Quantum Dot Molecules

Lecture Notes in Nanoscale Science and Technology

Volume 14

For further volumes:
http://www.springer.com/series/7544

Jiang Wu • Zhiming M. Wang
Editors

Quantum Dot Molecules

Springer

Editors
Jiang Wu
University of Electronic Science
and Technology of China
State Key Laboratory of Electronic
Thin Film and Integrated Devices
Chengdu, China, People's Republic

Zhiming M. Wang
University of Electronic Science
and Technology
State Key Laboratory of Electronic
Thin Film and Integrated Devices
Chengdu, China, People's Republic

ISSN 2195-2159 ISSN 2195-2167 (electronic)
ISBN 978-1-4939-4766-9 ISBN 978-1-4614-8130-0 (eBook)
DOI 10.1007/978-1-4614-8130-0
Springer New York Heidelberg Dordrecht London

Printed on acid-free paper

Springer is part of Springer Science+Business Media (www.springer.com)

Preface

Quantum dot molecules are artificial molecules made from coupled quantum dots. Quantum dot molecules have attracted significant attention since the first observation of coupled quantum dots with molecular wavefunctions. Such artificial molecules possess unique and fascinating properties, which make these coupled quantum systems promising in quantum information technologies. Over the last few years, quantum dot molecules have been used to implement quantum qubits, quantum gates, and exciton-spin memory. It is clear that quantum dot molecules will have great impact on next-generation information technologies. The main objective of this edited book is to provide a description of recent experimental and theoretical developments in quantum dot molecules.

The book consists of 11 chapters that present both experimental and theoretical considerations covering the fabrication, characterization, and properties of quantum dot molecules. The first chapter reviews two fabrication methods for quantum dot molecules. These methods involve of pre-patterning of substrates with either in situ droplet epitaxy "nanodrilling" or ex situ local oxidation nanolithography. In Chap. 2, the fabrication of ring-shaped InP quantum dot molecules is presented. This growth method employs droplet epitaxy to form self-assembled quantum dot molecules. Chapter 3 covers the growth of InGaAs quantum dot molecules by a partial-capping and regrowth technique. The nucleation mechanism and optical properties of lateral quantum dot molecules are also discussed. Chapter 4 presents a unique composite quantum system made of "quantum dot–dye" nanoassemblies. This chapter discusses "bottom-up," non-covalent self-assembly principles and exciton dynamics of nanoassemblies that contain organic colloidal semiconductor quantum dots and various inorganic dye molecules.

Chapters 5–10 are theoretical contributions focused on various types of quantum dot molecules. Chapter 5 presents a detailed analysis of the polarization response of multilayer quantum dot molecules. Chapter 6 presents a systematic theoretical investigation of the polarization potential in isolated and coupled $In_xGa_{1-x}N/GaN$ QDs and its effect on electronic structure and optical properties. Chapter 7 presents a theory of interference blockade and analyzes interference single electron transistors based on quantum dot molecules. In Chap. 8, the authors deal with two quantum interference-based phenomena, the Fano and Aharonov-Bohm effects in a ring of quantum dot molecules. Chapter 9 introduces phonon-assisted carrier dynamics in coupled quantum dots. Chapter 10 considers Förster resonant energy transfer in optically driven quantum dot molecules. In this chapter, the authors theoretically demonstrate a detection method of resonant energy transfer in quantum dot molecules by using level anticrossing spectroscopy.

Finally, in the last chapter, the authors report an experimental investigation of the Stark effect in quantum dot molecules.

The editors would like to thank all the contributors for their outstanding chapters. We would also like to acknowledge Dr. David Packer and Springer staff for their support. We are also grateful to Ms. Lina Wan for her helpful editorial assistance. The editors are also pleased to acknowledge the financial support from the National Basic Research Program (973) of China (2013CB933300).

Chengdu, People's Republic of China Jiang Wu
Chengdu, People's Republic of China Zhiming M. Wang

Contents

Contributors

Pablo Alonso-González Instituto de Microelectrónica de Madrid (CNM-CSIC), Madrid, Spain

Pornchai Changmoang Faculty of Engineering, Semiconductor Device Research Laboratory, Department of Electrical Engineering, Chulalongkorn University, Bangkok, Thailand

Diego Acosta Coden Facultad de Ciencias Exactas y Naturales y Agrimensura, Instituto de Modelado e Innovaci on Tecnol ogica (CONICET-UNNE), Corrientes, Argentina

Andrea Donarini Theoretische Physik, Universität Regensburg, Regensburg, Germany

Mauricio Garrido Department of Physics and Astronomy Clippinger Labs 364, Ohio University, Athens, OH, USA

Krzysztof Gawarecki Instiute of Physics, Wrocław University of Technology, Wrocław, Poland

Sergio S. Gomez Facultad de Ciencias Exactas y Naturales y Agrimensura, Instituto de Modelado e Innovaci on Tecnol ogica (CONICET-UNNE), Corrientes, Argentina

Milena Grifoni Theoretische Physik, Universität Regensburg, Regensburg, Germany

Wipakorn Jevasuwan Faculty of Engineering, The Semiconductor Device Research Laboratory, Electrical Engineering Department, Chulalongkorn University, Bangkok, Thailand

National Institute of Advanced Industrial Science and Technology (AIST), Ibaraki, Japan

S. Kanjanachuchai Semiconductor Device Research Laboratory, Department of Electrical Engineering , Chulalongkorn University, Bangkok, Thailand

Paweł Karwat Instiute of Physics, Wrocław University of Technology, Wrocław, Poland

Daniel A. Lovey Facultad de Ciencias Exactas y Naturales y Agrimensura, Instituto de Modelado e Innovaci on Tecnol ogica (CONICET-UNNE), Corrientes, Argentina

Paweł Machnikowski Instiute of Physics, Wrocław University of Technology, Wrocław, Poland

Javier Martín-Sánchez Instituto de Microelectrónica de Madrid (CNM-CSIC), Madrid, Spain

Eoin P. O'Reilly Tyndall National Institute, Cork, Ireland

Department of Physics, University College Cork, Cork, Ireland

Somsak Panyakeow Faculty of Engineering, The Semiconductor Device Research Laboratory, Electrical Engineering Department, Chulalongkorn University, Bangkok, Thailand

Nirat Patanasemakul Faculty of Engineering, Semiconductor Device Research Laboratory, Department of Electrical Engineering, Chulalongkorn University, Bangkok, Thailand

Swati Ramanathan Department of Physics and Astronomy Clippinger Labs 364, Ohio University, Athens, OH, USA

Somchai Ratanathammapan Faculty of Engineering, The Semiconductor Device Research Laboratory, Electrical Engineering Department, Chulalongkorn University, Bangkok, Thailand

Juan E. Rolon Department of Physics and Astronomy, University of North Carolina at Chapel Hill, Chapel Hill, NC, USA

Rodolfo H. Romero Facultad de Ciencias Exactas y Naturales y Agrimensura, Instituto de Modelado e Innovaci on Tecnol ogica (CONICET-UNNE), Corrientes, Argentina

Diego Sebastian Acosta Coden Facultad de Ciencias Exactas y Naturales y Agrimensura, Instituto de Modelado e Innovaci on Tecnol ogica (CONICET-UNNE), Corrientes, Argentina

Katarzyna Roszak Instiute of Physics, Wrocław University of Technology, Wrocław, Poland

Stefan Schulz Tyndall National Institute, Cork, Ireland

Naparat Siripitakchai Faculty of Engineering, Semiconductor Device Research Laboratory, Department of Electrical Engineering, Chulalongkorn University, Bangkok, Thailand

Anna Sitek Instiute of Physics, Wrocław University of Technology, Wrocław, Poland

Eric Stinaff Department of Physics and Astronomy Clippinger Labs 364, Ohio University, Athens, OH, USA

Supachok Thainoi Faculty of Engineering, Semiconductor Device Research Laboratory, Department of Electrical Engineering, Chulalongkorn University, Bangkok, Thailand

Natapong Thongkamkoon Faculty of Engineering, International School of Engineering, Chulalongkorn University, Bangkok, Thailand

Nitidet Thudsalingkarnsakul Faculty of Engineering, Semiconductor Device Research Laboratory, Department of Electrical Engineering, Chulalongkorn University, Bangkok, Thailand

Sergio E. Ulloa Department of Physics and Astronomy and Nanoscale and Quantum Phenomena Institute, Ohio University, Athens, OH, USA

Muhammad Usman Tyndall National Institute, Cork, Ireland

Christian von Borczyskowski Institute of Physics and Centre for Nanostructured Materials and Analytics (nanoMA), Chemnitz University of Technology, Chemnitz, Germany

Lina Wan State Key Laboratory of Electronic Thin Film and Integrated Devices, University of Electronic Science and Technology of China, Chengdu, People's Republic of China

Kushal Wijesundara Department of Physics and Astronomy Clippinger Labs 364, Ohio University, Athens, OH, USA

Eduard Zenkevich Department of Information Technologies and Robotics, National Technical University of Belarus, Minsk, Belarus

Chapter 1
Fabrication of Semiconductor Quantum Dot Molecules: Droplet Epitaxy and Local Oxidation Nanolithography Techniques

Pablo Alonso-González and Javier Martín-Sánchez

Abstract A semiconductor quantum dot molecule (QDM) composed of two interacting quantum dots (QDs) is the simplest coupled system formed by semiconductor quantum nanostructures. Potentially, a QDM is the ideal building block for the realization of a quantum computation device. However, the fabrication of QDMs is far from being a straightforward task, particularly if a precise control of QDs density, size, or spatial location is required. Recently, an important improvement in the control of these properties has been achieved by using patterned semiconductor substrates followed by preferential epitaxial growth. In this chapter we will overview two of such fabrication methods, which are based on: (1) in situ droplet epitaxy "nanodrilling" and (2) ex situ local oxidation nanolithography.

Abbreviations

AFM-LAO	Atomic force microscopy local anodic oxidation
ALMBE	Atomic layer molecular beam epitaxy
BEP	Beam equivalent pressure
CCD	Charge coupled device
MBE	Molecular beam epitaxy
ML	Monolayer
MSM	Metal–semiconductor–metal
PL	Photoluminescence
QD	Quantum dot
QDM	Quantum dot molecule

P. Alonso-González (✉) • J. Martín-Sánchez
Instituto de Microelectrónica de Madrid (CNM-CSIC), Isaac Newton 8 (PTM), 28760 Tres Cantos, Madrid, Spain
e-mail: p.alonso@nanogune.eu; Javier.martin.nano@gmail.com

J. Wu and Z.M. Wang (eds.), *Quantum Dot Molecules*, Lecture Notes in Nanoscale Science and Technology 14, DOI 10.1007/978-1-4614-8130-0_1,

RH	Relative humidity
RMS	Root mean square
TEM	Transmission electron microscopy
WL	Wetting layer

1.1 Introduction

A semiconductor quantum dot (QD) is a semiconductor nanostructure where the motion of electrons and holes are confined in all three spatial directions. A consequence of such three-dimensional (3D) carrier confinement is that, in analogy with atoms, QDs show a discrete quantized energy spectrum, which makes them to be generally referred to as "artificial atoms." However, semiconductor QDs have some important advantages with respect to real atoms in terms of applicability into devices. As they can be obtained in a solid host matrix with their position permanently defined, they can be further processed and functionalized using well-developed semiconductor technology. The integration of semiconductor QDs as active elements in quantum information processing and communication applications has been extensively studied during the last years. Electrically driven on-demand sources of single photons or entangled photon pairs have been realized by incorporating QDs in a device structure [1, 2]. Also, from a more fundamental point of view, a single QD embedded within an optical micro-resonator has been studied, forming an excellent playground to investigate cavity quantum electrodynamics phenomena in the weak and strong coupling regimes [3–5]. Despite all this progress, to take full advantage of the use of QDs in such applications or to develop novel functionalities, a precise control of QDs size, density, and spatial location is necessary.

Particularly interesting is the formation of ordered networks of well-defined QDs in close proximity. The simplest case is a quantum dot molecule (QDM) formed by two interacting semiconductor QDs [6], which can be considered as a basic building block for the realization of solid-state quantum computation devices [7]. Arranged in a vertical geometry (in the growth direction), QDMs have been deeply studied during the last years allowing for the direct observation of quantum-mechanical coupling and hybridization of energy levels between the QDs. Coupling in these molecules is mediated by coherent electron or hole tunneling through a thin intermediate layer, which is normally a few nanometers thick. The most common fabrication process of vertical QDMs is based on self-assembling processes and, more specifically, on the formation of two layers of self-assembled QDs separated by an epitaxial intermediate layer of another material, which acts as the tunneling potential barrier. The QDs of the second layer form mostly on top of the buried QDs of the first layer due to a strain relaxation process [8–10]. A drawback inherent to this self-assembling growth procedure is the low control of QDs size, strain, and composition, which makes the emission energy of the QDs forming the molecule to be generally mismatched and difficult to tune on demand. These growth inhomogeneities are often solved by either choosing a suitable thickness of the intermediate layer, which modifies the strain field distribution on the surface and therefore the size of the upper QDs, or by applying an external perturbation such as strain or/and an electric field in the growth direction [11–14].

Although the study of vertical QDMs has permitted the implementation of basic quantum information devices such as optical quantum gates [15], QDMs arranged in a lateral geometry (in the growth plane) is a more appealing design from the quantum computation technology point of view. Apart from facilitating the gating technology, it would easily permit to contact single QDMs one by one in the same device or to naturally increase the number of coupled QDs in the molecule, thus forming a more complex quantum network. From a more fundamental point of view, lateral QDMs would also allow for studying, for the first time, coupling and hybridization effects in a two-dimensional scenario. Experimentally, lateral coupling of two adjacent QDs has been barely tackled [16, 17]. Difficult to achieve by pure self-assembling processes, the most successful fabrication approaches to form lateral QDMs are based on the selective epitaxial growth of QDs on modified surface morphologies such as mounds or nanoholes [18–22]. Generally, the use of mounds takes advantage of its in situ self-assembled formation and consequent high optical quality of the semiconductor material [23–25]. However, this approach shows a limited control in the number of QDs forming the molecule. On the other hand, the main drawback encountered with the used of etched nanoholes is that the optical quality of the semiconductor material is often degraded by imperfections introduced during their fabrication, mainly carried out by ex situ lithographic processes [26, 27].

In this chapter, we will overview two fabrication methods that circumvent the majority of these growth-engineering problems, allowing for a controlled fabrication of QDMs with high optical quality.

In the first section, we focus on the in situ droplet epitaxy etching technique to obtain a template of low-density nanoholes on a GaAs surface through a self-assembling process [28]. Subsequent preferential InAs epitaxial growth into the nanoholes, when adequate growth conditions are used [29, 30], yields the formation of low-density vertical QDMs with deliberate emission properties [31]. The capability and versatility of this fabrication procedure is further demonstrated by showing the formation of lateral QDMs with high optical quality at the single nanostructure level [20]. Lateral quantum coupling in these QDMs is also demonstrated by micro-photoluminescence (micro-PL) studies of a single QDM as a function of a laterally applied electric field [17].

In the second section, we present an alternative way to produce lateral arrangements of QDs by ex situ atomic force microscopy local anodic oxidation (AFM-LAO) nanolithography followed by preferential epitaxial growth [22, 32, 33]. By means of this approach, not only the number of QDs in a lateral QDMs-like arrangement can be controlled but also the spatial positioning of the nanostructure on the substrate with nanometer resolution. The whole optimized fabrication process is demonstrated to be fully compatible with obtaining high optical quality nanostructures [34].

1.2 Fabrication of Vertical and Lateral QDMs by Droplet Epitaxy Technique

The droplet epitaxy growth technique was first used at the beginning of the 1990s by Koguchi and Ishige [35, 36]. It was initially proposed as a new molecular beam epitaxy (MBE) method for the fabrication of III–V semiconductor nanostructures on II–VI semiconductor substrates with nearly equal lattice constant. It basically consists in the initial deposition of atoms of Group-III element in absence of any supply of Group-V element, creating liquid metal droplets on the substrate. These metal droplets are, right afterwards, exposed to an atmosphere of Group-V element, which induces their crystallization into III–V nanostructures. On GaAs surfaces, and depending on the substrate temperature, this growth method yields different types of nanostructures. Particularly, QDs [37, 38] and quantum rings (QRs) complexes [39] are formed at relatively low temperatures of 200–300°C, whereas nanoholes surrounded by mounds [28] are obtained at high substrate temperatures of approximately 500°C. This relatively high temperature "nanodrilling" process has been recently emerged as a very promising in situ patterning technique. From a purely formation point of view, such a nanohole etching process is not a surprising result on GaAs, as long as it is well known that GaAs surfaces are unstable under Ga-rich conditions and As desorption occurs at a temperature of 500°C [40]. Specifically, when depositing Ga droplets on the GaAs surface, As atoms from the underlying GaAs may diffuse and escape to the vacuum chamber or incorporate into the droplet surroundings, which explains the formation of GaAs mounds around them even before any additional supply of As [28]. On the other hand, the remaining Ga atoms at the interface region after As desorption merge into the Ga droplet. As this process continues, there are more GaAs being dissolved below the droplet, more Ga atoms from the GaAs substrate being incorporated into the droplet, and more As atoms forming GaAs mounds on the surroundings. The Ga droplet acts then as a "nanodrill" into the GaAs surface inducing the formation of nanoholes. This etching process presents many advantages over other patterning techniques, as the density of nanoholes obtained on the surface is controllable by the initial Ga supply in the form of droplets, which can be as low as 10^7 cm^{-2}. Also, as the kinetics of this mechanism is strongly affected by the supply of As atoms that induces a fast Ga droplet crystallization (i.e., the "nanodrilling" process is stopped), the nanohole size can be accurately controlled. More importantly, due to its self-assembling nature and the high substrate temperatures used, this in situ patterning mechanism avoids any contamination or degradation of the substrate, thus ensuring an excellent optical quality of the semiconductor material [30].

1.2.1 Vertical QDMs

The potential and versatility of the droplet epitaxy etching for subsequent InAs preferential epitaxial growth is demonstrated in the following by the fabrication

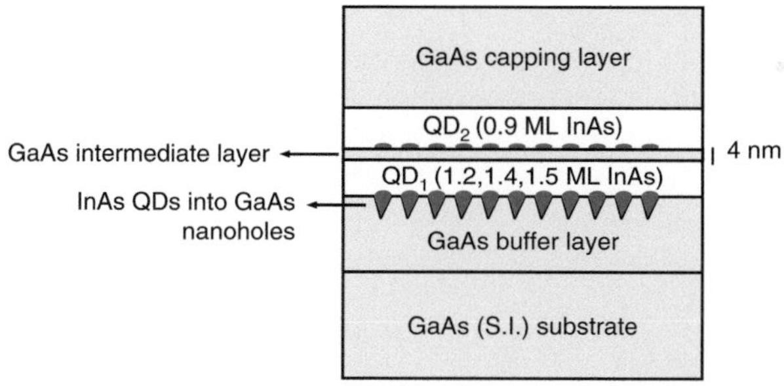

Fig. 1.1 Schematic diagram of the structure grown to obtain vertical QDMs. The first layer of nanostructures (QD_1) is formed through preferential InAs nucleation into a GaAs nanoholes template obtained by the droplet epitaxy "nanodrilling" technique. Three samples were fabricated by varying the amount of InAs deposited (1.2, 1.4, and 1.5 ML) to form QD_1. By the growth of a 4-nm-thick GaAs barrier layer and the deposition of 0.9 ML of InAs, a second layer of nanostructures (QD_2) is formed on top of QD_1 by a strain-driven nucleation process

of low density vertical In(Ga)As QDMs. Formed by two vertically aligned InAs QDs, a deliberate control of size/energy emission is achieved in these nanostructures [31]. Particularly, the emission energy of the first layer of QDs (hereafter referred to as QD_1) that forms the molecule is tuned by the preferential nucleation of controlled amounts of InAs into etched nanoholes. After this first deposition, a thin intermediate GaAs layer is grown in order to define a tunneling barrier that separates a second layer of InAs QDs (hereafter referred to as QD_2), which nucleate on top of the first ones via a strain-driven self-assembled process [8–10]. This fabrication procedure allows for obtaining either symmetric (the energy levels of both QDs forming the molecule are in resonance) or asymmetric (the energy levels of both QDs forming the molecule are out of resonance) vertically coupled structures where, without any further technological process, the type of carrier mediating the tunnel coupling in the QDMs can be controlled on demand.

The fabrication process for obtaining vertical InAs QDMs is illustrated in Fig. 1.1. Initially, a 0.5 μm-thick undoped GaAs buffer layer is grown at a substrate temperature of 580°C by MBE on a semi-insulating GaAs (0 0 1) substrate. The nanoholes template is fabricated at a substrate temperature of 500°C maintaining the Ga shutter opened during 10 s, providing a Ga flux equivalent to the growth of GaAs at a growth rate of 1 monolayer (ML) s^{-1}. During this time, the arsenic cell is opened periodically during 0.2 s every 0.8 s. The beam equivalent pressure (BEP) of As_4 is fixed during the process to $BEPAs_4 = 5 \times 10^{-7}$ Torr. This process, similar to an *arsenic-debt* atomic layer molecular beam epitaxy (ALMBE) growth [41], results in Ga droplets spread all over the surface with a density of 2.5×10^8 cm^{-2}. The sample is then annealed under As_4 flux during 6 min. Once irradiated with As_4, the "nanodrilling" etching process stops and the Ga droplets crystallize into GaAs, yielding a template of nanoholes surrounded by mounds. As shown in the atomic force microscopy (AFM) image of Fig. 1.2, these nanoholes show an elongated morphology along the GaAs [1 −1 0] crystallographic direction as a result

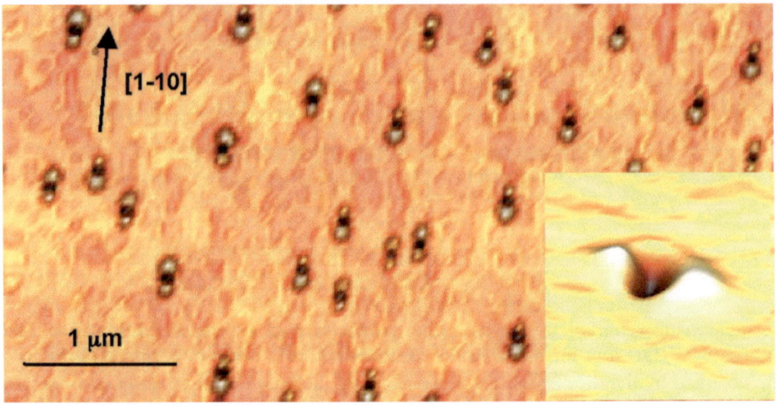

Fig. 1.2 AFM topography image of the nanoholes template formed by the droplet epitaxy "nanodrilling" technique. The *inset* shows a 3D rendered image of a single nanohole. Adapted from [42]. Copyright 2008 American Institute of Physics

of the higher Ga atoms diffusion in this direction. The mean nanoholes depth is 4.4 ± 0.8 nm with mean diameter values of 95 ± 5 nm and 43 ± 3 nm along the GaAs [1 1 0] and [1 −1 0] directions, respectively.

On this template of nanoholes, three samples are fabricated depending on the amount of InAs material deposited to form QD_1. Particularly, 1.2, 1.4, and 1.5 ML of InAs were deposited at a substrate temperature of 500°C, a growth rate of 0.01 ML s^{-1}, and $BEPAs_4 = 5 \times 10^{-7}$ Torr. As an intermediate tunneling barrier for carriers, a 4-nm-thick GaAs epitaxial layer is grown by ALMBE [41] at a low substrate temperature of 450°C, growth rate of 0.5 ML s^{-1}, and $BEPAs_4 = 2 \times 10^{-6}$ Torr. This growth mode allows for growing atomically flat GaAs epitaxial layers at low substrate temperatures, thus reducing possible material inter-diffusion processes (i.e., abrupt interfaces can be achieved). Finally, for completing the molecule structure, a second layer of QDs (QD_2) is grown on the surface by depositing 0.9 ML of InAs under the same growth conditions than QD_1. Note that due to the large strain field built-in on the surface by QD_1, the amount of InAs material needed for the formation of 3D nanostructures is much lower than the critical thickness for QDs formation in a self-assembled Stranski–Krastanov process. In order to study their optical emission, these molecules were finally capped with a 155 nm-thick GaAs layer.

The structure of the resulting nanostructures is shown in the dark field (cross-sections) transmission electron microscopy (TEM) images of Fig. 1.3 [43]. The contrast in the images is given by changes in material composition, being the bright areas InAs-rich zones. The formation of a double structure consisting of two InAs nuclei separated by a thin and darker GaAs intermediate layer is clearly observed. The corresponding InAs wetting layers (WLs) are also observed in the image. Interestingly, the nanostructure QD_1 forms at a lower level than its respective WL, contrarily to what is observed for the nanostructure QD_2 formed on top,

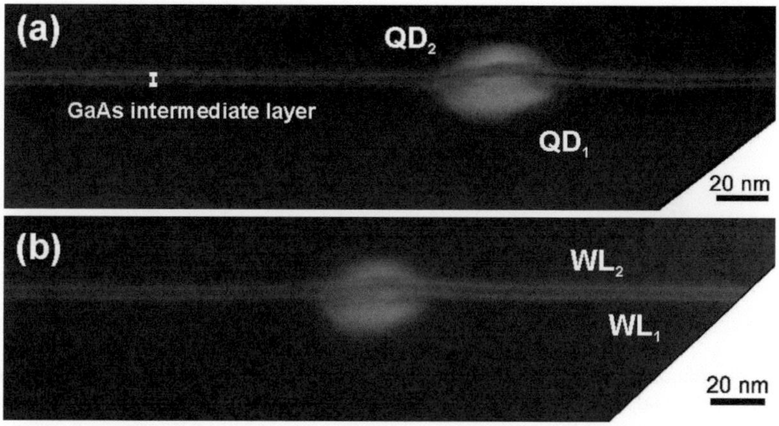

Fig. 1.3 Cross-section images taken by dark field transmission electron microscopy (TEM) of vertical QDMs formed when (**a**) 1.4 ML and (**b**) 1.2 ML of InAs are deposited on the nanohole template. QD_1 and QD_2 indicate the *lower* and *upper* QDs, respectively. While QD_2 forms above the wetting layer WL_2, QD_1 forms below the wetting layer WL_1, demonstrating the preferential nucleation of InAs into a nanohole. Adapted from [31]

whose formation is strain-driven. This is a clear demonstration that the mechanism involving the formation of QD_1 is the preferential nucleation of InAs material into the GaAs nanohole. Regarding the sizes of each QD, when depositing 1.4 ML of InAs (Fig. 1.3a) the lower QD_1 shows a bigger size than the upper QD_2. On the other hand, when depositing 1.2 ML of InAs (Fig. 1.3b) the buried QD_1 is similar in size to the top QD_2. These structural results provide a clear indication that the size of a QD into a vertical QDM structure can be tuned on demand by this fabrication procedure. However, it is noticeable that despite the engineered change in size of QD_1, the nucleation of InAs on top of these nanostructures yields QD_2 with similar dimensions in all cases. Importantly, this effect provides an independency in size between the QDs forming the molecule, which opens the door for a real controllability of a vertical QDM design.

The QDs size tunability in the QDMs is further studied by measuring the QDMs photoluminescence (PL) at low temperature ($T = 30$ K). Figure 1.4 shows the PL spectra for the three different samples fabricated. The convolution of two main Gaussian peaks is observed in all three cases. They are associated with two different QDs families in the molecules as verified by excitation power experiments where no relative saturation effects, corresponding to a given QD family with ground and excited states, were found in the PL peaks. By combining these optical emission results and the TEM images, a direct correspondence of the different emission peaks to the recombination of carriers can be established in either the lower or upper QDs (QD_1 and QD_2) forming the molecule. Figure 1.5a shows as filled areas the three PL emission spectra ascribed to the nanostructure QD_1 for the three different amounts of InAs material deposited on the nanoholes template. As the amount of

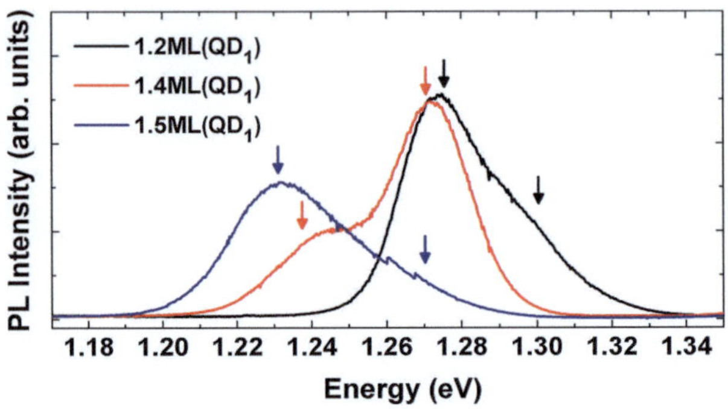

Fig. 1.4 Photoluminescence spectra for the three vertical QDMs designs fabricated. *Black*, *red*, and *blue curves* correspond to 1.2, 1.4, and 1.5 ML of InAs deposited into the nanoholes, respectively. The *arrows* in the figure point out the two families of QDs formed on each sample. Adapted from [31]

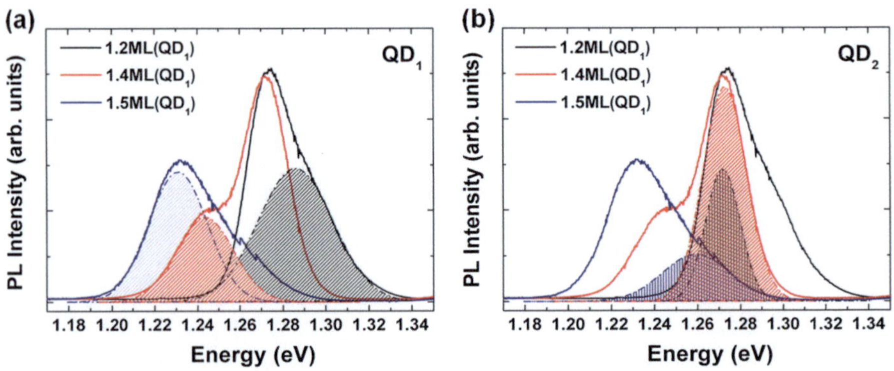

Fig. 1.5 Photoluminescence emission (*filled areas*) ascribed to (**a**) QD_1 and (**b**) QD_2 nanostructures. A clear tuning effect on the emission of QD_1 is obtained independently of the emission of QD_2, which shows similar emission for all the samples. Adapted from [31]

InAs increases, the emission energy of QD_1 decreases, further demonstrating the preferential nucleation of InAs into the nanoholes. Analogously, Fig. 1.5b shows as filled areas the three PL emission spectra ascribed to the nanostructure QD_2. In this case, the emission energy is similar in all cases indicating that the size of QD_2 remains constant despite the varying amounts of InAs material nucleated underneath. The emission energies ascribed to QD_1 and QD_2 in these PL curves are listed in Table 1.1. The arrow in this table highlights the tunability of the energy emission of QD_1 as a function of InAs material deposited into the nanoholes.

Altogether, these results show that the preferential nucleation of controllable amounts of InAs material into a nanoholes template formed by the droplet etching technique allows for a deliberate designing of symmetric or asymmetric vertical QDMs where the emission of one QD can be tuned with independency of the other.

Table 1.1 PL peak emission energies ascribed to QD_1 and QD_2 nanostructures

		Peak energy (eV)	
InAs (ML) QD_1	InAs (ML) QD_2	QD_1	QD_2
1.2	0.9	1.286 ↓	1.272
1.4	0.9	1.243 ↓	1.272
1.5	0.9	1.230 ↓	1.26

A clear dependency of the peak emission energy of QD_1 and the amount of InAs material deposited is demonstrated

Note that this fabrication procedure produces areal densities of nanostructures as low as 2×10^8 cm^{-2}, which would permit an optimal integration of these QDMs as active optical emitters in advanced photonic devices, where spectroscopic studies at the single nanostructure level are required.

1.2.2 Lateral QDMs

As commented in the introduction of this chapter, the fabrication of QDMs in a lateral geometry presents several technological advantages as compared to a vertical arrangement. In this subsection, we will describe a novel in situ fabrication process that, based on a preferential nucleation procedure of InAs material into GaAs nanoholes etched by the droplet "nanodrilling" technique, allows for the formation of lateral In(Ga)As QDMs with high optical quality [20]. Particularly, we will show that depending on the As_2 pressure at which InAs is deposited, the formation of single QDs or QD pairs into each of the nanoholes can be obtained.

Similarly to the previously described fabrication process of vertical QDMs, the formation of lateral QDMs starts with an initial flattening of the GaAs substrate by growing a 0.5 μm-thick buffer layer at a substrate temperature of 580°C by MBE. Right afterwards, the droplet etching "nanodrilling" process is performed in order to obtain a template of nanoholes similar to the one shown in Fig. 1.2. On this patterned surface, 1.5 ML of InAs are deposited at a substrate temperature of 510°C and following a growth sequence consisting in the deposition of 0.1 ML at a growth rate of 0.05 ML s^{-1} followed by a pause of 2 s under an As_2 flux. Keeping constant the rest of growth parameters, the As_2 pressure during this InAs deposition and annealing process was varied in two different samples taking values of 3.5×10^{-7} and 1×10^{-6} Torr, respectively. A (2×4) surface reconstruction was always observed at the used As_2 pressures. In both samples, for optical and topographic investigation purposes, capped with a 100 nm-thick GaAs layer and uncapped nanostructures were fabricated.

Figure 1.6 shows the AFM topography corresponding to the deposition of InAs on the nanoholes template for As_2 pressures of 3.5×10^{-7} Torr (Fig. 1.6a) and 1×10^{-6} Torr (Fig. 1.6b). When using an As_2 pressure of 5×10^{-7} Torr (Fig. 1.6a) a single QD is obtained inside the nanoholes with occupancy of 95%. On the other

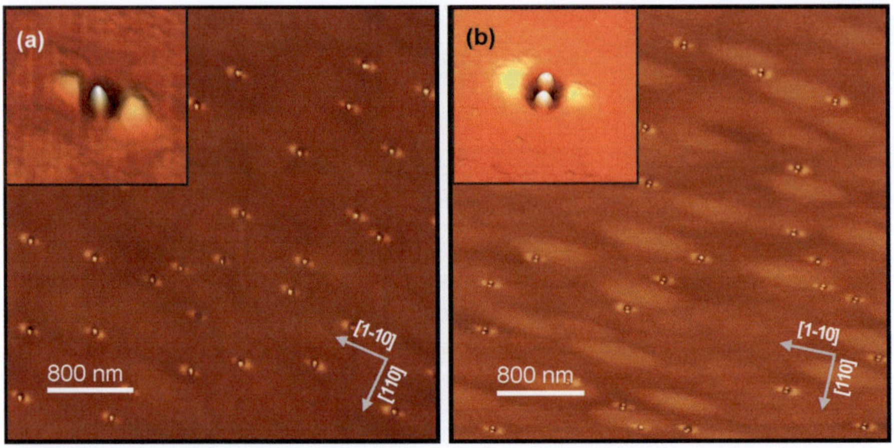

Fig. 1.6 Topography AFM images obtained after depositing 1.5 ML of InAs at an As_2 pressure of (**a**) 3.5×10^{-7} Torr and (**b**) 1×10^{-6} Torr. A direct correlation between the As_2 pressure and the formation of either a single QD or a QDM into each of the nanoholes is observed. Reprinted (adapted) with permission from [20]. Copyright 2009 American Chemical Society

hand, an InAs nucleation enhancement into the nanoholes is found when the As_2 pressure is increased to 1×10^{-6} Torr, as shown in Fig. 1.6b. In this case, QDMs aligned along the GaAs [1 1 0] crystallographic direction are formed inside the nanoholes with occupancy of 98%, which clearly indicates a higher preferential nucleation process into them. Detailed 3D rendered images for the initial nanohole, a single QD, and a QDM inside the nanohole are shown in Fig. 1.7, together with AFM profiles along the [1 1 0] direction (dotted lines).

As already mentioned, the nanoholes obtained by the droplet "nanodrilling" process show an elongated shape along the [1 1 0] direction and the presence of surrounding GaAs mounds aligned along the [1 −1 0] direction (Fig. 1.7a). In the case of single QDs (Figs. 1.6a and 1.7b), a monomodal distribution of QD sizes is obtained with a mean diameter of 58 ± 3 nm and a height of 7.4 ± 0.8 nm, both quantities referred to the flat surface plane. On the other hand, the QDMs (Figs. 1.6b and 1.7c) show systematic differences in the QDs sizes, as clearly observed in the distribution of heights represented in Fig. 1.8. A fit by two Gaussian peaks shows the presence of a bimodal distribution with height values of 5.3 nm $\pm$ 17% and 6.6 nm $\pm$ 24%. The separation edge to edge between QDs in the molecule is practically zero within AFM resolution, which yields an identical QDs mean diameter and center-to-center distance of 37 ± 4 nm. This QDs separation is short enough to potentially observe quantum coupling phenomena [18].

The PL characterization for single QDs and QDMs is shown in Fig. 1.9. Interestingly, although the AFM topography showed the single QDs bigger in size than the QDs in the QDM, a similar PL peak emission energy is found for both nanostructures. This effect is ascribed to a different In(Ga)As composition of the single QDs and QDMs, which can be explained by a different filling ratio of Group-

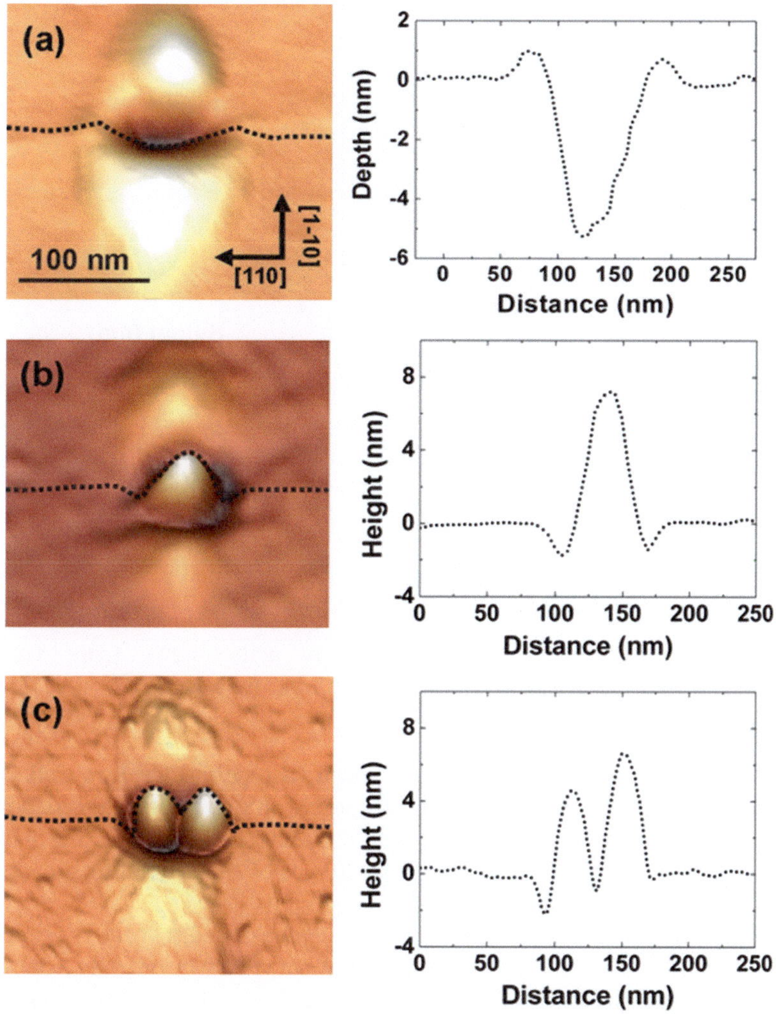

Fig. 1.7 3D rendered AFM images of a typical (**a**) nanohole, (**b**) single QD, and (**c**) QDM obtained by droplet epitaxy etching technique and posterior preferential nucleation of InAs material. Profiles on these nanostructures along the GaAs [1 1 0] direction are shown at the *right*. Reprinted (adapted) with permission from [20]. Copyright 2009 American Chemical Society

III atoms (Ga and In) into the nanoholes for the different As_2 pressures used in each case. In this regard, special relevancy has the crystalline orientation of the facets defined inside the nanoholes, which eventually conditions the nucleation process of In atoms [22]. On patterned substrates, apart from the GaAs (0 0 1) surface, two different surfaces can be identified: (0 0 1) disoriented surfaces towards [1 1 0] direction with step edges parallel to this direction that are Ga-terminated (A-type facets), and (0 0 1) disoriented surfaces towards [1 −1 0] direction with step edges

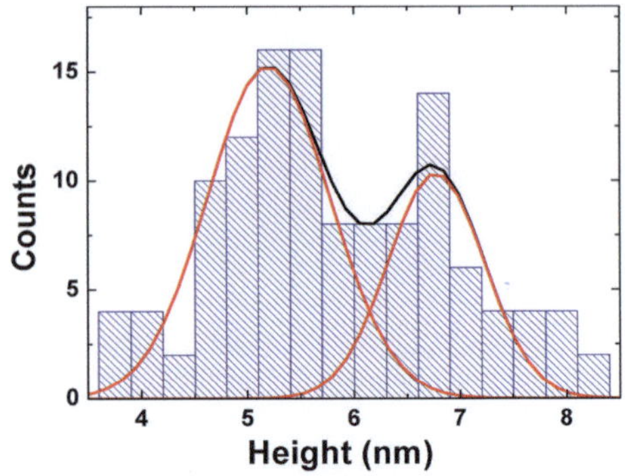

Fig. 1.8 Distribution of QD heights in the QDMs sample

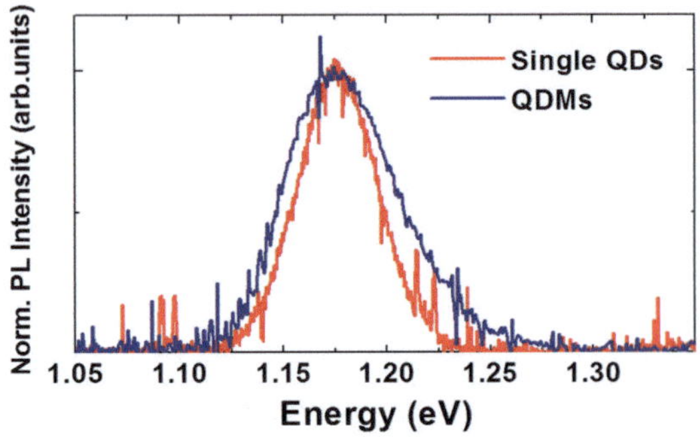

Fig. 1.9 Normalized PL spectra of single QDs (*red solid curve*), and QDMs (*blue solid curve*). Reprinted (adapted) with permission from [20]. Copyright 2009 American Chemical Society

parallel to this direction that are As-terminated (B-type facets). It is well known that the incorporation kinetics of Group-III atoms on each type of facets depends strongly on the experimental growth conditions and, specifically, on the substrate temperature [44], arsenic pressure [44–47], and type of As used during growth (As_2 or As_4) [48].

In conclusion, in this subsection we have demonstrated that the As_2 pressure used during InAs material deposition is a key parameter in order to control the formation of either single QDs or QDMs into previously in situ etched nanoholes. Under the appropriate conditions occupancy of 98% of the nanoholes with QDMs can be obtained. In general terms, this result demonstrates that the ability to fabricate QDMs into patterned nanoholes lies on the understanding of the preferential self-assembling growth process into them rather than the simply fabrication of a suitable motif.

1.2.3 Lateral Coupling in a Single QDM

Quantum-mechanical coupling in a lateral QDM has been barely studied experimentally [16, 17]. Previous indications of electronic coupling in a lateral QDM have been based on an analysis of the anomalous Stark shifts and photon correlation

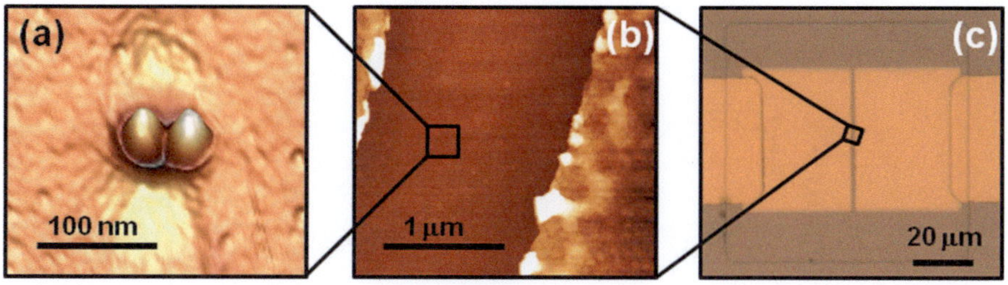

Fig. 1.10 Topography AFM images of (**a**) a single QDM, and (**b**) a 1.5-μm-wide GaAs channel. (**c**) Optical image of the Metal-Semiconductor-Metal (MSM) diode. Adapted from [17]. Copyright 2011 American Physical Society

statistics of the neutral exciton under a lateral electric field [16]. However, the observation of electrically tunable energy anticrossings in lateral QDMs remains a difficult task, mainly due to the exponential decrease of the tunnel coupling energy with the QD center-to-center distance [6, 16, 49]. In the following, we will show the emission spectrum by micro-PL of electrically tunable lateral QDMs with a varying number of electrons. Particularly, it is observed that for a QD center-to-center distance of 30–40 nm electron tunneling affects the negative trion emission energy in the molecule before clear exciton anticrossings may take place.

A typical lateral QDM used in this study shows a mean QDs center-to-center separation of 37 ± 4 nm (Fig. 1.10a). In order to apply an electric field along the QDs pair mutual axis, a metal–semiconductor–metal (MSM) diode is defined by evaporation of two metal contacts consisting of 15 nm Mo + 30 nm Au on top of 100-μm-square mesas. The contacts are separated by an 80 μm long × 1.5 μm wide un-doped GaAs channel embedding the nanostructures, as shown in Fig. 1.10a–c. The micro-PL of individual QDMs is collected at 5 K using a fiber-based confocal microscope excited with 785-nm continuous-wave laser light, dispersed by a 2 m × 0.3 m focal length double spectrograph and detected with a peltier cooled silicon charge coupled device (CCD) camera.

The emission of single semiconductor nanostructures in the presence of a lateral electric field has been studied in the last few years [50]. For moderate electric fields, or when the separation between the contacts is large, the changes observed in the QD emission spectrum have been related to the modulation of the carrier capture probability induced by the external field. On the other hand, for large enough electric fields or a small contact separation for a given lateral bias range, a modulation of the electronic confinement levels can also be obtained. In this regime, the exciton wave function can be directly modified, which leads to energy shifts, carrier tunneling, and fine-structure splitting reduction, among other effects [51, 52]. The MSM diodes used in this study were designed to have a channel width of only 1.5 μm, which allows for applying large electric fields along the GaAs [1 1 0] crystallographic direction (0–60 kV cm^{-1}). This is required to independently tune the exciton energy of the two QDs in the lateral QDM and, if their center-to-center separation is small enough, to observe resonant quantum tunneling phenomena [6, 16, 53].

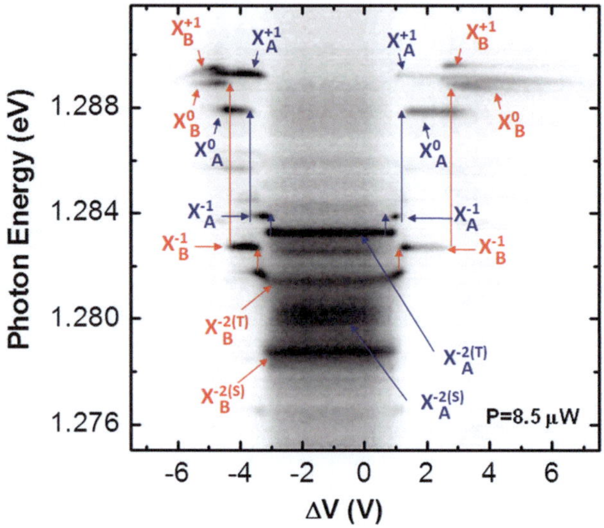

Fig. 1.11 Micro-PL contour plot map vs. lateral bias for a QDM. Adapted from [17]. Copyright 2011 American Physical Society

The contour map in Fig. 1.11 shows the evolution of the micro-PL spectrum as a function of the lateral bias ΔV applied to QDM. The number of spectral lines is twice the expected for a single QD [6, 16, 54]. The spectrum can be understood by the tentative assignment proposed in the figure, which is roughly that of two single QDs, A and B, with emission from the negative quarton X^{2-} to the negative trion X^{-}, to the neutral exciton X^0, and finally to the positive trion X^+. The two QDs are asymmetric in size, as expected (see Fig. 1.8), being A slightly bigger. Starting from the X^{2-} transitions (singlet and triplet), with increasing external field, the number of additional electrons is tuned from two to zero, and each QD in the molecule follows its own Coulomb staircase. This effect shows that net confined charge can be controlled in a QDM by applying a lateral electric field. The presence of negatively charged exciton complexes in the absence of an electric field is typical for QDs grown into nanoholes formed by the droplet "nanodrilling" technique [30]. This is explained by an incomplete crystallization of the Ga droplet, which induces the presence of arsenic vacancies in the surroundings of the nanohole. These localized states are close in energy to the electron-confined levels and are occupied by one or more electrons depending on their state of valence [55].

The energy levels ordering in the spectral assignment of Fig. 1.11 is the usual for isolated QDs. However, there is a remarkable anomaly. While the emission energy of X^0(A) is lower than that of X^0(B), indicating that QD A is bigger than QD B, the emission energy of X^-(A) is higher than that of X^-(B). Importantly, this result, difficult to explain in terms of isolated QDs, demonstrates electron tunneling coupling between the QDs in the QDM. A theoretical model explaining this anomaly trion emission in lateral QDMs has been recently published [49]. Overall, these results demonstrate the high potential of the droplet epitaxy "nanodrilling" technique to obtain high optical quality QDMs showing quantum mechanical coupling signatures in a lateral geometry.

1.3 Fabrication of Lateral Molecular Arrangements of QDs by Ex Situ Local Oxidation Nanolithography

In this section, we will demonstrate the suitability of ex situ AFM-LAO technique in combination with MBE growth technique to fabricate site-controlled single InAs QDs and lateral QDMs-like arrangements on GaAs (0 0 1) substrates. Similarly to the fabrication process presented in the previous section, this approach is based on the fabrication of patterned nanoholes that will act as preferential nucleation sites for QDs formation. Particularly, we will show that by controlling the size/shape of the patterned nanoholes and the MBE growth conditions of the subsequent epitaxial deposition, not only the number of QDs in a lateral molecular arrangement can be controlled but also their position on the substrate, which is also very interesting for their optimal integration in real and practical devices.

1.3.1 Fabrication of Patterned GaAs (0 0 1) Substrates by AFM-LAO

AFM-LAO is a well-known technique used for fabricating oxide structures on surfaces of a wide range of materials under ambient conditions [56, 57]. The AFM-induced oxidation mechanism is relatively fast (a few milliseconds) and spatially limited down to nanometric areas with high resolution in the lateral positioning and size/shape of the oxide structures on the substrate surface. The oxidation mechanism is a process induced by the application of an external voltage between the AFM tip and the substrate surface [in this case GaAs (0 0 1)], as sketched in Fig. 1.12.

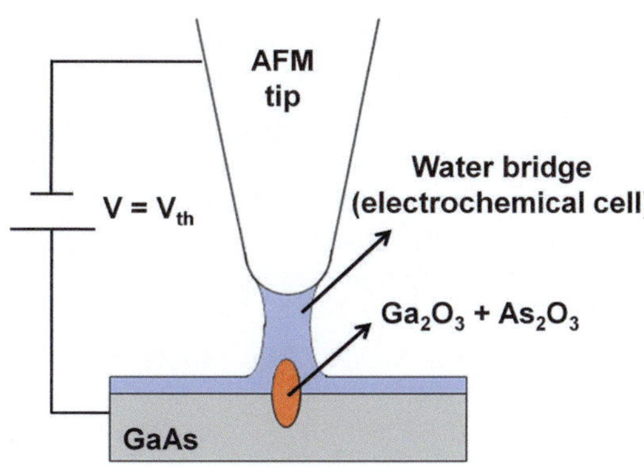

Fig. 1.12 Schematic representation of the AFM-LAO process. Oxides structures are formed on GaAs (0 0 1) substrates as a consequence of the reduction/oxidation processes through a water bridge

The electric field intensity can reach very high values of the order of 1 GV/m due to the reduced dimensions of the AFM tip (below 10 nm). Initially, at relative humidity (RH) values above zero and in thermal equilibrium, several MLs of water molecules are chemically absorbed on both substrate and AFM tip surfaces. When a strong electric field is applied, it induces the formation of a water bridge with nanometric dimensions that can be seen as a nanometric electrochemical cell. By applying a negative voltage above a threshold value (V_{th}) to the tip while the substrate is set to ground [58], oxidation and reduction processes take place on the substrate surface and AFM tip, respectively.

The eventual oxide size/shape and the oxidation kinetics depend mainly on the electric field intensity distribution on the substrate surface (i.e., oxidation experimental parameters). Given that the formation of the oxide structure is spatially confined within the water bridge, the maximum lateral extension of the oxide can be controlled by the AFM working conditions in non-contact-mode [59]. On the other hand, reproducibility in size/shape of the formed oxides depends strongly on the stability and uniformity of the water bridge during the oxidation process, which mainly depends on the surface roughness. In this regard, epitaxial GaAs (0 0 1) substrates are more convenient as they present a flat surface at the atomic level as measured by AFM (root mean square, RMS ~0.16 nm). In this work these surfaces were obtained by growing a 0.5-μm-thick GaAs layer by MBE at a growth rate of 1 ML s^{-1}, substrate temperature of 580°C, and a V(As)/III(Ga) flux ratio of 8.

In the case of AFM-LAO patterning of GaAs (0 0 1) surfaces, the resulting oxides are mainly composed of Ga_2O_3 and As_2O_3 species through the reaction $2GaAs + 12h^+ + 6OH^- \rightarrow Ga_2O_3 + As_2O_3 + 6h^+$ [60]. The volume per mol values for GaAs, Ga_2O_3, and As_2O_3 are 24.77, 65.74, and 68.04 $cm^3\ mol^{-1}$, respectively. Hence, the oxide volume is larger than that of the GaAs material and consequently, the oxides emerge above the substrate surface level. In order to obtain patterned substrates with nanoholes, these oxides can be easily removed by selective wet chemical etching with HF (49%). In Fig. 1.13, square arrays of oxide dots and the resulting nanoholes for applied voltages of $V_{ox} = -8$ V (Fig. 1.13a–c), and $V_{ox} = -14$ V (Fig. 1.13d–f) are shown. The oxides were fabricated in AFM non-contact-mode (RH = 30%) using n-doped Si tips ($K = 40$ N m^{-1}). A fixed oxidation time of 500 ms was used. For $V_{ox} = 8$ V the resulting oxides show a "simple conical structure" (Fig. 1.13c), whereas for $V_{ox} = -14$ V, an oxide with a broad and low base crown with a narrow spike is obtained (hereafter referred as oxides with "double structure") (Fig. 1.13f). This special morphology obtained at high applied voltages is ascribed to the vertical growth saturation of the oxide. Interestingly, the size/shape of the resulting nanoholes exactly replicates the size/shape of the emerging oxide with respect to the surface. Hence, customized nanoholes can be designed just by controlling the size/shape of the oxide structures, which eventually depends on the experimental oxidation parameters [61].

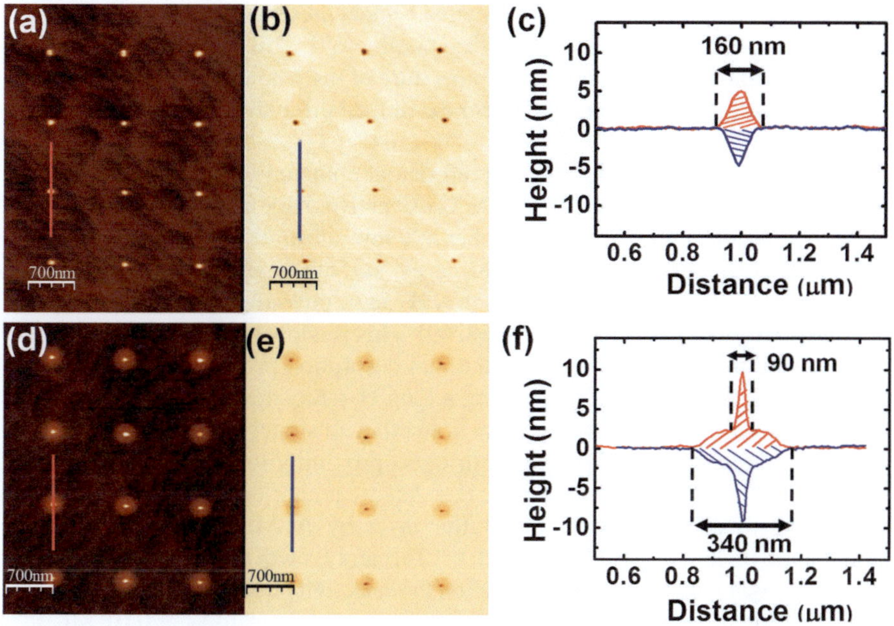

Fig. 1.13 AFM topography images of oxide structures fabricated on GaAs (0 0 1) by AFM-LAO for two different oxidation voltages and the corresponding nanoholes obtained after HF selective etching: "simple conical structure" for $V_{ox} = -8$ V (**a–c**); "double structure" for $V_{ox} = -14$ V (**d–f**). The profiles drawn on one of the oxides/nanoholes are shown at the *right* for both kinds of structures

1.3.2 Epitaxial Growth on Patterned GaAs (0 0 1) Substrates

The MBE Stranski–Krastanov growth of coherent self-assembled InAs QDs ensembles on un-patterned GaAs (0 0 1) substrates leads to randomly located QDs with a relatively broad size distribution. As shown in the previous section, the droplet epitaxy etching technique allows for a certain control of the QDs size and density into a nanohole. However, due to the random nature of Ga droplet formation on the surface, the QDs site formation cannot be controlled as desired. Many efforts have been dedicated to control the InAs/GaAs (0 0 1) QDs sites nucleation and study their formation process [18, 21, 22, 62–65], whereas less work has been realized for a simultaneous control of both spatial localization and QDs number in a lateral molecular-like arrangement by ex situ lithographic techniques [19, 22]. As shown in this chapter, the aim of such ex situ approaches is to define artificially preferential nucleation sites (nanoholes) to induce the InAs QDs self-assembling into them. The main constraints imposed by the use of patterned substrates are basically due to possible evolutions of the nanoholes size (a few nanometers) through: (1) Ga atom surface migration on GaAs (0 0 1) surfaces for substrate temperatures close to 500°C [66]; (2) growth of a relatively thick GaAs buffer layer, necessary to avoid proximity of InAs QDs to non-radiative recombination centers present at the interface with the substrate that can degrade the QDs optical quality [67]. This means that it would be desirable to maintain the maximum temperature of the whole fabrication process as close as possible to 500°C and the thickness of the buffer layer to a few nanometers thick.

Accordingly, to proceed with the preferential epitaxial growth of QDs into the fabricated nanoholes (Fig. 1.13b, e), the GaAs native oxide resulting after exposure to HF was removed in the MBE chamber at a low substrate temperature of 450°C. This process was carried out by exposing the surface to an atomic hydrogen (H) flux using a Ta H_2 thermal cracker with a H_2 base pressure of 10^{-5} Torr together with As_4 ($BEPAs_4 = 5 \times 10^{-7}$ Torr) [68, 69]. This is a low temperature process that avoids using the conventional oxide thermal desorption process at 600°C. On the other hand, it is known that performing a long $H + As_4$ treatment followed by the growth of a very thin GaAs buffer layer is an efficient strategy to obtain high optical emission efficiencies in QDs located in close proximity to the interface with the substrate [67]. Hence, in this work a $H + As_4$ treatment during 30 min preceded the growth of a 7-nm-thick GaAs buffer layer by ALMBE [41] at a substrate temperature of 450°C, $BEPAs_4 = 2 \times 10^{-6}$ Torr and growth rate of 0.5 ML s^{-1}. The effect of growing this GaAs buffer layer on the selectivity of the patterned nanoholes for InAs growth was investigated by depositing 1.5 ML of InAs at 0.01 ML s^{-1}, substrate temperature of 500°C and $BEPAs_4 = 5 \times 10^{-7}$ Torr [a (2×4) surface reconstruction was observed] on patterned samples with and without the growth of a GaAs buffer layer [22].

For an initial oxide with a "simple conical structure," Fig. 1.14 shows the AFM topography of the resulting nanoholes after HF wet chemical etching (Fig. 1.14a), after HF wet chemical etching and $H + As_4$ exposure (Fig. 1.14b), and after HF wet chemical etching, $H + As_4$ exposure, and the growth of a 7-nm-thick GaAs buffer layer (Fig. 1.14c). Profiles along the [1 1 0] and [1 −1 0] directions of the nanoholes are also shown in Fig. 1.14d. The nanohole evolution due to the previously described processes and before InAs deposition is summarized in Table 1.2. Particularly, the average relative differences in depth and width of the obtained nanohole along the [1 1 0] and [1 −1 0] directions with respect to the initial nanohole (after HF wet chemical etching) are shown. A positive (negative) value means an increase (decrease) of the specified nanohole dimension. The observed nanoholes depth increases after the treatment with atomic H is a well-known and reported effect [62].

After HF wet chemical etching (Fig. 1.14a), the nanohole shows a clear round shape with a diameter of 160 nm and a depth of 10 nm. This round shape evolves to a hexagonal one (Fig. 1.14b) after performing the atomic $H + As_4$ treatment, showing two sides parallel to the [1 1 0] direction that correspond to the intersection of the B-type facets (As-terminated) with the (0 0 1) surface plane [70]. When a 7-nm-thick GaAs buffer layer is grown, a clear nanohole enlargement along the [1 1 0] direction occurs (Fig. 1.14c). The observed decrease of the nanohole width along the [1 −1 0] direction is attributed to the presence of well-defined B-type

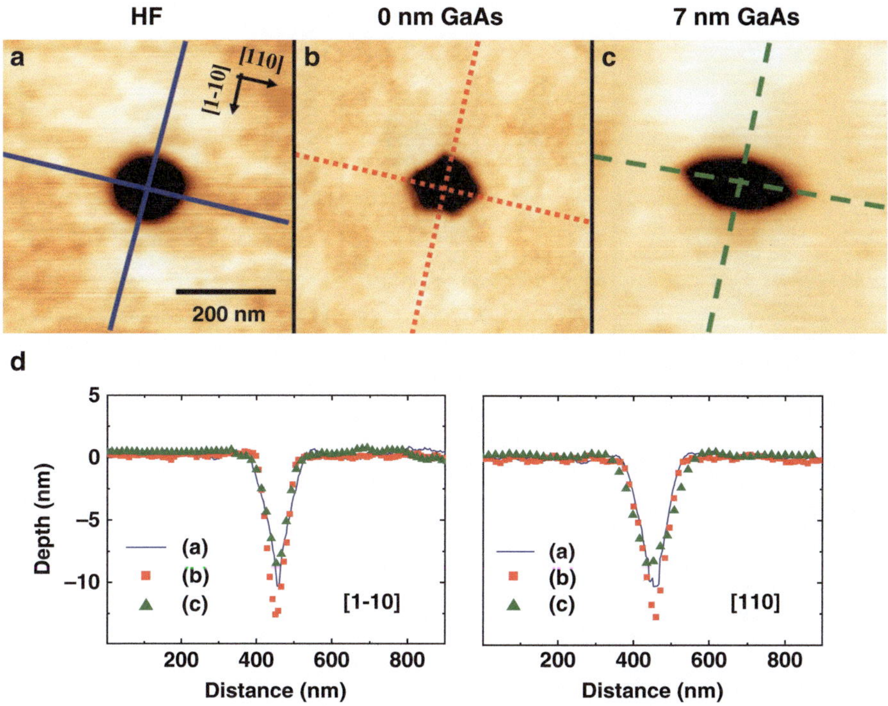

Fig. 1.14 AFM topography images of a nanohole after (**a**) HF treatment, (**b**) HF + (H + As_4) treatment, and (**c**) HF + (H + As_4) treatment followed by the growth of a 7-nm-think GaAs buffer layer. The nanoholes profiles along [1 −1 0] and [1 1 0] crystallographic directions are shown in (**d**). Adapted from [22]. Copyright 2009 Institute of Physics

Table 1.2 Average values of the relative differences in depth and widths along [1 1 0] and [1 −1 0] directions of the initial nanoholes (after HF treatment) and the nanoholes obtained with and without a 7-nm-thick buffer layer growth

GaAs buffer layer thickness (nm)	Width [1 1 0] (%)	Width [1 −1 0] (%)	Depth (%)
0	+1	−18	+20
7	+32	−13	−11

facets inside the nanoholes that, under certain growth conditions, incorporate Ga atoms preferentially with respect to (0 0 1) surfaces. Nevertheless, for both surface preparation processes (with or without GaAs buffer layer growth) the patterned nanoholes are preserved before the InAs deposition is performed. Similar qualitative results are obtained for nanoholes corresponding to initial oxides with a "double structure."

The dependence of InAs nucleation process on the crystallographic orientation of the nanoholes can be studied fabricating oxide lines (120 nm width, 500 nm length, and 8 nm height) with orientations along the [1 −1 0], [0 1 0] and [1 1 0]

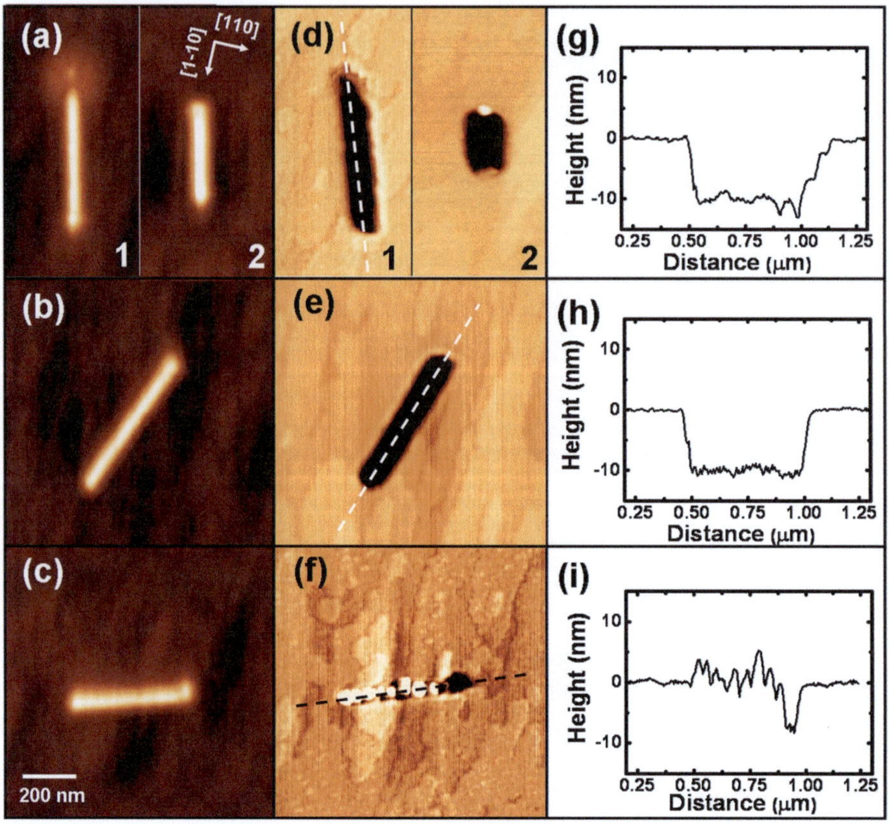

Fig. 1.15 AFM topography images of initial oxide lines fabricated closely along (**a**) [1 −1 0], (**b**) [0 1 0], and (**c**) [1 1 0] crystallographic directions. The results after the deposition of 1.5 ML of InAs at a substrate temperature of 500°C without GaAs buffer layer growth are reported in (**d–f**). The profiles on the AFM images for (**d1**), (**e**) and (**f**) are shown in (**g–i**), respectively. Adapted from [22]. Copyright 2009 Institute of Physics

crystallographic directions as shown in Fig. 1.15a(1), b, c. The results obtained after the deposition of 1.5 ML of InAs on samples without a GaAs buffer layer show that a chain of 3D InAs nuclei is formed in the stripes closely aligned along the [1 1 0] direction (see Fig. 1.15f). In contrast, no clear evidence of InAs nucleation is observed inside stripes aligned closely along the [1 −1 0], and [0 1 0] directions [Fig. 1.15d(1), e, g, h]. Interestingly, only when the width of the pattern stripealigned along the [1 −1 0] direction is increased [from 120 to 160 nm, see Fig. 1.15a(2)], do isolated 3D nuclei appear leaning on their [1 1 0] sides [see Fig. 1.15d(2)]. The extension of B-type facets is clearly directly related with the length of [1 1 0] sides in the nanoholes. These experimental results can be explained if the In atom incorporation, under the growth conditions used, is clearly higher on B-type facets (As-terminated) inside the patterned stripes. Therefore, by increasing

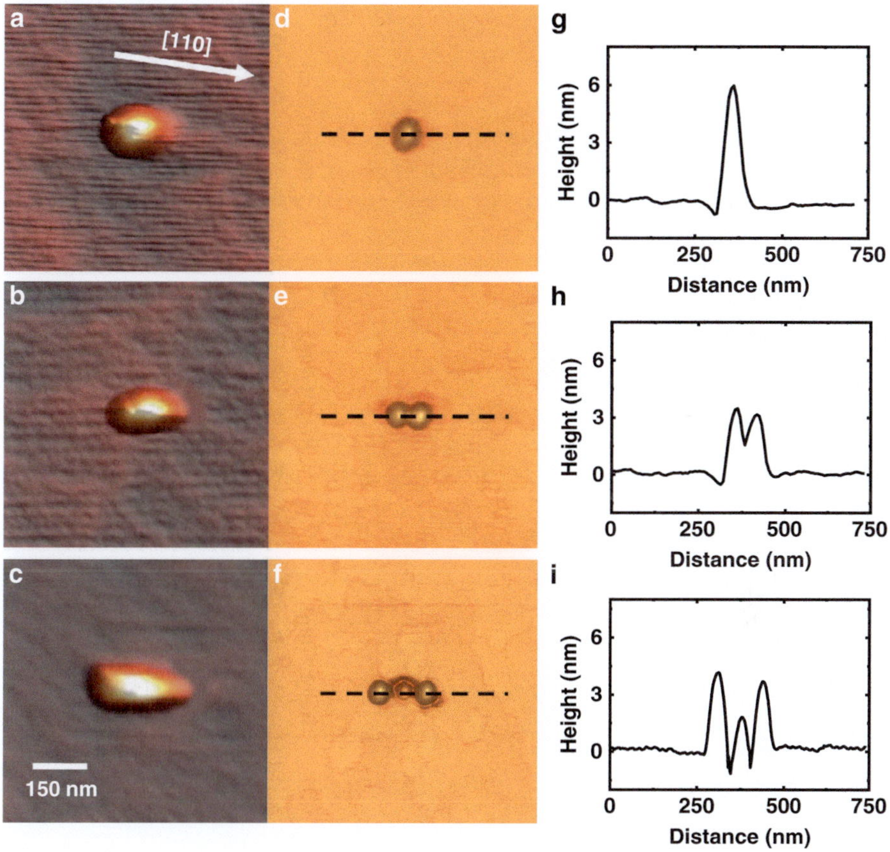

Fig. 1.16 AFM topography images of initial oxide dots with lateral dimensions 160 nm × 160 nm (**a**) and oxide lines fabricated closely along [1 1 0] direction with lateral dimensions 160 nm × 230 nm (**b**) and 160 nm × 270 nm (**c**). The obtained results after deposition of 1.5 ML of InAs on samples without GaAs buffer layer growth are shown in (**d–f**). The profiles along the lateral arrangement of QDs are depicted in (**g–i**). Adapted from [22]. Copyright 2009 Institute of Physics

the B-slope areas inside the stripes, the incorporation rate of In atoms inside this pattern increases, which finally determines the selectivity of the pattern for the InAs nucleation process. Taking into account these findings, it can be inferred that the amount of InAs (i.e., number of QDs formed inside the nanohole) can be controlled by previously designing the nanohole size and shape during the lithography process. Figure 1.16 shows AFM images of the initial oxide dots with lateral dimensions 160 nm × 160 nm (Fig. 1.16a), and oxide lines fabricated closely along the [1 1 0] direction with lateral dimensions 160 nm × 230 nm (Fig. 1.16b), and 160 nm × 270 nm (Fig. 1.16c). The height of both, the oxide dots and the lines, is about 8 nm. The results obtained after deposition of 1.5 ML of InAs on these

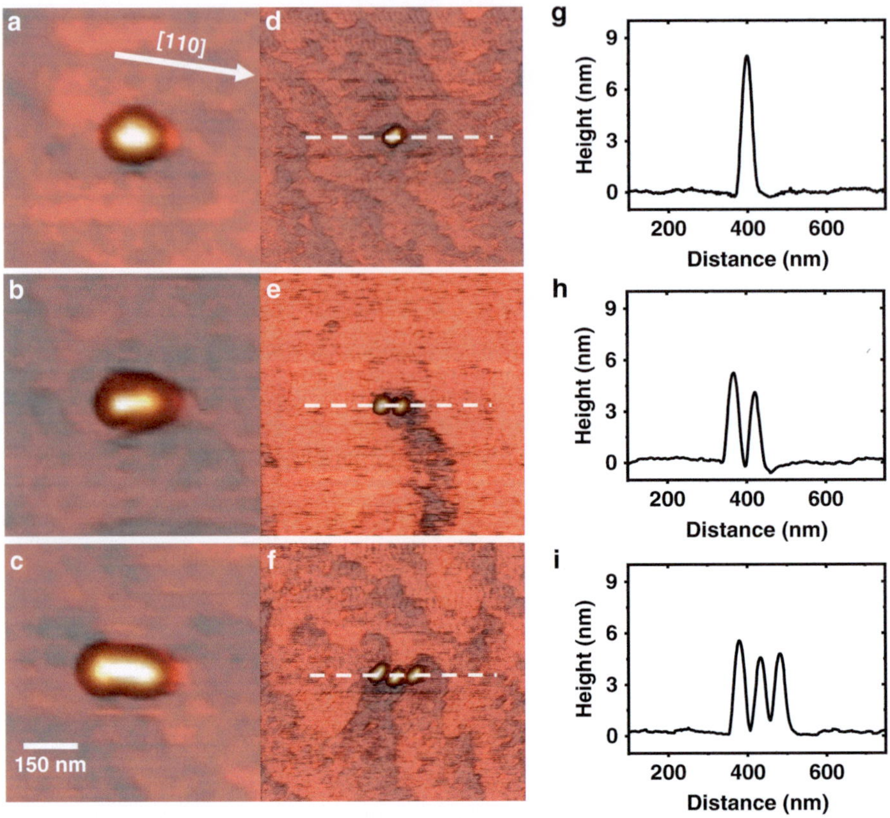

Fig. 1.17 AFM topography images of initial oxide dots with lateral dimensions 160 nm × 160 nm (**a**) and oxide lines fabricated closely along [1 1 0] direction with lateral dimensions 160 nm × 230 nm (**b**) and 160 nm × 270 nm (**c**). The obtained results after deposition of 1.5 ML of InAs on samples with a 7-nm-thick GaAs buffer layer growth are shown in (**d–f**). The profiles along the lateral arrangement of QDs are depicted in (**g–i**). Adapted from [22]. Copyright 2009 Institute of Physics

samples (after oxide etching) without a GaAs buffer layer growth demonstrate that, simply by varying the oxide size along the [1 1 0] direction from 160 to 270 nm, an isolated QD (Fig. 1.16d) or QDMs with two and three QDs in close proximity (Fig. 1.16e, f) can be obtained. These results are reproduced when a 7-nm-thick GaAs buffer layer is grown. An isolated single QD (Fig. 1.17d) is obtained for the initial round shaped oxide dot (Fig. 1.17a) and QDMs with two and three QDs in close proximity (Fig. 1.17e, f) for the initial oxide lines with lengths of 230 nm (Fig. 1.17b) and 270 nm (Fig. 1.17c) closely along the [1 1 0] direction. The average peak-to-peak QDs distance in these QDMs is 54 ± 3 nm, as measured on the AFM profiles shown in Fig. 1.17h, i. This separation is short enough to ensure charge carrier interaction between the adjacent QDs as previously reported [71].

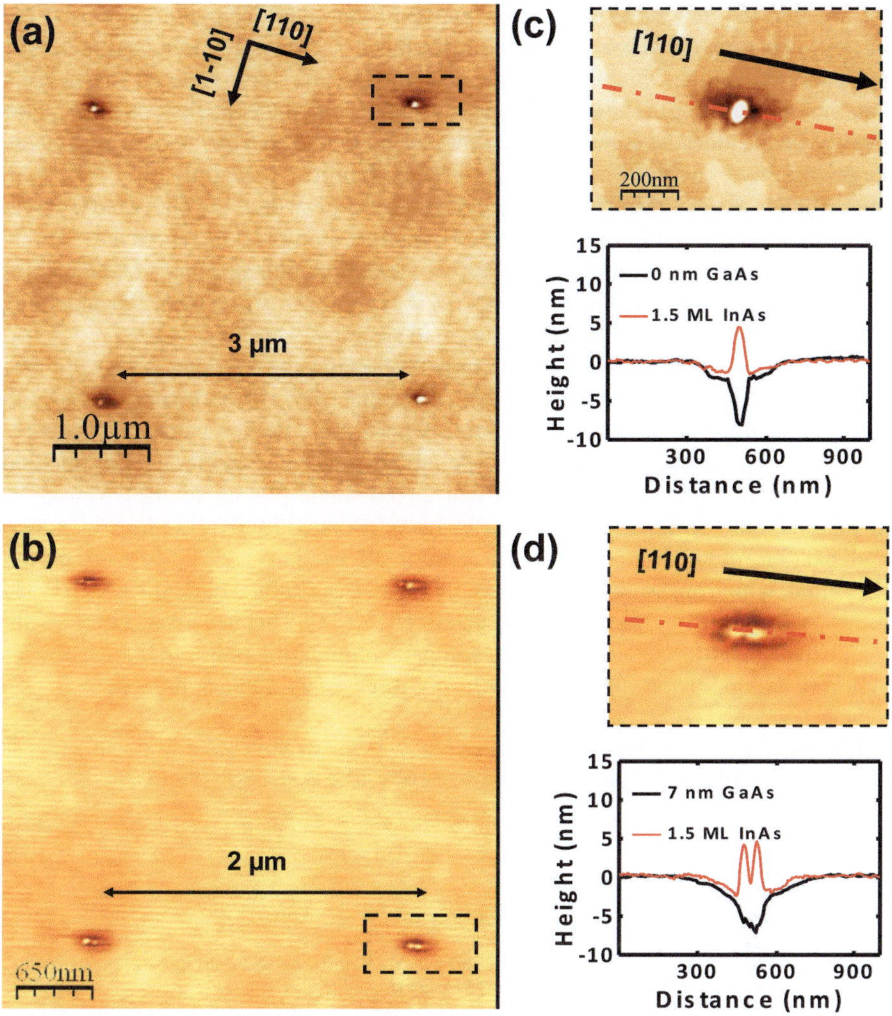

Fig. 1.18 Nucleation of site-controlled single QDs (**a**) and QDs pairs (**b**) after epitaxial growth into a "double structure" nanohole for samples without and with a 7-nm-thick GaAs buffer layer growth, respectively. The profiles drawn on the nanoholes along the [1 1 0] and [1 −1 0] directions are shown at the *right* for single QDs (**c**) and QDs pairs (**d**)

Importantly, these results show that the number of QDs nucleated inside patterned stripes can be controlled by simply varying the length of the initial oxide line along the [1 1 0] direction, with no limitations in QDs number.

In the case of initial oxide dots with a "double structure," single QDs (Fig. 1.18a) and QDs pairs with a peak-to-peak distance of 65 ± 18 nm (Fig. 1.18c) are obtained in a reproducible way for samples without and with GaAs buffer layer growth. The initial fabricated oxide structures are identical and therefore, the number of InAs

QDs inside the nanoholes is directly related to the nanohole evolution during the buffer layer growth. When no GaAs buffer layer is grown, the obtained nanohole preserves its initial "double structure" and single QDs are obtained inside each nanohole (Fig. 1.18b). These results clearly suggest that InAs nucleation takes place mainly in the deepest inner structure of the nanohole that presents a significant larger negative curvature (i.e., lower surface chemical potential) [27, 72]. More interesting is the evolution of the nanohole after the growth of the GaAs buffer layer leading to the formation of an elongated nanohole along [1 1 0] where the "double structure" is lost (Fig. 1.18d). The QDs pair formation in these elongated nanoholes can be explained considering an increase of B-type slopes areas inside the nanohole, similarly to the results obtained on elongated stripes.

The QDs obtained inside the nanoholes following the fabrication process presented in this section show a high optical quality [34]. In particular, the overall emission properties like emission efficiency and single photon emission capacities are not compromised by the nearby patterned interface, validating this method for the future development of laterally arranged single photon sources. Moreover, despite the QDs and the patterned surface are separated by only 7 nm, a biexciton and exciton lifetimes values of $\tau_{xx} = 665 \pm 20$ ps and $\tau_x = 1{,}180 \pm 40$ ps are measured, which are similar to those found for self-assembled quantum dots formed far away from the interface with the substrate [28, 73]. This is a satisfactory result, which must be attributed to the almost complete elimination of non-radiative centers achieved using the optimized re-growth conditions used in this approach.

Acknowledgments The authors would like to acknowledge the MBE group of the "Instituto de Microelectrónica de Madrid (IMM-CNM-CSIC)" where these works were carried out.

References

1. Benson, O., Santori, C., Pelton, M., Yamamoto, Y.: Regulated and entangled photons from a single quantum dot. Phys. Rev. Lett. **84**, 2513–2516 (2000)
2. Stevenson, R.M., Young, R.J., Atkinson, P., Cooper, K., Ritchie, D.A., Shields, A.J.: A semiconductor source of triggered entangled photon pairs. Nature **439**, 179–182 (2006)
3. Yoshie, T., Scherer, A., Hendrickson, J., Khitrova, G., Gibbs, H.M., Rupper, G., Ell, C., Shchekin, O.B., Deppe, D.G.: Vacuum Rabi splitting with a single quantum dot in a photonic crystal nanocavity. Nature **432**, 200–203 (2004)
4. Hennessy, K., Badolato, A., Winger, M., Gerace, D., Atature, M., Gulde, S., Falt, S., Hu, E.L., Imamoglu, A.: Quantum nature of a strongly coupled single quantum dot-cavity system. Nature **445**, 896–899 (2007)
5. Noda, S., Fujita, M., Asano, T.: Spontaneous-emission control by photonic crystals and nanocavities. Nat. Photonics **1**, 449–458 (2007)
6. Wang, L., Rastelli, A., Kiravittaya, S., Benyoucef, M., Schmidt, O.G.: Self-assembled quantum dot molecules. Adv. Mater. **21**, 2601–2618 (2009)
7. DiVincenzo, D.P.: Quantum computation. Science **270**, 255–261 (1995)
8. Xie, Q., Madhukar, A., Chen, P., Kobayashi, N.P.: Vertically self-organized InAs quantum box islands on GaAs(100). Phys. Rev. Lett. **75**, 2542–2545 (1995)

9. Wasilewski, Z.R., Fafard, S., McCaffrey, J.P.: Size and shape engineering of vertically stacked self-assembled quantum dots. J. Cryst. Growth **201–202**, 1131–1135 (1999)
10. Bayer, M., Hawrylak, P., Hinzer, K., Fafard, S., Korkusinski, M., Wasilewski, Z.R., Stern, O., Forchel, A.: Coupling and entangling of quantum states in quantum dot molecules. Science **291**, 451–453 (2001)
11. Ortner, G., Bayer, M., Lyanda-Geller, Y., Reinecke, T.L., Kress, A., Reithmaier, J.P., Forchel, A.: Control of vertically coupled InGaAs/GaAs quantum dots with electric fields. Phys. Rev. Lett. **94**, 157401 (2005)
12. Krenner, H.J., Sabathil, M., Clark, E.C., Kress, A., Schuh, D., Bichler, M., Abstreiter, G., Finley, J.J.: Direct observation of controlled coupling in an individual quantum dot molecule. Phys. Rev. Lett. **94**, 057402 (2005)
13. Stinaff, E.A., Scheibner, M., Bracker, A.S., Ponomarev, I.V., Korenev, V.L., Ware, M.E., Doty, M.F., Reinecke, T.L., Gammon, D.: Optical signatures of coupled quantum dots. Science **311**, 636–639 (2006)
14. Trotta, R., Zallo, E., Ortix, C., Atkinson, P., Plumhof, J.D., van den Brink, J., Rastelli, A., Schmidt, O.G.: Universal recovery of the energy-level degeneracy of bright excitons in InGaAs quantum dots without a structure symmetry. Phys. Rev. Lett **109**, 147401 (2012)
15. Robledo, L., Elzerman, J., Jundt, G., Atatüre, M., Högele, A., Fält, S., Imamoglu, A.: Conditional dynamics of interacting quantum dots. Science **320**, 772–775 (2008)
16. Beirne, G.J., Hermannstädter, C., Wang, L., Rastelli, A., Schmidt, O.G., Michler, P.: Quantum light emission of two lateral tunnel-coupled (In, Ga)As/GaAs quantum dots controlled by a tunable static electric field. Phys. Rev. Lett. **96**, 137401 (2006)
17. Muñoz-Matutano, G., Royo, M., Climente, J.I., Canet-Ferrer, J., Fuster, D., Alonso-González, P., Fernández-Martínez, I., Martínez-Pastor, J., González, Y., González, L., Briones, F., Alén, B.: Charge control in laterally coupled double quantum dots. Phys. Rev. B **84**, 041308 (2011)
18. Wang, L., Rastelli, A., Kiravittaya, S., Atkinson, P., Ding, F., Bufon, C.C.B., Hermannstädter, C., Witzany, M., Beirne, G.J., Michler, P., Schmidt, O.G.: Towards deterministically controlled InGaAs/GaAs lateral quantum dot molecules. New J. Phys. **10**, 045010 (2008)
19. Zallo, E., Atkinson, P., Wang, L., Rastelli, A., Schmidt, O.G.: Epitaxial growth of lateral quantum dot molecules. Phys. Status Solidi B **249**, 702–709 (2012)
20. Alonso-González, P., Martín-Sánchez, J., González, Y., Alén, B., Fuster, D., González, L.: Formation of lateral low density In(Ga)As quantum dot pairs in GaAs nanoholes. Cryst. Growth Design **9**, 2525–2528 (2009)
21. Songmuang, R., Kiravittaya, S., Schmidt, O.G.: Formation of lateral quantum dot molecules around self-assembled nanoholes. Appl. Phys. Lett. **82**, 2892–2894 (2003)
22. Martín-Sánchez, J., Alonso-González, P., Herranz, J., González, Y., González, L.: Site-controlled lateral arrangements of InAs quantum dots grown on GaAs (0 0 1) patterned substrates by atomic force microscopy local oxidation nanolithography. Nanotechnology **20**, 125302 (2009)
23. Lee, J.H., Wang, Z.M., Strom, N.W., Mazur, Y.I., Salamo, G.J.: InGaAs quantum dot molecules around self-assembled GaAs nanomound templates. Appl. Phys. Lett. **89**, 202101–202103 (2006)
24. Yakes, M.K., Cress, C.D., Tischler, J.G., Bracker, A.S.: Three-dimensional control of self-assembled quantum dot configurations. ACS Nano **4**, 3877–3882 (2010)
25. Alonso-Gonzalez, P., Gonzalez, L., Fuster, D., Martin-Sanchez, J., Gonzalez, Y.: Surface localization of buried III–V semiconductor nanostructures. Nanoscale Res. Lett. **4**, 873–877 (2009)
26. Alonso-González, P., González, L., González, Y., Fuster, D., Fernández-Martínez, I., Martín-Sánchez, J., Abelmann, L.: New process for high optical quality InAs quantum dots grown on patterned GaAs (0 0 1) substrates. Nanotechnology **18**, 355302 (2007)
27. Atkinson, P., Ward, M.B., Bremner, S.P., Anderson, D., Farrow, T., Jones, G.A.C., Shields, A.J., Ritchie, D.A.: Site-control of InAs quantum dots using ex-situ electron-beam lithographic patterning of GaAs substrates. Jpn. J. Appl. Phys. **45**, 2519 (2006)

28. Wang, Z.M., Liang, B.L., Sablon, K.A., Salamo, G.J.: Nanoholes fabricated by self-assembled gallium nanodrill on GaAs(1 0 0). Appl. Phys. Lett. **90**, 113120 (2007)
29. Liang, B.L., Wang, Z.M., Lee, J.H., Sablon, K., Mazur, Y.I., Salamo, G.J.: Low density InAs quantum dots grown on GaAs nanoholes. Appl. Phys. Lett. **89**, 043113 (2006)
30. Alonso-Gonzalez, P., Alen, B., Fuster, D., Gonzalez, Y., Gonzalez, L., Martinez-Pastor, J.: Formation and optical characterization of single InAs quantum dots grown on GaAs nanoholes. Appl. Phys. Lett. **91**, 163104 (2007)
31. Alonso-Gonzalez, P., Gonzalez, L., Martin-Sanchez, J., Gonzalez, Y., Fuster, D., Sales, D., Hernandez-Maldonado, D., Herrera, M., Molina, S.: Growth of low-density vertical quantum dot molecules with control in energy emission. Nanoscale Res. Lett. **5**, 1913–1916 (2010)
32. Martín-Sánchez, J., González, Y., González, L., Tello, M., García, R., Granados, D., García, J.M., Briones, F.: Ordered InAs quantum dots on pre-patterned GaAs (0 0 1) by local oxidation nanolithography. J. Cryst. Growth **284**, 313–318 (2005)
33. Kim, J.S., Kawabe, M., Koguchi, N.: Ordering of high-quality InAs quantum dots on defect-free nanoholes. Appl. Phys. Lett. **88**, 072107 (2006)
34. Martín-Sánchez, J., Muñoz-Matutano, G., Herranz, J., Canet-Ferrer, J., Alén, B., González, Y., Alonso-González, P., Fuster, D., González, L., Martínez-Pastor, J., Briones, F.: Single photon emission from site-controlled InAs quantum dots grown on GaAs (0 0 1) patterned substrates. ACS Nano **3**, 1513–1517 (2009)
35. Koguchi, N., Takahashi, S., Chikyow, T.: New MBE growth method for InSb quantum well boxes. J. Cryst. Growth **111**, 688–692 (1991)
36. Koguchi, N., Ishige, K., Takahashi, S.: New selective molecular-beam epitaxial growth method for direct formation of GaAs quantum dots. J. Vac. Sci. Technol. B **11**, 787–790 (1993)
37. Kim, J.S., Koguchi, N.: Near room temperature droplet epitaxy for fabrication of InAs quantum dots. Appl. Phys. Lett. **85**, 5893–5895 (2004)
38. Kim, J.S., Jeong, M.S., Byeon, C.C., Ko, D.-K., Lee, J., Kim, J.S., Kim, I.-S., Koguchi, N.: GaAs quantum dots with a high density on a GaAs (1 1 1)A substrate. Appl. Phys. Lett. **88**, 241911 (2006)
39. Mano, T., Kuroda, T., Sanguinetti, S., Ochiai, T., Tateno, T., Kim, J., Noda, T., Kawabe, M., Sakoda, K., Kido, G., Koguchi, N.: Self-assembly of concentric quantum double rings. Nano Lett. **5**, 425–428 (2005)
40. Cho, A.Y., Arthur, J.R.: Molecular beam epitaxy. Progr. Solid State Chem. **10**(Part 3), 157–191 (1975)
41. Briones, F., González, L., Ruiz, A.: Atomic layer molecular beam epitaxy (ALMBE) of III–V compounds: growth modes and applications. Appl. Phys. A **49**, 729–737 (1989)
42. Alonso-Gonzalez, P., Fuster, D., Gonzalez, L., Martin-Sanchez, J., Gonzalez, Y.: Low density InAs quantum dots with control in energy emission and top surface location. Appl. Phys. Lett. **93**, 183106 (2008)
43. Hernández-Maldonado, D., Herrera, M., Sales, D.L., Alonso-González, P., González, Y., González, L., Pizarro, J., Galindo, P.L., Molina, S.I.: Transmission electron microscopy study of vertical quantum dots molecules grown by droplet epitaxy. Appl. Surf. Sci. **256**, 5659–5661 (2010)
44. Koshiba, S., Nakamura, Y., Tsuchiya, M., Noge, H., Kano, H., Nagamune, Y., Noda, T., Sakaki, H.: Surface diffusion processes in molecular beam epitaxial growth of GaAs and AlAs as studied on GaAs (0 0 1)-(1 1 1)B facet structures. J. Appl. Phys. **76**, 4138 (1994)
45. Shen, X.-Q., Kishimoto, D., Nishinaga, T.: Arsenic pressure dependence of surface diffusion of Ga on nonplanar GaAs substrates. Jpn. J. Appl. Phys. **33**, 11 (1994)
46. Shitara, T., Zhang, J., Neave, J.H., Joyce, B.A.: As/Ga ratio dependence of Ga adatom incorporation kinetics at steps on vicinal GaAs (0 0 1) surfaces. J. Cryst. Growth **127**, 494 (1993)
47. Shen, X.Q., Nishinaga, T.: Arsenic pressure dependence of the surface diffusion in Molecular beam epitaxy on (1 1 1)B-(0 0 1) mesa-etched GaAs substrates studied by in situ scanning microprobe reflection high-energy electron diffraction. Jpn. J. Appl. Phys. **32**, L1117 (1993)
48. Hayakama, T., Morishima, M.: Surface reconstruction limited mechanism of molecular-beam epitaxial growth of AlGaAs on (1 1 1)B face. Appl. Phys. Lett. **59**, 3321 (1991)

49. Royo, M., Climente, J.I., Planelles, J.: Emission spectrum of quasiresonant laterally coupled quantum dots. Phys. Rev. B **84**, 235312 (2011)
50. Moskalenko, E.S., Larsson, M., Karlsson, K.F., Holtz, P.O., Monemar, B., Schoenfeld, W.V., Petroff, P.M.: Enhancement of the luminescence intensity of InAs/GaAs quantum dots induced by an external electric field. Nano Lett. **7**, 188–193 (2006)
51. Kowalik, K., Krebs, O., Lemaitre, A., Laurent, S., Senellart, P., Voisin, P., Gaj, J.A.: Influence of an in-plane electric field on exciton fine structure in InAs-GaAs self-assembled quantum dots. Appl. Phys. Lett. **86**, 041907 (2005)
52. Alén, B., Fuster, D., Fernández-Martínez, I., Martínez-Pastor, J., González, Y., Briones, F., González, L.: Electrical control of a laterally ordered InAs/InP quantum dash array. Nanotechnology **20**, 475202 (2009)
53. Peng, J., Bester, G.: Charged excitons and biexcitons in laterally coupled (In, Ga)As quantum dots. Phys. Rev. B **82**, 235314 (2010)
54. Climente, J.I., Bertoni, A., Goldoni, G.: Photoluminescence spectroscopy of trions in quantum dots: a theoretical description. Phys. Rev. B **78**, 155316 (2008)
55. Laasonen, K., Nieminen, R.M., Puska, M.J.: First-principles study of fully relaxed vacancies in GaAs. Phys. Rev. B **45**, 4122–4130 (1992)
56. Dagata, J.A., Schneir, J., Harary, H.H., Evans, C.J., Postek, M.T., Bennett, J.: Modification of hydrogen-passivated silicon by a scanning tunneling microscope operating in air. Appl. Phys. Lett. **56**, 2001–2003 (1990)
57. García, R., Calleja, M., Rohrer, H.: Patterning of silicon surfaces with noncontact atomic force microscopy: field-induced formation of nanometer-size water bridges. J. Appl. Phys. **86**, 1898–1903 (1999)
58. Gómez-Moñivas, S., Sáenz, J.J., Calleja, M., García, R.: Field-induced formation of nanometer-sized water bridges. Phys. Rev. Lett. **91**, 056101 (2003)
59. Wang, D., Tsau, L., Wang, K.L.: Nanometer-structure writing on Si(1 0 0) surfaces using a non-contact-mode atomic force microscope. Appl. Phys. Lett. **65**, 1415–1417 (1994)
60. Huang, W.P., Cheng, H.H., Jian, S.R., Chuu, D.S., Hsieh, J.Y., Lin, C.M., Chiang, M.S.: Localized electrochemical oxidation of p-GaAs (1 0 0) using atomic force microscopy with a carbon nanotube probe. Nanotechnology **17**, 3838–3843 (2006)
61. Calleja, M., García, R.: Nano-oxidation of silicon surfaces by noncontact atomic-force microscopy: size dependence on voltage and pulse duration. Appl. Phys. Lett. **76**, 3427 (2000)
62. Atkinson, P., Kiravittaya, S., Benyoucef, M., Rastelli, A., Schmidt, O.G.: Site-controlled growth and luminescence of InAs quantum dots using in situ Ga-assisted deoxidation of patterned substrates. Appl. Phys. Lett. **93**, 101908 (2008)
63. Schneider, C., Straub, M., Sünner, T., Huggenberger, A., Wiener, D., Reitzenstein, S., Kamp, M., Höfling, S., Forchel, A.: Lithographic alignment to site-controlled quantum dots for de-vice integration. Appl. Phys. Lett. **92**, 183101 (2008)
64. Kiravittaya, S., Songmuang, R., Rastelli, A., Heidemeyer, H., Schmidt, O.G.: Multi-scale ordering of self-assembled InAs/GaAs (0 0 1) quantum dots. Nanoscale Res. Lett. **1**, 1–10 (2006)
65. Pelucchi, E., Watanabe, S., Leifer, K., Zhu, Q., Dwir, B., De Los Rios, P., Kapon, E.: Mechanisms of quantum dot energy engineering by metalorganic vapor phase epitaxy on patterned nonplanar substrates. Nano Lett. **7**, 1282–1285 (2007)
66. Kiravittaya, S., Heidemeyer, H., Schmidt, O.G.: Growth of three-dimensional quantum dot crystals on patterned GaAs (0 0 1) substrates. Physica E **23**, 253–259 (2004)
67. Martín-Sánchez, J., González, Y., Alonso-González, P., González, L.: Improvement of InAs quantum dots optical properties in close proximity to GaAs (0 0 1) substrate surface. J. Cryst. Growth **310**, 4676–4680 (2008)
68. Sugaya, T., Kawabe, M.: Low-temperature cleaning of GaAs substrate by atomic hydrogen irradiation. Jpn. J. Appl. Phys. **30**, L402 (1991)
69. Tomkiewicz, P., Winkler, A., Krzywiecki, M., Chasse, T., Szuber, J.: Analysis of mechanism of carbon removal from GaAs (0 0 1) surface by atomic hydrogen. Appl. Surf. Sci. **254**, 8035 (2008)

70. Heidemeyer, H., Müller, C., Schmidt, O.G.: Highly ordered arrays of In(Ga)As quantum dots on patterned GaAs (0 0 1) substrates. J. Cryst. Growth **261**, 444–449 (2004)
71. Li, S.S., Xia, J.B., Liu, J.L., Yang, F.H., Niu, Z.C., Freng, S.L., Zheng, H.Z.: InAs/GaAs single-electron quantum dot qubit. J. Appl. Phys. **90**, 6151 (2001)
72. Yang, B., Liu, F., Lagally, M.G.: Local strain-mediated chemical potential control of quantum dot self-organization in heteroepitaxy. Phys. Rev. Lett. **92**, 025502 (2004)
73. Feucker, M., Seguin, R., Rodt, S., Hoffmann, A., Bimberg, D.: Decay dynamics of neutral and charged excitonic complexes in single InAs/GaAs quantum dots. Appl. Phys. Lett. **92**, 063116 (2008)

Chapter 2
InP Ring-Shaped Quantum Dot Molecules by Droplet Epitaxy

Wipakorn Jevasuwan, Somchai Ratanathammapan, and Somsak Panyakeow

Abstract Droplet epitaxy technique is a key fabrication method to create ring-shaped nanostructures. InP ring-shaped quantum dot molecules are grown on In0.5Ga0.5P/GaAs(0 0 1) due to lattice mismatch of 3.8% between InP and In0.5Ga0.5P and isotropic migration property of In atoms during the crystallization step of In droplets on In0.5Ga0.5P. The ring shape, density of the ring and number of dots on the ring are controlled by various growth parameters such as deposition and crystallization temperatures, In deposition rate and thickness. InP ring-shaped quantum dot molecules provide photoluminescence peak at 740 nm (1.66 eV) with FWHM of 45 meV at 20 K. Potential applications of ring-shaped quantum dot molecules in quantum cellular automata are discussed.

2.1 Introduction

Quantum dot molecules (QDMs) are equivalent to artificial molecules while quantum dots (QDs) are considered as artificial atoms. QDMs hence possess more variety of properties and functionalities than a single quantum nanostructure like an individual quantum dot. Their fundamental characteristics and performances are based on how the QDMs are formed in terms of number of quantum dots in each QDM and in terms of their configurations and shapes of QDMs. QDMs are packages of QDs either in vertical or in lateral configuration. Multi-stacks of QDs with narrow spacers grown by residual strain of underneath QD layers are a typical vertical alignment of QDs in this type of QDMs [1]. Multi-stacked QDMs are key nanostructures for laser and solar cell applications [2–5] because of their high dot density and intermediate band formation due to uniform dot size. However,

W. Jevasuwan • S. Ratanathammapan • S. Panyakeow (✉)
Faculty of Engineering, The Semiconductor Device Research Laboratory, Electrical Engineering Department, Chulalongkorn University, Bangkok 10330, Thailand
e-mail: wipakornj@gmail.com; somchai.r@chula.ac.th; s_panyakeow@yahoo.com

J. Wu and Z.M. Wang (eds.), *Quantum Dot Molecules*, Lecture Notes in Nanoscale Science and Technology 14, DOI 10.1007/978-1-4614-8130-0_2,

lateral formation of QDMs gives more variety of geometrical configurations of QDMs depending on their number of QDs per QDM like bi-QDMs [6], quadra-QDMs [7], QDM chains [8] which have potential applications for single electron transistors, spintronics, quantum computation based on quantum cellular automata concept [9 12].

Beside the 0D quantum dot nanostructure, quantum ring (QR) is another feature which exhibits uniqueness due to 1D circular electron transport in ring-shaped nanostructure for magnetic sensitive behavior [13]. Most of the quantum rings are grown by droplet epitaxy magnetic sensitive behavior technique where group III elements like Ga, In in droplets are deposited and are followed by crystallization under pressure of group V elements like As, P [14–16]. During the crystallization process, the droplets will be transformed to ring shape due to out diffusion of group III elements from the center of the droplets because the crystallization is preferably started from the outer rim of the droplets. This technique has a growth merit that it can be conducted for either lattice-matched or lattice-mismatched system. GaAs single quantum rings and GaAs double quantum rings are examples of lattice-matched nanostructures grown by droplet epitaxy [17, 18]. InGaAs quantum rings grown on GaAs are realized from lattice-mismatched system [19, 20].

Combination of QDs and QRs in the same nanostructure is quite interesting due to dual characteristics of both 0D and 1D electron behaviors. Therefore, ring-shaped QDMs become our research target in this article. In order to do so, isotropic ring structure is required to create good ring-shaped QDMs. Isotropic out diffusion of group III elements during crystallization under the pressure of group V elements is a key selection of this ring-shaped QDMs where circular and uniform ring robe can be obtained. It is found that GaAs quantum rings and InGaAs quantum rings exhibit anisotropic property and show noncircular ring shape having difference ring robes along [110] and [1–10] crystallographic directions [21, 22].

InGaP having 0.5 In and 0.5 Ga compositions is lattice matched to GaAs [23]. Conventional growth technique of Stranski–Kastanov (S–K) is used to prepare InGaP buffer layer on GaAs substrate. InP quantum rings are then created on InGaP buffer layer by droplet epitaxy. Isotropic and circular quantum rings are obtained using InP/InGaP material system. Symmetrical and uniform ring robe of quantum rings is very important condition to further growth of ring-shaped QDMs. InP and InGaP have 3.8% lattice mismatching [24], and this condition can lead to the formation of QDs on the ring robe of QRs. Ring-shaped QDMs are then possible when the growth process continues. Control of some growth parameters such as droplet deposition temperature, crystallization temperature, and growth rate can define the number of QDs on the QRs. The dimension as well as the density of ring-shaped QDMs is also affected from those growth parameters. Experimental results will be discussed in the following section of the article.

InP is a direct band-gap semiconductor having good thermal, optical, and electrical properties for high performances of high power photonic and electronic devices [25]. InP nanostructure would further improve the device performances by the quantum effects like carrier and optical confinements. InP QDs as well as InP ring-shaped QDMs can provide also narrow spectrum of red emission at 0.7 μm. Photoluminescence experiment of InP ring-shaped QDMs is also conducted and reported in this article.

2.2 The Formation Mechanism of InP Ring-Shaped QDMs

Semiconductor QDs are normally created by strain relaxation at the interface between lattice-mismatched semiconductor epitaxial layers such as InAs/GaAs. This growth technique is called Stranski–Krastanov (S–K) growth mode and is widely used in solid source molecular beam epitaxy (MBE) to prepare InAs QD nanostructure having good crystal quality for various nanoelectronic and nanophotonic applications.

This growth mechanism is self-assembly providing random individual InAs QDs spreading on the plane of epitaxial layer after the critical thickness of InAs wetting layer is reached (Fig. 2.1). InAs QDMs can be self-assembled by modification of growth process such as thin capping of initial InAs QDs leading to particular templates for InAs QDMs formation [26]. Partially capped InAs QDs can also provide QR templates for InAs bi-QDMs using gas source MBE [27]. The origin of InAs QDMs in these cases is based on elastic strain relaxation.

Droplet epitaxy is another alternative technique to prepare quantum nanostructures from both lattice-matched system such as GaAs on GaAs and small-lattice mismatched system such as InGaAs on GaAs. The growth process consists of two steps, i.e., droplet deposition and crystallization. The crystallization of the droplets begins at the outer edge of the droplets leading to the formation of QRs such as GaAs QRs and InGaAs QRs on GaAs substrates [28, 29]. Out diffusion of Ga and InGa atoms in the droplets under As pressure is the main growth mechanism of respective QRs (Fig. 2.2).

It is also found that there is some etching effect in the middle part of the Ga and InGa droplets in contact with the surface of GaAs buffer layer, which leads to deeper nanoholes inside the QRs. There will remain some residual strain field inside those deep nanoholes which originates InAs QD formation around and inside QRs [30, 31]. Anisotropy also occurs during InGaAs QR formation and gives birth to square-shaped nanoholes which also become templates for InAs quadra QDMs [7]. Combination of QR and QD formations in the same consecutive MBE growth

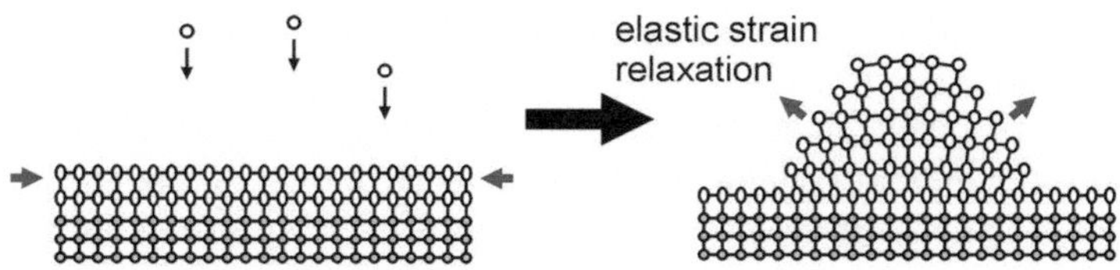

Fig. 2.1 Formation mechanism of self-assembled QDs by S–K growth mode

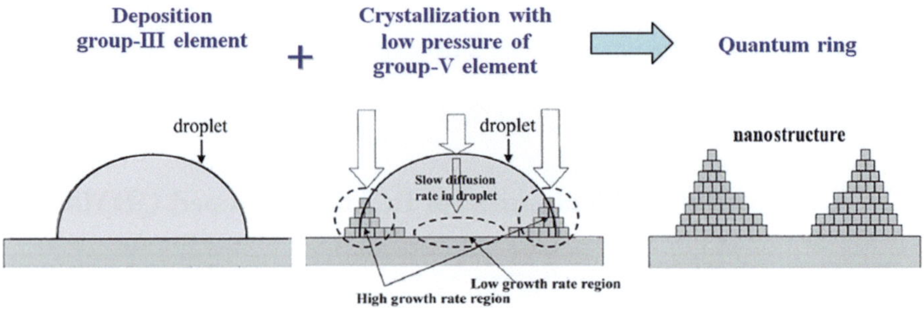

Fig. 2.2 Formation mechanism of self-assembled QRs by droplet epitaxy

process using both droplet epitaxy and strain relaxation would give more complex nanostructures of QDMs which are composed of both QD and QR at the same site.

In this article, we focus on the formation mechanism of InP ring-shaped QDMs which is quite different from the above-mentioned explanations in terms of both the material system and the outcome of combined QR/QD. In order to create InP quantum nanostructures, InGaP with In:Ga content of 50:50 layer is firstly prepared on GaAs substrate by conventional MBE growth method. This InGaP buffer layer is lattice matched to GaAs substrate; therefore, no strain exists in this bi-layer structure. Droplet epitaxy is applied for the whole process in creating InP ring-shaped QDMs. The very first step starts from In droplet deposition. Then, In droplets are exposed to P pressure at optimal crystallization temperature to form InP QRs. It is found that at small thickness of In droplets, only InP QRs are formed. These InP QRs have symmetrical circular shape due to isotropic crystallization in all crystallographic directions in the plane of growth. The ring robe of QR is uniform in terms of both width and height along the whole circumference of QR. When the thickness of In droplets becomes bigger, the InP QRs are transformed to InP ring-shaped QDMs during the crystallization process of those thick In droplets. The formation mechanism of InP QRs is similar to the explanation of InGaAs QRs. The only difference is their symmetrical circular shape of InP QRs compared to nonsymmetrical elliptical shape of InGaAs QRs. However, at thick In droplets, InP QRs are initially formed. The lattice mismatch between InP and InGaP is about 3.8%. This accumulates a strain in the InP QRs. When thickness of ring robe reaches a critical value, the same strain relaxation works and creates InP QD chain on the ring robe. The number of QDs on the ring robe depends on the droplet size which determines the dimension of ring diameter of InP ring-shaped QDMs. Smaller ring diameter has less number of QDs on the ring robe. The bigger ring diameter gives more QDs on the ring robe.

With single droplet epitaxy growth process, InP ring-shaped QDMs are created by the formation of QR as commonly being obtained in most of droplet epitaxy, and at the same time QDs are created on the ring robe by strain relaxation mechanism due to lattice mismatch between InP and InGaP epitaxial layers (Fig. 2.3).

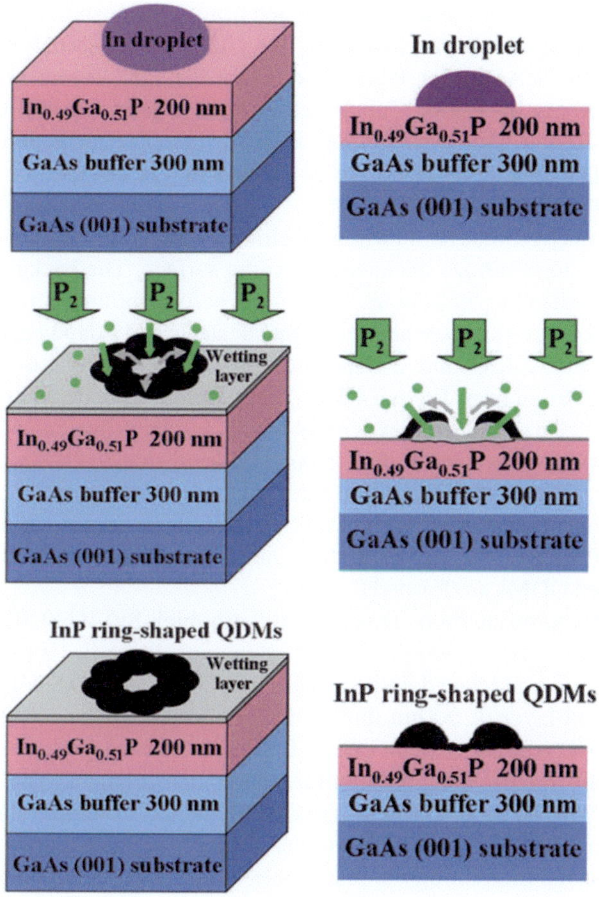

Fig. 2.3 Formation mechanism of InP ring-shaped QDMs by droplet epitaxy together with strain relaxation due to lattice mismatching of InP/InGaP

2.3 Experimental Works on Droplet Epitaxy of InP Ring-Shaped QDMs

InP ring-shaped QDMs are created by the droplet epitaxy technique. GaAs(0 0 1) substrate is used as the starting material. The GaAs wafer surface is cleaned by heating at 450°C for 1 h in the introduction chamber of MBE machine to eliminate the moisture. Then, the GaAs substrate is transferred to the growth chamber for surface oxide desorption under As pressure with BEP of 6×10^{-6} Torr at slow ramping of substrate temperature up to 610°C. After the oxide desorption process, GaAs surface becomes rough. It is necessary to grow 300 nm thick GaAs buffer layer on top to provide smooth GaAs surface for further MBE growth. Now, the environment in the growth chamber is changed from As atmosphere to P atmosphere for the preparation of P-based materials. We choose In0.5Ga0.5P as buffer layer being grown on GaAs buffer layer. These two epitaxial layers are well lattice matched. A 200 nm thick In0.5Ga0.5P buffer layer is grown at 470°C with a BEP V/III ratio of 10 and with a growth rate of 0.5 ML/s. The formations of these bi-buffer layers (In0.5Ga0.5P/GaAs) are confirmed by the observation of 2×4 and 2×1 RHEED patterns, respectively.

Droplet epitaxy process starts when substrate temperature is decreased to 250°C without P beam to minimize the excess P on the surface. The background pressure in the growth chamber is kept at 10^{-9} Torr before indium droplet deposition begins. Indium droplets are spread on the InGaP surface at a deposition temperature between 120 and 290°C and a deposition rate from 0.2 to 1.6 ML/s. Indium thicknesses are varied from 1.6 to 6.4 ML. These variations of deposition parameters define the droplet size and droplet density which will be prime condition to control the configuration and number of dots in each InP ring-shaped QDMs.

Later on, P beam is introduced into the growth chamber again for crystallization process of InP nanostructures. The crystallization temperature is varied from 150 to 300°C under P BEP of 4×10^{-6} Torr for 5 min. At appropriate growth parameters, InP ring-shaped QDMs are formed.

InP nanostructures are then capped by 100 nm thick In0.5Ga0.5P double layers grown by the two-step technique. The first capping layer is 10 nm thick InGaP grown by migration enhanced epitaxy (MEE) at 300°C with 0.5 ML/cycle growth rate. The second layer of 90 nm thick InGaP is grown by the conventional MBE process at 470°C with the growth rate of 0.5 ML/s. This capped InP nanostructure sample is ready for ex situ photoluminescence measurement. The same growth process of InP ring-shaped QDMs is repeated on the top of sample surface with the same growth parameters to produce another layer of InP ring-shaped QDMs. Now, the sample surface morphology is ready to be observed by a tapping mode AFM. Fractional part of this sample is also the cross section observed by TEM to confirm the ring-shaped nanostructure. Schematic sample structures at different processing steps are displayed in Fig. 2.4.

Major characterization of InP ring-shaped QDMs is conducted by photoluminescence (PL) measurement. The PL intensity reflects the crystal quality of the nanostructures as well as their density. PL spectra give specific identity of the nanostructures as well as their uniformity. Ar+ laser with emission line of 478 nm is used to excite the sample. The excitation power is varied from 10 to 80 mW. The laser beam is chopped and focused onto the sample placed in a cryostat with a cooling sample temperature from room temperature down to 20 K. The PL signal is collected and resolved by a 1-m monochromator. The resolved PL signal is detected by a liquid nitrogen cooled InGaAs detector and sent to a lock-in amplifier for PL data analysis.

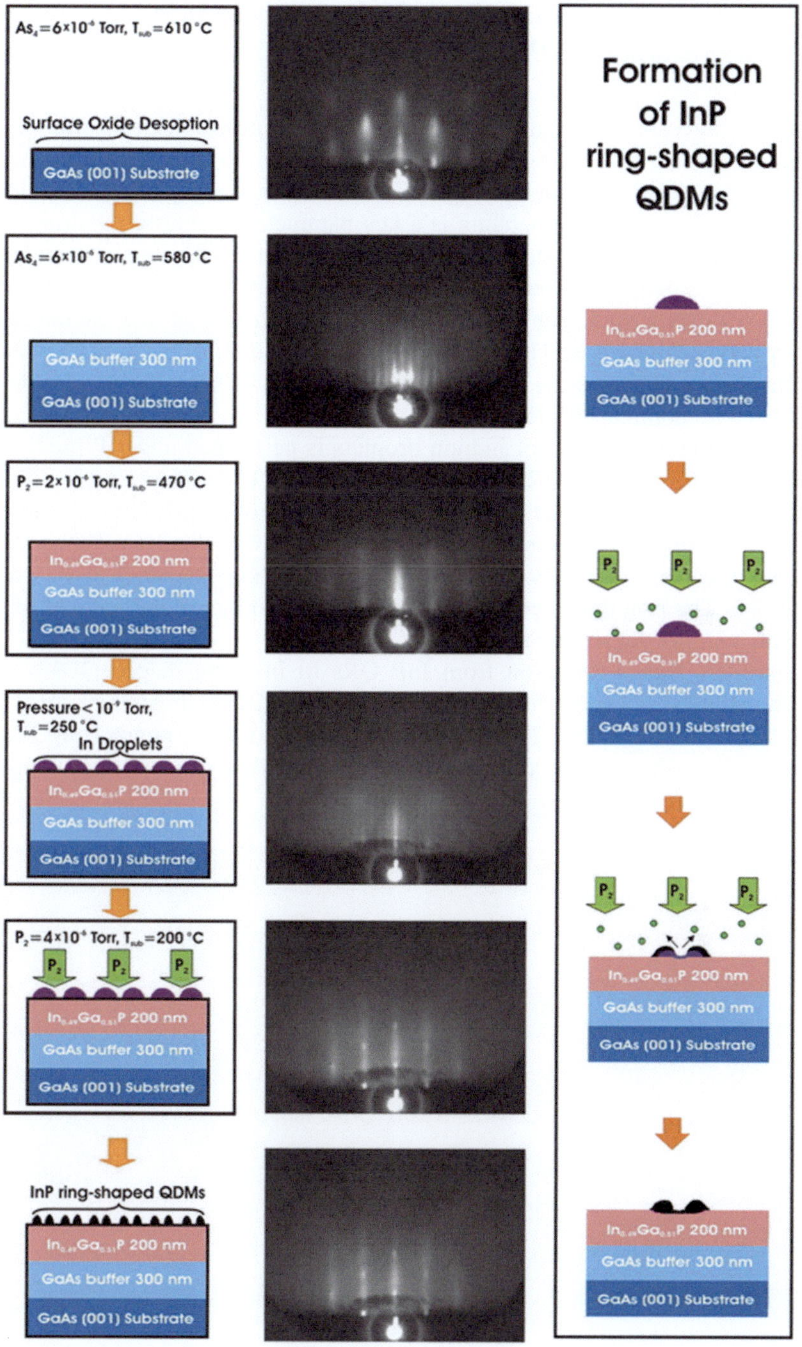

Fig. 2.4 Schematic diagrams of the sample structures grown at different steps and their in situ RHEED observations

2.4 Structural Evolution from In Droplets to InP Ring-Shaped QDMs

Indium droplets are starting elemental materials deposited on InGaP buffer layer at the beginning step of droplet epitaxy. Indium thickness (in ML), deposition rate (in ML/s), and deposition temperature (in °C) will define the droplet size and droplet density. The deposition temperature is a key parameter to control the droplet size and droplet density. At low deposition temperature, the droplet size is small but the droplet density is high. When the deposition temperature is higher, nearby small droplets will merge with each other resulting in a bigger droplet size, but the droplet density will become lower. The deposition rate will also affect the droplet evolution in a more complicated way. The indium thickness will keep the droplet density saturated with nominal droplet size.

When the crystallization of indium droplets is performed under P pressure, In droplets transform to InP QRs and InP ring-shaped QDMs, respectively. At small indium thickness, In droplets will be crystallized only to InP QRs as shown in Fig. 2.5. The ring has an isotropic shape and a small ring diameter. When indium thickness increases, InP QRs are transformed to InP ring-shaped QDMs as shown in Fig. 2.6. The ring diameter becomes longer with an increasing number of QDs and with a bigger dot size on the ring.

Figure 2.7 shows evolution from InP QRs to InP ring-shaped QDMs when indium thickness is varied from 1.6 to 6.4 ML (deposition temperature 250°C and crystallization temperature 200°C). Ring diameter, dot size, number of QDs per QDM, and density of ring-shaped QDMs are increased due to larger supply of

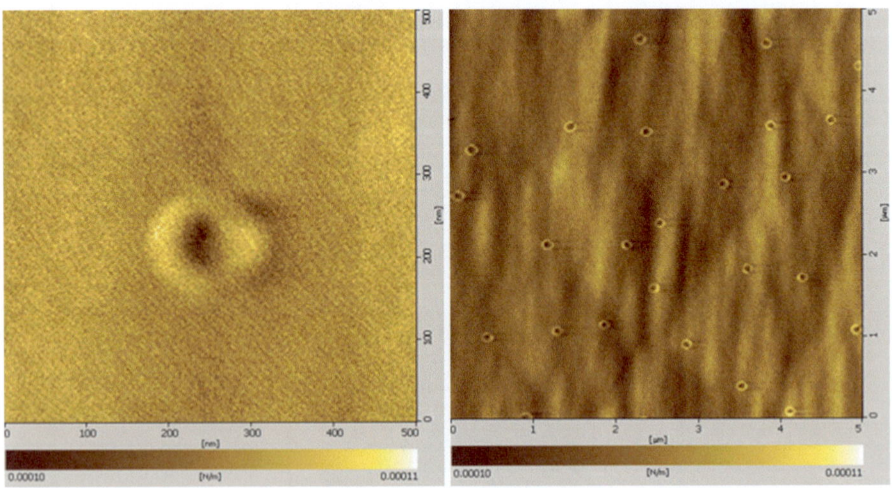

Fig. 2.5 AFM image of InP QRs grown from In thickness 1.6 ML (crystallization temperature 200°C)

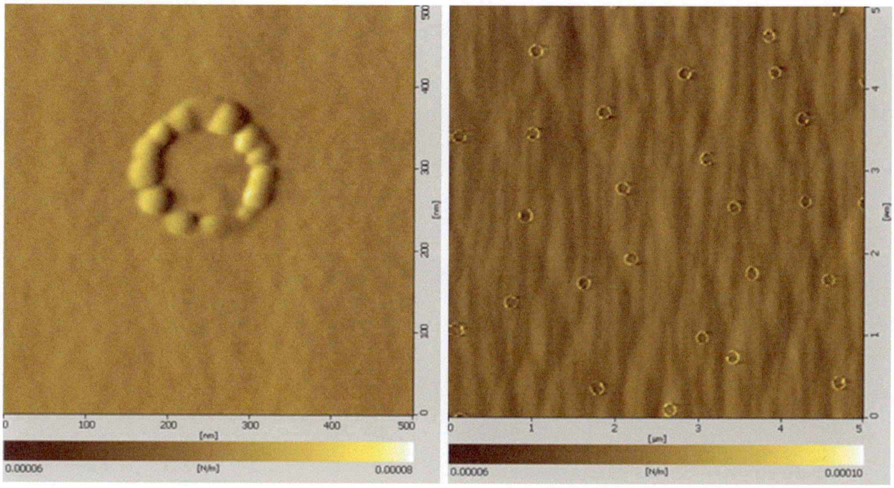

Fig. 2.6 AFM image of InP ring-shaped QDMs grown from In thickness 3.2 ML (crystallization temperature 200°C)

indium volume to each droplet spreading onto InGaP buffer layer. In this series of droplet epitaxy, the lowest indium thickness of 1.6 ML can provide only InP QR shape due to the limit volume of group III element. The ring robe is thin only around 1 nm, which does not give enough strain at InP/InGaP interface for QD formation. When thicker indium is deposited, QR becomes bigger in diameter. Strain is exhibited on the ring robe and then is relaxed by forming circular QD chain on the ring. QD uniformity is optimized at appropriate indium thickness. Further increase of indium thickness, QD size becomes larger and nonuniform.

Evolution of In droplets to InP ring-shaped QDMs by droplet epitaxy is shown step-wise in Fig. 2.8. The InP ring formation is originated from the crystallization at the outer peripheral of In droplets under P pressure together with the out diffusion of In atoms from the center of the droplets. At the same time, InP/InGaP lattice mismatch becomes critical when In thickness reaches a certain value. The strain relaxation happens at the ring robe and gives birth to circular QD chain formation on the ring.

Deposition temperature is a key parameter to control the ring diameter, dot size, and number of QDs on the ring robe. Deposition temperature is varied from 120 to 290°C for In deposition rate of 0.8 ML/s and In thickness of 3.6 ML. At a higher deposition temperature, a longer ring diameter, larger QD size, and larger number of QDs on the ring are obtained due to larger droplets being created by merging of small indium droplets. The density of InP ring-shaped QDMs is therefore decreased. The experimental results in this series of droplet epitaxy are displayed in Fig. 2.9. The uniformity of QD on the ring is optimized at a deposition temperature of 250°C. This value of deposition temperature is therefore widely used in other series of droplet epitaxy.

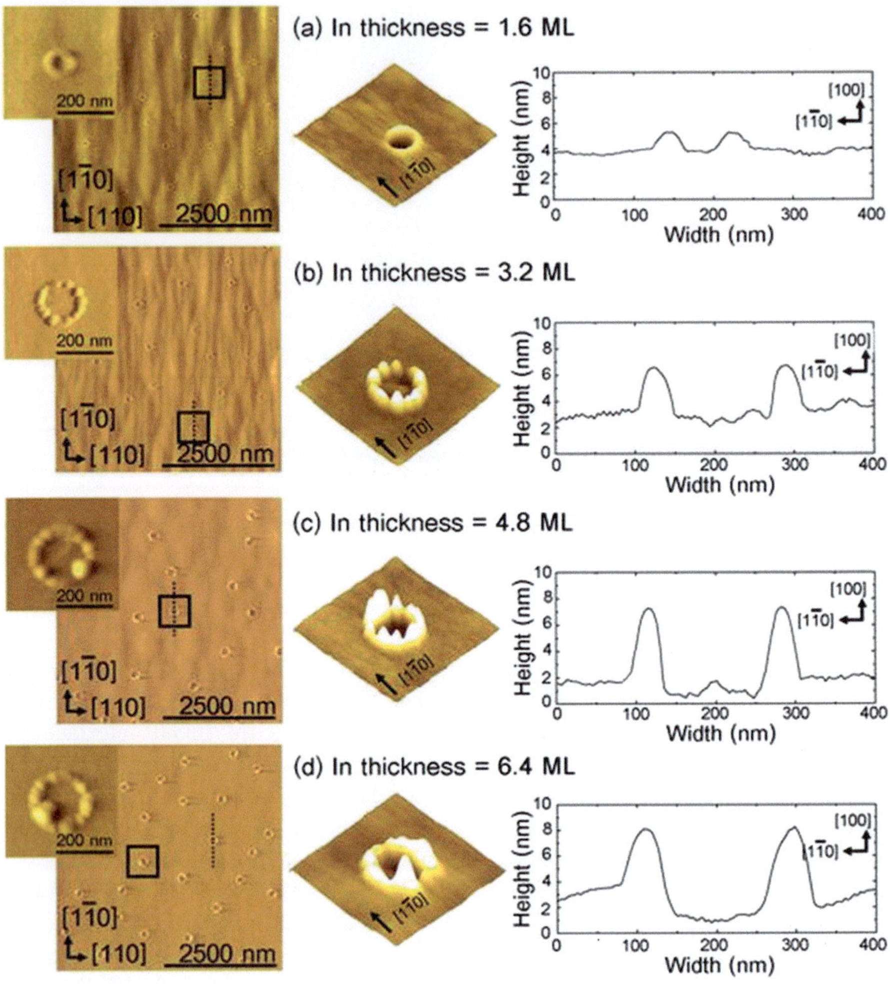

Fig. 2.7 In droplets are transformed to InP QRs and InP ring-shaped QDMs when more indium thickness is supplied from 1.6 to 6.4 ML during the deposition process. The deposition temperature and crystallization temperature are fixed at 250 and 200°C, respectively

When growth conditions at the deposition step are optimized and fixed like In thickness, In deposition rate and deposition temperature, in the next consecutive step of crystallization, InP ring-shaped QDMs will be affected by the crystallization temperature. At higher crystallization temperature, the ring diameter is longer but the QDM density and the number of QDs per QDM are smaller. Figure 2.10 shows the experimental results when crystallization temperature is varied from 150 to 300°C (deposition temperature 250°C, deposition rate 0.8 ML/s, indium thickness 3.2 ML). However, the effect of crystallization temperature on InP ring-shaped QDMs formation seems to be a secondary one. No large change in the nanostructure

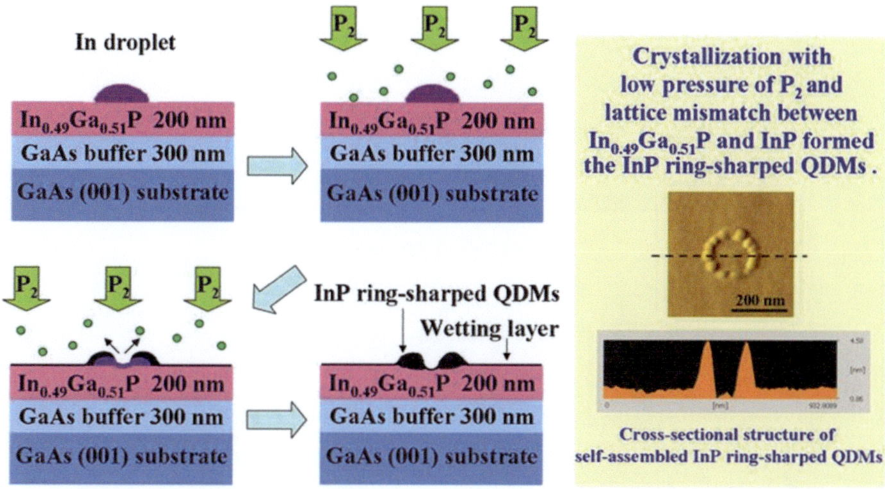

Fig. 2.8 Schematic displays of each processing steps of InP ring-shaped QDMs on InGaP/GaAs by droplet epitaxy technique

is observed, compared to the effect of primary step in the deposition process. Uniform QD is optimized at a crystallization temperature of 200°C.

In several series of experiments on droplet epitaxy, we can optimize the growth conditions to give fine InP ring-shaped QDMs by setting the deposition temperature and crystallization temperature at 250 and 200°C, respectively. The ring shape and density as well as the dot size are mainly defined by indium thickness during deposition step.

The number of QDs per QDM is another feature that is needed to be controlled for QD device applications. QDMs with eight QDs in ring shape are one of our aims in fabricating InP ring-shaped QDMs. This octa-quantum dot molecule will be a quantum dot set applicable for extended quantum dot cellular automata (EQCA). The concept of QCA will be discussed in detail in the following section.

When InP ring-shaped QDMs are capped by 100 nm thick InGaP for analytical photoluminescence measurement and for multi-stacked nanostructure, ring-shaped QDMs should be little affected by InP/InGaP matrix by using first InGaP capping layer grown by migration enhanced epitaxy technique at a low temperature of 300°C. To confirm the existence of ring-shaped QDMs, the sample is then prepared for TEM observation of their cross-sectional profiles. TEM image of InP ring-shaped QDMs grown at a deposition temperature of 250°C and a crystallization temperature of 200°C with indium deposition rate of 1.6 ML/s and indium thickness of 3.2 ML is shown in Fig. 2.11. TEM image can give actual dimensions of the nanostructure as follows. The outer and inner ring diameters are 140 and 80 nm, respectively. The basement of QD is approximately 40 nm with the dot height of few nm. These data are collected from the cross-sectional profile of the topmost nanostructure which has no thermal annealing effect by the capping process.

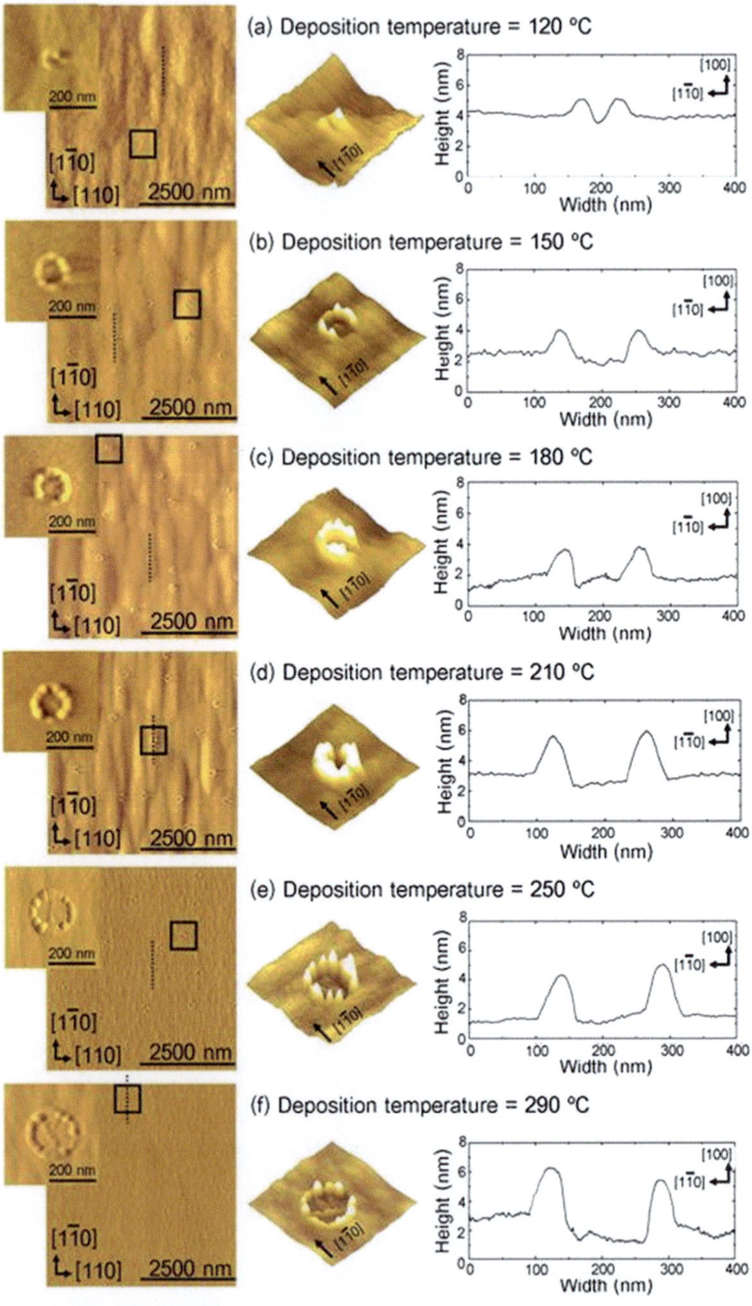

Fig. 2.9 When deposition temperature is varied from 120 to 290°C (crystallization temperature 200°C, In deposition rate 0.8 ML/s, In thickness 3.6 ML), InP ring-shaped QDMs become bigger in ring diameter, dot size, and number of dot on the ring with less QDM density due to merging of small droplets to larger droplet size leading to less droplet density at high deposition temperature

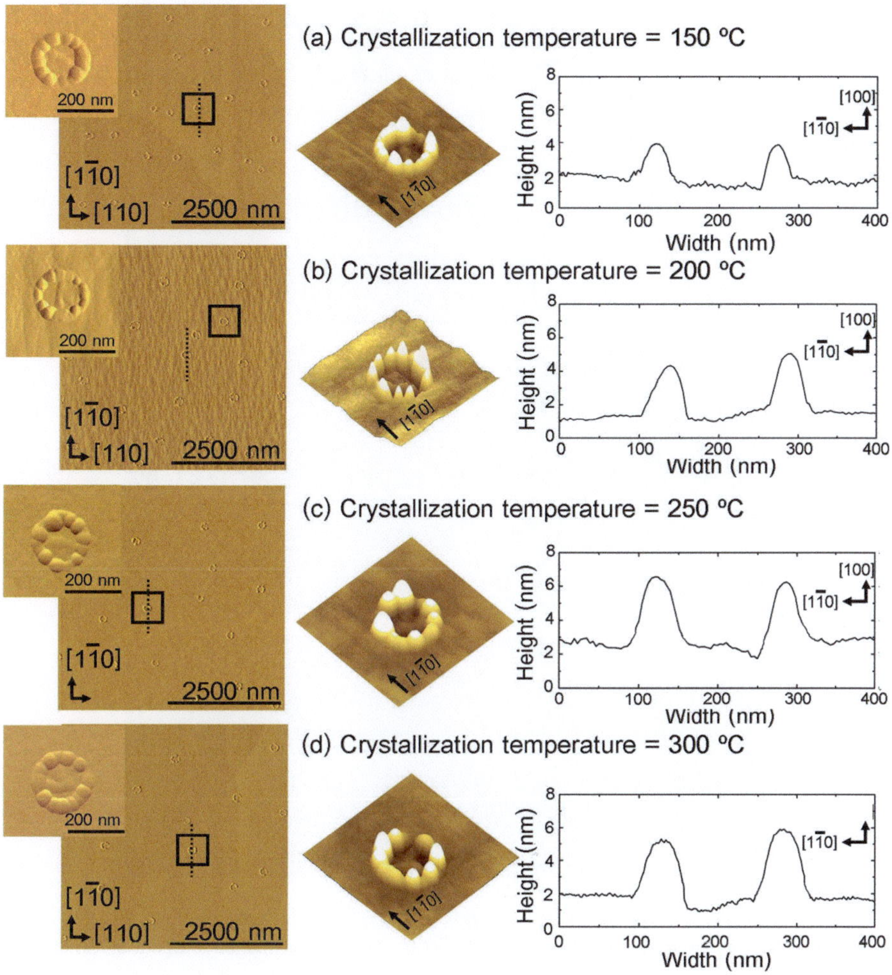

Fig. 2.10 InP ring-shaped QDMs being created at different crystallization temperatures from 150 to 300°C (In thickness 3.2 ML, In deposition rate 0.8 ML/s, and deposition temperature 250°C)

Capped samples of several InP ring-shaped nanostructures are also tested for their photoluminescence spectra. Their optical characteristics will be reported and discussed in the following section.

2.5 Photoluminescence Spectra of InP Ring-Shaped QDMs

InP is a direct bandgap (1.344 eV) semiconductor material. It gives strong light emission at a near infrared wavelength. When InP nanostructure is formed and is capped by a wider bandgap of InGaP, the emission wavelength is shifted to a shorter one in the red light region.

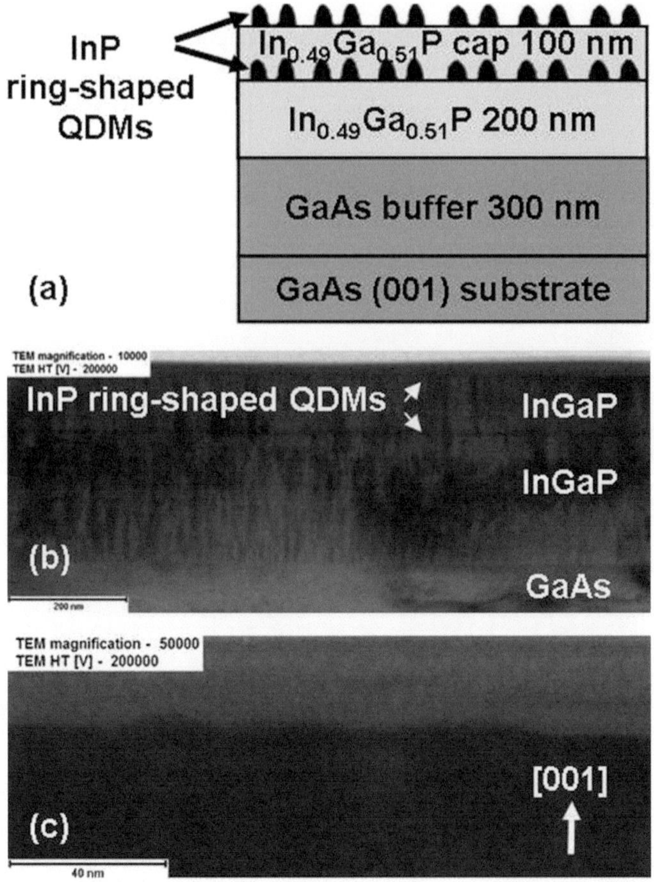

Fig. 2.11 TEM image of InP ring-shaped QDMs grown from In thickness 3.2 ML with deposition rate of 1.6 ML/s at a deposition temperature of 250°C and a crystallization temperature of 200°C

When InP ring-shaped QDMs are excited by blue laser, PL emission at red light will be detected.

We use Ar+ laser with a 478 nm wavelength and laser power up to 80 mW for our PL excitation. The sample is set up in a liquid Helium cryostat for cooling down to 20 K. PL signal is detected by either a CCD detector or a liquid nitrogen cooled InGaAs detector.

Photoluminescence (PL) is a basic analytical tool in studying the formation of InP ring-shaped QDMs throughout our whole experiment. InP ring-shaped QDMs are composed of QD chains on the ring robe. QDs on the ring are self-assembled. Therefore, they are not uniform in QD size. The difference in QD size leads to a broad PL spectrum. The PL spectroscopy is used to determine the homogeneity of the QDs on the ring as well as their crystal quality.

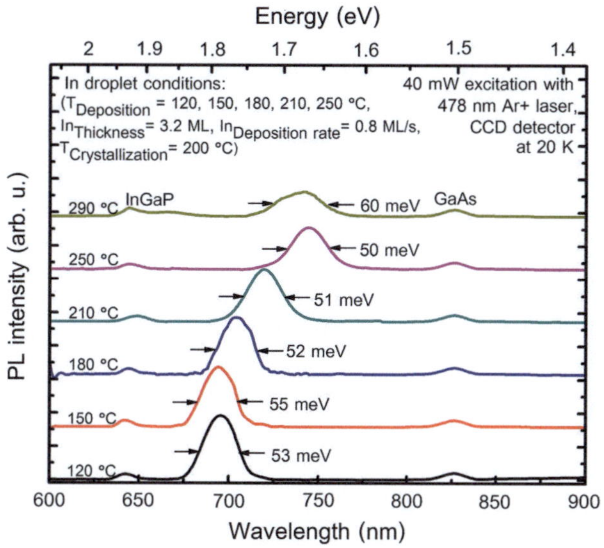

Fig. 2.12 PL spectra of InP ring-shaped QDMs with various deposition temperatures (120–290°C) excited by 40 mW Ar+ laser at 20 K

InP ring-shaped QDM samples grown with different deposition temperatures are investigated for their PL spectra. It is found that, at higher deposition temperatures, PL peaks are red-shifted due to a larger dot size. PL intensity becomes weakened due to lower QD density at a higher deposition temperature. FWHM is broader due to relaxed dot uniformity when deposition temperature is much higher than 250°C. (See details in Fig. 2.12.) This is a reason for fixing our deposition temperature at 250°C for the rest of our experiment.

At a fixed deposition temperature of 250°C, InP ring-shaped QDM samples are grown at different crystallization temperatures. Those samples are investigated for their PL spectroscopy. It is found from Fig. 2.13 that crystallization temperature is optimized at 200°C where PL peak becomes the highest and its FWHM is the narrowest.

From these two series of PL experiments, we find out that the deposition and crystallization temperatures have similar effects on the physical and optical properties of InP ring-shaped QDMs. The effect of deposition temperature is remarkably strong, since the QD and ring-shaped QDM sizes are determined during the deposition process prior to the following crystallization. We conclude that the deposition and crystallization temperatures are likely to be at 250 and 200°C, respectively. Under these conditions, PL peak is emitted at 750 nm with an FWHM of 50 meV at 20 K.

The next growth parameter to be investigated is In deposition rate. InP ring-shaped QDMs are grown at deposition and crystallization temperatures of 250 and 200°C with different In deposition rates (0.2–1.6 ML/s). Those samples are then evaluated for their PL spectroscopy as shown in Fig. 2.14.

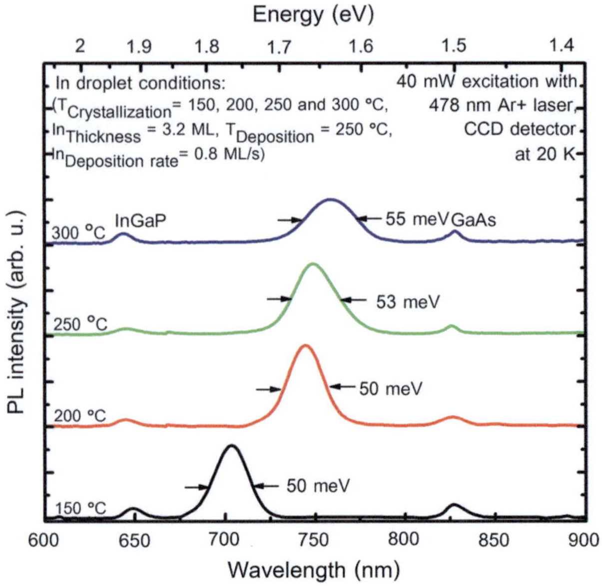

Fig. 2.13 PL spectra of InP ring-shaped QDMs with various crystallization temperatures (150–300°C) excited by 40 mW Ar+ laser at 20 K

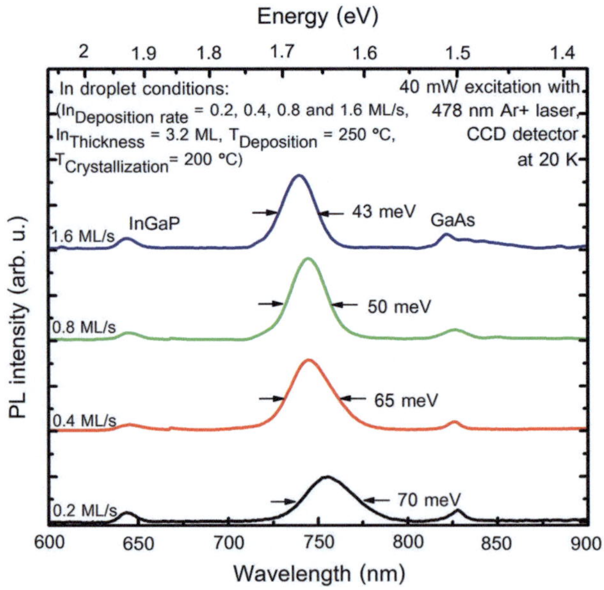

Fig. 2.14 PL spectra of InP ring-shaped QDMs with various indium deposition rates (0.2–1.6 ML/s) excited by 40 mW Ar+ laser at 20 K

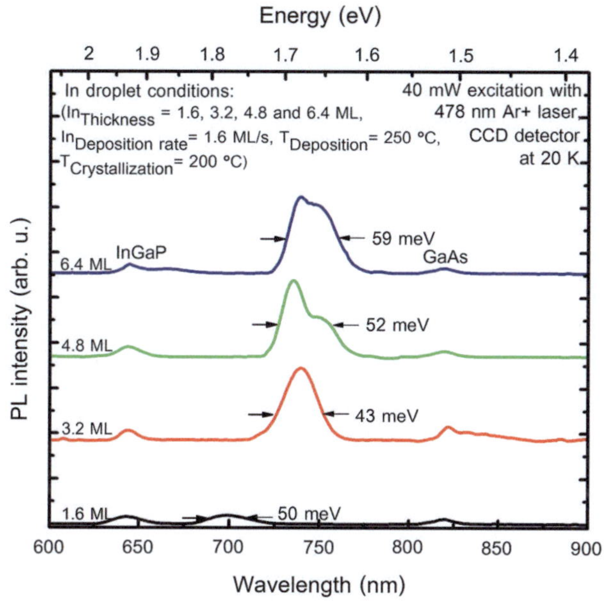

Fig. 2.15 PL spectra of InP QRs and InP ring-shaped QDMs with different indium thicknesses (1.6–6.4 ML) excited by 40 mW Ar+ laser at 20 K

It is found from this series of experiment that the higher indium deposition rate yields the better QD quality, and QD uniformity can be achieved. At 1.6 ML/s indium deposition rate, PL peak is blue-shifted to 740 nm wavelength with FWHM of 43 meV. We find also that, at this 1.6 ML/s deposition rate of indium, the percentage of creating octa-QDMs is as high as 46%. It is unfortunate that higher deposition rate than 1.6 ML/s could not be conducted due to the limitation of In cell temperature in our MBE equipment.

Final growth parameter to be optimized is indium thickness which gives more volume to the formation of InP ring-shaped QDMs. At a low indium thickness of 1.6 ML, only InP QRs are formed. Densities of QDs and of ring-shaped QDMs, number of QDs per ring-shaped QDMs and dimensions of QDs and of ring-shaped QDMs are increased with indium thickness greater than 1.6 ML. InP QRs and InP ring-shaped QDM samples with different indium thicknesses (1.6–6.4 ML) are evaluated for their crystal quality and homogeneity via PL measurement. At higher indium thickness than 3.2 ML, QDMs lose their dot uniformity. PL peak begins to split into two peaks because of the QD inhomogeneity attributed to two variations of QD height which can be confirmed from their respective AFM images. PL spectra have broader FWHM as shown in Fig. 2.15.

From the whole series of PL experiments, we optimize the growth conditions for InP ring-shaped QDM formation by droplet epitaxy as follows. The deposition

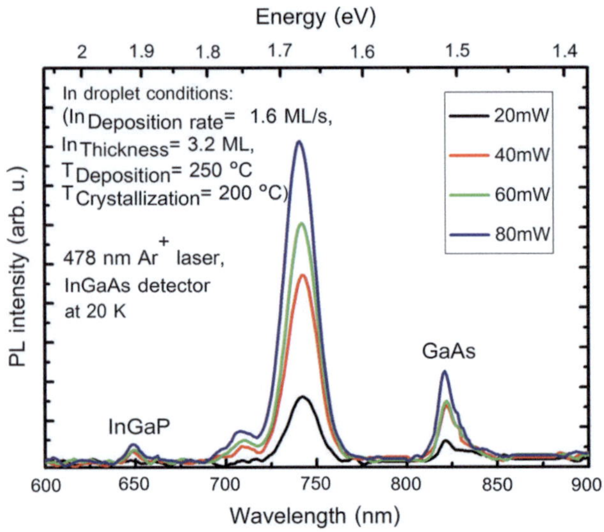

Fig. 2.16 Power dependence of PL emission from InP ring-shaped QDMs at 20 K

and crystallization temperatures are 250 and 200°C, respectively, with indium deposition rate of 1.6 ML/s and with indium thickness of 3.2 ML. Under these optimized growth conditions, the best QD size uniformity and the percentage of octa-QDMs as high as 46% can be maintained.

In order to understand the optical behavior of InP ring-shaped QDMs, further experiment on photoluminescence is continued by changing the excitation power and sample temperature. Figure 2.16 shows laser power dependence of PL emission from InP ring-shaped QDMs. The main PL peak is originated from the ground state having an emission wavelength of 740 nm, which is slightly blue-shifted at high excited laser power. PL peaks have FWHM values of 40–45 meV reflecting the uniformity of QDs on the rings. At high laser power, a weaker PL peak from the first excited state is also detected at 710 nm. In the PL spectra, PL emissions from InGaP and GaAs bulk materials are also observed at 650 and 850 nm, respectively.

When laser power excitation is fixed at 40 mW and sample temperature is varied from 20 to 100 K, temperature dependence of PL from InP ring-shaped QDMs is measured. The experimental result is shown in Fig. 2.17.

At a high sample temperature, PL peaks are red-shifted and FWHMs are broadened. PL signal is weakened due to thermal carrier escape from QDs to In0.5Ga0.5P capping layer. Defects in MEE grown In0.5Ga0.5P would trap those carriers leading to lower PL signal in a high temperature range.

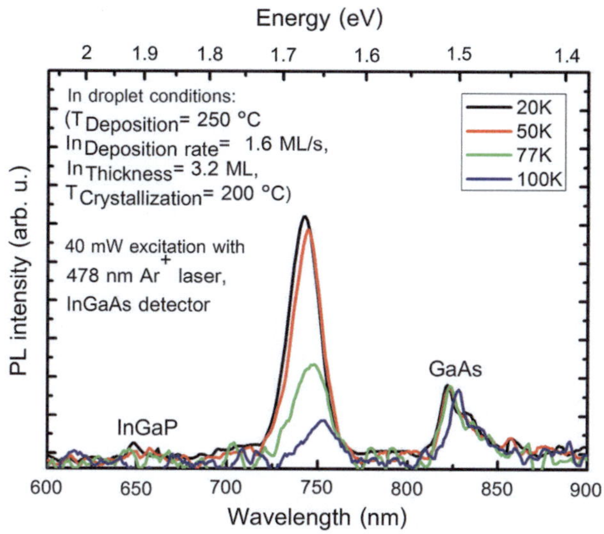

Fig. 2.17 Temperature dependence of PL emission from InP ring-shaped QDMs

2.6 Number of QDs per Ring and Potential Applications of InP Ring-Shaped QDMs

The growth parameters can modify QRs to ring-shaped QDMs. With limited volume of In thickness less than 1.6 ML, only QRs are realized. Increase in In deposition under appropriate crystallization condition with P pressure, InP QDs are created on the ring robe. The critical thickness for QD formation on the ring robe is in the range of 1.6–3.2 ML.

At a deposition temperature of 250°C and a crystallization temperature of 200°C, when In deposition rate of 1.6 ML/s and with In coverage of 3.2 ML, 46% of InP ring-shaped QDMs have eight dots per ring. These octa-QDMs are an interesting feature for extended quantum cellular automata application as shown in Fig. 2.18.

The basic idea of QCA is based on two electrons being located in quadra quantum dot set in diagonal positions due to Coulomb repulsion. The position of this electron pair can be either left or right diagonal configuration which represents either "1" or "0" in the binary system.

In octa QDMs having eight QDs per cell, the electron pair will have more stable positions, i.e., two diagonal configurations and more vertical and horizontal configurations. The EQCA gives more logic encodings as "1", "0", "+1/2," and "−1/2". This would enable QDMs to be utilized in future sophisticated quantum memory and computation.

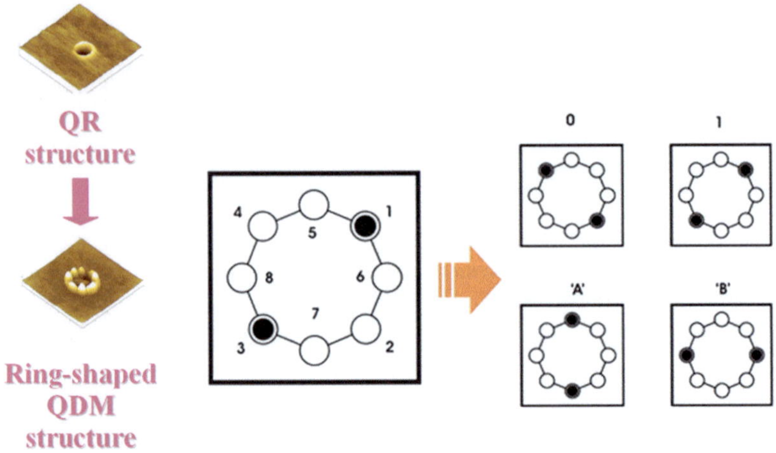

Fig. 2.18 Octa-QDMs can encode more logic values of 1, 0, and 1/2 for extended quantum dot cellular automata (EQCA) application

2.7 Summary

In this presentation, we intensively study the droplet epitaxial growth of InP ring-shaped QDMs. The quantum nanostructure is created on In0.5Ga0.5P/GaAs substrate based on the combination of droplet epitaxy technique for QR formation and strain-driven mechanism due to lattice mismatch between InP and In0.5Ga0.5P for QD formation on the ring robe when the critical thickness of indium coverage is reached.

Based on the AFM observation together with TEM confirmation and the experimental results on photoluminescence of the InP ring-shaped QDMs, we optimize the growth parameters as follows: the deposition and crystallization temperatures at 250 and 200°C with indium deposition rate of 1.6 ML/s and indium thickness of 3.2 ML. The QDMs with eight QDs per ring are our target in the fabrication which has a potential application in EQCA. The percentage of these ring-shaped octa-QDMs at the optimized growth condition is about 46%.

Acknowledgments This research article is an output which is supported by Thailand Research Fund (TRF) and Office of High Education Commission (OHEC) of Thailand in combined projects, i.e. Senior Research Scholar (RTA5080003), Distinguished Professor Grant (DPG5380002), the Royal Golden Jubilee Ph.D. program (Grant No. PHD/0040/2549). This research work is also supported by the Higher Education Research Promotion and National Research University Project of Thailand, Office of the Higher Education Commission (EN1180A-55) as well as by Nanotechnology Center of Thailand with counterpart fund from Chulalongkorn University.

References

1. Schmidt, O.G., Eberl, K.: Phys. Rev. B **61**, 13721 (2000)
2. Alfarov, Z.: Rev. Mod. Phys. **73**, 767 (2001)
3. Grundmann, M.: Physica E **5**, 167–184 (2000)
4. Tanabe, K., et al.: Appl. Phys. Lett. **100**, 193905 (2012)
5. Laouthaiwattana, K., et al.: Sol. Energy Mater. Sol. Cell **93**, 746–749 (2009)
6. Suraprapapich, S., et al.: J. Vac. Sci. Technol. B **24**, 1665 (2006)
7. Boonpeng, P., et al.: Microelectron. Eng. **86**, 853–856 (2009)
8. Wang, Z.M., et al.: Appl. Phys. Lett. **84**, 1931 (2004)
9. Lent, C.S., et al.: Nanotechnology **4**, 49–57 (1993)
10. Lent, C.S., Tougaw, P.D.: Proc. IEEE **85**, 491 (1997)
11. Porod, W.: J. Franklin Inst. **334B**(5/6), 1147–1175 (1997)
12. Bajec, I.L., et al.: Microelectron. Eng. **83**, 1826–1829 (2006)
13. Fisher, A.M.: Phys. Rev. Lett. **102**, 076405 (2009)
14. Watanabe, K., et al.: Jpn. J. Appl. Phys. **39**, 179–181 (2000)
15. Yamagiwa, M., et al.: Appl. Phys. Lett. **29**, 113115 (2006)
16. Sanguinetti, S., et al.: J. Appl. Phys. **104**, 113519 (2008)
17. Stemmann, A., et al.: J. Appl. Phys. **106**, 064315 (2009)
18. Mano, T., Mano, T., et al.: Thin Solid Films **515**, 531–534 (2006)
19. Heyn, C., et al.: Appl. Phys. Lett. **90**, 203105 (2007)
20. Strom, N.W., et al.: Nano Res. Lett. **2**, 112 (2007)
21. Esser, N., et al.: J. Vac. Sci. Technol. B **19**, 1756–1761 (2001)
22. Naraporn, P., et al.: J. Cryst. Growth **323**, 282–285 (2011)
23. Kurtenbach, A., et al.: J. Electron. Mater. **25**, 395–400 (1996)
24. Zundel, M.K., et al.: Appl. Phys. Lett. **73**, 1784–1786 (1998)
25. Lewis, G.M., et al.: Appl. Phys. Lett. **85**, 1904–1906 (2004)
26. Suraprapapich, S., et al.: Appl. Phys. Lett. **90**, 183112 (2003)
27. Suraprapapich, S., et al.: J. Cryst. Growth **302**, 735–739 (2007)
28. Mano, T., Kiguchi, N.: J. Cryst. Growth **278**, 108–112 (2005)
29. Lee, J.H., et al.: J. Appl. Phys. **106**, 073106 (2009)
30. Mazur, Y.I., et al.: Appl. Phys. Lett. **86**, 063102 (2005)
31. Somaschini, C., et al.: Nanotechnology **20**, 185602 (2011)

Chapter 3
Optical Properties of Lateral InGaAs Quantum Dot Molecules Single- and Bi-Layers

Songphol Kanjanachuchai, Nirat Patanasemakul, Natapong Thongkamkoon, Nitidet Thudsalingkarnsakul, Naparat Siripitakchai, Pornchai Changmoang, Supachok Thainoi, and Somsak Panyakeow

Abstract Growth of InGaAs nanostructures by molecular beam epitaxy using partial-cap and regrowth technique results in an ensemble of lateral quantum dot molecules (QDMs). Each QDM comprises a large, central quantum dot and several small, satellite quantum dots which emit at different energies and exhibit qualitatively different optical behaviors. This chapter reviews, explains, and discusses the various fundamental aspects of lateral QDM single layers: the nucleation mechanism, the photoluminescent spectra, and the unique bimodal optical characteristics. The chapter ends by demonstrating the usefulness of lateral QDM bi-layers as a promising broadband near-infrared material.

3.1 Introduction

Semiconductor quantum dots (QDs) have been intensively investigated for several decades since early pioneering work show them to be of high crystalline quality, capable of optical emission in the near-infrared region [1]. They are usually grown by metal-organic chemical vapor deposition (MOCVD) or molecular beam epitaxy (MBE) with elemental solid sources or compound gas sources. The standard route

---

S. Kanjanachuchai (✉) • N. Patanasemakul • N. Thudsalingkarnsakul • N. Siripitakchai
P. Changmoang • S. Thainoi • S. Panyakeow
Semiconductor Device Research Laboratory, Department of Electrical Engineering, Faculty of Engineering, Chulalongkorn University, Bangkok 10330, Thailand
e-mail: songphol.k@chula.ac.th; nirat.pat@gmail.com; smearsand@hotmail.com; s.naparat@gmail.com; pornchai.c@chula.ac.th; Supachok.t@chula.ac.th; s_panyakeow@yahoo.com

N. Thongkamkoon
International School of Engineering, Faculty of Engineering, Chulalongkorn University, Bangkok 10330, Thailand
e-mail: ixohoxi_35847@hotmail.com

J. Wu and Z.M. Wang (eds.), *Quantum Dot Molecules*, Lecture Notes in Nanoscale Science and Technology 14, DOI 10.1007/978-1-4614-8130-0_3,

for growing them is via the Stranski–Krastanow (SK) growth mode where lattice mismatch between the growing epilayer and the substrate is relaxed by the formation of a thin wetting layer (WL) and a three-dimensional (3D) QD structure whose size, density, and uniformity are dictated by growth conditions and are a subject of intensive investigations during the past decade [2, 3]. Another route for growing QDs follows a two-step, droplet epitaxial (DE) process where, typically, group-III metallic droplets are first grown on a III–V substrate and subsequently crystallized by a group-V element at low temperatures to preserve the original droplet geometry [4]. Both routes yield good quality QDs whose properties can be engineered. Optical properties in particular are tunable as they are sensitive to QD geometry; the latter dictates the degree of zero-dimensionality or quantum confinements. For InGaAs QDs, the effective mass for holes is much greater than those for electrons and, as a result, holes are confined within the QDs whereas extended states of electrons exist outside the QDs [5]. This offers the possibility of controlling the overlapping or coupling electron wavefunctions between two nearby QDs and is one of the main forces that drive the MBE and MOCVD communities towards the growth of QD molecules (QDMs).

Quantum coupling between or among constituent QDs in a QDM can occur vertically, in the growth direction, or laterally, in the growth plane [6]. Vertical geometry allows precise control of tunnel barrier thickness and it is this advantage that is critical to the demonstrated quantum coupling [7] and state entanglements [8] which form the foundation of quantum computation [9]. Unfortunately for the vertical geometry, access to the barrier, key to coupling control, is limited to top and bottom contacts. Lateral geometry, on the other hand, lacks precise control over barrier thickness but allows better electrostatic control of the tunnel barriers using top- and/or side gates which can be readily fabricated using well-established planar technology.

This chapter begins by giving a general overview of the growth procedures used to form lateral QDMs where coupling occurs in the growth plane. This includes a specific growth procedure central to this chapter called partial-cap and regrowth process. The optical properties of the resulting QDMs are then explained and compared with typical QDs as well as QDMs obtained from other techniques. The chapter ends by discussing stacked QDM structures in the form of bi-layers and demonstrating their broadband characteristics which are potentially useful for devices such as solar cells and superluminescent diodes.

3.2 Growth of Lateral QDMs

Lateral QDM is a term used to broadly describe ensembles of QDs that, by design, are closely connected in the horizontal direction or the growth plane. Many growth procedures have demonstrated successful formation of lateral QDMs with high crystalline quality; yet, there is still no consensus as to which procedure would yield QDMs with characteristics that best match requirements for quantum coupling

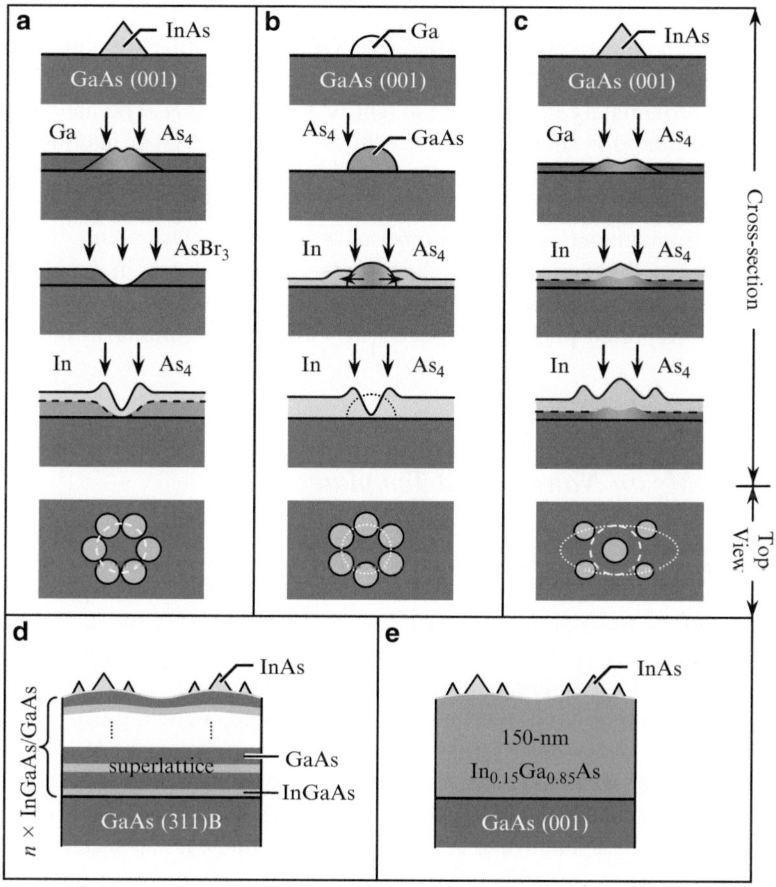

Fig. 3.1 Growth process and evolution schematics of lateral QDMs grown on various templates: (**a**) nanohole, (**b**) nanomound, and (**c**) nanohole-and-mound. Schematic cross-sectional structures of lateral QDMs formed on (**d**) InGaAs/GaAs-(3 1 1)B superlattice and (**e**) InGaAs/GaAs-(0 0 1) cross-hatch pattern templates

and coupling control. Most procedures reported so far are based on InGaAs/GaAs systems using some kinds of templates and employ standard (0 0 1)-GaAs substrates as summarized in Fig. 3.1 and described below.

3.2.1 QDMs on Nanohole Templates

This procedure was first reported by Songmuang et al. in 2003 [10] and shown in Fig. 3.1a. The first step is the formation of standard InAs QDs on a flat GaAs substrate. The seed QDs are then capped with 10-nm GaAs, resulting in QDs collapsing and elongating along the [1 −1 0] direction. The surface is then exposed to $AsBr_3$ which selectively etches the seed QDs away while the GaAs surface remains relatively unscathed. Holes formed from the etched QDs are typically 5-nm deep, approximately the nominal height of the seed QDs. The nanoholes are used as a template for subsequent regrowth of InAs, but In–Ga intermixing and Ga diffusion result in InGaAs QDMs. During regrowth, adatoms arriving at the surface preferentially attach to the nanohole rim. By varying the deposition amount and the substrate temperature, the number of constituent QDs making up a QDM along the rim can be varied between 2 and 6; the latter is shown in the plane-view schematic drawing in the bottom of Fig. 3.1a. These QDMs are of high crystalline quality as the reported photoluminescent (PL) intensity is similar to the original QDs [10].

3.2.2 *QDMs on Nanomound Templates*

Nanomound templates can also be used to form lateral InGaAs QDMs as reported by Lee et al. in 2006 [11]. This approach is best described as a hybrid process between droplet homoepitaxy and standard SK heteroepitaxy as summarized in Fig. 3.1b. The procedure begins by the formation of a few monolayers (MLs) Ga droplets at high substrate temperatures without the presence of arsenic. The large lattice mismatch between the metallic Ga and the semiconducting GaAs substrate means that Ga droplets readily form on the flat GaAs surface via the Volmer–Weber growth mode. The substrate temperature is then lowered, typically to below 200°C, and the surface exposed to As_4 which then crystallizes the Ga droplets into GaAs nanomounds with base length and height around 100 and 50 nm, respectively. These nanomounds serve as a template for subsequent overgrowth in a similar fashion to the nanohole approach described above. The substrate temperature, kept low during crystallization to preserve the original mound geometry, is raised prior to the deposition of InAs. By varying the deposition amount the number of InAs QDs nucleated around the GaAs nanomounds can be varied between 2 and 6, similar to the previous approach. GaAs nanomounds not only serve as a spatial template but also act as a Ga reservoir since Ga atoms get incorporated into the growing In(Ga)As QDs. These QDMs also have high crystalline quality as shown by strong PL.

A variant of the above hybrid DE–SK approach is also reported by Strom et al. where, using different conditions, the starting Ga droplets are crystallized into nanorings and subsequent overgrowth results in the formation of InAs QDs inside and outside the rings with a significant amount of GaAs rings still remaining after crystallization [12]. While the QDs which nucleate outside the nanorings are closely spaced and likely be coupled, those nucleated inside are completely decoupled as they are isolated from outside QDs by the thick GaAs nanorings.

3.2.3 QDMs on Nanohole-and-Mound Templates

Similar to the two approaches above, this third approach by Suraprapapich et al. employs a unique template shaped like the back of a camel (side-view) or a two-blade propeller (top-view) and is thus originally referred to as a camel-like nanostructure and nanopropeller [13]. Overgrowth of InAs on this template create a unique lateral QDM geometry as summarized in Fig. 3.1c. The process starts from the deposition of 1.8–2 ML of InAs QDs at a slow growth rate of 0.01 ML/s, followed by a short growth interruption to reduce the substrate temperature from 500 to 470°C before capping the seed QDs by GaAs. The thin GaAs cap layer does not completely cover the original QDs. The apex is hence thermodynamically unstable, resulting in In out-diffusion and a shallow dip, or nanohole, in the middle. The out-diffusing In atoms then react with the incoming Ga and the existing Ga atoms on the surface to form InGaAs nanomounds with a pronounced elongation along the [1 −1 0] direction. The nanoholes are the most favorable locations for QD formation, followed by the edges of the surrounding nanomounds. Subsequent regrowth of InAs always yields InAs QDs inside the nanoholes before the nucleation and growth on nanomounds. While the number of QDs inside the nanohole is limited to one, those along the edges of the nanomounds can be varied from 1 to as many as 13 simply by changing the deposition amount [14]. These QDMs are also of high crystalline quality, exhibiting room-temperature luminescence with noticeable polarization anisotropy [15].

3.2.4 QDMs on Strain-Modulated Templates

This approach is fundamentally different from the above approaches by its use of strain-modulated templates where strains are global in nature, i.e. present on a wafer scale, as opposed to local strains effective only around nanoholes or nanomounds.

One implementation of this approach by Lippen et al. and with schematic cross section in Fig. 3.1d uses a superlattice (SL) template of InGaAs/GaAs on (3 1 1)B-GaAs to form a slowly modulated strained network that upon overgrowth results in a lattice of ordered QDMs [16]. These QDMs are optically active up to room temperature and exhibit a complex PL behaviors [17].

Another implementation by Thet et al. and with schematic cross section in Fig. 3.1e uses as template a network of orthogonal, interfacial dislocations which under optical microscopy appears as a cross-hatch pattern (CHP) to form a slowly varying surface undulations that also supports the growth of QDM lattice [18]. Despite the presence of dislocation, QD chains grown on these kinds of templates have been shown to exhibit high crystalline quality [19, 20]. QDMs on these kinds of templates are expected to have high crystalline quality as well though no PL results have been reported so far. Recently, Seravelli et al. reported PL results of QDs grown on thick metamorphic InGaAs/GaAs templates and found QDs to be of high crystalline quality [21].

3.3 Structure, Morphology, and Evolution

The cross-sectional schematic showing the QDM-embedded structures grown on the nanohole-and-mound template described above is shown in Fig. 3.2. All growth takes place in a Riber's 32P solid-source MBE chamber. Growth on epi-ready (0 0 1)-GaAs substrate starts from thermal desorption of the surface oxide at 580°C, growth of a 300-nm GaAs smoothening layer at 610°C, a brief pause to bring down the substrate temperature to 500°C before the growth of x ML of InAs QD seed layer at a rate of 0.01 ML/s where x is between 1.8 and 2. This results in a QD ensemble with typical base width of 30–50 nm and height of 5 nm. The onset of QD formation, monitored in situ via the reflection high-energy electron diffraction (RHEED) pattern, is confirmed by the appearance of the transmission spots which subsequently develop into chevrons [22]. The substrate temperature is then brought down further to 470°C before the InAs QDs are partially capped with y ML of GaAs where y is typically between 6 and 25. This ensures a successful formation of the nanoholes whose depth can be varied approximately between 0.2 and 1 nm. Each nanohole sits atop a nanomound elongated along the [1 −1 0] direction as shown in the lower atomic force microscopy (AFM) image in Fig. 3.2. The elongation observed is a result of asymmetrical In adatom diffusion lengths along the ⟨1 1 0⟩ directions which can be enhanced or suppressed by simply increasing or decreasing the capping temperature [15]. To form the QDMs at the same temperature used for capping, z ML of InAs is regrown on top of the nanohole-and-mound template where z is typically between 1 and 2.5 for fully-developed molecules. For the sake of convenience and precision, a QDM ensemble formed from x-ML InAs seed QDs, y-ML GaAs partial capping, and z-ML InAs overgrowth is hereafter referred to as $x/y/z$ QDMs.

The QDMs are then capped with 100-nm GaAs where the first 10 nm is grown at 470°C and the remaining 90 nm is grown while ramping the temperature up to 500°C. For a single-layered QDM structure, the sample is quenched after completing the 100-nm GaAs growth. For a bi-layer QDM structure, another cycle of $x/y/z$ QDM (with different values of x, y, and/or z from the first QDM layer) and

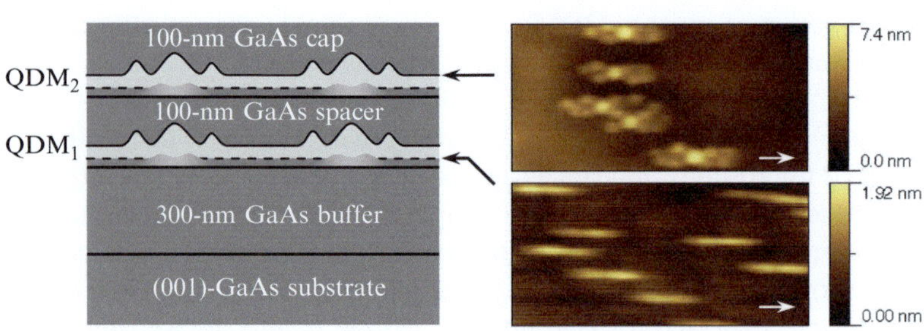

Fig. 3.2 (*Left*) Cross-sectional structures of lateral QDMs grown on the nanohole-and-mound template via the partial-cap and regrowth process. (*Right*) 1μm × 0.5μm AFM images of (*lower image*) nanohole template and (*upper*) lateral QDMs. The arrows point along the [1 −1 0] direction

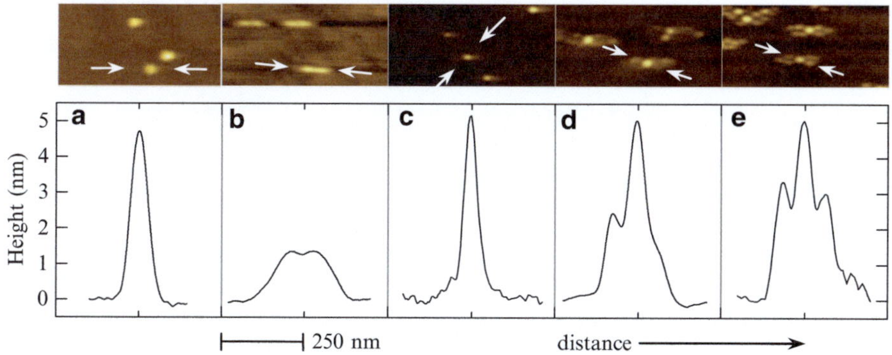

Fig. 3.3 AFM images (1μm × 0.5μm) and line scans elucidating the evolution of lateral QDMs grown on the nanohole-and-mound template: (**a**) seed InAs QDs, (**b**) nanoholes and mounds—formed after partial capping the seed InAs QDs with GaAs, (**c**) cQDs—overgrown on the nanohole-and-mound template, (**d**) partially-formed QDMs comprising saturated cQDs and non-saturated sQD, and (**e**) fully-formed QDMs comprising saturated cQDs and sQDs. For the AFM images, the horizontal direction is the [1 −1 0] direction, and the *arrows* indicate the sections where line scans are taken

100-nm GaAs is repeated. After quenching and removing from the growth chamber, the sample is mounted in a variable temperature optical cryostat and characterized by PL using the 476.5-nm line of an Ar^+ laser as the excitation source. The PL signal is dispersed by a 1-m monochromator and collected by a liquid nitrogen cooled InGaAs detector using standard lock-in detection technique.

The morphology of the QDMs is obtained by quenching the sample as soon as the regrowth of InAs on the nanohole-and-mound template finishes and probing the surface in air using AFM in the tapping mode. A typical QDM morphology is shown in the upper AFM image in Fig. 3.2.

By varying the regrowth thickness (z) it is possible to observe the various stages of QDM growth. The evolution of the QDM structure—from the seed QD layer to the complete molecule formation for 2/6/z QDMs—is shown in a series of AFM images and line profiles in Fig. 3.3. Right after the seed QD formation, Fig. 3.3a shows that the base length of a typical QD is approximately 50 nm while the height is 5 nm. Upon partial capping, the QD height collapses, the base elongates towards the [1 −1 0] direction, and a shallow dimple ~0.2-nm deep appears in the center as shown in Fig. 3.3b. After only 0.5-ML InAs regrowth, an indication that a QD is being formed at the center of each nanohole is shown in Fig. 3.3c. Usually this sub-monolayer amount is taken up as a WL on flat GaAs surfaces. This clearly shows the much lower chemical potentials associated with the nanohole and mound. With a thicker 1-ML InAs regrowth, a few QDs are formed in addition to the central QD as shown in Fig. 3.3d. The QD at the center is referred to as the central QD (cQD) while those nucleated around it are called satellite QDs (sQDs). With 1-ML regrowth, cQDs are well saturated while sQDs are only starting to form. And, finally, with 1.4-ML InAs regrowth, cQDs and sQDs are saturated, forming fully-grown QDMs as shown in Fig. 3.3e. Exceeding the saturation, free-standing QDs will be formed on the previously denuded regions between QDMs. The cQDs and sQDs exhibit fundamentally different optical behaviors and can interact due to their proximity as described next.

3.4 Optical Properties of QDMs

The macro-PL results described in this section are obtained at 20 K under a low excitation power density of 0.45 W/cm^2, unless otherwise stated. The QDMs are of high crystalline quality with clear, separate emission peaks from the cQDs and sQDs as described in Sect. 3.4.1. Transfers of luminescent intensity between cQDs and sQDs as temperature varies are reported in Sect. 3.4.2 and explained in terms of phonon-assisted carrier transfer between the constituent QDs. The QDM ensemble exhibits a unique, bimodal *optical* characteristics explained in Sect. 3.4.3 which differ from bimodal *size* distributions generally reported for intermediate sized QDs.

3.4.1 cQDs and sQDs Emissions

This sub-section describes the PL of six QDM samples. Each sample is grown using a different QDM recipe, i.e. different x, y, and/or z, with identical procedures for the rest of the structure as described in Sect. 3.3. The first series of three samples is grown with the same nanohole template obtained from 2-ML InAs seed QDs and 25-ML GaAs partial-cap layer. The nanoholes are filled with either 1, 2, or 2.5-ML InAs, resulting in 2/25/1, 2/25/2 or 2/25/2.5 QDMs, respectively. The second series of three samples is grown with the same 2-ML InAs seed QDs and 1.4-ML InAs regrown QDMs, but with different GaAs capping layer thickness of 6, 10, or 25 ML, resulting in 2/6/1.4, 2/10/1.4, or 2/25/1.4 QDMs, respectively. Systematic variation of parameters in the two series enables unambiguous assignment of PL peaks to the constituent cQDs and sQDs in the QDMs [23].

Figure 3.4a shows, from bottom to top, the PL spectra from the first sample series which contain 2/25/z QDMs where $z = 1$, 2, and 2.5 ML, respectively. The 2/25/1 QDMs emission is dominated by a single peak at 1.056 eV with a narrow full-width at half-maximum (FWHM) of 24 meV. The 2/25/2 QDMs emission shows a low-energy peak also at 1.068 eV with FWHM of 40 meV, and a high-energy peak at 1.168 eV with FWHM of 68 meV. The 2/25/2.5 QDMs emission show the same low-energy peak at 1.068 eV but with a slightly narrower FWHM of 25 meV, and a high-energy peak at 1.150 eV with FWHM of 73 meV. The double peak feature of the latter two samples go up and down together under excitation-dependent experiments, indicating that the high-energy peak is not an excited state (ES) arising from the saturation of the low-energy, ground-state (GS) peak. In addition to confirming GS emission, it is necessary to identify the sources and predict the characteristics of the two peaks if the QDMs are to be optimized and employed in devices.

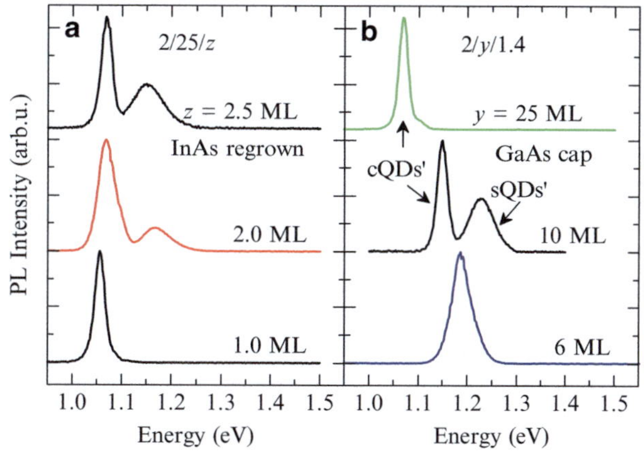

Fig. 3.4 Twenty-Kelvin PL spectra of (**a**) 2/25/z QDMs with $z = 1$, 2 or 2.5 ML, and (**b**) 2/y/1.4 QDMs with $y = 6$, 10 or 25 ML. Spectra are offset and rescaled for clarity. Reproduced from [23]

By comparing the PL spectra with surface morphologies obtained from AFM images, the origins of the low- and high-energy PL peaks can be attributed to the cQDs and sQDs, respectively. A typical 2/25/1 QDM contains just a single dot nucleated inside the nanohole, or cQD, similar to Fig. 3.3c while 2/25/2 QDM has in addition another dot or two outside the nanohole, or sQDs, similar to Fig. 3.3d, and 2/25/2.5 QDM has at least four sQDs, forming a fully-developed molecule similar to Fig. 3.3e. The presence of cQDs and absence of sQDs in 2/25/1 QDM ensemble is a definitive confirmation that the 1.056-eV peak originates from the cQDs. All three samples share almost identical cQD GS energy at around 1.06 eV which is not surprising because they all share the same nanohole template and similar size cQDs. After the nanoholes are saturated, sQDs start nucleating which coincides with the appearance of the high-energy PL peak. The 1.168-eV peak for 2/25/2 QDMs and 1.15-eV peak for 2/25/2.5 QDMs must therefore originate from the GS of sQDs as excitation-dependent experiments already rule them out as an ES.

The energetic position, the intensity, and the FWHM of these high-energy GS peaks agree with surface morphologies and reaffirm their assignments to sQDs. Figure 3.3 shows that cQDs must first be saturated before sQDs are nucleated. It is important to note that the base of cQDs is limited by the nanohole template, while the nominal height of cQDs is always higher than sQDs even though a greater amount of deposited materials seems to be taken up by the sQDs. This is simply due to the greater sQD density: four (or more) sQDs per QDM as opposed to one cQD per QDM. The maximum regrowth thickness for the three samples is 2.5 ML, with at least 0.5 ML taken up entirely by cQDs and at most 2 ML by sQDs, the per-dot material accumulated by an sQD is thus much lower than by a cQD. This translates to smaller sQD and higher GS energy than cQD. It is informative to compare the energetic positions and integrated intensities of the sQDs in 2/25/2 and 2/25/2.5 QDMs. The sQD PL in 2/25/2 and 2/25/2.5 QDMs peak at 1.168 and 1.150 eV, respectively. This is simply due to the latter gaining more material, becoming larger, and thus emitting at a lower GS energy. As more sQDs are nucleated, the relative intensity of sQDs with respect to cQDs also increases as obvious from the progressively greater proportion of the sQD peaks in Fig. 3.4a.

The FWHM of the 1.168-eV peak in the case of 2/25/2 QDMs (68 meV) and the 1.150-eV peak in the case of 2/25/2.5 QDMs (73 meV) are higher than those of the 1.068-eV peak (25 meV), indicating that the inhomogeneous broadening of sQDs is worse than those of cQDs. This is expected, considering the almost simultaneous nucleation of cQDs in the nanoholes across the wafer, and the sequential, random nucleation of sQDs only after the nanoholes are saturated with cQDs.

The PL spectra from the second sample series affirm the peak assignments above and elucidate the role of nanohole depth. Figure 3.4b shows, from bottom to top, the PL spectra of 2/y/1.4 QDMs where $y = 6$, 10, and 25 ML, respectively. The 2/6/1.4 QDMs emit a single PL peak at 1.186 eV, the 2/10/1.4 QDMs a double peak at 1.150 and 1.226 eV, and the 2/25/1.4 QDMs a single peak at 1.068 eV. Since the thicknesses of the seed QDs x and regrown QDs z are identical, the differences in the three spectra originate from the GaAs capping thickness y. The GaAs capping thickness y dictates the degree of In out-diffusion from the seed InAs QDs and thus the nanohole depth which has been carefully characterized and found to vary almost linearly from 0.4 to 1 nm as y increases from 6 to 25 ML [14]. The shallower the nanohole, the easier for it to be saturated, or the smaller the cQDs.

In the case of 2/6/1.4 QDMs with the most shallow nanoholes, both cQDs and sQDs co-exist and the average sizes of cQDs and sQDs do not much differ, resulting in a near-merged spectra. Though the bottom spectrum in Fig. 3.4b shows a single peak, closer examination of its semi-logarithmic plot reveals different rise and fall characteristics, indicating different origins. In the case of 2/25/1.4 QDMs with deepest nanoholes, on the other hand, only cQDs exist, resulting in a single Gaussian peak at 1.068 eV shown in the top spectrum of Fig. 3.4b. This peak is located at the same energetic position as the 2/25/1 QDMs from the first series but is narrower (FWHM = 21 meV), possibly due to the better uniformity achieved when approaching full saturation condition at a thicker regrowth thickness of 1.4 ML. Finally, in the case of 2/10/1.4 QDMs with nanohole depth intermediate between the two extreme cases above, both cQDs and sQDs co-exist and their average sizes are distinctly difference, resulting in two well-separated PL peaks with the low-energy portion attributed to cQDs and the high-energy to sQDs.

3.4.2 cQD–sQD Coupling

The presence of the WL and the proximity between cQDs and sQDs bring with them the possibility of tunnel coupling, especially in view of the fact that electron wavefunctions are not well confined and the existent of extended states has been confirmed [5, 17]. Microscopically, coupling gives rise to bonding and anti-bonding molecular states, resulting in energetic separation or anti-crossings when the luminescence is measured under varying electromagnetic fields [24, 25]. Macroscopically, coupling gives rise to electron transfer between QDs, resulting in PL intensity transfer between or among related energy levels. Tunnel coupling in

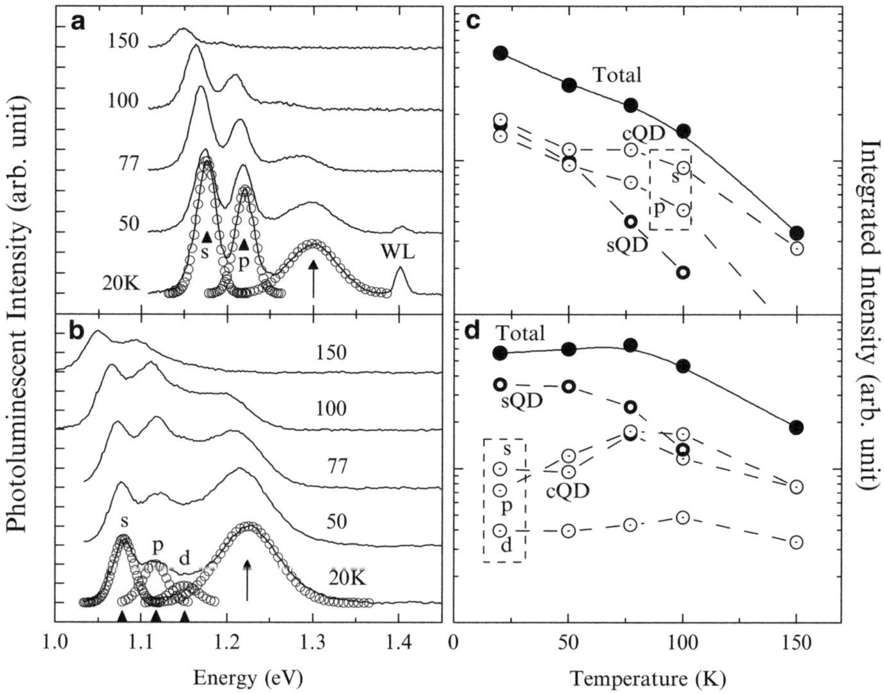

Fig. 3.5 Temperature-dependent PL spectra of (**a**) 1.8/15/1.2 and (**b**) 1.8/25/1.5 QDMs. Text labels (s, p, and d) indicate cQD emissions; the *arrow*, sQD emissions. The *dotted lines* are multiple Gaussian function fits to the 20 K data. Graphs are offset for clarity. Temperature-dependent integrated intensities of constituent peaks of (**c**) 1.8/15/1.2 QDMs extracted from (**a**), and (**d**) 1.8/25/1.5 QDMs extracted from (**b**). The *dotted lines* are guide to the eye. Adapted from [27] with permission from Elsevier

lateral QDMs are due to electrons only; holes are well confined in the QDs due to high effective mass [26]. The macro-PL setup used in our experiments only allow the observation of the macroscopic PL intensity transfer.

Two QDM samples are grown to study the coupling effects: One sample contains a single layer of 1.8/15/1.2 QDMs, the other a single layer of 1.8/25/1.5 QDMs [27]. Temperature-dependent PL spectra of the two samples illustrate the differences in tunneling strength, and the competition between tunneling and thermionic emissions. The former maintains the total integrated intensity whereas the latter quenches it due to carrier losses to non-radiative recombination (NRR) channels. Both samples are excited at a high excitation power density of 20 W/cm^2 sufficient for the observation of ES as tunneling into ES is easier than into GS due to the wider spread of ES wavefunctions.

The PL spectra of the 1.8/15/1.2 and 1.8/25/1.5 QDMs are shown in Fig. 3.5a, b, respectively. The broad, high-energy peaks indicated by arrows in both figures are related to sQDs as explained previously. The narrow, low-energy peaks are due to cQDs. The labels s, p, and d denote the ground, first, and second excited states, respectively. The nature of excited states has been confirmed by excitation-dependent experiments. For 1.8/15/1.2 QDMs in Fig. 3.5a, the spacing between the ground and first excited states, or the s–p spacing, is 45 meV. For 1.8/25/1.5 QDMs in Fig. 3.5b, the s–p and p–d spacings are approximately 37 meV. These values are within the expected range for QDs of similar size [28]. The equi-distance of the s–p and p–d spacings results from the harmonic oscillator-type potentials in the growth plane and is also expected [29]. The overall spectra can be fitted with multi-Gaussian functions which are shown as examples in the 20-K spectra in both figures. The fits allow the FWHM of all the spectral peaks to be accurately determined which together with the associated peak intensity yield the integrated intensity (II) vs. temperature plots of the 1.8/15/1.2 and 1.8/25/1.5 QDMs in Fig. 3.5c, d, respectively. Non-monotonous variations of integrated intensity with temperature are evident in both cases.

For ensembles of isolated QDs, a simple, monotonous decrease of II with increasing temperature is expected due to the presence of NRR channels [30] or reduced PL yields [31]. For laterally-coupled QDMs, the temperature dependency of the PL spectra is complicated by tunnel coupling which has many possible routes and is also temperature dependent. The overarching monotonous II reduction with increasing temperature for 1.8/15/1.2 QDMs in Fig. 3.5c results from carriers gaining sufficient energy, escaping into the WL and/or GaAs matrix before recombining with NRR centers/channels as is typical for InAs/GaAs QD systems [32]. Close examination of the constituent IIs shows that the monotonous reduction of the total II results from a rapid, monotonous decrease of sQDs II and a non-monotonous decrease of cQDs II. In fact, a slight increase of the cQDs GS is registered at around 75 K. This increase can be interpreted as typical carrier redistribution between QD ensembles of different nominal size, or bimodal QDs where carriers in sQDs may escape, diffuse towards the cQD, and be captured. Or it can be interpreted as resulting from direct tunnel coupling. The underlying mechanism(s) that governs the temperature dependency of the II for this particular QDMs is not clear.

The 1.8/25/1.5 QDMs exhibit a qualitatively different II temperature dependency as shown in Fig. 3.5d. This is due to the bigger cQDs and sQDs as compared to the 1.8/15/1.2 QDMs above. The total II in this case is virtually constant from 20 to 75 K, above which it decreases monotonously. The overall change results from the complex behaviors of the s and p peaks of the cQDs while the sQDs exhibit a simple, monotonous decrease in intensity. As the temperature increases from 20 to 50 K, the II of the p peak significantly increases. The intensity gain cannot originate from carrier redistribution as the temperature is still too low for excitons to appreciably escape the potential barriers. The most likely mechanism is thus tunnel coupling, from sQDs to cQDs. AFM images show that the cQDs and sQDs in 1.8/25/1.5 QDMs almost merge which supports the conclusion. The II of the p peaks keeps on increasing with temperature up to 75 K before carrier loss from sQDs to NRR channels begin to aversely impact the availability of carriers tunneling to cQDs, and thus the intensities of the s, p, and d peaks. The II of the s peak is approximately constant between 20 and 50 K, indicating saturation. It, however, increases as the temperature reaches 75 K which contradicts the saturation interpretation. Closer inspection of the PL spectra shows that the II increases not by increasing peak intensity but by broadening. The saturation conclusion thus remains valid.

The observation of intensity transfer and the implied underlying mechanism via tunnel coupling in our lateral QDMs are at first surprising, considering the significant dot-to-dot separation in the order of 50 nm. Tunnel coupling decreases exponentially with distance and Szafran and Peeters predicted that coupling is negligible for a dot-to-dot distance of 20 nm [33]. This, however, is strictly true for isolated QDs with a thin WL acting as the main coupling path. Our lateral QDMs, especially the sQDs, form on the nanomound template which in itself can act as a low barrier region connecting the constituent QDs in a QDM. A similar "basin" has recently been demonstrated to be acting as a coupling channel for double InGaAs QDs spaced as far as 40 nm apart [34].

3.4.3 Bimodal Optical Characteristics

Depending on growth procedures, lateral QDMs can take shape in many different geometries, with varying degrees of uniformity and size distribution. The latter can be categorized into a mono-, bi-, or multi-modal size distribution. This section describes a novel bimodal *optical* characteristics which results from the bimodal *size* distribution of QDMs. The converse is not necessarily true. The unique bimodal optical characteristics stems from specific spatial arrangement of constituent QDs resulting from the partial-cap and regrowth process.

To demonstrate the unique bimodal optical properties, temperature-dependent GS PL spectra of the 1.8/25/1.2 and 1.8/25/1.5 QDMs are shown in Figs. 3.6a and 3.7a, respectively [35]. The two QDM ensembles mainly differ in the nominal size of sQDs, and consequently the degree of cQD–sQD coupling. The excitation power density at 2 W/cm^2 is sufficiently low to avoid filling up the GS and complicating peak analyses with ES. The spectra can be fitted with double Gaussian functions; examples are shown as the dashed lines in the 20-K spectra of both figures. The fittings allow temperature variations of peak position, intensity, FWHM, and hence II to be accurately determined.

The cQD and sQD peak energy variations with temperature of the 1.8/25/1.2 and 1.8/25/1.5 QDMs are shown in Figs. 3.6b and 3.7b, respectively. In both cases, the cQDs and sQDs exhibit fundamentally different behaviors: the cQDs-related peak energies exhibit a slow red-shift with increasing temperature throughout the experimental temperatures, while the sQDs-related peak energies exhibit a slow red-shift only up to a certain temperature ($\sim$75 K for 1.8/25/1.2 QDMs and 100 K for 1.8/25/1.5 QDMs) before they rapidly decrease at a rate of 1 meV/K. Dashed (dotted) lines in the figures approximate the slow (fast) red-shift of peak energies.

The cQDs's energy slow red-shift can be readily explained in terms of bandgap variations with temperature. The dashed lines in the figures are obtained from Varshni's equation [36] using bulk InAs bandgap parameters, shifted up the energy

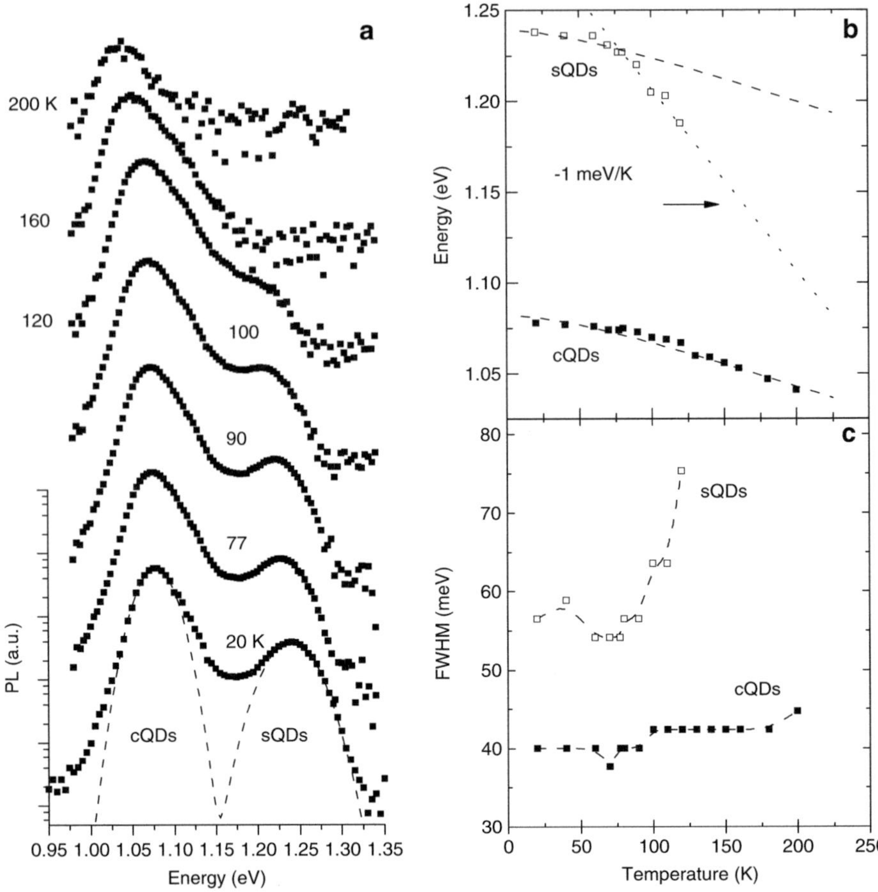

Fig. 3.6 Optical properties of 1.8/25/1.2 QDMs: (**a**) PL spectra at various temperatures, shifted vertically for clarity; temperature dependencies of (**b**) peak energy positions and (**c**) FWHM of cQDs- and sQDs-related spectra. The *dashed curves* in (**a**) are double Gaussian functions fit. The *filled-square* and *open-square* symbols in (**b**) and (**c**) are measured values for cQDs- and sQDs-related spectra, respectively. The *lower* and *upper dashed lines* in (**b**) are bulk InAs bandgap temperature variation according to Varshni's equation, shifted along the energy axis by 0.667 and 0.824 eV, respectively. The *dashed lines* in (**c**) are guide to the eye. Reproduced from [35] with permission from Elsevier

scale to aid visual comparison. The agreement between the experimental trend and the trend calculated based on Varshni's equation is good, except for the small diversion at low temperatures which can be further improved by the Fan model [37].

The sQDs's energy fast red-shift, usually referred to as sigmoidal [38], has been reported by many groups [32, 39–41] and is attributed to carrier escape and redistribution among inhomogeneous QDs. This occur concurrently with an anomalous FWHM reduction which is indeed observed here and best explained together.

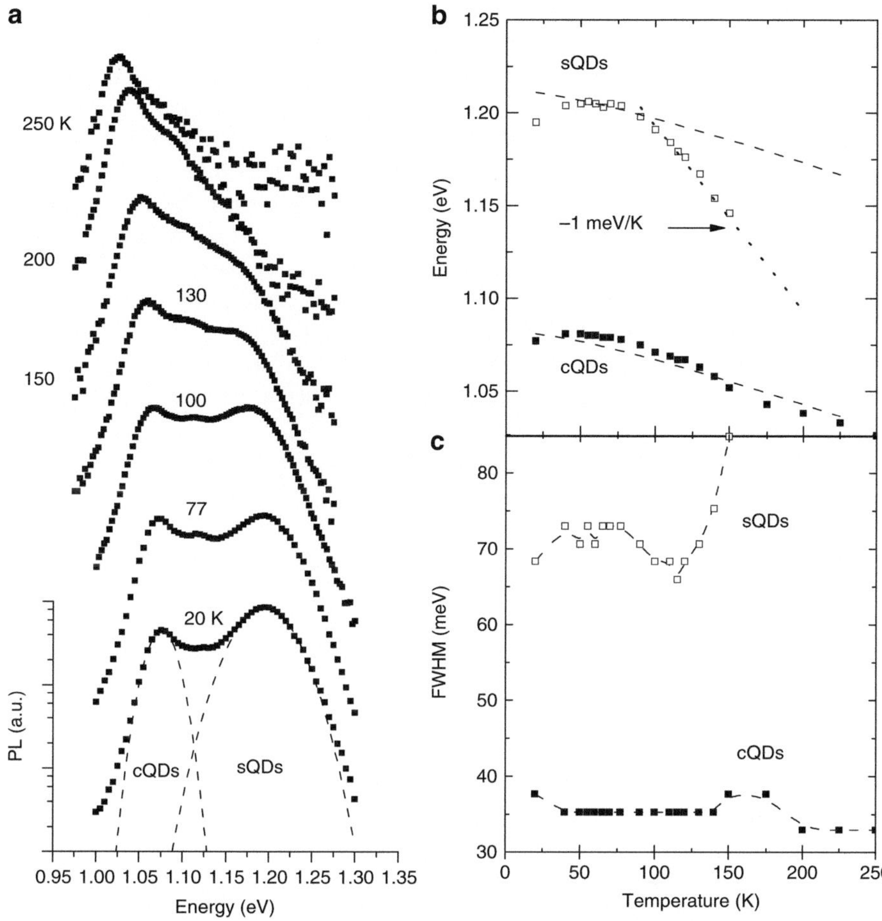

Fig. 3.7 Optical properties of 1.8/25/1.5 QDMs: (**a**) PL spectra at various temperatures, shifted vertically for clarity; temperature dependencies of (**b**) peak energy positions and (**c**) FWHM of cQDs- and sQDs-related spectra. The *dashed curves* in (**a**) are double Gaussian functions fit. The *filled-square* and *open-square* symbols in (**b**) and (**c**) are measured values for cQDs- and sQDs-related spectra, respectively. The *lower* and *upper dashed lines* in (**b**) are bulk InAs bandgap temperature variation according to Varshni's equation, shifted along the energy axis by 0.667 and 0.797 eV, respectively. The *dashed lines* in (**c**) are guide to the eye. Reproduced from [35] with permission from Elsevier

The cQD and sQD FWHM variations with temperature of the 1.8/25/1.2 and 1.8/25/1.5 QDMs are shown in Figs. 3.6c and 3.7c, respectively. Again, the variations due to cQDs and sQDs are fundamentally different: the cQDs-related FWHM are approximately the same and are almost constant (40 and 35 meV for the 1.8/25/1.2 and 1.8/25/1.5 QDMs, respectively) while the sQDs-related FWHM decrease towards a minimum at intermediate temperatures before increasing again at high temperatures. The minimum for 1.8/25/1.2 QDMs occurs at around 75 K while those for 1.8/25/1.5 QDMs occurs at around 100 K.

The FWHM reduction at intermediate temperatures is well explained by thermal escape and carrier redistribution. At low temperatures, excitons are created, confined, and recombined radiatively in individual QDs. The FWHM reflects QD size distribution. For uniform ensembles, the FWHM can be lower than 30 meV [42] while for nonuniform ones the FWHM can be greater than 100 meV [43]. As sample temperature increases, carriers are thermally activated out of individual QDs, into the WL and/or the GaAs barriers where they can subsequently be re-trapped by nearby QD. Since confined electron levels in small dots are higher than in large dots, or the thermal energy required by the electrons to escape into the adjacent WL is smaller for small dots than for large dots, it is more probable for carriers to be distributed from small to large dots. As temperature increases, the dominant emitting structure thus shifts from small to large dots, resulting in a peak energy shift much faster than the rate due to bulk bandgap reduction with temperature. This explains the dotted lines in Figs. 3.6b and 3.7b.

The carrier redistribution occurs at a slightly lower temperature in the 1.8/25/1.2 QDMs ($\sim$75 K) than in the 1.8/25/1.5 QDMs (100 K). This is possibly due to the smaller sQDs in the former because the regrown InAs thickness is 1.2 ML, lower than 1.5 ML in the latter. The smaller sQDs emit at 1.240 eV while the bigger sQDs emit at 1.195 eV. The smaller sQDs thus require a slightly lower thermal energy to escape to the 1.4-eV WL previously identified in Fig. 3.5a.

The co-existence of slow, Varshni type, and fast sigmoidal changes in the same sample is unusual. The slow red-shift that follows Varshni's equation is usually observed in QD ensembles which are carefully grown to achieve low FWHM, especially if they are later to be fabricated into lasers [42]. On the other hand, the fast red-shift is usually observed in highly inhomogeneous QD ensembles, particularly if the FWHM is greater than 80 meV [39, 40] and if they are later to be fabricated into superluminescent diodes [44]. The two types of QD ensembles are achieved with conflicting growth parameters and not usually observed on the same sample even when clear bimodal size distribution exists. Kissel et al., for example, reported the growth of intermediate sized QD ensembles with clear bimodal size distribution but both exhibit essentially the same sigmoidal behavior [45].

Carrier redistribution from small to large sQDs is effective up to a certain temperature beyond which thermal broadening will increase the FWHM and carrier loss to NRR channels will quench the overall PL signal. The more complicated growth procedure of lateral QDMs as compared to standard SK QDs makes it likely that additional NRR channels/centers maybe present and limit the usefulness of QDMs.

Using Arrhenius plots and simple, single activation energy fittings we prove that these concerns are unwarranted as the main loss mechanism is caused by the WL, as is the case for typical SK QDs. Figure 3.8a, b shows the IIs vs. inverse temperature plots for the 1.8/25/1.2 and 1.8/25/1.5 QDMs, respectively. The dashed lines are fits to the equation $I = I_0/[1 + A\exp(-E_A/k_BT)]$ where I stands for the integrated intensity, I_0 is the low-temperature integrated intensity, A is the pre-exponential factor, E_A is the activation energy, k_B is the Boltzmann's constant, and T is the temperature. The best-fit values for E_A in the 1.8/25/1.2 QDMs are 250 meV for

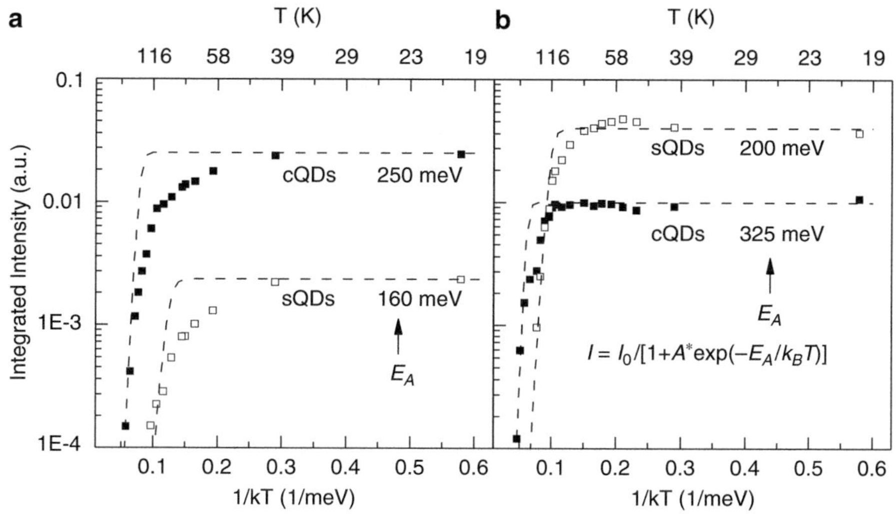

Fig. 3.8 Integrated intensities of cQDs- and sQDs-related spectra for (**a**) 1.8/25/1.2 and (**b**) 1.8/25/1.5 QDMs, plotted as a function of temperature T and $1/k_BT$. The *filled-square* and *open-square* symbols are measured values for cQDs- and sQDs-related spectra, respectively. The *dashed lines* are single activation energy fittings with pre-exponential factor $A = 3 \times 10^8$ and E_A as indicated. Reproduced from [35] with permission from Elsevier

cQDs and 160 meV for sQDs, and in the 1.8/25/1.5 QDMs are 325 meV for cQDs and 200 meV for sQDs. These values agree with the energetic difference between the related peak energy and the WL energy. The fits for the 1.8/25/1.5 QDMs are excellent: the cQD peak at 1.075 eV and the sQD peak at 1.195 eV are lower than the 1.4-eV WL energy by 325 and 205 meV, respectively. The matches are almost exact. The fits for the 1.8/25/1.2 QDMs are satisfactory for sQDs only: the cQD peak at 1.075 eV and the sQD peak at 1.240 eV are lower than the 1.4-eV WL energy by 325 and 160 meV, respectively. The corresponding best fits values are 250 and 150 meV. It is unclear what causes the large discrepancy in the case of cQDs but possible causes include the presence of some small percentage of excited-state excitons (which lowers the effective potential barrier or activation energy) or defects in or around the nanohole-and-mound template (which introduces NRR centers/channels).

3.5 Stacked QDMs

The existence of two ground-state energies makes lateral QDMs a promising candidate as a broadband material for near-infrared (NIR) emissions, especially if more than one active layer is present. It is typical for quantum wells (QWs) or QDs active layers employed in superluminescent diodes (SLDs) to be chirped, i.e. stacked in the growth direction where each layer is grown differently [46, 47]. Broadband SLDs find applications in various fields, the most prominent of which is possibly as the broadband source at the heart of optical coherence tomography (OCT) [48]. The resolution of images acquired from OCT is inversely proportional to the luminescent linewidth (FWHM) of the NIR source [49]. We proposed and demonstrated that lateral QDMs are well suited as an active material for broadband devices and systems. The main advantage over conventional QDs or QWs-based chirped structures is the broader FWHM for the same number of stacked layer, or smaller number of stacked layers for the same FWHM. This section describes chirping schemes based on lateral QDMs bi-layers as the active material. The bi-layers comprise four nominally different sub-ensembles; the PL in each of which has a predictable temperature dependency, allowing easy design and optimization of structures with a greater number of stacks.

3.5.1 *Chirped Bi-Layers*

A single layer of lateral QDMs exhibits two GS energies: a low-energy, narrow emission from cQDs and a high-energy, broad emission from sQDs. A lateral QDMs bi-layer thus exhibits four GS energies which, in order to maximize the FWHM for broadband applications, can be designed to overlap in accordance with one of the three schemes shown in Fig. 3.9a–c [23].

The straddled scheme or Type-I chirp depicted in Fig. 3.9a makes use of a wide separation between the two GS energies of one layer (cQD_1 and sQD_1 in the figure, unshaded curves) to straddle or sandwich the narrow separation of the other (cQD_2 and sQD_2, shaded). The wide separation can be achieved by a relatively thick capping and regrowth, a condition where cQDs are filled but sQDs are forming and still far from saturation. The 1.8/25/1.5 QDMs described in Sect. 3.4.2, for example, meet this criteria. The spectrum shown in Fig. 3.5b indicates GS separation as wide as 145 meV. The narrow separation, on the other hand, can be achieved by a relatively thin capping and thick regrowth. The 2/6/1.4 QDMs described in Sect. 3.4.1 with PL spectrum in Fig. 3.4b, for example, show the separation of almost zero as the cQD and sQD GS peaks are unresolved. Alternatively, a single cQD peak can be employed. By growing a bi-layer of 2/26/2 QDM_1 and 2/26/1.4 QDM_2, a Type-I chirp spectrum can be obtained. (Subscripts 1 and 2 indicate, respectively, the lower and upper QDM layers in the growth sequence.) Figure 3.9d shows the PL spectra of the bi-layer (upper spectrum) with respect to the controlled, single layer 2/25/1.4 QDMs. The latter, previously shown as a linear plot in Fig. 3.4b, exhibits a single PL peak since the deep nanoholes have yet to be saturated. The former exhibits three GS peaks: the minimum at 1.048 eV and the maximum at 1.214 eV are from the 2/26/2 QDM_1, whereas the intermediate peak at 1.086 eV is from the 2/26/1.4 QDM_2. Multi-Gaussian function fits (dashed lines) show that the lower two peak energies are narrow, indicative of cQDs-based origin, and the high peak energy is broad, indicative of sQDs-based origin. The intermediate peak energy from the

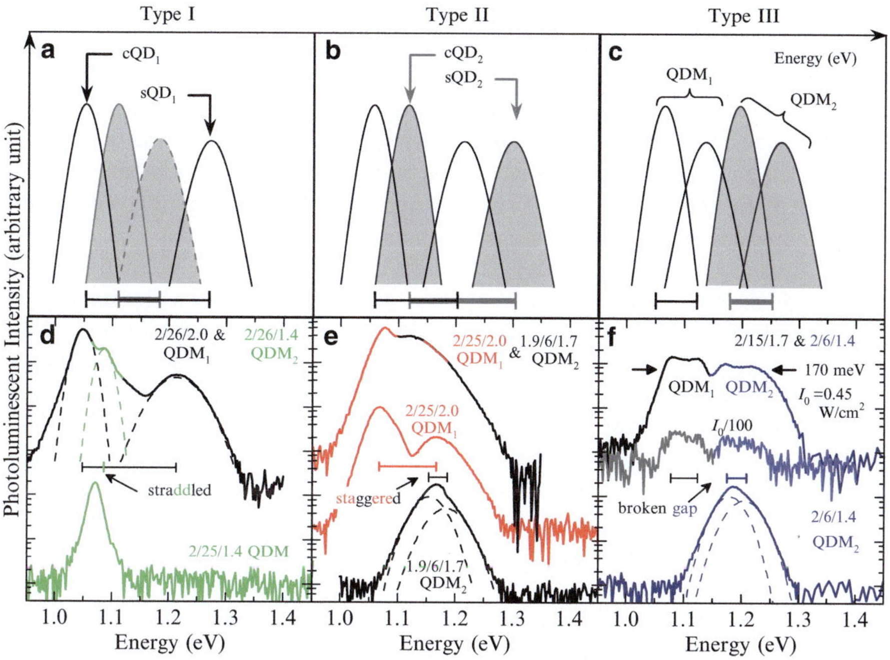

Fig. 3.9 Schematic spectral superposition of (**a**) straddled or Type-I, (**b**) staggered or Type-II, and (**c**) broken-gap or Type-III chirped QDM bi-layers. The *black* (*gray*) *bar* marks the spectral range between the cQD and sQD peaks of QDM_1 (QDM_2). Subscripts 1 and 2 denote the lower and upper QDM layers, respectively. Measured 20 K PL spectra of samples containing single QDM layer or chirped QDM bi-layer as active layers: (**d**) single 2/25/1.4 QDMs reference (*lower plot*), and chirped 2/26/2.0 QDM_1 bottom layer and 2/26/1.4 QDM_2 top layer (*upper*); (**e**) single 1.9/6/1.7 QDM_2 (*lower*), single 2/25/2.0 QDM_1 (*middle*), and chirped QDM_1/QDM_2 bi-layer (*upper*); and (**f**) single 2/6/1.4 QDM_2 (*lower*), and chirped 2/15/1.7 QDM_1 and 2/6/1.4 QDM_2 bi-layer under nominal (*upper*) and reduced excitations (*middle*). *Dashed lines* in (**d–f**) are multiple Gaussian function fits. Spectra are offset for clarity. Adapted from [23]

2/26/1.4 QDM_2 is sandwiched between the minimum and maximum from the 2/26/2 QDM_1 as expected and provides a smoothening effect of the whole spectrum. The bi-layer spectrum clearly demonstrates the superposition of individual PL peaks, indicating that reabsorptions are insignificant.

The staggered scheme or Type-II chirp depicted in Fig. 3.9b makes use of two QDM ensembles with similarly wide energetic separations and which are offset along the energy scale. One possible implementation as shown in Fig. 3.9b positions cQD_2 between cQD_1 and sQD_1, and sQD_1 between cQD_2 and sQD_2. This can be achieved by designing the bi-layers to have different capping thickness, to ensure separated cQD peak energies, but similar regrowth thickness. Figure 3.9e shows the PL spectra of the 2/25/2 QDM_1 and 1.9/6/1.7 QDM_2 bi-layer (upper spectrum) with respect to the controlled, single layers of 2/25/2 QDM_1 (middle) and 1.9/6/1.7 QDM_2 (lower). The lower spectrum shows the almost merged cQDs and sQDs peaks, similar to the 2/6/1.4 QDMs spectrum in Fig. 3.4b. The Gaussian fits (dashed lines in the lower spectrum) reveal the constituent cQDs and sQDs peaks. The middle spectrum shows two well-separated cQD and sQD peaks. The four peaks from the lower and middle spectra are staggered in accordance with the design in Fig. 3.9b. The dip in the middle of QDM_1's spectrum is made up by the rapid rise at the same energetic position of QDM_2's spectrum, resulting in a smoothened overall spectrum in the QDMs bi-layer. The bi-layer spectrum demonstrates that with proper design a broad Gaussian spectrum can be obtained.

The broken-gap scheme or Type-III chirp depicted in Fig. 3.9c makes use of two QDM ensembles with similarly narrow energetic separations where the highest peak energy of one QDM ensemble is lower than the lowest peak energy of the other. This can be achieved by designing the bi-layers to have different capping thickness, to ensure separated cQD peak energies as in Type II, and similarly thick regrowth, to ensure narrow cQD–sQD separation. Figure 3.9f shows the PL spectra of the 2/15/1.7 QDM_1 and 2/6/1.4 QDM_2 bi-layer (upper spectrum) with respect to the controlled, single layer of 2/6/1.4 QDM_2 (lower). The existent of multiple peaks in the upper spectrum of the bi-layers begs the question as to whether all these four peaks are GS. To answer this we reduce the excitation power density by two orders of magnitude, observe the linear decrease of the four peaks down to almost the noise floor as shown in the middle spectrum, and thus confirm that all the peaks in the upper spectrum are indeed GS peaks. Ignoring the small dip in the middle, the spectrum has a broad FWHM of 170 meV. This non-optimized value by chirping *two* layers of lateral QDMs compares favorably with 125 meV obtained from chirping *four* layers of QDs [47], or 200 meV from *sixty* stacks of strain-compensated structure [50]. Lateral QDM bi-layers thus provide the best active layer in terms of cost-performance: a broader FWHM can be achieved for the same stack number, or the smallest stack number is required for the same FWHM.

3.5.2 Temperature Dependencies

The optical properties of a lateral QDM *single* layer has been shown to follow the bimodal optical characteristics explained in Sect. 3.4.3, it is thus expected that the *bi*-layer should follow the same temperature dependencies since both layers are separated by a thick 100-nm GaAs spacer layer and hence should be optically uncoupled. Though reabsorptions (of QDM_1 emissions by QDM_2, and vice versa) are a concern, low-temperature PL spectra in the three chirping schemes above indicate that they do not qualitatively affect the wavelength superposition. Variable temperature PL spectra in this section additionally indicate that they do not qualitatively affect the underlying carrier escape and redistribution from and between cQDs and sQDs either.

The temperature-dependent PL spectra of the Types I–III chirped samples above are shown in Fig. 3.10a–c, respectively. The overall spectra are similarly quenched as the temperature increases, and beyond 250 K no luminescence can be measured.

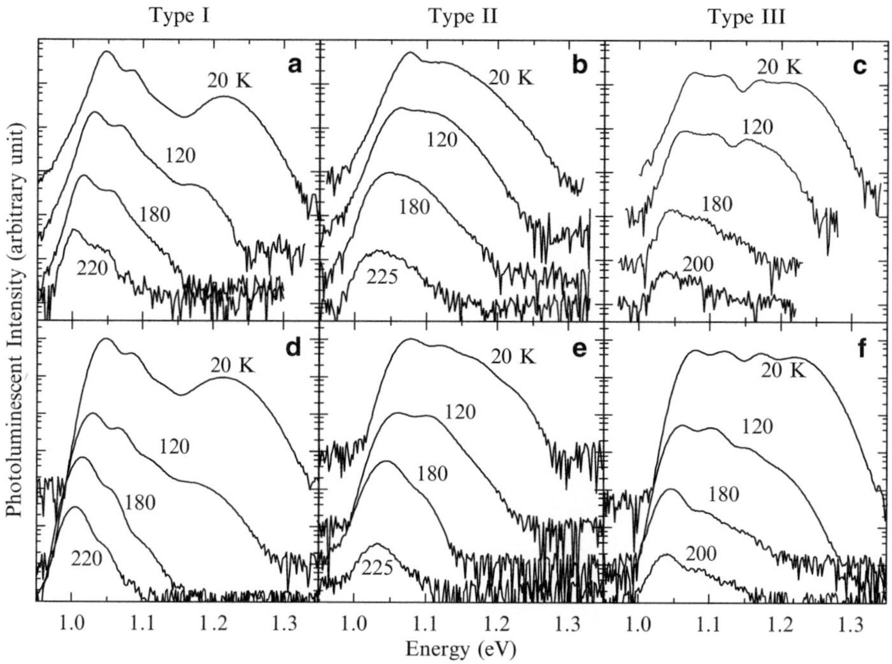

Fig. 3.10 Temperature-dependent PL of chirped QDM bi-layers: measured PL spectra of Types (**a**) I, (**b**) II, and (**c**) III; simulated PL spectra of Types (**d**) I, (**e**) II, and (**f**) III. Line spectra in (**a–f**) are offset for clarity. Simulated line spectra in (**d–f**) are performed at the same temperatures as the measured spectra in (**a–c**), respectively. Adapted from [23]

This is not to be taken as a limiting factor for room-temperature operations as the structures have yet to be optimized. The overarching trend in all samples is the subsequent quenching from the high-energy ends. In Fig. 3.10a, for example, the highest-energy peak at 1.214 eV is the first to be quenched, followed by the next immediate peak at 1.086 eV, and finally by the lowest-energy peak at 1.048 eV. Such orderly quenching is characteristic of thermal activation of carriers out of QDs into the adjacent WL and/or GaAs matrix where carriers recombine non-radiatively. The multiplicity of luminescent peaks in the QDM bi-layers makes it difficult to identify the NRR channels and associated activation energies without prior knowledge from controlled single QDM layer structures. If our hypothesis of optical independence between the QDM bi-layer is correct, the main escape channel should be the same as QDM single layers, i.e. the WL as identified by the Arrhenius plots in Fig. 3.8.

In order to identify the NRR channels and to understand the temperature dependencies of the three chirp structures, the spectra are fitted to the equation:

$$I = \sum_{i=1}^{4} \frac{I_i \exp(E - E_i)^2 / \Gamma_i^2}{1 + A \exp -(E - E_{\mathrm{WL}}) / \eta_i k_{\mathrm{B}} T}$$

Table 3.1 Simulation parameters for the PL maps and spectra of chirped QDM bi-layers

	QDM_1		QDM_2	
	cQD_1	sQD_1	cQD_2	sQD_2
Type I—straddled				
I_i	1.000	0.094	0.358	0.071
E_i (eV)	1.048	1.214*	1.085	1.114*
FWHM (meV)	30.6	77.7*	28.3	51.8*
η_i	1.0–1.6	1.1–1.9	1.0–1.6	1.1–1.9
Type II—staggered				
I_i	1.000	0.315	0.591	0.044
E_i (eV)	1.077	1.160*	1.120	1.213*
FWHM (meV)	40.0	53.0*	40.0	49.5*
η_i	1–1.4	1–1.8	1–1.4	1–1.8
Type III—broken-gap				
I_i	1.0	1.0	0.7	0.7
E_i (eV)	1.078	1.121	1.170	1.220*
FWHM (meV)	33.0	42.4	33.0	65.9*
η_i	1.1–2	1.1–2	1.1–2	1.0–2

Twenty-Kelvin peak energy position E_i, relative intensity I_i, and FWHM of cQDs and sQDs ensembles of Types I–III chirped QDM bi-layers extracted from Fig. 3.10a–c, respectively. The ideality factor η_i varies linearly with temperature from the lower limit value at 20 K to the upper limit value at 300 K. Subscripts 1 and 2 represent the lower and upper QDM layers, respectively. E_i's temperature dependency follows Varshni's equation unless marked by * where it instead follows the sigmoidal behavior. FWHM is assumed constant unless marked by * where it follows the anomalous temperature behavior. The FWHM is related to the standard deviation of the Gaussian distribution or the broadening parameter Γ through the relationship: FWHM (meV) = 1,665.11 $\times$ Γ. Reproduced from [23]

where the overall intensity I at energy E and temperature T is a summation of constituent intensities from QDM_1 (cQD_1 and sQD_1) and QDM_2 (cQD_2 and sQD_2), hence the summation from $i = 1$ to 4. Each constituent's luminescence has a low-temperature intensity I_i at a peak energy E_i, a broadening parameter Γ_i and is quenched by thermal escape to the WL level E_{WL}. A is the pre-exponential factor, k_B is the Boltzmann's constant and η_i is the ideality factor indicating the dominance of the WL over other NRR channel(s). If the WL is the sole factor responsible for quenching, then $\eta_i = 1$. If other NRR channel(s) co-exist and acting in parallel, then $\eta_i > 1$. The further η_i is from 1, the less significant the WL is as excitons escape route. The fitting parameters extracted from the measured spectra are summarized in Table 3.1.

The simulated line spectra of Types I–III chirps at selected temperatures are shown underneath the measured spectra in Fig. 3.10d–f, respectively. A white noise is added to the simulated data to reflect the actual noise levels in our setup. The full simulations of Types I–III chirps covering the 20–300 K temperature range are shown in the PL maps $I(E, T)$ in Fig. 3.11a–c, respectively. The dashed lines

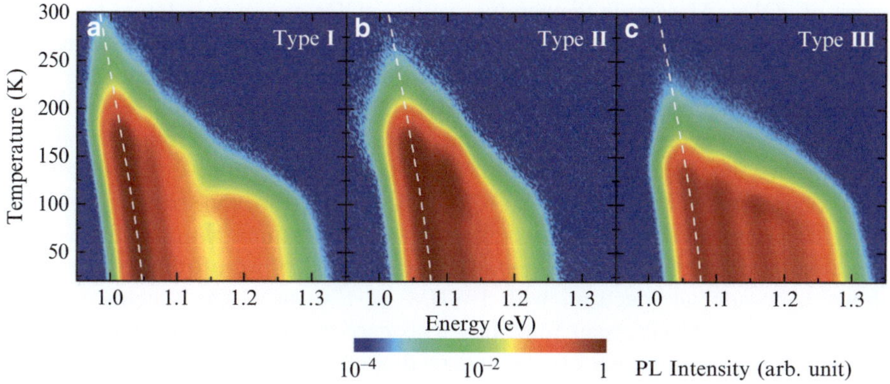

Fig. 3.11 Simulated PL maps $I(E, T)$ of Types (**a**) I, (**b**) II, and (**c**) III chirped QDM bi-layers. *Dashed lines* are bandgap variations with temperature according to Varshni's equation with bulk InAs parameters, linearly shifted to match the lowest-energy cQD peaks. Adapted from [23]

in the PL maps are calculated temperature variations of the cQD peak based on Varshni's equation and bulk InAs parameters; the trend is followed only by cQDs-related peaks. Despite the simplicity of the approach, the fits predict all the major characteristics of the measured spectra at all experimental temperatures. Though we can further improve the simulations by incorporating the recently reported parameters for the Fan model [37], it does not change the qualitative nature of our conclusions that the WL is the main escape path and that the activation energy falls somewhere between the ideal case where $\eta_i = 1$ and the limiting case where $\eta_i = 2$.

Though the chirping schemes described in Sect. 3.5.1 and the demonstrated optical characteristics in Sect. 3.5.2 employ a QDM bi-layer as the active layer, the same fundamental concepts can be readily extended to structures with number of stacks greater than two, or to other material systems in order to increase the bandwidth or to shift the nominal wavelength to other region of the electromagnetic spectrum, respectively.

3.6 Conclusion

Lateral InGaAs quantum dot molecules are grown by solid-source MBE via the partial-cap and regrowth process using the nanohole-and-mound template. Each QDM comprises two types of QDs: a cQD located at the nanohole center, and sQDs surrounding it. Differences in nucleation locations and dynamics result in cQDs and sQDs having different sizes and degrees of homogeneity: cQDs are generally taller and more uniform than sQDs. Temperature-dependent photoluminescent spectra reflect the geometrical differences: cQDs typically emit at a low GS energy around 1.05 eV with a narrow FWHM below 30 meV whereas sQDs emit at about 150 meV higher with a broad FWHM that more than doubles those of cQDs. As sample temperature varies, cQDs emission is found to simply follow the InAs bulk bandgap variation in accordance with Varshni's equation. On the contrary, sQDs exhibit a sigmoidal temperature behavior resulting from carrier redistribution among inhomogeneous QDs. The two qualitatively different temperature-dependent PL or the bimodal *optical* property results from the intrinsic bimodal *size* distribution of our lateral QDM ensemble. With this unique property we proposed and demonstrated a QDM bi-layer structure that exhibits four GS energies whose spectra can be arranged to overlap in three basic configurations: straddled, staggered, and broken-gap. A non-optimized, proof-of-principle structure shows a broadband spectrum with FWHM of 170 meV. The spectra are well explained by multi-Gaussian functions with carrier redistributions among sQDs and quenching via thermal escape and recombination via the wetting layer and non-radiative recombination centers/channels acting in parallel. We introduced an ideality factor to indicate the dominance of the WL as the quenching channel. Well-understood optical properties of lateral QDM single- and bi-layers are necessary if they are to serve as an active material for devices destined for broadband absorption such as solar cells or for broadband emissions such as superluminescent diodes.

Acknowledgments AFM data analyses are performed using Gwyddion. This work is supported by Thailand Research Fund (RSA5580015, DPG5380002); Nanotec; Integrated Innovation Academic Center (IIAC), Chulalongkorn University Centenary Academic Development Project; and the Higher Education Research Promotion and National Research University Project of Thailand, Office of the Higher Education Commission (CU56-EN09).

References

1. Leonard, D., Krishnamurthy, M., Reaves, C.M., Denbaars, S.P., Petroff, P.M.: Appl. Phys. Lett. **63**, 3203 (1993)
2. Bimberg, D., Grundmann, M., Ledentsov, N.N.: Quantum Dot Heterostructures. Wiley, Chichester (1999)
3. Wang, Z.M. (ed.): Self-Assembled Quantum Dots. Lecture Notes in Nanoscale Science and Technology. Springer, New York (2007)
4. Watanabe, K., Koguchi, N., Gotoh, Y.: Jpn. J. Appl. Phys. **39**, L79 (2000)
5. Wang, L., et al.: New J. Phys. **10**, 045010 (2008)
6. Wang, L., Rastelli, A., Kiravittaya, S., Benyoucef, M., Schmidt, O.G.: Adv. Mater. **21**, 2601 (2009)
7. Schedelbeck, G., Wegscheider, W., Bichler, M., Abstreiter, G.: Science **278**, 1792 (1997)
8. Bayer, M., Hawrylak, P., Hinzer, K., Fafard, S., Korkusinski, M., Wasilewski, Z.R., Stern, O., Forchel, A.: Science **291**, 451 (2001)
9. Loss, D., DiVincenzo, D.P.: Phys. Rev. A **57**, 120 (1998)
10. Songmuang, R., Kiravittaya, S., Schmidt, O.G.: Appl. Phys. Lett. **82**, 2892 (2003)
11. Lee, J.H., Wang, Z.M., Strom, N.W., Mazur, Y.I., Salamo, G.J.: Appl. Phys. Lett. **89**, 202101 (2006)
12. Strom, N.W., Wang, Z.M., Lee, J.H., AbuWaar, Z.Y., Mazur, Y.I., Salamo, G.J.: Nanoscale Res. Lett. **2**, 112 (2007)
13. Suraprapapich, S., Thainoi, S., Kanjanachuchai, S., Panyakeow, S.: J. Vac. Sci. Technol. B **23**, 1217 (2005) [Ibid. **24**, 1665 (2006)]

14. Siripitakchai, N., Suraprapapich, S., Thainoi, S., Kanjanachuchai, S., Panyakeow, S.: J. Cryst. Growth **301**, 812 (2007)
15. Suraprapapich, S., Kanjanachuchai, S., Thainoi, S., Panyakeow, S.: Microelectron. Eng. **83**, 1526 (2006)
16. Lippen, T.v., Nötzel, R., Hamhuis, G., Wolter, J.: Appl. Phys. Lett. **85**, 118 (2004)
17. Lippen, T.v., Silov, A.Y., Notzel, R.: Phys. Rev. B **75**, 115414 (2007)
18. Thet, C.C., Sanorpim, S., Panyakeow, S., Kanjanachuchai, S.: Semicond. Sci. Technol. **23**, 055007 (2008)
19. Welsch, H., Kipp, T., Köppen, T., Heyn, C., Hansen, W.: Semicond. Sci. Technol. **23**, 045016 (2008)
20. Himwas, C., Panyakeow, S., Kanjanachuchai, S.: Nanoscale Res. Lett. **6**, 496 (2011)
21. Seravalli, L., Frigeri, P., Nasi, L., Trevisi, G., Bocchi, C.: J. Appl. Phys. **108**, 064324 (2010)
22. Nabetani, Y., Ishikawa, T., Noda, S., Sasaki, A.: J. Appl. Phys. **76**, 347 (1994)
23. Patanasemakul, N., Panyakeow, S., Kanjanachuchai, S.: Nanoscale Res. Lett. **7**, 207 (2012)
24. Krenner, H.J., Sabathil, M., Clark, E.C., Kress, A., Schuh, D., Bichler, M., Abstreiter, G., Finley, J.J.: Phys. Rev. Lett. **94**, 057402 (2005)
25. Scheibner, M., Yakes, M., Bracker, A.S., Ponomarev, I.V., Doty, M.F., Hellberg, C.S., Whitman, L.J., Reinecke, T.L., Gammon, D.: Nat. Phys. **4**, 291 (2008)
26. Beirne, G.J., Hermannstädter, C., Wang, L., Rastelli, A., Schmidt, O.G., Michler, P.: Phys. Rev. Lett. **96**, 137401 (2006)
27. Kanjanachuchai, S., Thudsalingkarnsakul, N., Siripitakchai, N., Changmoang, P., Thainoi, S., Panyakeow, S.: Microelectron. Eng. **87**, 1352 (2010)
28. Grundmann, M., Ledentsov, N.N., Stier, O., Bimberg, D., Ustinov, V.M., Kop'ev, P.S., Alferov, Z.I.: Appl. Phys. Lett. **68**, 979 (1996)
29. Mukai, K., Ohtsuka, N., Shoji, H., Sugawara, M.: Appl. Phys. Lett. **68**, 3013 (1996)
30. Lambkin, J.D., Dunstan, D.J., Homewood, K.P., Howard, L.K., Emeny, M.T.: Appl. Phys. Lett. **57**, 1986 (1990)
31. Heitz, R., Mukhametzhanov, I., Madhukar, A., Hoffmann, A., Bimberg, D.: J. Electron. Mater. **28**, 520 (1999)
32. Fafard, S., Raymond, S., Wang, G., Leon, R., Leonard, D., Charbonneau, S., Merz, J.L., Petroff, P.M., Bowers, J.E.: Surf. Sci. **361–362**, 778 (1996)
33. Szafran, B., Peeters, F.M.: Phys. Rev. B **76**, 195442 (2007)
34. Peng, J., Hermannstädter, C., Witzany, M., Heldmaier, M., Wang, L., Kiravittaya, S., Rastelli, A., Schmidt, O.G., Michler, P., Bester, G.: Phys. Rev. B **81**, 205315 (2010)
35. Thongkamkoon, N., Patanasemakul, N., Siripitakchai, N., Thainoi, S., Panyakeow, S., Kanjanachuchai, S.: J. Cryst. Growth **323**, 206 (2011)
36. Varshni, Y.P.: Physica **34**, 149 (1967)
37. Yeo, I., Dong Song, J., Lee, J.: Appl. Phys. Lett. **99**, 151909 (2011)
38. Brusaferri, L., et al.: Appl. Phys. Lett. **69**, 3354 (1996)
39. Lubyshev, D.I., Gonzalez-Borrero Jr., P.P., Marega, E., Petitprez Jr., E., La Scala, N., Basmaji, P.: Appl. Phys. Lett. **68**, 205 (1996)
40. Polimeni, A., Patane, A., Henini, M., Eaves, L., Main, P.C.: Phys. Rev. B **59**, 5064 (1999)
41. Xu, Z.Y., et al.: Phys. Rev. B **54**, 11528 (1996)
42. Yang, T., Tatebayashi, J., Tsukamoto, S., Nishioka, M., Arakawa, Y.: Appl. Phys. Lett. **84**, 2817 (2004)
43. Chia, C.K., Chua, S.J., Dong, J.R., Teo, S.L.: Appl. Phys. Lett. **90**, 061101 (2007)
44. Ngo, C.Y., Yoon, S.F., Fan, W.J., Chua, S.J.: Appl. Phys. Lett. **90**, 113103 (2007)
45. Kissel, H., Muller, U., Walther, C., Masselink, W.T., Mazur, Y.I., Tarasov, G.G., Lisitsa, M.P.: Phys. Rev. B **62**, 7213 (2000)
46. Lin, C.-F., Lee, B.-L.: Appl. Phys. Lett. **71**, 1598 (1997)
47. Haffouz, S., Raymond, S., Lu, Z.G., Barrios, P.J., Roy-Guay, D., Wu, X., Liu, J.R., Poitras, D., Wasilewski, Z.R.: J. Cryst. Growth **311**, 1803 (2009)
48. Greenwood, P.D.L., et al.: IEEE J. Sel. Topics Quantum Electron. **16**, 1015 (2010)
49. Schmitt, J.M.: IEEE J. Sel. Topics Quantum Electron. **5**, 1205 (1999)
50. Akahane, K., Yamamoto, N.: J. Cryst. Growth **323**, 154 (2011)

Chapter 4
Formation Principles and Exciton Relaxation in Semiconductor Quantum Dot–Dye Nanoassemblies

Christian von Borczyskowski and Eduard Zenkevich

Abstract In this chapter we discuss "bottom-up" non-covalent self-assembly principles which define a strategy for the formation of organic–inorganic nanoassemblies containing colloidal semiconductor quantum dots (QD) of different types (based on a CdSe core) and various heterocyclic molecules (dyes) with functionalized anchoring side substituents (*meso*-pyridyl substituted porphyrins and perylene diimides). Using a combination of ensemble and single molecule spectroscopy of "QD–Dye" nanoassemblies, we show that single functionalized molecules can be considered as extremely sensitive probes for studying the complex interface physics and chemistry (influence of the embedding environment and temperature) and related exciton relaxation processes in QDs. It will be quantitatively laid out that the major part of the observed QD photoluminescence (PL) quenching in nanoassemblies can be understood, on the one hand, in terms of exciton wave function tunneling under the condition of quantum confinement and, on the other hand, by the influence of ligand dynamics. In nanoassemblies, photoinduced Foerster-type energy transfer (FRET) QD $\rightarrow$ Dye is often only a small contribution to the PL quenching and is effectively suppressed already in slightly polar solvents which is often overlooked in literature. Finally we would like to point out that properties of "QD–Dye" nanoassemblies are not only interesting in themselves but also provide a valuable tool to study surface-related phenomena in

---

C. von Borczyskowski (✉)
Institute of Physics and Centre for Nanostructured Materials and Analytics (nanoMA), Chemnitz University of Technology, Reichenhainerstr. 70, Chemnitz, 09107 Germany
e-mail: borczyskowski@physik.tu-chemnitz.de

E. Zenkevich
Department of Information Technologies and Robotics, National Technical University of Belarus, Prospect Nezavisimosti 65, Minsk, 220013 Belarus
e-mail: zenkev@tut.by

J. Wu and Z.M. Wang (eds.), *Quantum Dot Molecules*, Lecture Notes in Nanoscale Science and Technology 14, DOI 10.1007/978-1-4614-8130-0_4,

QDs on an extremely low level of surface modification, thus providing the data for a further development of defined multi-component structures for exploitation as artificial light-harvesting complexes, electro- and photochemical devices or nanosensors.

4.1 Introduction to Nanoassemblies

At present, nanotechnology is emerging as an integrated research field aimed at investigating and controlling nanomaterials by combining concepts from chemistry, physics, biology, and engineering. Nanostructured materials with tunable morphology have attracted exceptional interest because of their unique architectures, tailored physicochemical properties, and central role in fabricating nano-electronics, and potential applications in nano-biomedicine [1–7]. Very often, the bottom-up formation of nanomaterials is based on self-assembly approaches being the fundamental phenomenon that generates structural organization on all scales and may be realized in solutions and solid state via various basic interactions: hydrogen bonding, coordination bonding, electrostatic and donor–acceptor interactions, or metal–ion binding [8]. The designed self-assembled process, the information necessary to initiate such a process, and the algorithm behind must be stored in the components and must be operative via selective/specific interactions [9].

One direction in the field of nanoscience and nanotechnology is connected with the study and applications of inorganic semiconductor materials of nanoscale dimensions. Semiconductor nanocrystals [often referred to as quantum dots (QD), e.g. CdSe or CdSe/ZnS and other II–VI compounds] represent a specific class of matter between atomic clusters and bulk materials with well-defined size-dependent tunable photophysical properties [10–14]. Moreover, based on self-assembly ideas discussed above, the anchoring of functional organic molecules, molecular complexes, and biostructures to QDs is of considerable scientific and a wide practical interest including material science and biomedical applications [15–24]. In this respect, there are few a principal aspects which should be taken into account upon analysis of photoluminescence (PL) characteristics for QDs being a part of heterogeneous nanoassemblies.

Optical properties of colloidal semiconductor QDs have been investigated intensively during the last two decades [10–14, 25, 26] including design strategies of, e.g. of core–shell systems [27], ligand chemistry [28–34], and surface functionalization [23, 35–40]. Tuneable band position and PL high quantum yields are of crucial importance for envisaged applications [41, 42]. Notably, because of the increased surface-to-volume ratio relative to bulk materials, QD surfaces are subject to chemical and structural disorder. Thus, along with size distribution, surface chemistry is the major source of heterogeneity in the optical properties of QDs both in time, which manifests as PL intermittency or "blinking" for single QDs, and within an ensemble of particles. It means that the heterogeneity and dynamics of QD surface complicate efforts to understand the mechanisms by which surface chemistry may influence the optical and dynamic properties of QDs. Correspondingly, PL characteristics for colloidal QDs in solution are subject to various dynamic processes which are related to QD interface properties: (1) the adsorption of spurious molecules [28, 29], (2) attachment and detachment of protective electrostatically bound [43, 44], chemisorbed [39, 40] or chelating [35] ligands, and (3) the anchoring of functionalized dye molecules [23, 36–38, 45] or protein complexes [46, 47] to QD surfaces in hybrid nanoassemblies. In general, a surface reconstruction due to "mobile traps" [48] can be invoked by any of the ligands mentioned. Typically, changes in solvent or surface-bound ligands have been found to affect these surface traps and thereby influence steady-state [49–53] and time-resolved [54, 55] PL of QDs. It has also been theoretically shown that surface states split off from the band edge of the valence and conduction band in case of non-passivated QDs [56]. In case of "QD–organic ligand" nanoassemblies the formation of OD-ligand delocalized (hybridized) surface states may take place which are mostly inactive in absorption spectra but have significantly influence on QD PL dynamics [57, 58]. Thus, the kind and concentration of ligands in nanoassemblies may control the optical energies of surface states and their influence on the optical properties of QDs which is important upon analysis of QD PL quenching mechanisms.

Basically there are several routes to realize "QD–Dye" nanoassemblies in liquid solutions and polymeric matrices in the form of (1) blends [59], (2) QD–polymer–dye composites [60, 61], and (3) self-assembled QD–Dye assemblies via functional groups [36, 38, 45, 62, 63]. With respect to formation of self-assembled "QD–Dye" nanoassemblies in the liquid phase (besides attachment/detachment of dye molecules and the presence/formation of various surface trap states mentioned above) the competitive interplay of dye molecule attachment and capping ligand exchange dynamics [e.g., tri-*n*-octyl phosphine oxide (TOPO) or long chain amines (AM)] [28–34, 64, 65] is of essential importance both from fundamental principles and from possible applications. In addition, this competitive interplay dynamic is influenced by temperature [66] and the dielectric properties of the solvent [67]. In the past a few studies have been devoted to the elucidation of thiol-type ligand exchange at relative low concentrations between one and ten relative to the QD concentration [33, 68]. Recently, "QD–Dye" nanoassembly formation has been studied under the conditions of extremely low concentrations of QDs and dye molecules and a microscopic description of QD PL quenching has been presented [64, 65]. In contrast to the number of bulk measurements on large ensembles of QDs, there have been far fewer studies on the effects of surface ligands on the photoluminescence of CdSe QDs at the single-nanocrystal level focusing predominantly on photoluminescence intermittency (or blinking) investigations [69–73, 124].

However, a better understanding of the effects of ligand binding on single QD PL is critical to interprete existing ligand binding data derived from solution photoluminescence measurements. In this respect, the combination of bulk and single molecule/single nanoassembly experiments [74, 75] is a tool to precisely identify the interaction of exactly one QD with one dye molecule leading to a microscopic understanding of the formation (including ligand dynamics) and related mechanisms of PL quenching dynamics for "QD–Dye" nanoassemblies. Knowledge of the ligand dynamics and surface functionalization can play an important role in various technological fields, e.g. for the fabrication of nanostructured inks for solution-processed photovoltaics [76] or printed semiconductor layers in flexible electronics [77].

As far as the quantum efficiency and PL dynamics for QDs depend drastically on the environment it is a demanding task to control the environment in a systematic and well-defined way, e.g. by changing the temperature in a controlled manner and observing the corresponding influence on PL for, at least, individual QDs, single "QD–Dye" nanoassemblies and bulk solutions. From this point of view, experiments in a wide temperature range (300–2 K) are of urgent interest because of two reasons: (1) temperature- and size-dependence of the exciton decay of individual QDs provides a direct way to characterize the QD energy band gap [33, 68, 78–80], as well as the elucidation of the specificity of the interactions between QDs and organic ligands and (2) the phase transition (or reconstruction) of the capping ligand shell [81, 82] or rearrangement of organic ligands upon temperature changes may affect the PL behavior of QDs. With respect to the band gap even for pure QDs reported results and interpretations are, however, not fully consistent with each other, while for "QD–Dye" nanoassemblies the corresponding data are even missing. The crucial role of the solvent and ligands on PL properties upon temperature variation has also been demonstrated for individual QDs [80, 81]. Nevertheless, there are no studies on temperature effects in "QD–Dye" nanoassemblies. In the last case, the controlled use of dyes, which quench the PL of the QD, is a useful tool to follow indirectly the thermodynamics of ligand–QD interactions at various temperatures. In addition, due to the simultaneous observation and analysis of PL quenching effects, which serve as an indicator for the formation of an "QD–Dye" nanoassembly, there is a direct access to surface-related processes as well as to the elucidation of PL quenching mechanisms [83]. Interestingly, recent in situ diffraction data show that the TOPO- or hexadecylamine-capped CdSe QDs in toluene exhibit predominantly wurtzite crystal structure, which undergoes a phase transformation to zinc blende crystal structure following drop-casting on Si, thus showing environment-induced phase transformation of even the CdSe core [84].

In most cases the formation of "QD–Dye" nanoassemblies is followed by QD PL quenching, which is studied both in bulk solutions and on a single particle detection level. Commonly, this PL quenching is interpreted as being due to photoinduced charge transfer (CT) [45, 85–88] and/or energy transfer processes QD $\rightarrow$ Dye [21, 36, 38, 46, 60, 62, 74, 89–95]. To date, though for lot of systems ample qualitative evidence for the presence of such quenching processes is given, only a limited number of papers unravel quantitatively whether the PL quenching (full or in some cases partly, at least) can uniquely be assigned to CT [59, 96] or Foerster resonance energy transfer (FRET) for bulk solutions [60, 62, 90, 91, 97–99] and for single "QD–Dye" nanoassemblies [74, 94, 100]. On the other hand, PL quenching may be induced by other non-FRET processes [36, 96, 101–103] and be related to the involvement of QD surface states [23, 56, 58, 83] or photoinduced self-trapping of charges in the dielectric medium of the environment of QDs [67, 104, 105, 124].

Finally, the dynamically controlled attachment process may be an additional reason for PL quenching in QD–Dye nanoassemblies [30, 31, 33, 39, 44, 106]. It turned out that in case of FRET in "QD–Dye" nanoassemblies, the quantitative verification of an energy transfer process as the dominant reason for QD PL quenching is only provided by the comparison of FRET efficiencies obtained, on the one hand, via the donor (QD) PL quenching and, on the other hand, via the sensitization of the acceptor (dye) fluorescence. This is often missing in many publications resulting in incorrect assignments of processes and erroneous data evaluation. In this respect, experimental evidence for such QD PL quenching processes without a corresponding fluorescence enhancement of dye molecules attached to QD surfaces was also reported for complexes of QDs and tetramethylrhodamine-labeled streptavidin [46], but mechanisms have not been quantitatively discussed.

With these ideas in mind, we will present in this contribution a comparative and quantitative analysis of recent quenching and sensitization results for "QD–Dye" nanoassemblies based on steady-state and PL picosecond time-resolved measurements showing the existence of a strong competitive non-FRET quenching process together with the evaluation of the related mechanism. This contribution is organized as follows. Available data based on a combination of ensemble and single molecule/particle spectroscopy for "QD–Dye" nanoassemblies will be discussed. In Sect. 4.2 the basic formation principles for self-assembled "QD–Dye" nanoassemblies (based on TOPO- and amine-capped CdSe or CdSe/ZnS QDs and functionalized dyes such as pyridyl substituted porphyrins and perylene diimids) as well as their spectral-kinetic properties will be highlighted. Section 4.3 is devoted to the analysis of exciton relaxation pathways and QD photoluminescence quenching in "QD–Dye" nanoassemblies including FRET and non-FRET processes. Temporal dynamics of ligand exchange as well as the influence of the solvent and temperature effects will be discussed in Sects. 4.4 and 4.5. It will be demonstrated that the QD surface is inhomogeneous with respect to the involved attachment and detachment processes, i.e. the formation of "QD–Dye" nanoassemblies is in competition with exchange dynamics of TOPO or amine ligands and functionalized organic dye molecules. It will be shown also that very few or even only one attached dye molecule change the distribution and/or presence of dye-related surface trap states considerably. The "decoration" of QDs by dye molecules makes a phase transition of the QD capping ligand shell (at low temperatures) highly visible or even amplifies this phase transition.

We would like to emphazise that this contribution should be viewed as a review of a likewise comparative characterization of the non-covalent self-assembly possibilities and the influence of interface properties on the excitation dynamics in "QD–Dye" nanoassemblies. It may be pointed out that properties of "QD–Dye" nanoassemblies are not only interesting in themselves but also provide a valuable tool to investigate surface-related phenomena in QDs on an extremely low level of surface modification, thus providing the data for a further development of defined multi-component structures for exploitation such as artificial light-harvesting complexes, electro- and photochemical devices or nanosensors.

4.2 Basic Principles of "QD–Dye" Nanoassembly Formation

4.2.1 Characterization of Assembly Constituents

For ensemble and single particle experiments highly monodisperse colloidal core/shell CdSe/ZnS QDs passivated by TOPO or long-chain amines (AM) as well as uncapped CdSe QDs have been used (Evident Technologies, Inc, Troy, NY, USA) (Fig. 4.1). The main structural and optical properties of QDs are presented in our earlier publications [62, 63, 65, 75, 90].

With respect to the formation of "QD–Dye" nanoassemblies, our focus is on dye molecules of two classes: *meso*-pyridyl substituted porphyrins, $(m\text{-Pyr})_n\text{-H}_2P$ or H_2P (synthesis and properties are described in [107, 108]) (Fig. 4.2), and high photostable pyridyl functionalized perylene diimide molecules, PDI [109, 110] (Fig. 4.3).

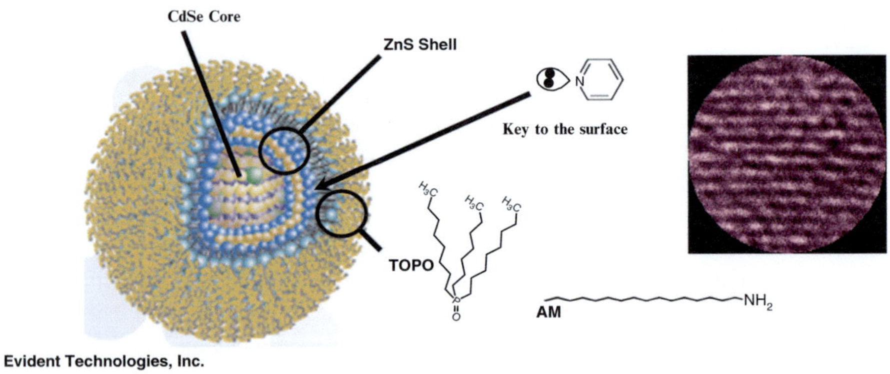

Fig. 4.1 Schematic structural presentation and transmission electron microscope image of CdSe/ZnS QD. Tri-*n*-octyl phosphine oxide (TOPO) and long-chain amine (AM) ligands are also shown. Key to the surface presents the basic scheme of functional anchor for a dye molecule

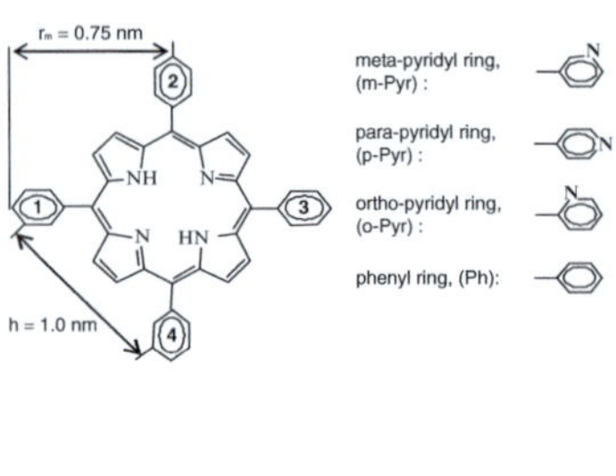

Porphyrin abbreviation	Meso-substituents	
$(m\text{-Pyr})_1\text{-H}_2P$	1: (*m*-Pyr)	2, 3, 4: (Ph)
$(m\text{-Pyr})_2\text{-H}_2P$	1, 3: (*m*-Pyr), opposite	2, 4: (Ph)
$(m\wedge\text{Pyr})_2\text{-H}_2P$	1, 2: (*m*^Pyr), adjacent	3,4: (Ph)
$(m\text{-Pyr})_3\text{-H}_2P$	1, 2, 3: (*m*-Pyr)	4: (Ph)

Fig. 4.2 Chemical structure, abbreviations, and positions of pyridyl-substituents for *meso*-pyridyl substituted porphyrins (H_2P). The same scheme applies to para- or ortho pyridyl substituted H_2P

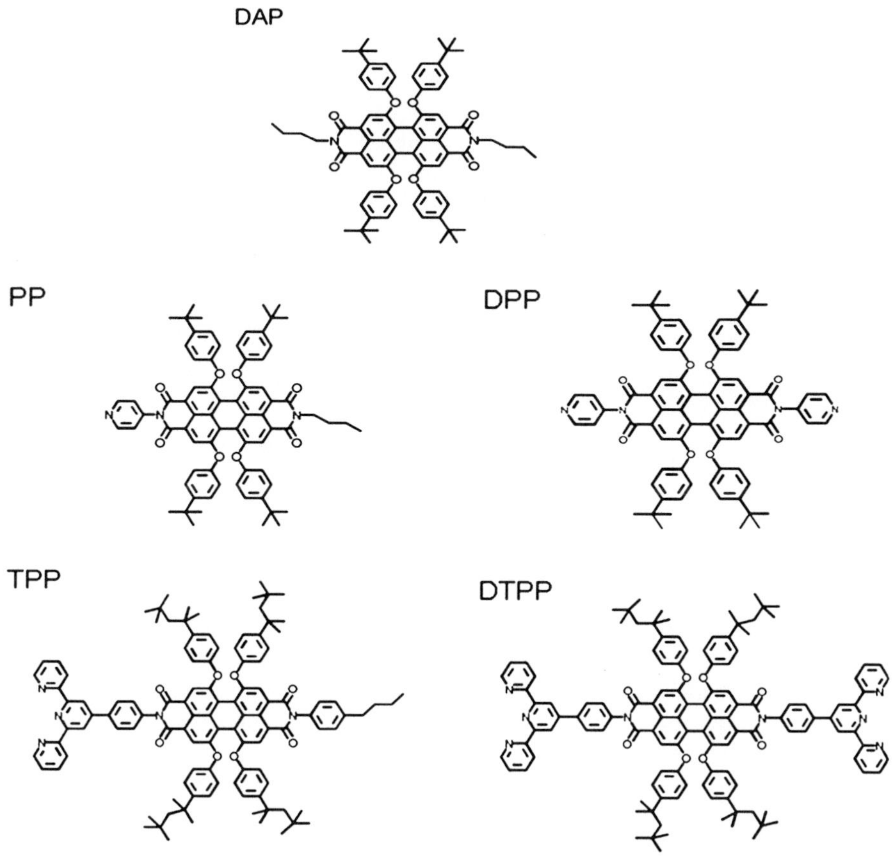

Fig. 4.3 Chemical structure and abbreviations for perylene diimides (PDI). *DAP* dialkyl-PDI, *PP* $(\text{pyridyl})_1$-PDI, *DPP* $(\text{pyridyl})_2$-PDI, *TPP* $(\text{ter-pyridyl})_1$-PDI, *DTPP* $(\text{ter-pyridyl})_2$-PDI

Like in the case of multiporphyrin complexes previously studied in our group [107, 108, 111–113], the existence of substituted pyridyl rings in these organic ligands offers the possibility for a controllable formation of "QD–Dye" nanoassemblies via a non-covalent coordination "key-hole" principle during titration experiments in solution [62, 63, 65, 75, 90, 101]. It is well known for chemical reasons that the transition metal ions Zn^{2+} or Cd^{2+} have an empty $3d^{10}$ orbital while the heteroatom *N*-pyr of the porphyrin *meso*-pyridyl ring is a very good e-donor having an electron lone pair orbital. Correspondingly, in this case a "key-hole" principle is realized via one- or two-fold non-covalent coordination $Zn^{2+}\cdots\cdot N$-pyr or $Cd^{2+}\cdots\cdot N$-pyr. Recent calculations with respect to the bonding conditions of ligands [106] have shown that the bond strength for the ligands can be described in most cases by electrostatic interactions between the partial charges of the respective surface ions and the ones of the ligand anchoring group. Similar conditions also hold for the formation of nanoassemblies with participation of H_2P and PDI molecules via pyridyl anchoring groups.

A versatile scheme for the preparation of "QD–Dye" nanoassemblies in non-polar liquids or low polar environments at ambient conditions for bulk experiments is based on quantitative titration steps: into low concentrated ($C_{QD} \sim 10^{-7}$ M) QD solutions dye molecules are added sequentially in steps of 10 μL from a highly concentrated ($C_{Dye} \sim 10^{-5}$ M) stock solution up to the wanted molar ratios $x = [C_{Dye}]/[C_{QD}]$ (where C_{Dye} and C_{QD} denote the concentration of the species), thus varying the number of dye molecules per QD. Details of the experimental technique, steady-state, and time-resolved measurements for bulk solutions and single nanoassemblies as well as theoretical approaches are presented in corresponding papers [62, 65, 74, 90, 94, 114, 115].

4.2.2 *Spectral-Kinetic Manifestations of Nanoassembly Formation*

4.2.2.1 Semiconductor QDs and Porphyrin Molecules

It should be mentioned that the formation of "QD–Dye" nanoassemblies with various subunits shows some common spectral-kinetic features which may vary depending on the physicochemical properties of interacting QDs and attached dye molecules. In the case of nanoassemblies based on CdSe/ZnS QDs and functional porphyrin molecules, typical spectral-kinetic data of titration experiments are collected in Figs. 4.4 and 4.5. It is seen from Fig. 4.4 that upon molar ratio x increase, besides QD absorption bands ($\lambda_{max} = 556$ nm) a linear increase of the porphyrin absorption bands takes place. Noteworthy, in all cases QD absorption remains constant, while the PL (at $\lambda_{max} = 585$ nm) is considerably quenched upon titration by $(Pyr)_4H_2P$ molecules. No such effects are observed when using various porphyrins without pyridyl rings.

Time-resolved PL measurements show that emission of CdSe or CdSe/ZnS QDs in the absence of dyes is characterized by a non-exponential decay [116–120]. The decay time distribution shows in the given example a strong and broad contribution centered at 19–22 ns which is related to the "intrinsic" PL of CdSe QDs. The center and width of this intrinsic decay times depend noticeably on the excitation and detection wavelength and/or the corresponding spectral bandwidths.

In this respect, time-resolved fluorometry (TRF) was applied also to compare the excited-state depopulation kinetics of semiconductor QDs in absence and presence of attached $(m\text{-Pyr})_4H_2P$ molecules. Figure 4.5 summarizes the TRF results obtained at the near-band-edge emission peak wavelength of the QD ($\lambda_{em} = 580$ nm) and the $Q_x(0, 0)$-band emission wavelength of porphyrin ($\lambda_{em} = 653$ nm) at molar ratios $x = 1$, 2, and 4.

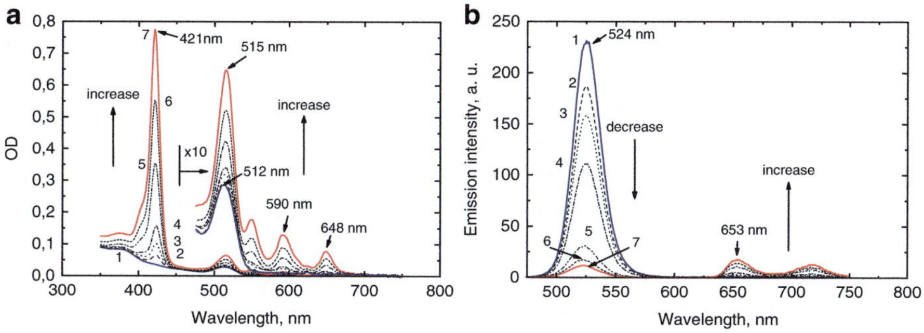

Fig. 4.4 Absorption (**a**) and emission (**b**, $\lambda_{ex} = 465$ nm) spectra of CdSe/ZnS QD (CdSe core diameter $d_{CdSe} = 2.5$ nm, number of ZnS monolayers $n_{ZnS} = 2$) and (*m*-Pyr)$_3$-H_2P porphyrin molecules in toluene at 295 K upon increase of the molar ratio $x = [C_{Porph}]/[C_{QD}]$: 0.0(1); 0.19 (2); 0.43 (3); 0.83 (4); 2.15 (5); 3.5 (6); 4.9 (7). The corresponding band maxima are shown according to [62]

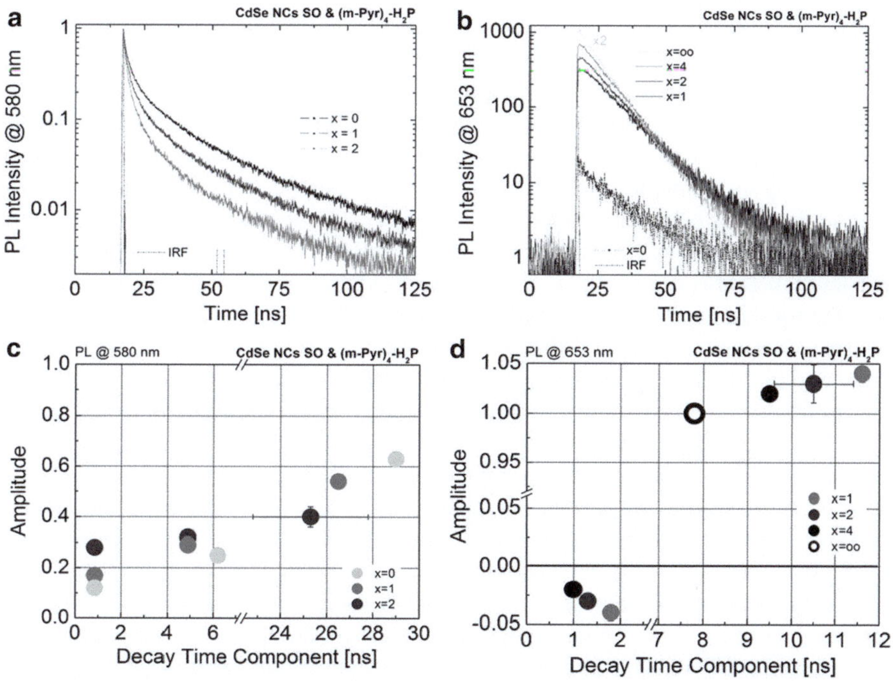

Fig. 4.5 Time-resolved emission traces (**a**, **b**) and fitted decay times and amplitudes (**c**, **d**) for CdSe QD (**a**, **c**; $\lambda_{em} = 580$ nm) and (m-Pyr)$_4$-H_2P molecules (**b**, **d**; $\lambda_{em} = 653$ nm) at various molar ratios $x = [C_{Porph}]/[C_{QD}]$ in toluene at 295 K at $\lambda_{exc} = 532$ nm. $C_{QD} = 4 \times 10^{-7}$ M, $d_{CdSe} = 3.2$ nm, $n_{ZnS} = 0$. IRF corresponds to the system response. (**b**) $x = 0$ corresponds to the spurious PL of the QD at the (m-Pyr)$_4$-H_2P emission wavelength and $x = \infty$ to a sample with only (m-Pyr)$_4$-H_2P. Figure adapted from [123]

The TRF experiments indicate a change of the (3-exponentially fitted) emission decay times of both QD and (m-Pyr)$_4$-H_2P porphyrin upon nanoassemblies formation. The main conclusion is that in cases of both CdSe/ZnS QDs and CdSe QDs [60, 62], the attachment of dye molecules leads to a PL mean decay time shortening accompanied by a reduction of the PL efficiency. For "QD–H_2P" nanoassemblies, the initial two components of 19–22 and 7 ns are shortened, while the fast component at ~700 ps remains nearly constant [62]. These shortened components reflect that there is an (static or dynamic) inhomogeneous distribution of QDs, some of them showing non-radiative shortening of the intrinsic PL decay, which is related to assembly formation. It turns also out from Fig. 4.5b, d that, especially at $x = 1$, the fluorescence decay time of porphyrin in the presence of QDs is longer ($\tau = 11.7 \pm 1.1$ ns) than that for pure porphyrin ($\tau_0 = 9.1 \pm 0.2$ ns). With increasing molar ratio x, the characteristic decay time approaches the one of the porphyrin. Further, for (m-Pyr)$_4$-H_2P molecules being attached to the surface of CdSe QDs, a weak rising component (with amplitudes below 5%) is seen, the characteristic time of which varies within the range of 1–2 ns. The longer decay for porphyrin molecules being attached to CdSe QD surface at small x ratio may be explained by two reasons: (1) the influence of QD on the radiationless rate constants of the porphyrin molecule in the nanoassemblies [62] and (2) the reduction of the porphyrin excited singlet state quenching by molecular oxygen due to the relative screening effect caused by QD with a high hydrodynamic volume like it has been observed upon interaction of porphyrin triads and pentads with molecular oxygen in non-degassed liquid solutions at 295 K [112]. In its turn, the detection of a weak rising component for (m-Pyr)$_4$-H_2P molecules in nanoassemblies reflects the existence of energy transfer QD $\rightarrow$ porphyrin. Taken together, all these facts indicate that the formation of "QD–H_2P" nanoassemblies manifests itself in PL quenching of the CdSe counterpart, and this quenching is a dynamic process caused by the increased non-radiative relaxation channels in QD excited states.

The detailed analysis of the main reasons (mechanisms) leading to the observed QD PL quenching in various "QD–Dye" nanoassemblies will be presented below. In this section, with respect to porphyrin moieties being used for the QD surface attachment (see Fig. 4.2), it is reasonable to outline the results of systematic titration experiments [62, 65] (which will be of interest for the following analysis of the basic principles for self-assembled nanoassemblies). Like it has been done previously with self-assembled multipophyrin arrays [107, 108, 112, 113], the strategy was: (1) to vary the number of pyridyl-rings from 1 to 4 including the two variants (*m*-Pyr)$_2$-H_2P and (*m*^Pyr)$_2$-H_2P and (2) to replace the type of nitrogen (N) position within the pyridyl ring from the *meta*-(*m*), to *ortho*-(*o*), and *para*-(*p*) N position in the case of the fourfold *meso*-pyridyl-substituted H_2P molecules. The data presented in Fig. 4.6a show that for CdSe/ZnS QDs under the same titration conditions, the observed QD PL quenching depends strongly on the number and type of pyridyl substituents. The most obvious observation is that (*o*-Pyr)$_4$-H_2P does almost not quench the PL, while the quenching is strongest for (*p*-Pyr)$_4$-H_2P closely followed by (*m*-Pyr)$_4$-H_2P. In addition, within the (*m*-Pyr)$_n$-H_2P manifold there is a systematic increase of quenching efficiency depending on the number of pyridyl

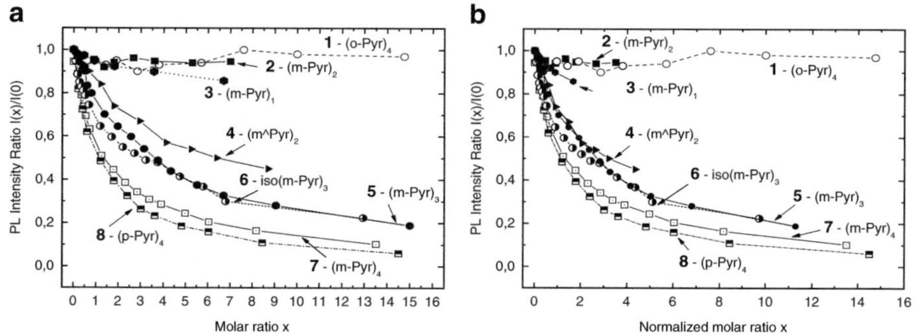

Fig. 4.6 Relative PL intensity changes (quenching) $I(x)/I(0)$ as function of the molar ratio $x = [C_{\mathrm{Porph}}]/[C_{\mathrm{QD}}]$ (**a**) and the normalized (to the number of pyridyl groups) molar ratio x (**b**) for CdSe/ZnS QD ($d_{\mathrm{CdSe}} = 2.5$ nm, $n_{\mathrm{ZnS}} = 2$, $C_{\mathrm{QD}} = 4 \times 10^{-7}$ M) and various porphyrin molecules: (*1*) (*o*-Pyr)$_4$-H_2P; (*2*) (*m*-Pyr)$_2$(Ph)$_2$-H_2P; (*3*) (*m*-Pyr)$_1$-H_2P; (*4*) (*m*^Pyr)$_2$-H_2P; (*5*) (*m*-Pyr)$_3$-H_2P; (*6*) (iso)-(*m*-Pyr)$_3$-H_2P; (*7*) (*m*-Pyr)$_4$-H_2P; (*8*) (*p*-Pyr)$_4$-H_2P. Toluene, 295 K. Adapted from [62]

substitutes when changing from (*m*-Pyr)$_1$ to (*m*-Pyr)$_4$, with one exception, namely (*m*-Pyr)$_2$ which shows an overall behavior more similar to (*m*-Pyr)$_1$, whereas (*m*^Pyr)$_2$ shows a much more efficient quenching similar, e.g., to (*m*-Pyr)$_3$.

As shown in Fig. 4.6a, the QD PL quenching efficiency and thus the probability to form QD–H_2P nanoassemblies is decreased with a decreasing number of pyridyl rings. Assuming that the probability of the nanoassembly formation is linearly proportional to the number of pyridyl rings, one can define an effective molar ratio $x_{\mathrm{pyr}} = x(N/4)$ that scales with N, where N is the number of pyridyl rings for a given H_2P molecule. Correspondingly, x_{pyr} becomes smaller with a decreasing number of pyridyl rings. Doing so, we obtain a rescaling of the QD PL quenching efficiency for every H_2P molecule [62] (depicted in Fig. 4.6b). In the result, all of the quenching curves besides those for (*m*-Pyr)$_1$-H_2P, (*m*-Pyr)$_2$(Ph)$_2$-H_2P and (*o*-Pyr)$_4$-H_2P are shifted towards one single curve. The overall result is a kind of "master" curve for the quenching efficiency. In case that only one pyridyl ring can be anchored effectively, the agreement with the master curve becomes less satisfactory. It follows from this behavior that the assumption relating the probability to form a QD–H_2P nanoassembly with the number of pyridyl rings having access to the QD surfaces is correct. The stability of a two-point interaction will be at least a factor of 2 stronger than a one-point interaction, as can be deduced from the pronounced mismatch of the (scaled) one-point interaction curves for (*m*-Pyr)$_1$-H_2P, (*m*-Pyr)$_2$(Ph)$_2$-H_2P as compared to the master curve. The importance of a two-point interaction has also been demonstrated for CdSe/ZnS QD–protein complexes [121]. The variation of the QD PL quenching efficiency with respect to the number, kind, and position of pyridyl rings in H_2P molecules points toward a dynamic equilibrium between QD–H_2P nanoassemblies and free entities, as has also been observed for multiporphyrin arrays [107, 108, 112, 113]. The equilibrium is dynamic, since

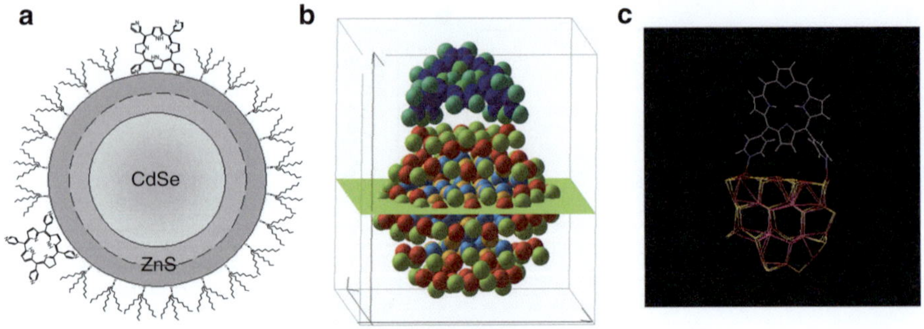

Fig. 4.7 Schematic presentation of "CdSe/ZnS QD–H_2P" nanoassemblies. (**a**) the scales of CdSe core ($d_{CdSe} = 2.5$ nm), ZnS shell, (*m*-Pyr)$_4$-H_2P and TOPO molecules as well as intercenter distances correspond to relative sizes of the main components of the assemblies [62]: the ZnS shell thickness for QDs was estimated on the basis of one ZnS layer thickness $l = 0.5$ nm; parameters for conical TOPO molecules $r_{bottom} = 0.55$ nm, $h_{con} = 0.99$ nm were used; for (*m*-Pyr)$_4$-H_2P molecule (see Fig. 4.2). $r_m = 0.75$ nm is the radius of a porphyrin molecule with opposite pyridyl rings having nitrogens in meta-positions, $h = 10$ Å is the mean distance between meta-nitrogens of adjacent pyridyl rings (HyperChem software, release 4.0 geometry optimization with semiempirical PM3 method). (**b**, **c**) optimized geometry for $Cd_{33}Se_{33} + (m^Pyr)_2$-$H_2P$ (optimization by HyperChem 7.0; simulations by ab initio density functional theory, DFT, with the VASP code [115])

assuming an infinitely strong coupling would not result in a dependence of the quenching on the number of pyridyl rings.

The above presented results lead to the conclusion that in "QD–H_2P" nanoassemblies, H_2P molecules anchor on the CdSe/ZnS surface in a nearly perpendicular fashion with two nitrogen lone pair orbitals (at most) forming coordination bonds with the surface. From geometric arguments, the QD PL weak quenching behavior observed for H_2P molecules with opposite pyridyl rings can thus be easily rationalized because a contact of opposite pyridyl rings to the surface is impossible due to geometric (steric) reasons in the case of a parallel orientation of the porphyrin macrocycle with respect to the QD surface. Theoretical simulations (ab initio DFT with the VASP code [115]) have shown also that for the optimized geometry of QD–H_2P nanoassemblies, the mutual arrangement of (*m*-Pyr)$_4$-H_2P molecules is perpendicular relative to QD surface. Considering space-filling molecular entities for such nanoassemblies, in the case of competitive exchange of TOPO capping molecules by attaching porphyrin ligands of one H_2P molecule may replace about two to three TOPO molecules or, alternatively, fills a free volume corresponding to two to three TOPO molecules. The overall structural information for QD–H_2P nanoassemblies is shown in Fig. 4.7.

The determination of the number of dye molecules per QD over the course of titration experiments is difficult, since the overall PL quenching depends both on the (a priori unknown) quenching efficiency and on the number of dye molecules on the QD surface. To separate these two effects, one needs the independent identification of the number of attached H_2P molecules merely from spectroscopic means. For the

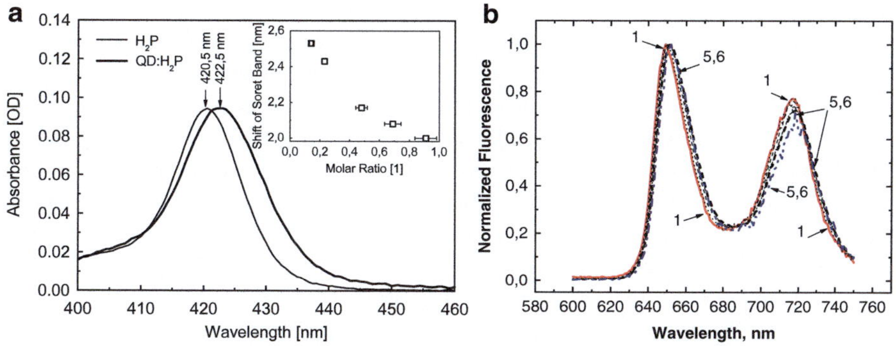

Fig. 4.8 Soret band absorption and fluorescence spectra (normalized to the maximum) for porphyrin (m-Pyr)$_4$H$_2$P molecules in absence and presence of CdSe (**a**) or CdSe/ZnS (**b**, $\lambda_{exc} = 420$ nm) QDs in toluene at 295 K. *Inset* in **a**: position of the Soret band (with respect to the Soret band of pure H$_2$P) in presence of CdSe QDs at small molar ratios *x*. (**b**) 1 ($x = 0.52$, $\lambda_{max} = 650$ nm), 2 ($x = 1.03$, $\lambda_{max} = 650$ nm), 3 ($x = 2.04$, $\lambda_{max} = 651$ nm), 4 ($x = 4.02$, $\lambda_{max} = 651$ nm), 5 ($x = 7.5$, $\lambda_{max} = 652$ nm), 6 (pure porphyrin without QD, $\lambda_{max} = 653$ nm)

"QD–H$_2$P" nanoassembly formation, the spectra show that only a small part of H$_2$P molecules added during the titration procedure becomes attached to the QD surface, indicated by the missing of an isosbestic point [64]. It should be mentioned that noticeable spectral shifts in the PL of the QDs were not observable upon nanoassembly formation in any case.

On the other hand, for the H$_2$P molecules being attached on QD surface, spectral red shifts of both the Q- and Soret absorption bands and a blue shift of the fluorescence Q-band accompanied by a slight change in the Franck–Condon envelope of the overall spectrum are observed [62, 101] (Fig. 4.8).

These shifts and the slight inhomogeneous broadening of the porphyrin Soret band indicate that at low molar ratios *x* most of the H$_2$P molecules are conjecturally complexed in "QD–H$_2$P" nanoassemblies. There are several possible explanations for the observed spectral shifts. In principle, the spectral properties depend on several factors such as the molecular geometry, QD surface composition, the overall dielectric function, and polarization effects via the *meso*-pyridyl rings upon attachment to the QD surface. A red shift (both for absorption and for emission) is expected in relation to an increase of the average dielectric constant which is approximately $\varepsilon = 9.7$ for CdSe and $\varepsilon = 2.4$ for toluene. On the contrary, the Q-band emission exhibits a blue shift which cannot be explained by a dielectric model. The Stokes shift between Q-band absorption and emission becomes smaller upon assembly formation. At the same time, the electronic Q(0,0)-band intensity of H$_2$P ($\lambda = 653$ nm) is reduced relative to Q(0,1) vibronic one (at 720 nm) which indicates that the Franck–Condon factor is changing upon assembly formation. It is known that the Q-bands in H$_2$P correspond to symmetry-forbidden transitions [122], which will become partly allowed by distortion of the H$_2$P ring upon attachment to the QD. These facts induce an increased Stokes shift and an increase of the formerly forbidden (vibration-less) transition at high energy. We assume that the same processes hold for the situation of H_2P in toluene solution. Assembly formation obviously "restores" the intrinsic H_2P symmetry (a planar H_2P ring) which has a symmetry-forbidden optical transition. Probably the planarity is enforced when the H_2P molecules coordinate with the QD surface and interact with the closely packed TOPO shell in the vicinity. The observed shifts depend obviously on the type of QD and decrease with increasing molar ratio x [64]. For large molar ratios, the shifts vanish from which we conclude that the spectroscopic properties are on average determined by non-assembled H_2P molecules. The critical molar ratio at which this happens is in the order of $x_c = 1–10$ and increases with increasing QD size.

Finally, it may be concluded that spectral shifts proof assembly formation immediately after the first titration step at $x < x_c$, conducted in our case for molar ratios as low as $x \sim 0.1$. At molar ratios $x > x_c$ (with $x_c = 1–10$ depending on QD size [64]), the porphyrins become less and less complexed. It follows from studies on single QD–PDI nanoassemblies [74, 94] that one dye molecule causes a PL quenching per QD that is at most two times stronger than in solution at $x = 1$. This implies that only a smaller number of dyes (about one in four [74]) becomes attached to the QD surface at comparable molar ratios in solution. As it was shown in Fig. 4.5, for "QD–H_2P" nanoassemblies, fluorescence lifetime measurements support a complexation efficiency that is of the same order of magnitude: for individual (m-Pyr)$_n$-H_2P molecules in absence of QDs in toluene, the fluorescence decay is measured to be 8.4–9.3 ns [62, 111–113]. At the first titration step at $x \sim 0.25$, the mono-exponential lifetime is increased to 11 ns [123] due to FRET from the long-lived CdSe/ZnS QDs and possible diminishing of H_2P fluorescence quenching by molecular oxygen because of QD screening action. This fact implies that, at these low molar ratios, most of the H_2P molecules are attached to a QD, whereas the complexation efficiency reduces considerably when the molar ratio is increased. All these experiments show that already one attached dye molecule can quench the PL intensity of a QD considerably. Further information on the number of H_2P per QD will be presented below from an analysis of the PL quenching following a detailed Stern–Volmer description.

As was outlined above, the attachment of functionalized porphyrin molecules to a QD surface manifests itself in noticeable QD PL quenching as well as in related complex interface dynamics caused by non-radiative relaxation channels for the exciton. Here, we demonstrate that QD PL quenching (as a manifestation of the nanoassemblies formation) is also visible in experiments with single nanoobjects. Figure 4.9 shows the comparison of blinking statistics for two samples in spin-coated toluene solution at 295 K: CdSe/ZnS QDs and "QD–(m-Pyr)$_4$-H_2P" nanoassemblies both having the same initial QD concentration and being excited within the QD first excitonic absorption band. Nanoassemblies were prepared at a molar ratio $x = [C_{\mathrm{Porphyrin}}]/[C_{\mathrm{QD}}] = 10$, at which the bulk QD PL quenching is about 40% [62].

It is seen from Fig. 4.9a, b that for both cases blinking statistics show a power law distribution for "on-" and "off-"times. Dark QD states are usually explained by charged nanocrystals [67], and the heterogeneity (power law behavior [124]) is

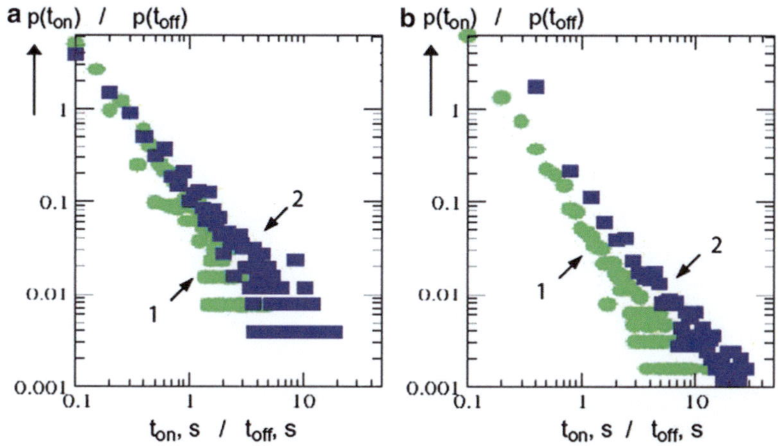

Fig. 4.9 Blinking statistics for single TOPO capped CdSe/ZnS QDs (**a**) and for single "QD–QDs in the presence of (m-Pyr)$_4$-H_2P" at molar ratio $x = 10$ (**b**). QD parameters: $C_{QD} = 1.8 \times 10^{-9}$ M, $d_{CdSe} = 3.2$ nm, $n_{ZnS} = 3$; laser excitation at $\lambda_{exc} = 514.5$ nm, $P = 250$ μW. Samples have been prepared by spin coating from toluene solution onto quartz surface at 295 K. The presented data were averaged for at least 20 individual objects from free and porphyrin-assembled QDs. Distributions of on-times (1, *green*) and off-times (2, *blue*) are indicated in each graph. Adapted from [75]

inherent to broadly distributed (de-)population processes of the dark state. In case of nanoassemblies, values for $\langle t_{on} \rangle = 0.18$ s do not change with respect to those measured for QD ($\langle t_{on} \rangle = 0.18$ s), while a substantial increase of the "off"-times is observed for QD in the presence of porphyrin molecules (1.2 s in comparison with 0.75 s). This elongation of dark periods is equivalent to PL quenching. These findings are considered as a proof of QD–porphyrin interactions leading to QD PL quenching also on a single assembly level.

Finally, when numerically analyzing QD PL quenching data for various porphyrin molecules in order to evaluate the corresponding complexation constants K_c one should take into account few aspects: ligand exchange dynamics (depending on TOPO concentration and solvent properties) and the number of H_2P molecules on the QD surface. Without consideration of the dynamic equilibrium, it is reasonable to assume that the PL quenching rate for a given QD with n attached chromophores is proportional to n, whatever the quenching mechanism is. Like it has been done earlier [125] as well as in our [62] and in recent publications [23, 126], the number of QD–porphyrin (or, in general, QD–Dye) assemblies may be described by a Poisson distribution

$$P(n) = x^n \exp(-x)/n!, \tag{4.1}$$

where x is the average number of chromophores per one QD (estimated from a molar ratio ($x = [H_2P]/[QD]$)), n is the number of attached chromophores on a given QD. Correspondingly, the QD PL intensity ratio $I(x)/I_0$ may be written as [62, 101, 122]

Table 4.1 Complexation constants K_c for "QD–porphyrin" nanoassemblies based on CdSe/ZnS QD ($d_{CdSe} = 2.5$ nm, $n_{ZnS} = 2$) and $(Pyr)_n$-H_2P ligands with $n = 1$, 2, 3 and 4

Ligand	$(m\text{-Pyr})_1H_2P$	$(m\hat{}\text{Pyr})_2H_2P$	$(m\text{-Pyr})_3H_2P$	$(m\text{-Pyr})_4H_2P$
K_c ($\times 10^7$ M^{-1})	<0.015	0.34	0.78	2.6

$C_{QD} = 5 \times 10^{-8}$ M in toluene at 295 K

$$I(x)/I_0 = e^{-x} \sum x^n/n!(1 + nk_Q/k_D) \text{ [122]} \tag{4.2}$$

or

$$\frac{I(x)}{I_0} = \sum_{n=o}^{\infty} \frac{k_D}{k_D + nk_Q} x^n \frac{\exp(-x)}{n!} \text{ [62, 101]}, \tag{4.3}$$

where k_D is the total QD PL intrinsic decay rate and k_Q the QD PL quenching rate.

The experimental PL quenching data $I(x)/I(0)$ of Fig. 4.6b presented as function of the rescaled molar ratio x demonstrate that nk_Q is approximately the same for most of the arrays based on QD. This implies that all variants of $(Pyr)_nH_2P$ molecules exhibit the same quenching mechanism upon assembling on QD. In addition, the dynamic equilibrium in the "QD–H_2P" nanoassemblies does not favor a strong complexation of $(Pyr)_n$-H_2P on the QD surface or that there are only a limited number of accessible sites on the QD surface. In fact, the maximum number of anchored H_2P molecules depends on several factors, such as the number and kinds of "empty places" on the QD surface which are not occupied by TOPO ligands, the competitive dynamic equilibrium between TOPO and H_2P molecules depending on both complexation constants K_c for TOPO and porphyrin molecules and their relative concentrations. It follows from detailed spectral observations [101, 127] that at low molar ratios x all $(Pyr)_n$-H_2P molecules especially those with four pyridyl rings are attached to the CdSe/ZnS surface. In addition, since the well-defined QD PL quenching effects are even observed at molar ratios $x < 1$ (see curves 7 and 8 in Fig. 4.6a), one can neglect (following statistical arguments) nanoassemblies with more than one $(Pyr)_nH_2P$ molecules per QD. Thus, the experimental titration data can be discussed on the basis of a bi-molecular reaction scheme valid for a dynamic equilibrium between self-assembled and free constituents as has been usually applied in numerous experiments on the formation of multipophyrin assemblies [107, 108]. Following this idea and using the values of the complexation constant (see Table 4.1) K_c may be estimated for solutions of QD as a function of various $(Pyr)_nH_2P$ ligands [101] according to

$$K_c = (1 - \beta) \times \frac{1}{C_{D0} \cdot \beta\,(x + \beta - 1)}, \tag{4.4}$$

where C_{D0} is the initial concentration of QD; $\beta = I/I_0$ corresponds to the QD PL ratio as a function of the molar ratios at $x = [H_2P]/[QD]$ and $x = 0$, respectively.

The above results clearly show that K_c values are increased with the number of pyridyl rings, thus reflecting a dynamic equilibrium between the QD and $(Pyr)_n$-H_2P. Comparing K_c values with those obtained for various multiporphyrin complexes [107, 108, 112, 113] shows that with respect to twofold coordination they are in a comparable range but being much larger as compared to the corresponding onefold coordination. Like for porphyrin self-assembled triads, the importance of the designing adapted coordination schemes has thus been also demonstrated for formation of the CdSe/ZnS QD–porphyrin nanoassemblies. It is seen also that for "QD–H_2P" nanoassemblies the increase of K_c values with the number n of pyridyl rings is less dramatic with respect to K_c differences of two to three orders of magnitude found between two- or one-fold coordination in the case of porphyrin triads and pentads. It means that in the case of "QD–H_2P" nanoassemblies the higher K_c values for two-point interaction reflect rather the competition with one-point interaction of TOPO ligands, thus leading to ligand exchange. In the latter case, allosteric effects seem to be not important, while they play a dominant role for porphyrin triad and pentad formation. An analysis of this dynamic exchange equilibrium will be discussed further on with respect to the quenching efficiency as a function of the molar ratio x.

4.2.2.2 Nanoassemblies from QDs and Perylene Diimide Molecules

The principle of molecular self-assembly was transcribed also to "QD–PDI" nanoassemblies based on amine-capped CdSe/ZnS QDs (QD–AM) and pyridyl functionalized perylene bisimides (dye structures and corresponding abbreviations are presented in Fig. 4.3), implementing a detailed investigation by single molecule spectroscopy (due to the high quantum yield of PDI and its large photostability). The observation of QD PL quenching caused by FRET and non-FRET processes was analyzed, even on a single assembly level, and the detailed analysis of this quenching has been presented recently [74].

Figure 4.10 shows spectral properties and results of titration experiments for a given CdSe/ZnS QD–AM and PP dye. Optical spectra of the five PDI dyes used in this study are almost identical. It is clearly seen from Fig. 4.10 that upon nanoassembly formation, QD–AM PL intensities change while spectral shapes remain the same. This finding is very similar to the one discussed for "QD–porphyrin" nanoassemblies. Titration of solution by various PP dye results, as shown in Fig. 4.10b, in instantaneous PL quenching of QD–AM.

Figure 4.11 collects experimental results on PL quenching for QD–AM upon titration by all PDI dyes as a function of molar ratio x.

The strongest quenching is observed in case of DPP. Quenching is quantitatively similar for TPP and DTPP and somewhat stronger than for PP and DAP. The ratio I_x/I_0 at $x = 1$ becomes 0.88, 0.82, and 0.66 for PP, DTPP, and DPP, respectively (see dashed lines in Fig. 4.11). The surprising result is that DAP (without any pyridyl group) quenches the PL almost to the same extent as, e.g., PP. In this case, we apt to believe that the attachment might be accomplished for dyes without pyridyl groups

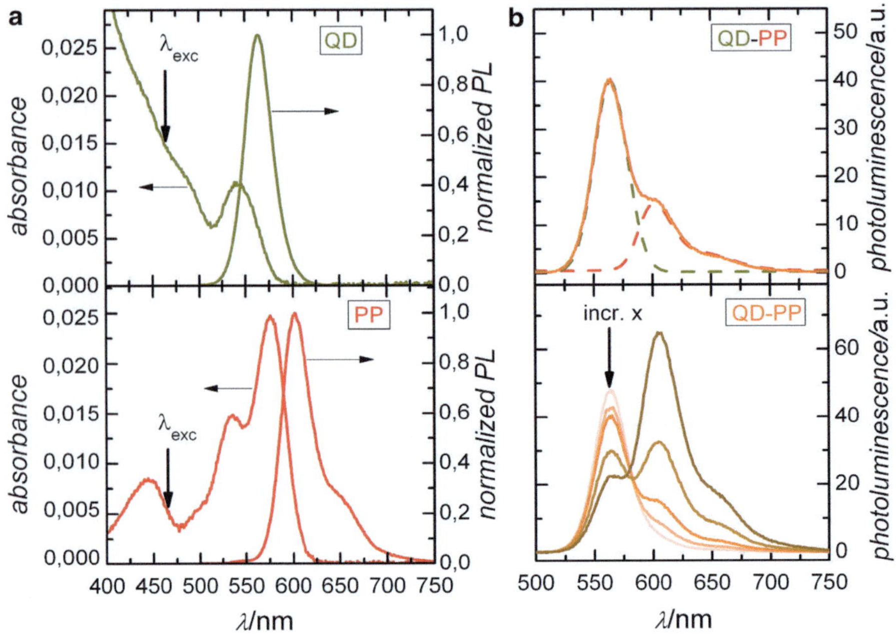

Fig. 4.10 (**a**) Absorption and emission spectra ($\lambda_{exc} = 465$ nm) of 0.1 μM QD–AM (*top*) and 0.55 μM PP (*bottom*) in toluene at 295 K; (**b**) photoluminescence of "QD–PP" nanoassemblies at $x = 1$. *Broken lines* show a deconvolution of the spectra of QD and PP, respectively (*top*). QD–AM photoluminescence quenching upon titration by PP at molar ratios $x = 0.2, 0.5, 0.9, 2.5$, and 5 (line color from *orange* to *brown*). CdSe/ZnS QD–AM parameters: $d_{CdSe} = 3.0$ nm, $n_{ZnS} = 2$. Adapted from [74]

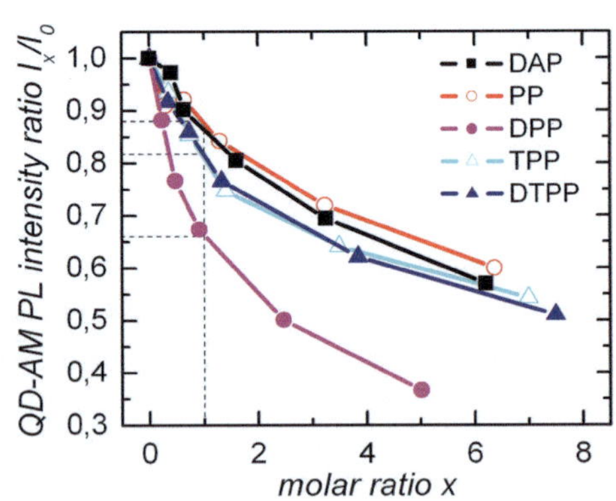

Fig. 4.11 Relative I_x/I_0 PL quenching for QD–AM where I_0 and I_x are PL intensities for pure QDs and at given molar ratio x, correspondingly. For *dashed lines* refer to discussion. Adapted from [74]

by intercalation of the aliphatic tale of DAP into the AM layer. A similar observation has been discussed recently [36, 37].

It is interestingly to note also that using PDI dyes with two different pyridyl-type functional groups (namely DPP, and DTPP molecules shown in Fig. 4.3), it is possible to realize geometrically well-defined "QD–PDI" nanoassemblies,

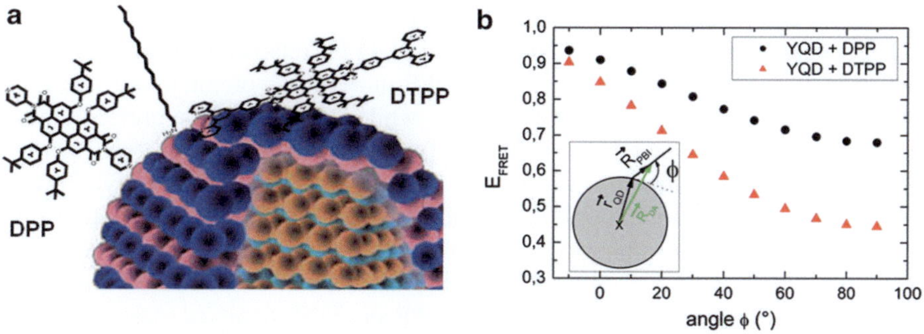

Fig. 4.12 Scheme of surface attachment of DPP and DTPP molecules to a CdSe/ZnS QD surface (**a**) and calculated FRET QD → Dye efficiency for DPP and DTPP as a function of angle ϕ (**b**). The *inset* represents a scheme of the geometrical model being used in calculations. From geometrical constraints the orientation for DTPP molecule with respect to QD surface corresponds to $\phi \approx 0°$ and for DPP molecule to $\phi \approx 90°$, respectively. Simple calculations of FRET efficiency (E_{FRET}) have been carried out using the standard Foerster model [128] $E_{\mathrm{FRET}} = [1 + (R_{\mathrm{DA}}/R_0)^6]^{-1}$, where R_0 corresponds to the Foerster transfer radius and R_{DA} is the center-to-center donor–acceptor distance. While almost all relevant parameters are nearly the same with respect to DPP and DTPP, they differ considerably in $R_{\mathrm{DA}} = R_{\mathrm{PDI}}/2 + r_{\mathrm{QD}}$, the (vectorial) distance between the centers of the QD and the dye ($R_{\mathrm{PDI}}/2$ corresponds to the distance of the coordinating point to the center of PDI chromophore while r_{QD} is the radius of CdSe/ZnS QD). Adapted from [162]

which allows to determine the dependence of dynamic QD–Dye interactions upon variation of the spatial arrangement of the dye molecule with respect to QD surface (e.g., distance and orientation) [74, 94]. Indeed, it is seen from Fig. 4.11 that at the same molar ratio x QD–AM PL quenching is stronger for DTPP molecules with respect to that found for the DPP dye. In order to analyze this difference one should look at Fig. 4.12a, which shows different attachment angles for DPP and DTPP molecules coordinating via pyridyl and terpyridyl substituents to the ZnS shell of the QD. The parallel orientation of DTPP on the QD surface is reasonable, since the attachment of the terpyridyl unit to the QD surface needs the coordination of the two ortho-pyridyl groups, which can only be achieved in a flat and slightly drilled configuration of the terpyridyl unit. A direct consequence of this coordination to the surface is a spectral red shift observed for DTPP in case of the reduction of excess of TOPO from the solution [94]. It follows from geometrical considerations that for a hypothetical perpendicular position of both DPP and DTPP molecules with respect to QD surface the distances become $R_{\mathrm{DA}}(\mathrm{DPP}) = 3.6$ nm and $R_{\mathrm{DA}}(\mathrm{DTPP}) = 4.75$ nm. This implies that FRET should be stronger for DPP than for DTPP, which is quite opposite to what is observed experimentally (see Fig. 4.11). However, taking into account the steric restrictions discussed above we calculated the FRET efficiency as a function of the orientation ϕ of the long molecular axis with respect to the QD surface as is shown in the inset of Fig. 4.12a. Now E_{FRET} becomes 0.85 for DTPP at $\phi = 0°$ and 0.68 for DPP at $\phi = 90°$ which is in rough agreement with the experimental result with respect to relative FRET efficiencies.

From a comparison of calculated FRET efficiencies for "QD–DPP" and "QD–DTPP" nanoassemblies (Fig. 4.12b) as well as QD PL relative quenching for these two dyes it has been concluded that the DPP molecule is oriented with the long axis nearly perpendicular to the QD surface, while DTPP molecule prefers a more tangential orientation [74]. Thus, it follows from the presented results that PDI dye molecules may be sorted into two classes, namely DPP-type one-pyridyl, DTPP-type three-pyridyl coordination via nitrogen lone pairs or electrostatic interaction between partial negative charge densities at the N-atom of pyridyl with surface Zn^{2+} ions [106], like for "QD–porphyrin" nanoassemblies, and a third namely the DAP-type intercalation into the AM capping shell. PP and TPP are mixed cases since they are subject to a competition between pyridyl-type coordination and intercalation. However, it became also evident from the experimental data [74, 94] that the FRET efficiency is quite low. Though relative FRET efficiencies can convincingly be related to various QD–Dye assembly geometric structures and to different relative orientations of the PBI dyes with respect to the surface, the surprisingly low FRET efficiency in these nanoassemblies will be discussed separately.

With respect to "QD–PDI" nanoassemblies, we would like to show how the spectral detection of single objects may give information about distinct nanoassembly geometries and conformational mobility of PDI molecules attached to the QD surface. Figure 4.13 shows spectra of single PP and TOPO-capped CdSe/ZnS QD obtained via confocal microscopy for spin-coated samples [162]. The spectra can be divided into three classes, namely QD PL (Fig. 4.13a), PP fluorescence (Fig. 4.13c), and the sum of the two spectra (Fig. 4.13). All types of spectra show typical single quantum object fingerprints, that is luminescence intermittency (blinking) [124] and spectral fluctuations (shown at the bottom). The experimental luminescence decay times averaged over 500 ms are shown as a function of observation time on the right side of each spectrum. In case of PP a nearly mono-exponential decay varying between 4 and 6 ns is detected for different single molecules, while the decay time for QDs is fluctuating between 50 and 5 ns (experimental lower limit in time resolution). The observation of the sum of the two spectra shown in the middle has been assigned to the formation of "QD–PP" nanoassemblies.

From Fig. 4.13b it is seen that at $t \approx 57$ s PP fluorescence is bleached while the QD remains emitting because of higher photostability in comparison with an organic dye. In addition, the PL band of this quantum dot is blue shifted during observation time which might be caused by photooxidation of the QD core and is enhanced by the attached dye molecule [75, 94, 129]. While the fluorescence of single PP molecules decays almost mono-exponentially, the PL of single QD–AM (independent whether free or assembled) decays clearly multi-exponential as has been reported for many similar cases [130–133]. It should be noted also that PL average decay times $\langle \tau_D \rangle$ of single QD (fitted by a stretched exponential function) show a tendency to decrease with increasing observation time, and this behavior is related to the spectral blue shift of the PL. The average decay time for single uncomplexed QD–AM was measured to be $\langle \tau_D \rangle \approx 20.5\,\text{ns}$ (based on measurements for 20 QDs), while for QD in single nanoassemblies (QD–AM)–PP we obtain

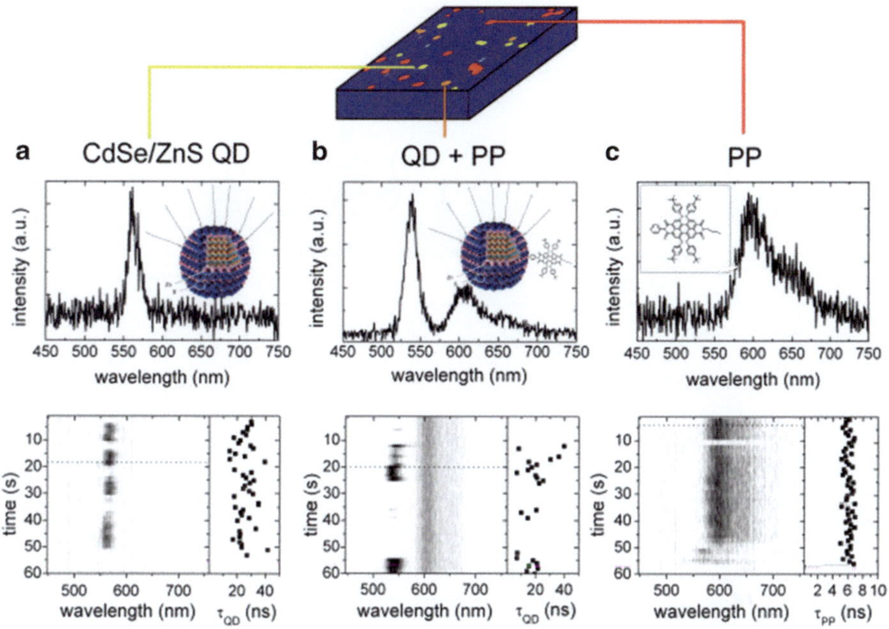

Fig. 4.13 Emission spectra of single TOPO-capped CdSe/ZnS QD ($d_{CdSe} = 3.0$ nm, $n_{ZnS} = 2$) and (pyridyl)$_1$-PDI molecule (PP). Spectra are related to various confocal spots of the spin-coated sample. It was possible to detect PL spectra (1 s binning time) of single QD (**a**), single PP (**c**), and single nanoassemblies with PP and QD spectra simultaneously (**b**). Typical luminescence intensities, spectral positions, and decay times are shown as a function of observation time of 60 s and longer. $\lambda_{exc} = 465$ nm; $P = 0.6$ kW cm^{-2}. A confocal scan of a spin-coated sample with solution ratio $x = 1$ is shown on *top*. Adapted from [162]

$\langle\tau_{DA}\rangle \approx 16$ns (based on measurements for 18 nanoassemblies). The shortening of the decay time is clearly related to the observed decrease of the QD PL intensity upon nanoassembly formation. While in a given sample the fluorescence intensity of PP is nearly (see discussion of FRET) indistinguishable between isolated PP and assembled PP, the QD PL intensity of single QD–PP assemblies is quenched on average by 50% as compared to isolated QD in the same sample. This has to be compared to the PL reduction to 0.88 observed in ensemble experiments at $x = 1$ (see Fig. 4.11). This indicates that at least the majority of the simultaneously detected QD–PP spectra are in fact due to assembly formation. Taken together, these facts prove that QD PL quenching is also observed on a single nanoassemblies level and reflects directly the existence of an additional non-radiative channel for QD exciton relaxation in these nanoassemblies.

Finally, the spectral distribution of the fluorescence of many single PP molecules with respect to the one of many single "QD–PP" nanoassemblies is related to a reduced conformational mobility of PP molecules on a QD surface [94]. It is known [130, 134] that individual PP molecules have different fluorescence spectra due to

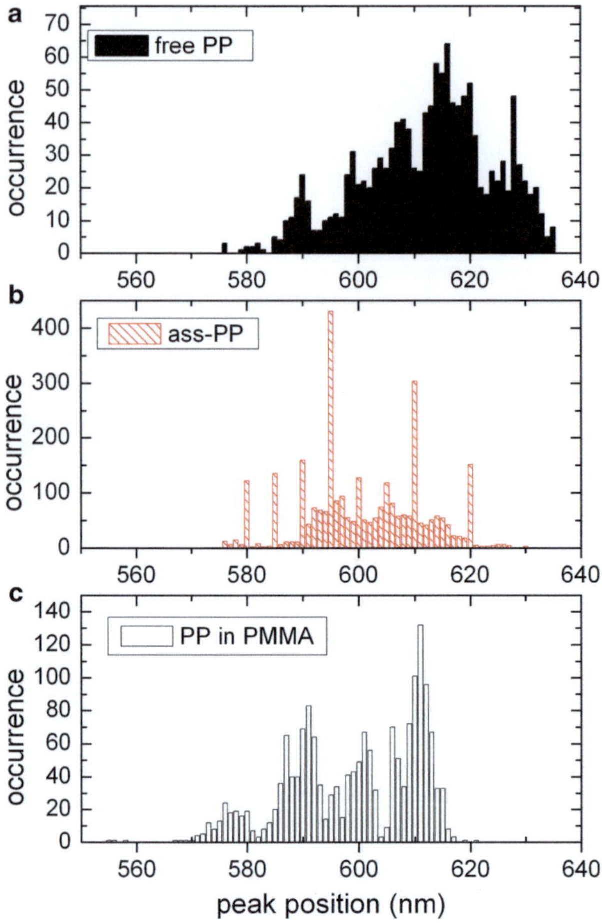

Fig. 4.14 Distribution of fluorescence maxima of single PP molecules on SiO_2 (**a**) and PP in in nanoassemblies (**b**). For comparison, distributions are shown for PP in a spin-coated PMMA film (**c**). Fluorescence maxima have been obtained by fitting the PP fluorescence spectra by two Gaussian lines. Adapted from [94] with permission from the Intern. Journal of Mol. Sciences @ 2009

various conformations of the bay groups (see Fig. 4.3) resulting in a broad distribution of the fluorescence maxima. The corresponding distribution of conformations and thus the fluorescence spectra depends on the local environment such as the embedding matrix. Figure 4.14 shows typical distributions of fluorescence maxima for PP molecules on an SiO_2 surface, in QD–PP nanoassemblies and in a PMMA polymer film, correspondingly. Comparison of these distributions reveals that the QD–PP nanoassembly formation favors conformations in the short wavelength range becoming more equivalent to the distribution observed for PMMA, which is a rigid matrix. Therefore the comparative spectral range of nanoassemblies reflects a reduced conformational range of bay group orientation. Nevertheless as is evident from Fig. 4.14 fluorescence spectra of individual PP molecules show that spectral fluctuations still occur.

Notably, single molecule data provide more specific information on the structure of the QD–PP nanoassemblies. According to Fig. 4.14 the spectral distribution of PP fluorescence is blue shifted upon nanoassemblies formation. Such a shift indicates that those conformations of PP are favored, for which the bay groups are most extended as this situation is related to short fluorescence wavelengths [94, 130, 134]. This corresponds to a situation where PP bay groups are as much within the molecular plane as possible. Such a situation is reasonable since PP has to be intercalated into the ligand shell which will be more easily accomplished for a nearly flat molecule, since less ligands have to be excluded from the QD surface. Thus, the distribution is shifted from a "free" PP molecule on SiO_2 surface to a "matrix-isolated" type such as in a PMMA film (see Fig. 4.14c). It is known also [94] that spectral fluctuations are enhanced for assembled PP as compared to PP in a PMMA film. From this it follows that the ligand shell imposes more flexibility than in a PMMA film allowing for remaining conformational changes of PP molecules. An alternative explanation is that the curved QD surface allows for more conformational flexibility than on an SiO_2 substrate. The high flexibility of the PP phenoxy side groups on the QD surface is only enabled in absence of steric hindrance. Therefore (1) a nearly perpendicular geometry has to be established and (2) the ligand density close to PP has to be low. In conclusion, the identification of PP conformations (1) has been used to confirm the surface geometry already suggested by FRET results and (2) serves as probe for the ligand density and dynamics on single quantum dots.

4.3 Exciton Relaxation Pathways and QD Photoluminescence Quenching in "QD–Dye" Nanoassemblies

The results presented in Sect. 4.2 clearly show the formation of "QD–Dye" nanoassemblies accompanied by the PL quenching of the QD counterpart, both from bulk experiments and single nanoassemblies detection. As it was mentioned in the introduction, the QD PL quenching being observed in numerous "QD–Dye" nanoassemblies needs a thorough analysis in order to evaluate the increased non-radiative relaxation channels in the excited states of QD with attached dye molecules. Such an approach has been carried out recently by us for "QD–porphyrin" [62–65, 114, 123] and "QD–perylene diimide" nanoassemblies [64, 65, 74, 94] and we will summarize only the basic findings.

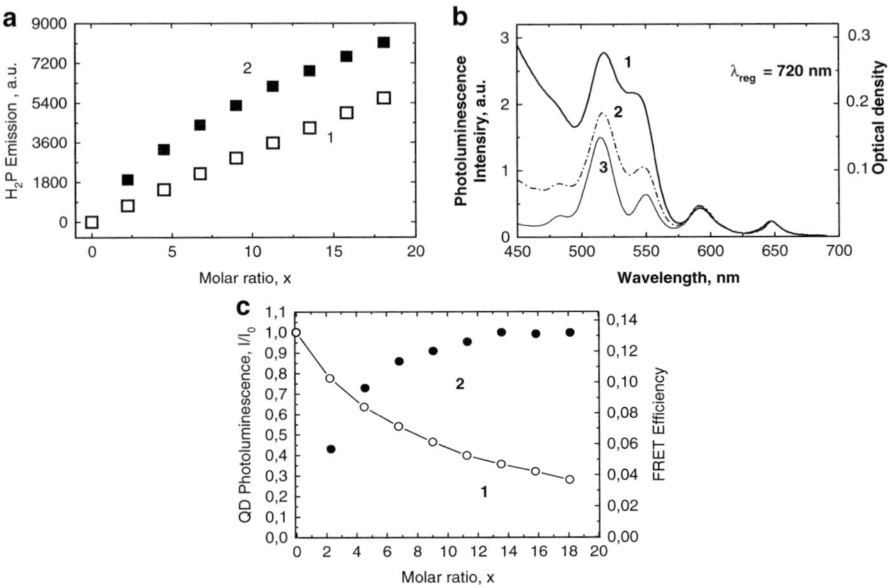

Fig. 4.15 Experimental comparison of TOPO-capped CdSe/ZnS QD PL quenching and (m-Pyr)$_4$-H_2P fluorescence enhancement in toluene at 295 K. (**a**) Integrated fluorescence intensity of (m-Pyr)$_4$-H_2P (*1*) and enhanced integrated fluorescence intensity of (m-Pyr)$_4$-H_2P in the presence of CdSe/ZnS QDs ($d_{CdSe} = 3.00$ nm, $n_{ZnS} = 2$ monolayers) (*2*) at the same molar concentrations of porphyrin and excitation conditions ($\lambda_{exc} = 465$ nm) for both cases. (**b**) Comparison of the absorption spectrum (*1*) and porphyrin fluorescence excitation spectrum (*2*, $\lambda_{reg} = 720$ nm, corrected for excitation spectral response) for a solution of CdSe/ZnS QD with (m-Pyr)$_4$-H_2P molecules at molar ratio $x = 4.0$; (*3*) is fluorescence excitation spectrum for alone porphyrin at the same porphyrin concentration and experimental conditions as for case (*2*). Spectra (*1*), (*2*), and (*3*) are normalized with respect to the intensity of Q(0,0) porphyrin band ($\lambda = 647$ nm). (**c**) QD relative integrated PL intensities I/I_0 (*1, left scale*) and efficiency of Foerster resonance energy transfer, based on the porphyrin fluorescence enhancement (*2, right scale*) for CdSe/ZnS QDs upon titration by (m-Pyr)$_4$-H_2P molecules as function of the molar ratio x

4.3.1 Foerster Resonant Energy Transfer

Usually, for FRET "QD → Dye" in nanoassemblies, the direct verification of the energy transfer process as a real reason of QD PL quenching should be done via the direct comparison of the experimental quenching efficiency of the donor (QD) PL and the efficiency of the acceptor (dye) fluorescence enhancement (sensitization).

Figure 4.15 collects the corresponding results of this comparison carried out for TOPO-capped CdSe/ZnS QDs and (m-Pyr)$_4$-H_2P porphyrin molecules upon variation of the molar ratio x.

It should be mentioned that the fluorescence quantum efficiency (φ) for attached (m-Pyr)$_4$-H_2P molecules remains almost the same as that for free porphyrin [62]. At the same time, Fig. 4.15a shows that upon increasing the molar ratio x in solutions of "QD–porphyrin" nanoassemblies the integrated fluorescence intensities of (m-Pyr)$_4$-H$_2$P are noticeably larger than those being measured for individual porphyrin molecules for the same excitation conditions ($\lambda_{ex} = 465$ nm, the donor QD strong absorption). This finding reflects the presence of sensitization (enhancement) that is typical for FRET. In addition, the fluorescence excitation spectrum of (m-Pyr)$_4$H$_2$P in nanoassemblies ($\lambda_{reg} = 719$ nm) provides direct information about the QD $\rightarrow$ porphyrin FRET process (Fig. 4.15b). Nevertheless at a given molar ratio x, there is no full coincidence between absorption and fluorescence excitation spectra for "QD–porphyrin" nanoassemblies. It follows also from TRF experiments (see Fig. 4.5d) that only a small build-up component (1–2 ns) is observed [62, 123] which indicates an only weak role of a population mechanism of porphyrin molecules via excited states of QD in nanoassemblies. Thus, the realization of FRET with a quantum efficiency of 100% seems not to be the case for the above system.

Experimental results [60, 117, 135] and theoretical considerations [115, 136] suggest that nanocrystalline materials and organic–inorganic hybrids are best described in the weak coupling regime. This allows the application of Foerster theory [137], which typically is used in the limit of "localized oscillators" as a result of the electric dipole approximation. The corresponding theoretical analysis in the frame of Foerster model shows [62, 115] that for the "QD–porphyrin" nanoassemblies under study, the energy transfer "QD $\rightarrow$ porphyrin" should take place within few nanoseconds time scale with the FRET efficiency $E_{FRET}^{theor} \sim 0.7$–$0.8$. This conclusion is in contradiction to our findings but shows the following relation between experimental and theoretical FRET rate constants: $k_{DA}{}^{exper} \ll k_{DA}{}^{theor}$.

Thus, we have carried out a direct comparison of experimental quenching results for QD emission and sensitization data for porphyrin fluorescence using a complete set of titration points (Fig. 4.15c). Based on the ideas presented in [62–64, 138, 139], the analysis of FRET efficiencies via sensitization effect has been carried at every titration step upon variation of the x ratio using the following equation:

$$E_{FRET} = [\varepsilon_A(\lambda_{ex})F_{AD} - F_A]/[\varepsilon_D(\lambda_{ex})F_A \cdot n], \tag{4.5}$$

where F_{AD} corresponds to the A fluorescence intensity measured at λ_{em} [651 nm, Q(0,0) band maximum of H$_2$P] while exciting at the strong donor absorption $\lambda_{ex} = 465$ nm. F_D corresponds to H$_2$P (acceptor) emission at the same λ_{em} and λ_{ex} in the absence of QD (donor). n is the number of A molecules per D, ε_A and ε_D are molar extinction coefficients of A and D, respectively, at the excitation wavelength λ_{ex}. It follows from Fig. 4.15c that at every molar ratio x, E_{FRET} values calculated from fluorescence enhancement (right scale) are significantly smaller than those estimated from QD PL quenching efficiency [$E_{QUENCH} = 1 - I(x)/I_0$]. It is seen also that beginning from $x > 6$ there is a saturation tendency for experimental E_{FRET} values.

In fact, the experimental situation, when $E_{FRET} \ll E_{QUENCH}$, has been observed not only for "QD–porphyrin" nanoassemblies with various CdSe/ZnS QDs [62–64] but also for "QD–PDI" systems [74, 94]. In order to verify the question whether the

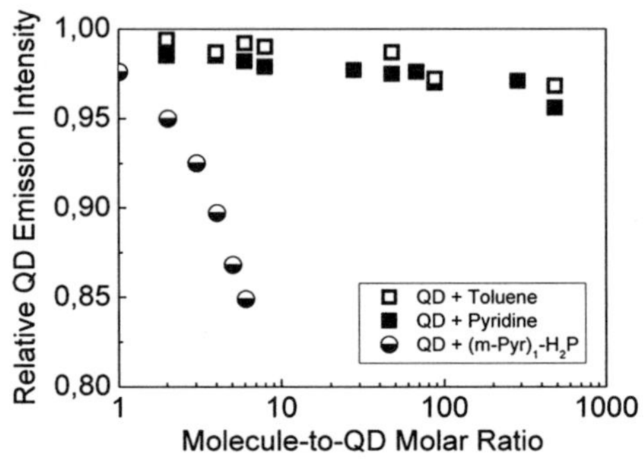

Fig. 4.16 Comparison of PL quenching of colloidal CdSe/ZnS QD ($d_{CdSe} = 2.14$ nm, $n_{ZnS} = 2$) induced by (m-Pyr)$_1$-H_2P (*half circles*) and pyridine (*solid squares*). *Open squares* show the PL quenching by addition of toluene only. Note that the PL change due to concentration change upon addition of the aliquot is already taken into account. Adapted from [123]

PL quenching of the QD really needs the attachment of the H_2P or PDI chromophore or merely the pyridine substituent, a comparison has been performed of PL quenching induced by titration with pyridine and with mono-pyridyl-substituted porphyrin molecules (m-Pyr)$_1$-H_2P containing only one *meso*-pyridyl ring which is necessary for the QD attachment (Fig. 4.16). It turns out that an only small (2–3%) decrease of the PL quantum efficiency of the QD is due to dilution by toluene (open squares). Practically the same tendency is observed for titration by pyridine (solid squares), which probably reflects a negligible quenching by this molecule (pyridine is a hole acceptor like amines or thiols [39, 44, 106]). In comparison, for molar ratios $x = 1-8$, the PL quenching of QDs by mono-pyridyl substituted porphyrin (m-Pyr)$_1$-H_2P is essentially stronger (half circles) in comparison with the above two cases.

Several reasons for this behavior have to be considered. (1) Van-der-Waals interactions between TOPO molecules and the (in comparison with pyridine) large porphyrin molecule may favor a strong incorporation of a porphyrin molecule into the QD interface (with respect to the QD surface perpendicular geometry, see Fig. 4.7). (2) Weak solubility of (m-Pyr)$_1$-H_2P in toluene may also favor a fixation on the QD surface. (3) Electronic properties of the π-conjugated porphyrin macrocycle having *meso*-pyridyl rings may play a specific role in the localization of the electron–hole pair at the QD surface [90]. Concluding, these experiments reveal that there is almost no PL quenching of CdSe/ZnS by pyridine even at 100 times increased relative molar ratios x. This observation prompts us to believe that the porphyrin chromophore itself is of central importance although the specific electronic structure is of much less influence [101]. Investigations on similarly attached perylene diimide chromophores support this statement [74, 94]. However, we have in case of H_2P dyes varied the electronic nature by fluorination of the molecules without observing a pronounced influence on the PL quenching [90]. Interestingly, in the case of CdSe/ZnS QDs ($d = 2.6$ nm and $n_{ZnS} = 2$ monolayers), upon the replacement of (m-Pyr)$_4$-H_2P molecules by the Cu-containing derivative (m-Pyr)$_4$-CuP, and tetrahydroporphyrin (m-Pyr)$_4$-THP (thus, changing the overlap integral $J(\nu) = \int_0^\infty f_D(\nu)\varepsilon_A(\nu)\frac{d\nu}{\nu^4}$ values by factor of 2.5) the quenching efficiency of QD PL remains nearly the same [90, 101]. So, the contribution of FRET QD $\rightarrow$ porphyrin to the total QD PL quenching seems to be minor and hence is negligible in most cases. For this reason we believe that the marked difference between pyridines and dyes relates to thermodynamic and/or volume effects as will be discussed later in detail.

The disagreement between PL quenching and FRET enhancement is also observed via single nanoassembly detection. While the ensemble experiments allow for a dynamic nanoassembly formation, this can be almost certainly ruled out in single nanoobject experiments, since an effective diffusion of dyes or QD on SiO_2 surfaces has not been observed experimentally [65, 74, 75, 94]. Figure 4.13b (bottom) presents for a single nanoassemblies typical luminescence spectra and PL decay times as a function of observation time. It is seen that in most cases both types of spectra (QD and PP, respectively) show uncorrelated blinking and bleaching events, which are a direct proof of the presence of in each case only one emitter [124]. The observation of uncorrelated blinking is a strong indication that the electronic interaction between QD and PP by either charge transfer or FRET is quite weak.

Since the PL intensity of the QD is reduced on average by 50% [74, 94] for each identified single nanoassembly, the corresponding dye-induced quenching rate should be (assuming $n = 1$) nearly equal to the (average) decay rate of the QD. A close inspection of the time-resolved QD PL decay of single nanoassemblies provides evidence for the shortening of the QD PL decay time. An average decay time of the multi-exponential QD PL can be deduced applying a stretched exponential fit (with $(0 < \beta \leq 1)$) according to

$$I(t) = I_0 \cdot \exp\left(-\left(\frac{t}{\tau}\right)^\beta\right). \tag{4.6}$$

Correspondingly, an average decay time τ_D for a single QD can be calculated according to

$$\tau_D = \frac{\tau}{\beta}\Gamma\left(\frac{1}{\beta}\right). \tag{4.7}$$

Calculated τ_D values for single QDs and single "QD–PP" nanoassemblies as a function of the observation time are shown as black circles in Fig. 4.13a, b. The spectral fingerprints of 90 QDs from 3 different samples have been used to compare the decay times of free and assembled QDs [74]. Normalizing the histograms for τ, β, and PL intensity I_{PL} to the number of emitters results as shown in Fig. 4.17 in probability distributions of τ_D, β_D, and I_{PL}^D for isolated QDs and those for the QDs in single "QD–PP" nanoassemblies, respectively. It is seen from Fig. 4.17 that the resulting decay time probability density of "QD–PP" nanoassemblies is as

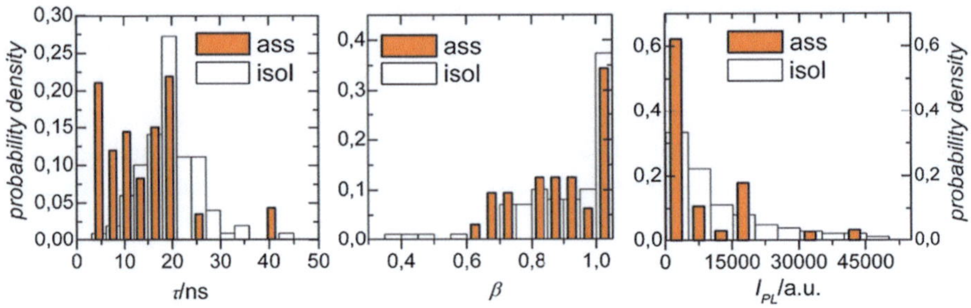

Fig. 4.17 Probability densities ($t_{\mathrm{bin}} = 20$ s) of τ, β, of s streched exponential function and PL intensity I_{PL} for 90 single isolated (isol) QDs and single 30 "QD–PP" nanoassemblies (ass). Adapted from [74]

compared to isolated QDs clearly shifted to shorter decay times τ. Similar effects are observed for the probability density of intensities I_{PL} for isolated QDs and "QD–PP" nanoassemblies. The related quenching efficiency of one dye molecule per QD is obviously broadly distributed between weak and near complete PL quenching.

Based on this statistical analysis, calculation of the mean value of $\bar{\tau}_{\mathrm{D}}$ by averaging over all nanoassemblies leads to the value $\bar{\tau}_{\mathrm{D}}^{\mathrm{A}} \approx 13.1$ ns for the decay time for QD PL in single "QD–PP" nanoassemblies while it is $\bar{\tau}_D^D \approx 21.3$ ns for isolated QDs. From this we obtain an average effective quenching rate $\bar{k}_{\mathrm{DA}}$ to be

$$k_{\mathrm{DA}} = \frac{1}{\overline{\tau_{\mathrm{DA}}}} = \frac{1}{\overline{\tau_{\mathrm{D}}^{\mathrm{A}}}} - \frac{1}{\overline{\tau_{\mathrm{DD}}}} \tag{4.8}$$

which results in $1/\bar{\tau}_{\mathrm{D}}^{\mathrm{A}} = \bar{k}_{\mathrm{DA}} = 2.9 \times 10^7\ \mathrm{s}^{-1}$.

This result is very close to $k_{\mathrm{q}} = 3.5 \times 10^7\ \mathrm{s}^{-1}$ obtained in bulk experiments for the corresponding "QD–PP" nanoassemblies in titration experiments at $x = 1$ in toluene [74, 94]. From this we conclude that the nature of quenching processes (including low efficiencies of FRET) is very similar in ensemble and single nanoassembly experiments. Both experimental approaches being applied for "QD–H_2P" and "QD–PDI" nanoassemblies evidently show the existence of strong non-FRET quenching pathways competing effectively with FRET.

It should be noted that for given CdSe/ZnS or CdSe QDs, ligand dynamics and non-FRET processes depend critically on the kind of solvent [62–65, 74, 75, 90, 94, 101, 127]. Most of the FRET investigations on "QD–Dye" nanoassemblies discussed above have been performed in toluene. Adding polar acetone increases considerably the QD PL quenching [62], which indicates the involvement of charged QD states. On the other hand, quenching has been effectively suppressed in non-polar n-octane [64]. To show the influence of solvent properties qualitatively we discuss here in short the properties of "QD–PDI" nanoassemblies based on TOPO-capped CdSe/ZnS QDs and DTPP molecules (see structure in Fig. 4.3) in a thin film of the solvent TEHOS (tetrakis-2-ethylhexoxy-silane has long aliphatic chains, see the structure on the top of Fig. 4.18) since efficient FRET has been observed

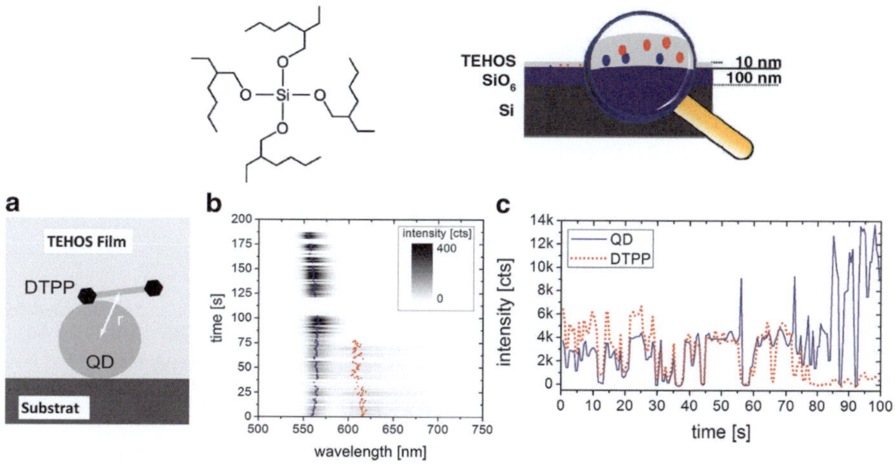

Fig. 4.18 *Bottom*: Scheme of DTPP orientation on the QD surface (**a**), spectral (**b**), and intensity (**c**) time traces of "QD–DTPP" nanoassemblies. *Top*: Scheme of TEHOS and of the sample composition

for single QD–DTPP nanoassemblies under these conditions [140]. Samples were prepared by dip-coating the silicon substrates (with 100 nm thick thermal oxide, Crystec, Berlin) into solutions of TEHOS in hexane with highly diluted CdSe/ZnS QDs and DTPP molecules resulting in 10 nm thick films of TEHOS with nanomolar concentrated CdSeZnS QDs and DTPP at $\lambda_{\text{exc}} = 488$ nm.

Similar to PP and CdSe/ZnS in toluene [64, 74], mixing CdSe/ZnS and DTPP in a thin TEHOS film results both in assemblies and in single constituents. Thus, luminescent spots observed by confocal microscopy show three types of emission spectra related to either nanoassemblies, QDs or DTPP. About 10% of the luminescent spots show spectra of both constituents. Figure 4.18 shows the intensity fluctuations of luminescence for CdSe/ZnS and DTPP as a function of an observation time of several minutes. Typically, strong intensity intermittency (blinking [124]) for QD PL in combination with spectral fluctuations (spectral diffusion) is quite obvious [141]. In addition, an enhanced blinking of DTPP is observed in contrast to the almost negligible blinking known for isolated single molecules [110].

A closer inspection reveals that the respective QD and DTPP emission intensities are strongly (positively) correlated, as can be directly seen in Fig. 4.18c, when plotting the corresponding intensities on top of each other: the strong blinking of QD is imposed onto DTPP, which now blinks very often. DTPP fluorescence is nearly completely "off" as soon as the QD photoluminescence is turned "off" according to a blinking process. When DTPP is photobleached or detached from the QD (at 77 s) the QD PL increases on average by about a factor of 3. There are only three times at which intensity correlation fails, namely at 56, 66, and 72 s, respectively, for which DTPP fluorescence decreases while the QD photoluminescence increases. This can

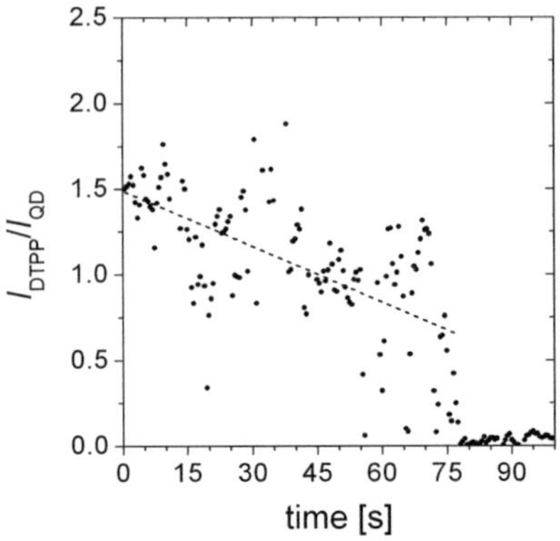

Fig. 4.19 Luminescence intensity ratio $R = I_{DTPP}/I_{QD}$ as a function of observation time for "QD–DTPP" nanoassemblies in TEHOS

be assigned to one of the rare blinking events of DTPP. The overall interpretation is that the PL of the QD is quenched due to FRET according to QD → DTPP and that DTPP fluorescence is predominantly caused via FRET.

Alternative to FRET, intensity correlation might be caused by photoinduced electron transfer from the QD to DTPP. In that case both constituents are charged and at least the luminescence of DTPP will be completely quenched. However, such an explanation can be ruled out because of the intrinsic nature of QD blinking which does not change upon aggregate formation. If charge transfer would apply, the PL of the dye will increase when the QD is "off" and cannot act as a charge donor.

To obtain a more quantitative description of FRET we have plotted the luminescence intensity ratio $R = I_{DTPP}/I_{QD}$ in Fig. 4.19. It is seen that the correlation clearly depends on the observation time and changes by more than a factor of 3. A detailed analysis (with normalized intensities for each constituent) shows that the linear correlation of intensities is quite constant during the first 30 s, but changes in the course of time. This can be interpreted as being due to time-dependent changes of the geometry of the assembly or the conformation of DTTP.

From the quenching data of the donor (QD) PL intensity I_{QD} in the single "QD–DTPP" nanoassembly, FRET efficiency can be calculated according to [139] to be

$$E_{FRET} = 1 - \frac{I_{QD-DTTP}}{I_{QD}}. \tag{4.9}$$

It is seen from Fig. 4.18b that DTPP is photobleached for times longer than 77 s. Taking the maximal PL intensity I_{QD} in that range and the maximum of the intensity $I_{QD-DTTP}$ during the time DTPP is not photobleached (Fig. 4.18c) we obtain according to Eq. (4.9) $E_{FRET} \approx 0.69$. This can be compared to the situation at 57 and 72 s when DTPP is "off" due to blinking providing I_{QD} and compare it to $I_{QD-DTPP}$ when DTPP is "on" again. According to Eq. (4.9) this results in $E_{FRET} \approx 0.55$. Both values are in reasonable agreement with $E_{FRET} \approx 0.85$ calculated within Foerster model [94] (see Sect. 4.2.2.2).

Obviously, strong variations of E_{FRET} are observed for on one and the same assembly, since I_{QD} depends on quenching processes other than FRET. Notably, FRET values (obtained in TEHOS) are essentially higher with respect to those ($E_{FRET} \leq 0.1$) obtained for the same "QD–DPI" nanoassemblies both from ensemble (toluene) and from single nanoobject detection (on an SiO_2 surface) [74, 94] as well as for "QD–porphyrin" nanoassemblies (in toluene) [62–64]. We have related these previous findings of low FRET efficiencies to the presence of competing non-FRET processes, which open a quenching pathway other than energy transfer by merely attaching a dye molecule to the QD surface [62–65, 74, 75, 90, 94, 101, 127]. This will be discussed in more detail later on.

Finally, in the hypothetical case of a photoinduced charge (hole or electron) transfer process in "QD–porphyrin" or "QD–PDI" nanoassemblies the fluorescence of the dye should be also quenched. However, fluorescence parameters (efficiency φ_F and decay time τ) for (m-Pyr)$_4$-H_2P and PDI molecules upon complexation with QDs remain practically the same with respect to those measured for individual constituents under the same conditions [62, 101]. In addition, the titration of identical QD solutions by (m-Pyr)$_4$-H_2P and (m-Pyr)$_4$THP (hole acceptor) as well as by H_2P(m^Pyr)$_2$(Ph)$_2$ and electron acceptors (m^Pyr)$_2H_2P$(5FPh)$_2$ or (m^Pyr)$_2H_2P$(Anthraquinone)$_2$ gives the same result for PL quenching [62, 90]. Thus, the independence of PL quenching efficiency on redox properties of porphyrin and the absence of the porphyrin fluorescence quenching in "QD–porphyrin" nanoassemblies rules out a dominant role of photoinduced charge transfer processes with participation of molecular orbitals of the porphyrin macrocycle for PL quenching for the systems under study.

4.3.2 Non-FRET Photoluminescence Quenching and Quantum Confinement

It is well documented that because of quantum confinement [10–13] the exciton in a QD is highly sensitive to local charges or distortions of the charge distribution. Therefore, the PL quantum efficiency is sensitive to the involved interfaces, such as inorganic shell structures [142, 143], surfactants [83, 144–146] and solvent molecules [147] as long as the exciton wave function extends beyond the QD core. In this respect, it would be tempting to probe the interactions locally by distorting the interfacial properties with a local probe. Such controlled distortions will unravel microscopic details of either non-radiative processes occurring at the QD interface or the influence of chemical bonds at the QD surface. Organic dye molecules attached to the surface might act as such probes. Several experiments upon the effect of organic molecules exchanging the surfactant layer have been reported [144–146]. In this section we show that direct surface labeling of colloidal semiconductor QD with only a few pyridyl-functionalized molecules reveals the nature of only recently identified PL quenching mechanism which has been identified to be dominant for

"QD–H_2P" and "QD–PDI" nanoassemblies in toluene [63, 74]. This non-FRET quenching mechanism has not been considered in most of the related publications and is connected with the extension of the wave function of the exciton to the outside of the QD core.

In such a study, the experimental background was the comparative analysis of the QD PL quenching by one type of porphyrin (m-Pyr)$_4$-H_2P molecule for QDs of different sizes as well as having different ZnS capping layers. For the analysis of the PL quenching as a function of the number of H_2P molecules per QD for various QD sizes, the well-known Stern–Volmer formalism [139] was modified. In our more generalized approach, the PL quenching can be described by

$$\frac{I_0}{I} = 1 + \int_0^\infty K(x) \cdot dx, \tag{4.10}$$

where $I(x)$ and I_0 represent the QD PL emission intensity in the presence and absence of quenchers, respectively. The corresponding results are presented in Fig. 4.20a.

In this approach, the Stern–Volmer function $K(x)$ depends explicitly on the molar ratio x and is expressed as the first derivative of the experimental data plotted in a Stern–Volmer representation. Further, the Stern–Volmer "constant" $K(x)$ can be written as

$$K(x) = k_q(x) \cdot \tau_0, \tag{4.11}$$

where k_q corresponds to the total quenching rate induced by the number of attached quencher molecules and τ_0 is the intrinsic PL lifetime of the QD in absence of quencher molecules (known from measurements [63] or extrapolated from literature [117]). It is seen from Fig. 4.20a that $I_0/I(x)$ does not show a linear correlation with x over the total H_2P concentration range. The double logarithmic plot presented for $K(x)$ values shows the clear dependence of PL quenching on QD size (Fig. 4.20b). It can be seen that $K(x)$ indeed initially constant but becomes smaller around a critical molar ratio (which we name x_c [63, 64]). Additionally, this critical molar ratio x_c increases systematically with the diameter of the QD. A non-constant $K(x)$ indicates eviation from a normal Stern–Volmer relationship in case of a bimolecular reaction.

The overall interpretation of the above findings presented recently in [64] is that QD PL quenching upon titration by porphyrin molecules occurs in two steps: (1) Immediately after titration, "QD–H_2P" nanoassemblies are effectively formed, which results in both PL quenching of the QD PL and low-effective FRET to the adsorbed dye molecules [63]. (2) More dye molecules become attached during a following-up waiting time. However, increased PL quenching and FRET do not follow the same dependence on x. Clearly, the titration (assembly formation) causes predominantly other quenching mechanisms than those related to FRET. As it is seen from Fig. 4.20c, the maximal FRET efficiency does not exceed ~12–15% over the total titration range, and becomes sequentially lower with increasing QD size. This makes us to believe that the dynamics initiated by a single titration step are not

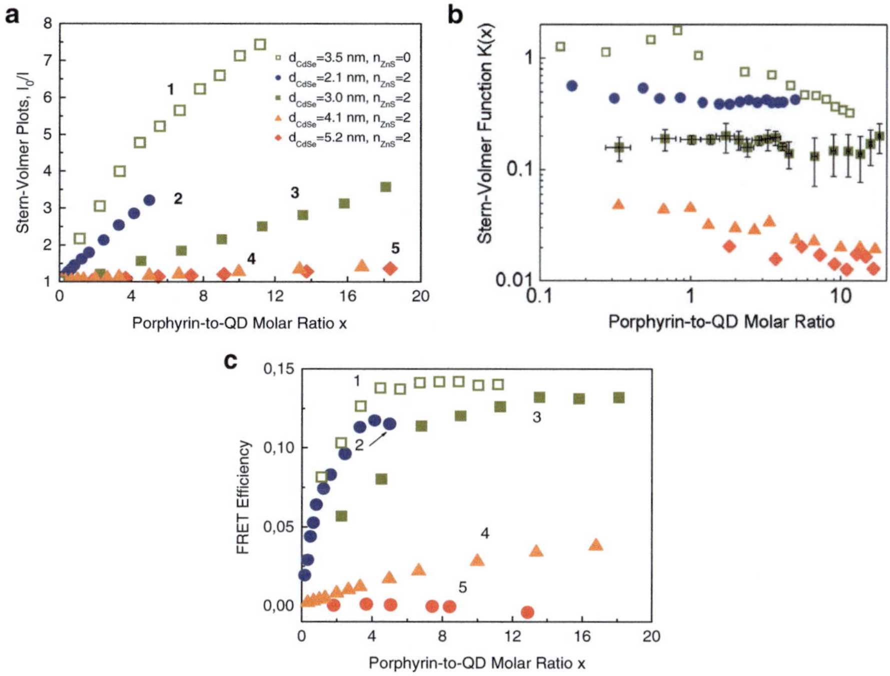

Fig. 4.20 Stern–Volmer plots $I_0/I(x)$ (**a**), calculated Stern–Volmer function $K(x)$ (**b**) and FRET efficiencies calculated using Eq. (4.2) (**c**) for an uncapped CdSe QD (*open markers*) and CdSe/ZnS-capped QDs (*solid markers*) of various sizes upon titration by (m-Pyr)$_4$-H_2P as function of the molar ratio $x = [C_{\text{Porphyrin}}]/[C_{\text{QD}}]$ in toluene at 295 K : *1* (*open green square*)—$d_{\text{CdSe}} = 3.5$ nm, $n_{\text{ZnS}} = 0$; *2* (*blue circle*)—$d_{\text{CdSe}} = 2.1$ nm, $n_{\text{ZnS}} = 2$; *3* (*filled green square*)—$d_{\text{CdSe}} = 3.0$ nm, $n_{\text{ZnS}} = 2$; 4 (*orange triangle*)—$d_{\text{CdSe}} = 4.1$ nm, $n_{\text{ZnS}} = 2$; 5 (*red diamond*)—$d_{\text{CdSe}} = 5.2$ nm, $n_{\text{ZnS}} = 2$. Adapted from [63]

Table 4.2 Mean $\langle K_{\text{SV}}(x)\rangle$ values and PL quenching rate constants k_q for QDs of various CdSe core diameters d_{CdSe} upon titration by (m-Pyr)$_4$-H_2P (toluene, 295 K)

d_{CdSe} (nm)	4.3	5.2	6.3	7.3
$\langle K_{\text{SV}}(x)\rangle$	0.65 ± 0.10	0.115 ± 0.015	0.055 ± 0.007	0.020 ± 0.003
k_q (ns^{-1})	0.041 ± 0.016	0.0057 ± 0.0023	0.0027 ± 0.0016	0.0015 ± 0.0009

only due to the molecular properties of H_2P but also due to a local change in the ligand shell on the QD surface upon assembly formation.

Experimental data for mean $\langle K_{\text{SV}}(x)\rangle$ values and PL quenching rate constants k_q evaluated from the results presented in Fig. 4.20 are collected in Table 4.2. As far as the QD PL quenching in "QD–H_2P" nanoassemblies is due to at least two contributions (FRET and non-FRET), $\langle K_{\text{SV}}(x)\rangle$ (presented also in Fig. 4.20b) and k_q values have been corrected for the FRET contribution (shown in Fig. 4.20c but which are relatively small).

Data presented in Table 4.2 show that for CdSe/ZnS quantum dots with two ZnS monolayers k_q values follow a monotonous function drastically decaying with the QD core diameter. For comparison, CdSe QDs without a ZnS shell show a much stronger quenching than the respective core–shell analogue. With respect to porphyrins, it was found [62, 90, 101, 127] that the observed QD PL quenching in "QD–H_2P" nanoassemblies does not show a remarkable dependence on the electronic structure and properties of attached dyes, and a non-specified quenching mechanism has to be postulated, which does not depend on the electronic structure of the attached dye. Therefore we concentrate only on those aspects that are related directly to the outer interface of a QD as a function of size and shell thickness.

At the outer interface of the "QD-H_2P" nanoassembly two contributions are important: (1) the presence of "binding" orbitals such as the lone pair of a pyridyl ring of the porphyrin or pyridine itself and (2) the evanescent wave function of the confined exciton "leaking" out of the core (and the ZnS shell) of the QD. From a physicochemical point of view it means that upon interaction of (m-Pyr)$_4$-H_2P with the QD surface, the electron wave function may be locally modified (e.g., via inductive and/or mesomeric effects [90]) forming a surface local state capable to trap the electron of the photogenerated electron–hole pair. So, as in any related quantum mechanical problem the wave function of the delocalized electron may extend beyond an imposed barrier or, in other words, the corresponding particle may tunnel through the barrier. With respect to QDs, quantum confinement of the excitonic wave function in its most simple version confines the wave function of the exciton (electron–hole pair) to a spherical box [148].

The general scheme of our model is shown in Fig. 4.21a. In addition to the coordination of a pyridyl ring to the surface the electron wave function at the outer interface of the QD is presented schematically. Correspondingly, all observed size dependence of the PL quenching (as expressed by the quenching rate k_q) should be directly related to the amplitude of the electron wave function at the interface. The calculation of changes of charge densities is therefore reduced to an evaluation of the radial probability function $r^2\psi^2(r)$ for the electron as a function of the CdSe core diameter and ZnS shell thickness. Now the QD PL quenching rate becomes

$$k_q(r) = C\psi^2(r). \tag{4.12}$$

The single-carrier envelope wave functions ψ_a in a spherical core/shell QD are determined by the Schrödinger equation (the particle-in-a-box model [149]) by

$$\left(\frac{\hbar^2}{2}\frac{\partial^2}{\partial r^2}\left(\frac{1}{m_a{}^*(r)}+V_a(r)\right)\right)\psi_a = E_a\psi_a, \tag{4.13}$$

where the index $a = e, h$ represents electron and hole, respectively. Solutions were obtained by applying continuity relations for the single-carrier wave functions at the interfaces between the CdSe core ($i = 1$), ZnS shell ($i, j = 2$), and matrix ($j = 2$), respectively, by assuming that

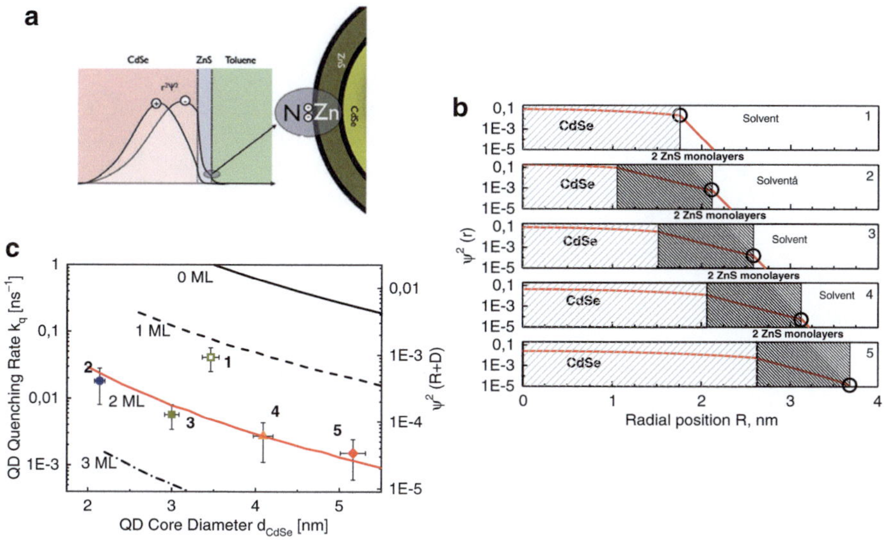

Fig. 4.21 General scheme of the QD PL quenching model and comparison of experimental results and theoretical calculations for "QD–H_2P" nanoassemblies: (**a**) QD–H_2P interaction model. (**b**) Logarithmic presentation of $\Psi^2(r)$ for a 1 s electron in a core–shell spherical potential for five different sizes of QD with $n_{ZnS} = 0$ or 2. (**c**) Comparison of the experimentally determined quenching rate constants k_q (symbols with *error bars*, *left axis*) and calculated probability density functions $\psi^2(r = R + D)$ (*lines*, *right axis*) of a 1 s electron at the outer interface ($r = R + D$) between the ZnS shell and the environment as a function of core diameter d and ZnS shell thickness D. The constant C in equation $k_q(r) = C\psi^2(r)$ has been adjusted with respect to $\psi^2(R + D)$ to fit the experimental value at $d_{CdSe} = 4.1$ nm, $n_{ZnS} = 2$. All theoretical lines correspond to calculations of $\psi^2(R)$ with the same proportionality constant C. QD parameters are the same as in Fig. 4.20. Adapted from [63]

$$\psi_i(r_{ji}) = \psi_j(r_{ji}) \quad \text{and} \quad \frac{1}{m^*_i}\frac{d}{dr}\psi_i(r)\bigg|_{r=r_{ji}} = \frac{1}{m^*_j}\frac{d}{dr}\psi_j(r)\bigg|_{r=r_{ji}}. \tag{4.14}$$

Figure 4.21b shows the calculated radial part of $\psi^2(r)$ of an *s*-type electron wave function for four CdSe/ZnS QDs with two ZnS monolayers and one CdSe QD without a ZnS monolayer. It can be seen that $\psi^2(r)$ becomes smaller at the outer interface (marked by a circle) upon increasing QD diameter. The corresponding value is largest for the uncapped CdSe QD ($R = 1.73$ nm).

It is seen from Fig. 4.21c that the comparison of rate constants k_q (symbols) and the calculated (and scaled) probability densities $\psi^2(r)$ (lines) shows a good correlation. Thus, one point-like charge density perturbation caused by an organic linker group or chromophore at the QD interface forces the electron of the delocalized exciton of the QD to become localized. Qualitatively the same physicochemical picture is characteristic also for "QD–PDI" nanoassemblies [64, 74, 94]. In fact, the dependence of QD PL quenching induced by even only one single molecule on the

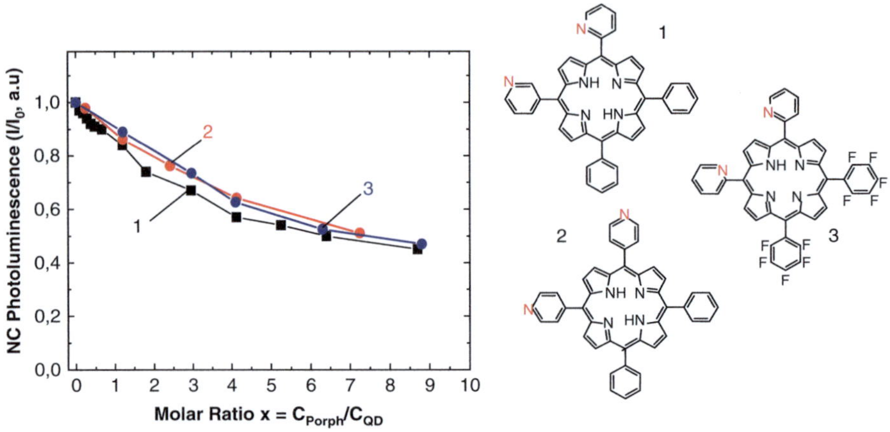

Fig. 4.22 Relative PL intensity changes (quenching) $I(x)/I(0)$ as function of the molar ratio $x = [C_{\mathrm{Porph}}]/[C_{\mathrm{QD}}]$ for CdSe/ZnS QD ($d_{\mathrm{CdSe}} = 2.5$ nm, $n_{\mathrm{ZnS}} = 2$, $C_{\mathrm{QD}} = 5 \times 10^{-7}$ M) and various porphyrin molecules. *1*: $(m^\wedge\mathrm{Pyr})_2(\mathrm{Ph})_2$-$H_2P$; *2*: $(p^\wedge\mathrm{Pyr})_2(\mathrm{Ph})_2$-$H_2P$; *3*: $(m^\wedge\mathrm{Pyr})_2(5\mathrm{FPh})_2$-$H_2P$ in toluene at 295 K ($\lambda_{\mathrm{exc}} = 465$ nm, $\lambda_{\mathrm{reg}} = 522$ nm). Structures of porphyrins 1, 3, and 3 are also presented on the right

QD clearly resembles the tunneling of an electron (through the ZnS barrier) to the outer interface of the QD. Such tunneling is followed by the (self-)localization of the electron–hole pair. This corresponds to the creation or modification of trap states in the semiconductor band gap. Such traps might be subject to non-radiative channels, e.g. via enhanced electron–phonon coupling. Thus, changes in local charge densities result in changes in the quantum efficiency of the QD which have to be related to PL lifetime changes and modification of blinking in case of single QDs [124, 131, 151]. These results reveal also that single functionalized porphyrin molecules can be considered as a probe for the complex interface physics and dynamics of colloidal semiconductor quantum dots [63, 64, 74, 75, 94, 150]. It is tempting to conclude that the quenching is merely due to the influence of the pyridyl anchor since the electronic properties of the porphyrins are obviously negligible [127] (see Fig. 4.22). Additionally, the PL quenching caused by PDI molecules is similar to the one observed for porphyrins as long as assemblies are formed in toluene solvent. However, obviously the anchor is not the main source for quenching, since titration by orders of magnitude larger amounts of pyridine as compared to porhyrins does not result in noticeable quenching as can be seen from Fig. 4.16.

Concluding, we suggest that the non-FRET quenching mechanism is related to the extension of the electronic wave function of the electron hole pair to the outer interface of the (capped) QD influencing the formation of near band edge states with new non-radiative features. However, though dye molecules heavily induce PL quenching, this quenching is in many cases—as reported here—only to a minor extent related to photoinduced charge and/or energy transfer. Nevertheless, the identification of FRET is at least a proof that nanoassemblies are formed and that even information on the geometry and stability of the assembly can be obtained. Obviously, non-FRET quenching does not depend on the electronic details of the attached dye molecules but on the properties of the solvent or embedding environment. These observations prompt us to suggest that dye molecules influence and modify the structure and dynamics of the (TOPO or amine) surfactant (ligand) shell. Since ligands saturate the dangling bonds at the surface [48, 56–58] they also modify the intra-band states or near band edge states of a QD [48, 56–58]. This implies that we have to investigate the competition of dye attachment and ligand coordination. We will approach this task in the following section.

4.4 Porphyrin and Ligand Exchange Dynamics in Nanoassemblies

A comprehensive description of dye attachment dynamics in relation to PL quenching processes is of crucial importance for the elucidation of mechanisms of photoinduced processes in "QD–H_2P" nanoassemblies (as well as for other dyes) as a step to an investigation of the chemical topography of a QD surface. In fact, dye molecules that quench the PL of the QD in a controlled way are an indirect measure for the ligand dynamics at the QD surface as well as they may influence surface-related processes.

While the formation of assemblies is manifested via QD PL quenching three issues remain open: (1) the identification of the average number of dye molecules attached to the QD surface, (2) the formation dynamics of such nanoassemblies, and (3) the closely related dynamics of ligands. In each case, the dynamics of nanoassembly formation is closely related to the ligand dynamics, since the dye molecule either has to find freely accessible sites on the QD surface or partly replaces the existing ligand shell. Papers on ligand dynamics [32–34, 106, 125–128] mostly deal with uncapped CdSe QDs, since the effects on the PL quenching following ligand dynamics are much more pronounced for uncapped than for capped systems, e.g. CdSe/ZnS QDs. In the present section, we present results on capped CdSe/ZnS QDs which generally exhibit much less variation of PL intensities. Moreover, the application of dyes offers the possibility to investigate surface attachments on an extremely low concentration level not accessible by conventional experiments on ligand dynamics. Additionally, as will be shown later, the use of dyes allows for an investigation of the microscopic aspects of the quenching process itself.

Here, we focus on nanoassemblies based on TOPO-capped CdSe/ZnS QDs (as well as CdSe QDs in some cases) and (*m*-Pyr)$_4$-H_2P molecules showing (like (*p*-Pyr)$_4$-H_2P) among a series of *meso*-pyridyl substituted free-base porphyrins the most effective PL quenching of QDs at the same titration conditions (Sect. 4.2.1, Fig. 4.6a). Typically, in titration experiments that study the PL intensity of CdSe or CdSe/ZnS QDs as a function of the added amount of H_2P molecules, the spectra were obtained within a time of 60 s after every addition of the aliquot. To prove this operational procedure, long-term stability of the applied chemical systems including the solvent quality has been checked within the time scale from 60 s to hours as well as upon variation of the solvent (toluene, TEHOS or *n*-octane) [64]. The detailed analysis of a whole series of these experiments leads to the following main results and conclusions.

4.4.1 Dependence of PL Quenching of QDs and QD–Dye Nanoassemblies on Long Observation Times

4.4.1.1 Development of Intrinsic PL Quenching with Time in Solvents of Various Qualities

After sample preparation via diluting the originally concentrated stock solution of QDs a decrease of PL intensity is observed with time. This decrease turns out to depend on the kind of solvent, the respective purity as well as on the absolute concentration of QDs and surface ligands in solution. Figure 4.23 shows for uncapped CdSe QDs the role of different toluene purities and the influence of the addition of (m-Pyr)$_4$-H_2P molecules at molar ratio $x = 1.2$.

It is clearly observed from Fig. 4.23a that the amount of water governs the stability of the PL over time. The poorly dried toluene results in an almost complete PL quenching within 30 min. We also find that PL quenching of the QDs is reduced upon an increased content of TOPO in the solution (molar ratio $y = [C_{\mathrm{TOPO}}]/[C_{\mathrm{QD}}] = 800$). However, the most stable PL is observed following an extensive drying of the solvent, which reduces PL quenching considerably. The related PL decrease is in all cases nearly exponential at early times. The corresponding decay time constants are in the range 240–3,000 s for the given examples.

The question regarding the long-term stability of the formed "QD–H_2P" nanoassemblies in solution was investigated by adding H_2P in one step at $x = 1.2$ to the respective samples of uncapped CdSe QDs (see Fig. 4.23b). Here, for all solvent qualities, similar results were found: addition of the H_2P aliquot to the master solution of QDs results in an immediate decrease of the PL intensity faster than our time resolution of about 60 s. In each of the samples, the initial fast PL decrease is followed by a decrease of the PL intensities on similar time scales, but it is no longer close to an exponential decay. As compared to the time behavior of the corresponding samples before the titration step (see Fig. 4.23a) the decay times seem to be broadly distributed (similar to a power law) reflecting the presence of different quenching processes, namely those already inherent in the initial QD sample and those imposed by addition of the H_2P resulting in assembly formation. In this respect, the slower QD PL quenching in time for "QD–H_2P" nanoassemblies

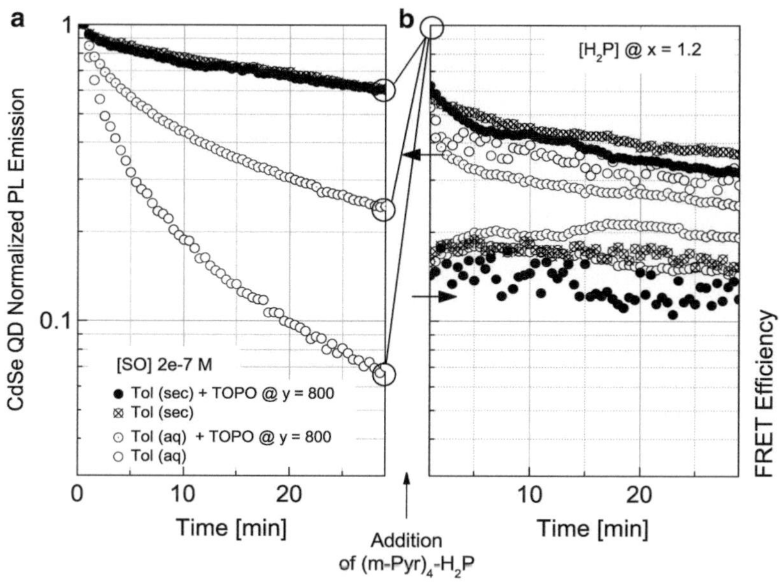

Fig. 4.23 Influence of sample conditions on the time-dependent PL quenching for uncapped CdSe QDs ($d_{CdSe} = 3.5$ nm) in toluene. (**a**) Influence of solvent composition. Tol (aq) and Tol (sec) correspond to toluene spectroscopic grade as-supplied and 48 h dried over a molecular sieve, respectively. The addition of tri-*n*-octyl phosphane oxide (TOPO) in large excess (molar ratio of $y = [C_{TOPO}]/[C_{QD}] = 800$). (**b**) Influence of the addition of (m-Pyr)$_4$-H_2P at molar ratio $x = 1.2$. PL intensities of CdSe are re-normalized with respect to the time of addition of H_2P at about 30 min. This time was set to zero. The time-dependent increase of the H_2P fluorescence (FRET efficiency, relative units, *right axis*) is also shown (*lower curves*) which proves the formation of "QD–H_2P" nanoassemblies. The absolute concentration C of uncapped CdSe QDs is $C = 2 \times 10^{-7}$ M. Adapted from [64]

in comparison with that for merely QDs may be explained by effective PL quenching in nanoassemblies competing with that caused by solvent impurities.

The increase of the H_2P fluorescence (Fig. 4.22b) upon titration can be clearly explained by FRET from the QD to the attached H_2P molecule. It levels off as soon as it cannot compete any more with other ingrowing quenching processes. Hence, aside from the confirmation conducted in earlier publications [62, 63, 101, 127], FRET is a measure for the formation kinetics of "QD–H_2P" nanoassemblies. It should be mentioned also that for CdSe QD+H_2P solutions of the same molar ratio $x = 1.2$ prepared at different initial concentrations of QDs ($C_{QD} = 1 \times 10^{-7}$, 2×10^{-7} and 4×10^{-7} M) the PL of the QDs is continuously decreasing, with the quenching most pronounced for the most dilute solutions.

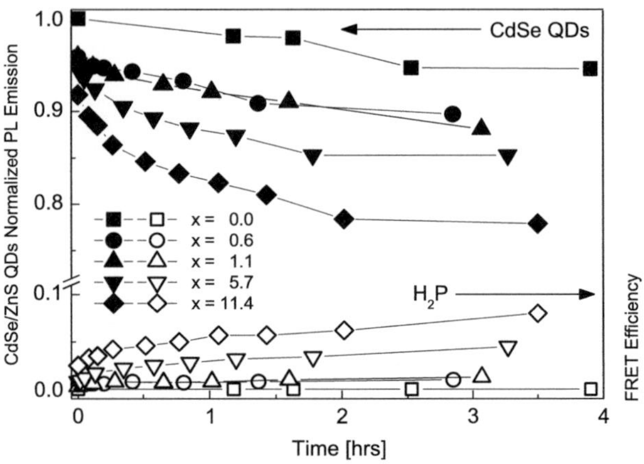

Fig. 4.24 Long-time development of the PL of CdSe/ZnS QDs ($d_{CdSe} = 3.0$ nm, $n_{ZnS} = 2$) and the FRET efficiency in presence of (m-Pyr)$_4$-H_2P molecules for "QD–H_2P" nanoassemblies at different molar ratios. The values $I(x)/I_0$ on the *left axis* were normalized to the PL of the individual sample prior titration. The related FRET efficiency (*right axis*) is plotted at the bottom. Adapted from [64]

4.4.1.2 One-Step PL Quenching of CdSe/ZnS QDs in "QD–H_2P" Nanoassemblies at Various Molar Ratios *x*

This example concerns the behavior of the PL intensity as a function of the waiting time after one titration step followed by the formation of "QD–H_2P" nanoassemblies at a given ratio *x*. In these experiments, the long-term stability of the PL of core–shell CdSe/ZnS QDs was investigated at various H_2P concentrations (Fig. 4.24). Here, samples at different molar ratios *x* were prepared in one single titration step which results in a PL quenching similar to the discussion above for the uncapped QDs. The data for this quasi-static titration are presented as a ratio of $I(x)$, the intensity after titration, to $I_{0,}$ the intensity just before the first titration event. It is clearly seen that PL quenching increases with observation time. This effect is most pronounced for high molar ratios (e.g., additional quenching of 15% over 3.5 h at $x = 11$), whereas solutions at molar ratios $x = 0$–6 remain stable in PL intensity within the range of 5–8% over 3.5 h.

The absolute efficiency of quenching increases with *x*, although again some quenching is present in the absence of H_2P. The comparison of the PL quenching with the FRET build-up during time shows that the "QD–H_2P" nanoassembly formation has multiple (at least three) macroscopic time scales. All these effects should be taken into account upon quantitative analysis of titration results with respect to the efficiency of FRET and non-FRET quenching processes in "QD–Dye" nanoassemblies.

4.4.1.3 Dye Attachment Dynamics and PL Quenching for CdSe/ZnS QDs of Various Sizes

As was outlined in Sect. 4.3.2 for "QD–H_2P" nanoassemblies based on CdSe/ZnS-capped QDs of various sizes and (m-Pyr)$_4$-H_2P molecules, the Stern–Volmer quenching constant $K(x)$ is initially constant but becomes smaller close to a "critical" molar ratio x_c (see Fig. 4.20c) [63, 64], while the x_c value increases systematically with the diameter of the QD. This can be explained by the fact that "QD–H_2P" nanoassemblies are formed in a dynamic equilibrium of attachment and detachment of porphyrin molecules to the QD surface. It follows from Fig. 4.23 that the time to reach the initial equilibrium is faster than our experimental observation time of about 60 s. This implies that there have to be some free or very easily accessible sites [28–34] for the dye attachment on the QD surface, since ligand replacement occurs typically on time scales of minutes or even longer. From the results presented in Fig. 4.20c we conclude that assembly formation can be tentatively divided into a low-molar-ratio and high-molar-ratio range which is separated by a "critical" molar ratio $x_{c.}$. Above the critical molar ratio $x_{c,}$ the probability of attaching dye molecules to the QD surfaces is considerably reduced. In our generalized approach [63, 64], the Stern–Volmer quenching constant K depends on x if, e.g. dye attachment is limited by the number of accessible sites n_{max} on a QD surface. This number can also vary according to the particular conditions, e.g. if a re-arrangement of the ligand shell or the surface topography [48] has to happen in order to allow for further dye attachment. Now let us assume that, at low x, the QDs have free or easily accessible attachment sites available for assembly formation with dye molecules. At these low concentrations, quenching relates to "quasi stable assemblies," and it is very similar to doping-dependent transfer processes in molecular crystals [152]. In that low molar ratio range, $K_{x<x_c}$ is independent of x and can be described (according to Eq. 4.11) by $K_{x<x_c} = K_c{}^r = k_q \cdot \tau_0$.

However, both the absolute value of K and the critical molar ratio x_c depend on the size of QDs. For the low-molar-ratio range $x < x_c$, the dependence of the magnitude of $K_{x<x_c} = K_c{}^r$ on the size is in agreement with the results of the quantum-mechanical considerations (see Fig. 4.21c) [63]. With respect to x_c value, it was shown [64] that the QD surface is far from being totally covered with dye molecules, but this value represents a landmark where the major processes responsible for assembly formation change. Based on the generalized Stern–Volmer formalism and the results of Fig. 4.20c, these x_c values (which should be related to the maximum number of freely accessible binding sites n_{max}) as well as the modified equilibrium constants $K_{x>x_c} = K_c^d$ for the high-molar-ratio range have been evaluated for a spherical approximation for QDs of different sizes [64] and are presented here in Fig. 4.24 as a function of the QD surface area.

The linear behavior of x_c as a function of the QD surface area for the same absolute QD concentration (presented in Fig. 4.25a) reveals that the number of freely accessible sites is proportional to the surface area of the respective QD. From the slope of the linear relation we obtain that (normalized by the QD surface unit) $n_0 \leq 0.032 \pm 0.006$ H_2P/nm^2 can freely be attached to the CdSe QD surface. Hence, the average number of attached molecules n_{max} corresponds via x_c to the QD size. According to this linear relationship, n_{max} is between 0.8 (for the smallest species: uncapped CdSe QDs with core diameter 3.0 nm) and 5.3 (for the largest CdSe/ZnS QDs with core diameter 5.2 nm) H_2P molecules per QD. This is only an upper limit and the real number of attached dye molecules n may be by far smaller than n_{max} as

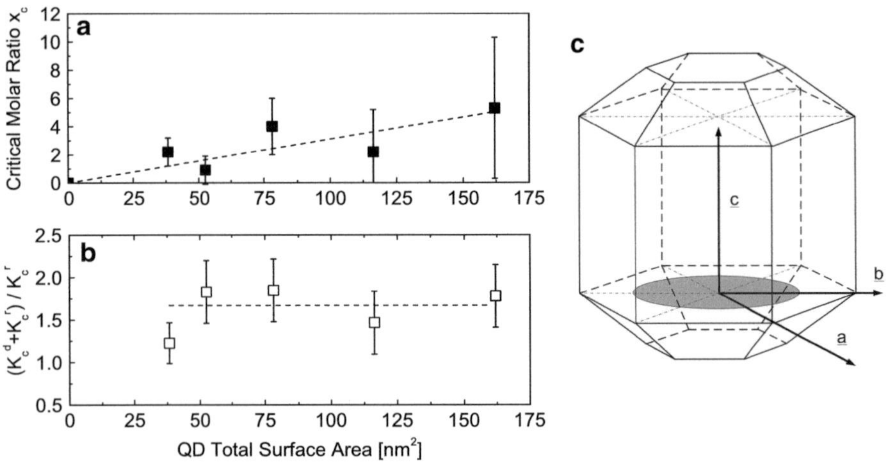

Fig. 4.25 (**a**) Critical molar ratio x_c for various types of QDs as a function of the QD surface area $4\pi(R+D)^2$ calculated from the radius R of the CdSe core and eventually the thickness D of the ZnS shell. (**b**) The ratio $(K_c^d + K_c^r)/K_c^r$ (*dots*) as a function of QD surface area. (**c**) Geometrical model of a CdSe quantum dot core (hexagonal wurtzite unit cell) including different crystal facets with polar and non-polar orientation. Adapted from [64]

there is not a direct measure for this number. Again, according to the experiments on PDI, this number may be a factor of up to 5 smaller than n_{max} [74, 94]. But even if n can be estimated from the corresponding absorption coefficients for QD and dye after precipitation and re-dissolution of the nanoassemblies [45], competitive attachment–detachment processes may take place.

The important conclusion from Fig. 4.25a is that n is almost linearly related to the QD surface area. Our interpretation is that only selected parts of the QD surface area allow an immediate attachment of the dye molecules as it is inhomogeneous with respect to its ligand dynamics. This argument is obvious from theoretical investigations of the growth mechanisms of semiconductor nanocrystals leading to facet structures [153, 154] as well as experiments about a complex optical transition moment attributed to strain within the crystal or an anisotropic growth of QDs during synthesis [155] (see simplified sketch in Fig. 4.25c). The relations of specific surface facets and ligand binding dynamics have been discussed recently in numerical simulations [48]. In the view of this picture, we see a relationship to the fact that the particular crystal facets of the QD nanocrystals or even the edges of these facets have a preferential chemical binding probability for an instantaneous attachment of dye molecules: only sub-areas with a particular structure contribute to the instanteneous attachment without significant competition with the original ligands (e.g., TOPO). When these specific attachment sites become occupied (in our picture, this happens around $x = x_c$), further PL quenching is controlled by more complex processes that are based on a competition between the dye molecules and the original ligands (TOPO) to attach on the QD surface.

When plotting $K_c^d + K_c^r / K_c^r$ as a function of QD surface area (see Fig. 4.24b), we obtain a nearly constant ratio between K_c^d and K_c^r (slope of 1.67 ± 0.09). It follows from this observation that the number of sites accessible, e.g. via a reconstruction of the ligand shell [48] is also related to the surface area. Hence there are reasonable arguments that surface attachment takes place in two steps: (1) direct attachment and (2) attachment following surface or surface-ligand reorganization. The crossover between these two routes (around x_c) depends on the total QD surface area and hence also on the QD size.

Finally, the strong difference in PL quenching observed for uncapped CdSe and core/shell CdSe/ZnS QDs may be explained by a few reasons: (1) the ZnS shell is not closed and effectively only one monolayer thick or less. This will enhance the probability of the electron wave function ψ^2 of the exciton at the attachment site and thus PL quenching in relation to the uncapped QD by more than one order of magnitude [63]. (2) The microscopic conditions are different for capped and uncapped QD. One reason can be the formation of different (induced) surface states [118]. Note that in the range of $x = 1$–1000 pyridine molecules do not quench the PL in case of CdSe/ZnS QDs (see Fig. 4.16) but very effectively in case of uncapped CdSe QDs. This fact might indicate that even the wave function of the hole has be considered in case of uncapped QD since pyridine is a strong hole acceptor [39, 44, 105, 106, 156]. (3) The binding strength of pyridine with Zn^{2+} or Cd^{2+} and thus the attachment probability for H_2P might be different for capped and uncapped QDs.

Concluding the results presented here show that it is possible to probe the interface dynamics of QDs with dye molecules in solution even from ensemble experiments, observing that a simplified picture of a spherical QD does not fully explain the details. This is intriguing in spite of the clear correlation of the initial quenching for small x with the amplitude of the exciton wave function at the QD surface.

4.4.1.4 Ligand Exchange: Rates, Dynamics, Geometrical Considerations

As has been shown previously in several papers by our group [74] and others [30–34, 106, 157–160] ligand dynamics play an essential role with respect to optical properties of QDs and have considerable impact on the QD quantum yield. In case of uncapped CdSe QDs ($d_{CdSe} = 3.5$ nm) detachment rates $k_{det} = 0.01\ s^{-1}$ and attachment rates $k_{att} = 0.5\ M^{-1}\ s^{-1}$ have been determined for octylamine ligands (OA) at concentrations [OA] = 0.01–0.1 M. The corresponding equilibrium constant $K^L = k_{att}/k_{det}$ for the ligands is about 50 M^{-1}. Compared to that, not only slower [159, 160] but also faster time scales have been reported [31]. According to these results, ligands experience a fast exchange between surface attached ones and those in the solvent. In case of octylamine on CdSe QDs, half of the ligand shell is exchanged within 70 s [30].

In our experiments at extremely low concentration of CdSe/ZnS QDs capped with amine ligands, a decrease of the PL is observed with a decay time of about

1.2 s [74], which is considerably longer than observed for uncapped CdSe QDs [30]. This difference might be due the addition of a ZnS shell which exhibits a stronger ligand binding [106] as is also imposed by our findings by comparing the quenching efficiencies for CdSe and CdSe/ZnS QDs. However, for TOPO ligands, this time may be considerably longer since the binding strength of TOPO is much higher than the one of amines [106]. In any case, the reduction of coverage of the QD surface by removing ligands [30] or upon dilution [34] results in PL quenching. This quenching is either due to the creation of unsaturated bonds causing a lower PL quantum yield or due to an increased attachment of impurities and/or titrated quenchers, respectively. In case that a fast ligand exchange is also present in our experiments, a titration step will be followed by two processes, that is an increased ligand desorption due to dilution and the replacement of ligands (L) by quenchers (M), in our case H_2P.

A fast ligand exchange rate would be consistent with our titration experiments, since PL quenching is observed within less than 60 s after a single titration step. However, in case that ligand exchange is much slower, the immediate PL response after titration points towards the presence of a few "free" attachment sites, which correspond to the volume of about 2–3 TOPO ligands per H_2P. Depending on the binding strength and the number of anchoring functional groups the equilibrium constant $K = k_{\text{att}}/k_{\text{det}}$ will be different for ligands (K^{L}) and quencher molecules (K^{M}). This has been clearly proven in recent experiments [62, 74] which revealed that at a given molar ratio the quenching efficiency depends drastically on the number and kind of (*ortho-*, *meta-*, *para-*) pyridyl substituents of H_2P (see Fig. 4.6). The complexation of (m-Pyr)$_4$$H_2P$ with CdSe/ZnS QD is estimated by the constant $K^{\text{M}} \approx 10^5$–$10^7$ M^{-1} [101], which is by at least three orders of magnitude larger than $K^{\text{L}} = 50$ M^{-1} observed for octylamine ligands [30]. Due to the high value of K^{M}, most of H_2P molecules will in the course of time be attached to the QD surface, at least at reasonable low concentrations. In fact, upon titration at molar ratios $x < 1$, the spectral shifts of the H_2P absorption and fluorescence spectra (see Fig. 4.8) as well as changes of the H_2P emission decay times (see Fig. 4.5b, d) are in accordance with the above considerations.

Since $K^{\text{M}} \gg K^{\text{L}}$, we would expect that upon the presence of a dye molecule at the QD surface the ligand shell is replaced at least proportional to the effective volume of the molecule, at least if the QD surface is assumed to be homogenously covered by ligands. In case of a completely closed ligand shell, for any QD with $d_{\text{CdSe}} = 3.5$ nm, the total average number of TOPO ligands per QD will be close to 120 [94]. Taking into account H_2P and TOPO known volumes (see Fig. 4.7) [62] we expect that H_2P will replace about three TOPO molecules, which would result in a maximum number of about 40 H_2P on the respective QD surface. However, as can be seen from Fig. 4.20a, PL quenching is still increasing up to $x \approx 20$, although the spectral shifts have saturated and the fluorescent decay times stay constant [62, 123] indicating a local replacement of TOPO by H_2P as follows from theoretical considerations [48]. Notably, it is known from a comparison of ensemble and single-molecule experiments on PDI molecules [74, 94] that the number of attached PDI molecules to a QD is much less than that given by the molar ratio x, and we assume that a similar ratio also holds for H_2P. This means that at very high H_2P concentrations and long waiting times the TOPO ligand shell becomes nearly completely replaced by H_2P and the related K^M is much larger than K^L.

Thus, the influence of the ligand exchange dynamics on the QD PL efficiency (quenching) may be outlined as follows. Sample preparation is accompanied by diluting the original QD–toluene solution followed by a new equilibrium of TOPO coverage in case that there is a relatively low concentration of TOPO in the sample. This implies an overall reduction of the QD surface coverage by TOPO ligands resulting in a decrease of the PL quantum yield due to the formation of more unsaturated surface states [30, 56]. Such surface states have even without additional quenchers different PL quantum yields as compared to the near band edge or excitonic states [56]. Additionally, this will allow for further attachment of quenchers, either those which are already inherently present or are added during a titration procedure. The time scale of such processes is in the present experiments typically between 200 and 2000 s (see Figs. 4.23 and 4.24) and depends critically on the experimental conditions such as the type of QD, TOPO, (unknown) impurities and/or dye concentration. However, besides processes on a time scale below 2000 s, a further reorganization of the surfactant shell (accompanied by increased quenching of QD PL) occurs on much longer time scales which can be still observed. It might be related to the formation of new TOPO structures such as surfactant islands on the QD surface or TOPO micelles in the solution. In principle, the configuration of even parts of the ligands on the surface of semiconductor nanocrystals plays a critical role for their stability [161]. Another plausible explanation might be related to an oxidative degradation of the QD surface on these long time scales, especially in the case of amine ligands and PDI molecules [30, 31, 74, 162]. We would like to emphasize that also at these long time scales a still increasing attachment of (m-Pyr)$_4$-H_2P as identified by further increasing FRET has been observed (see Fig. 4.23).

From all these facts and discussions it follows that that non-FRET quenching in "QD–Dye" nanoassemblies might be related to depletion of ligands by the respective dye molecules. Following the argument that non-coordinated Zn^{2+} atoms (and the corresponding surface, intra-band states) decrease the PL intensity according to their absolute number [30] replacement of ligands by dye molecules will increase the number of uncoordinated Zn^{2+} since pyridyl anchors of the given dye molecules can at most saturate two bonds. Since dye molecules replace several ligands due to their large volume the PL quenching will increase. However, the related number of removed ligands necessary to explain the observed non-FRET efficiency was much too large with respect to the volume of the dye molecule. Here, we suggest that this "unusual quenching mechanism" could be related to the crystal structure of the QD which results with respect to the nature of ligand sites in different binding strengths and/or different luminescent properties and/or charge localization features of intra-band states. In case that there are various attachment sites with varying binding strengths it is reasonable to assume that these sites are in a different way sensitive to dye attachment either due to a formation of (dye specific) trap states, which quench the QD PL effectively, or are more easily exchanged by dye molecules. The presence of various in part ligand-related trap states of various nature has recently been suggested by numerical calculations [46, 56–58].

Summarizing all these arguments, non-FRET quenching corresponds to the removal of ligands at specific surface sites accompanied by the formation of intra-band states and possibly the creation of new (PL quenching) trap states. Hence, for tuning the PL properties, it does not need a mayor exchange of the ligand shell, but already very few dye molecules can induce very effectively PL quenching. This conclusion is supported by results [71] showing that the luminescence quantum yields of QDs and the photobrightening are determined primarily by the surface stoichiometry and by the resultant selective surface adsorption of passivants. At this point we conclude from the long-time behavior of PL quenching and its dependence on the amplitude of the electronic wave function at the outer interface, that both processes play an essential role in PL quenching. If ligand depletion in combination with intra-band state formation plays the leading effect for non-FRET quenching it is not surprising that the solvent has an essential impact on PL quenching as will be discussed in more detail in the following section.

4.4.2 Influence of Solvent Composition on FRET and Non-FRET Contributions to PL Quenching

As we have discussed in detail, ligand dynamics and non-FRET processes depend critically on the kind of the solvent. Most of the FRET investigations on QD–Dye nanoassemblies have been performed in toluene. Addition of polar acetone increases considerably the PL quenching [62]. This is understood according to recent calculations, which show that ligand binding occurs via electrostatic forces [106].

On the other hand, quenching is effectively suppressed in non-polar *n*-octane [64]. Figure 4.26 shows the PL quenching of core–shell QDs CdSe/ZnS for the two solvents, toluene and *n*-octane, as a function of the molar ratio *x*. It is clearly seen that PL quenching is only observed in toluene, whereas the QD PL intensity stays more or less constant in *n*-octane indicating that in the latter case the ligand shell is stabilized against the H_2P molecules. This is easily understood since the solubility of TOPO is largely reduced in *n*-octane while that of H_2P is increased as compared to toluene. This will shift the equilibrium constants K^M and K^L towards each other. In that case the ligand shell remains intact preventing H_2P becoming attached effectively.

To show the influence of solvent properties, we analyze the formation "QD–Dye" nanoassemblies based on TOPO-capped CdSe/ZnS QDs and DTPP molecules (dye structure is shown in Fig. 4.3) in TEHOS (structure is presented in Fig. 4.18) since an efficient FRET QD → Dye is as shown in Fig. 4.18 observed for single nanoassemblies in this solvent [140]. Figure 4.27 shows QD PL quenching as a function of waiting time following titration by DTPP at various molar ratios *x*. It is

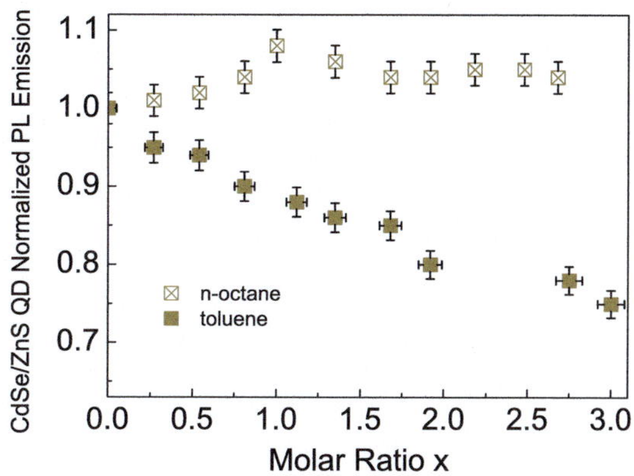

Fig. 4.26 Normalized PL intensities of CdSe/ZnS QDs upon titration of (m-Pyr)$_4$-H_2P as a function of the molar ratio $x = [H_2P]/[QD]$ in *n*-octane (*crossed markers*) and toluene (*solid markers*). Adapted from [64]

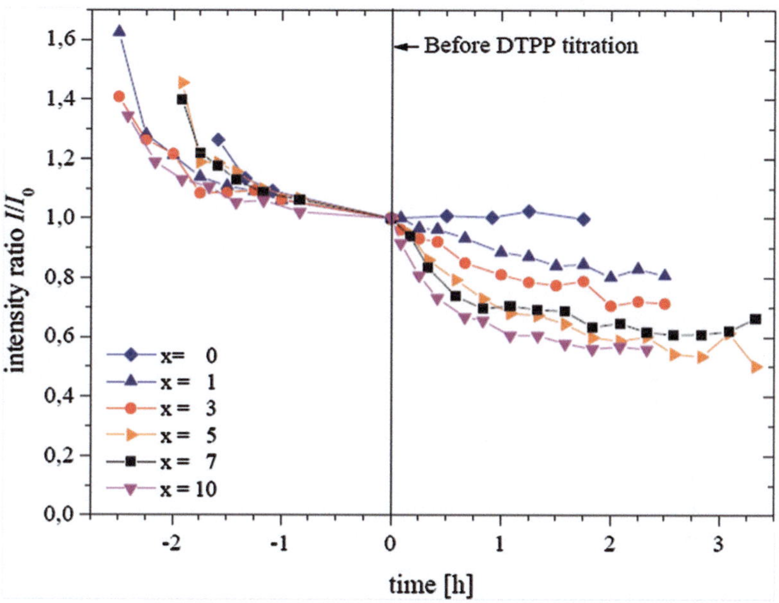

Fig. 4.27 Titration of CdSe/ZnS QDs in TEHOS for various relative DTPP molar ratios x at $t = 0$. The first spectra are taken 1 min after titration. Before $t = 0$ all samples come nearly to an equilibrium with $x = 0$ in each case. Spectra have been normalized at $t = 0$

evident that the quenching immediately after the respective titration step is very small but increases within hours as a function of x. The quenching time becomes shorter with increasing x.

A more detailed view on the time dependence of the PL quenching is depicted for $x = 7$ in Fig. 4.28, which shows a comparable increase of DTPP fluorescence on the same time scale.

Table 4.3 summarizes for a few ratios x the fit results for bi- or mono-exponential functions for the QD PL quenching and DTPP fluorescence enhancement,

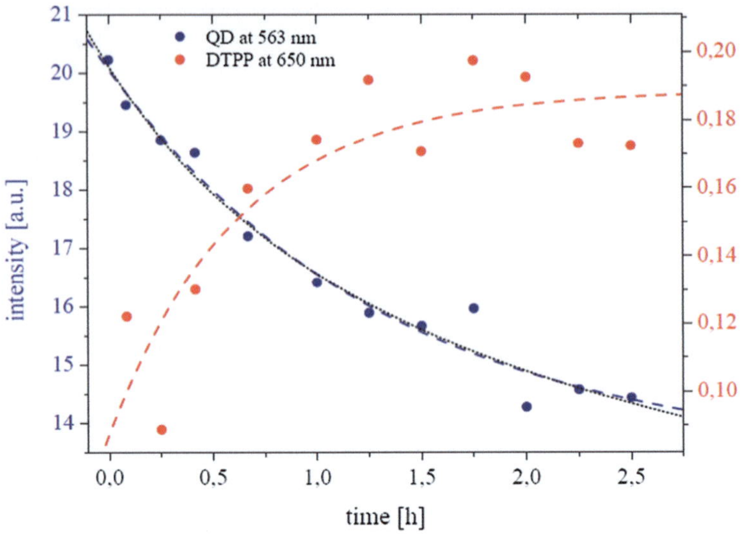

Fig. 4.28 Time-dependent luminescence intensities of CdSe/ZnS QDs and DTPP at $x = 7$. The first DTPP spectra have been taken about 1 min after titration. Lines are bi-exponential fits for CdSe/ZnS and mono-exponential for DTTP

Table 4.3 Comparison of time-dependent luminescence intensities of CdSe/ZnS (QD) and DTPP for variable concentration ratios in TEHOS (A) and in toluene: TEHOS mixture at different volume ratio R (B)

x	QD/DTPP	I_{end} (a.u.)	A_1 (a.u.)	τ_1 (min)	A_2 (a.u.)	τ_2 (min)
(A) Pure TEHOS, $x = [C_{Dye}]/[C_{QD}]$						
1	QD	17 ± 1	5.9 ± 0.9	103 ± 31		
	DTPP	0.13 ± 0.02	0.07 ± 0.01	137 ± 36		
3	QD	11.8 ± 0.9	1.9 ± 0.4	35 ± 10	6.4 ± 0.5	157 ± 43
	DTPP	0.19 ± 0.02	0.10 ± 0.02	38 ± 12		
5	QD	11.2 ± 0.5	10.1 ± 0.5	63 ± 9		
	DTPP	0.46 ± 0.09	0.31 ± 0.08	150 ± 90		
7	QD	$11.5 \pm 0, 2$	30 ± 1.4	62 ± 21	4.4 ± 1.7	24 ± 6
	DTPP	0.28 ± 0.01	0.09 ± 0.01	58 ± 18		
10	QD	14.1 ± 0.6	7.2 ± 4.1	47 ± 28	4.7 ± 1.9	14 ± 9
	DTPP	0.24 ± 0.05	0.04 ± 0.01	54 ± 33		
(B) $x = [C_{Dye}]/[C_{QD}] = 5$, volume ratio $R =$ toluene:TEHOS						
1	QD	17 ± 1	5.9 ± 0.9	103 ± 31		
	DTPP	0.13 ± 0.02	0.07 ± 0.01	137 ± 36		
3	QD	11.8 ± 0.9	1.9 ± 0.4	35 ± 10	6.4 ± 0.5	157 ± 43
	DTPP	0.19 ± 0.02	0.10 ± 0.02	38 ± 12		

respectively, during the observation time. The comparison shows that one of the PL quenching times is very close to the DTPP increase. Therefore we assign this component of the PL quenching to FRET-related quenching, whereas the second component is related to non-FRET quenching.

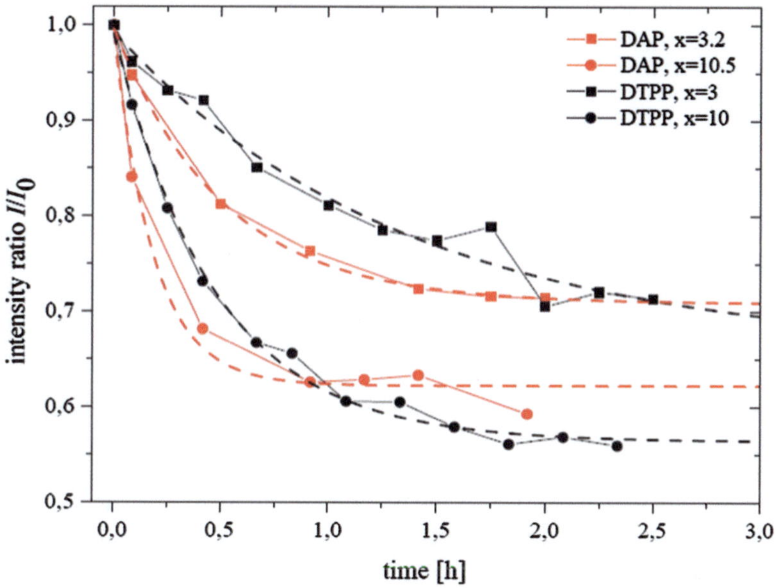

Fig. 4.29 Comparison of time-dependent luminescence intensities for DAP and DTTP each at two ratios x. *Broken lines* correspond to mono-exponential fits

To test whether PL quenching dynamics depend on the type of dye molecule we have performed similar titration experiments with DAP (see the dye structure in Fig. 4.3). The results are shown for two DAP concentrations in Fig. 4.29.

A comparison with data for DTPP (presented also in Fig. 4.29) reveals that the quenching dynamics for DAP are by about a factor of 2–3 faster than those for DTPP, while the final degree of quenching is somewhat larger for DTPP as compared to DAP. We have recently reported [74] that DAP (without any functional pyridyl group) is also able to cause QD PL quenching (see Fig. 4.11) since it intercalates into the ligand shell probably in an upright position of the long molecular axis with respect to the QD surface. On the other hand, DTPP is oriented perpendicular to the previous orientation (see scheme in Fig. 4.12). To realize such an orientation more ligands will have to be replaced in case of DTPP as compared to DAP, which will take a longer time as can be seen in the experiment and as is summarized in Table 4.4. However, FRET will be stronger for DTPP since the Foerster-related FRET efficiency is larger for a parallel as compared to a perpendicular orientation [74, 94, 162] as is also seen in the experiment.

It follows from data presented in Table 4.3 that on average about 42% of the QD PL quenching in "CdSe/ZnS QD–DTPP" nanoassemblies in TEHOS can be assigned to FRET. This is much more as compared to the same kind of nanoassemblies in toluene for which the observed FRET efficiency is less than 10% [74, 162]. It should be noted that FRET experiments in TEHOS on single nanoassemblies of the same composition resulted also in a large FRET efficiency of

Table 4.4 Comparison of time-dependent luminescence intensities for DAP and DTTP (structures are presented in Fig. 4.3) for variable concentration ratios x. I_{end} corresponds to the relative luminescence intensity for long times, τ_q is the time to reach I_{end}

Dye molecule	x	I_{end}	τ_q [min]
DAP	1.6	0.75 ± 0.02	43 ± 8
	3.2	0.71 ± 0.01	30 ± 2
	5.4	0.67 ± 0.02	20 ± 5
	7.5	0.59 ± 0.02	17 ± 3
	10.5	0.62 ± 0.02	11 ± 2
DTPP	1	0.74 ± 0.04	103 ± 31
	3	0.66 ± 0.04	81 ± 22
	5	0.53 ± 0.02	62 ± 9
	7	0.63 ± 0.01	34 ± 5
	10	0.56 ± 0.01	28 ± 2

up to 66% [140]. To investigate this strongly solvent-dependent FRET efficiency in more detail, we have carried out a series of experiments for these nanoassemblies in a toluene/TEHOS mixture at different relative ratios R (Figs. 4.30 and 4.31).

From the comparison of results (Fig. 4.31) obtained after titration with DTPP at $x = 5$–7 and presented in Fig. 4.30 two observations are immediately obvious: (1) the time scales of QD PL quenching and DTPP fluorescence increase are much faster for the mixture of the solvents ($R = 50{:}50$) as for TEHOS samples discussed before and (2) the DTPP fluorescence first increases strongly during about 9 min and thereafter decreases again on a much longer time scale. The letter process is accompanied by a red shift of the DTPP fluorescence band maximum by about 5 nm (see Fig. 4.30). This red shift equals the shift of the fluorescence maximum when measuring DTPP in TEHOS or in toluene. This observation might be explained by a time-dependent separation of the two solvent components in such a way that at the beginning "QD–Dye" nanoassemblies are predominantly solvated in a TEHOS and later on in a more toluene rich environment. This is supported by the finding that FRET efficiency decreases as is expected for a nanoassembly in toluene. For a solvent mixture at $R = 20{:}80$ DTPP fluorescence properties are more closely to those in TEHOS, while they are more close to a dye fluorescence in toluene for $R = 80{:}20$. Correspondingly, we find a strong "overshoot" of luminescence intensity for $R = 20{:}80$ which is completely lost for $R = 80{:}20$.

In titration experiments the excitation wavelength was $\lambda_{ex} = 412$ nm, which corresponds to an absorption minimum for DTPP and a 100 times stronger QD absorption. For this reason directly excited DTPP emission is very low and almost only established via FRET. Figure 4.31 shows a comparison of emission intensities of "QD–DTPP" nanoassemblies at different times after the titration step as a function of the solvent mixture ratio R.

It can be clearly seen from Fig. 4.30 that the DTPP fluorescence is very low both for $R = 100{:}0$ and for $R = 0{:}100$ even after long waiting times. On the contrary, strong DTPP emission at $R = 50{:}50$ is immediately present after titration and further increases in time. Comparing the related time scales of increasing FRET in Fig. 4.31a–c reveals that they are similar as already shown in Table 4.3 for QD PL quenching and DTPP fluorescence increase. For $R = 100{:}0$ (only toluene)

Fig. 4.30 Comparison of time-dependent luminescence spectra of Qd and DTPP as a function of solvent ratio R = toluene: TEHOS. (**a**) $R = 0{:}100$ ($x = 7$); (**b**) $R = 50{:}50$ ($x = 5$); (**c**) $R = 100{:}0$ ($x = 5$). In titration experiments we used $\lambda_{ex} = 412$ nm, which corresponds to an absorption minimum for DTPP and a 100 times stronger QD absorption. For this reason direct excitation of DTPP molecules is effectively low but might be realized via FRET

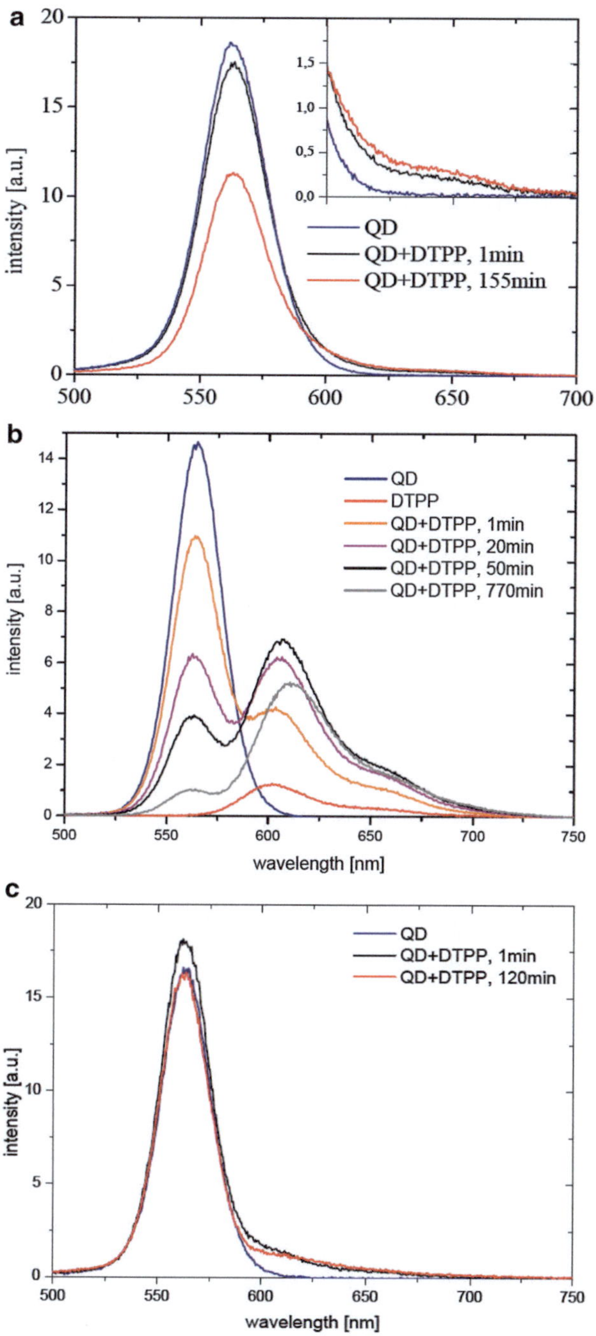

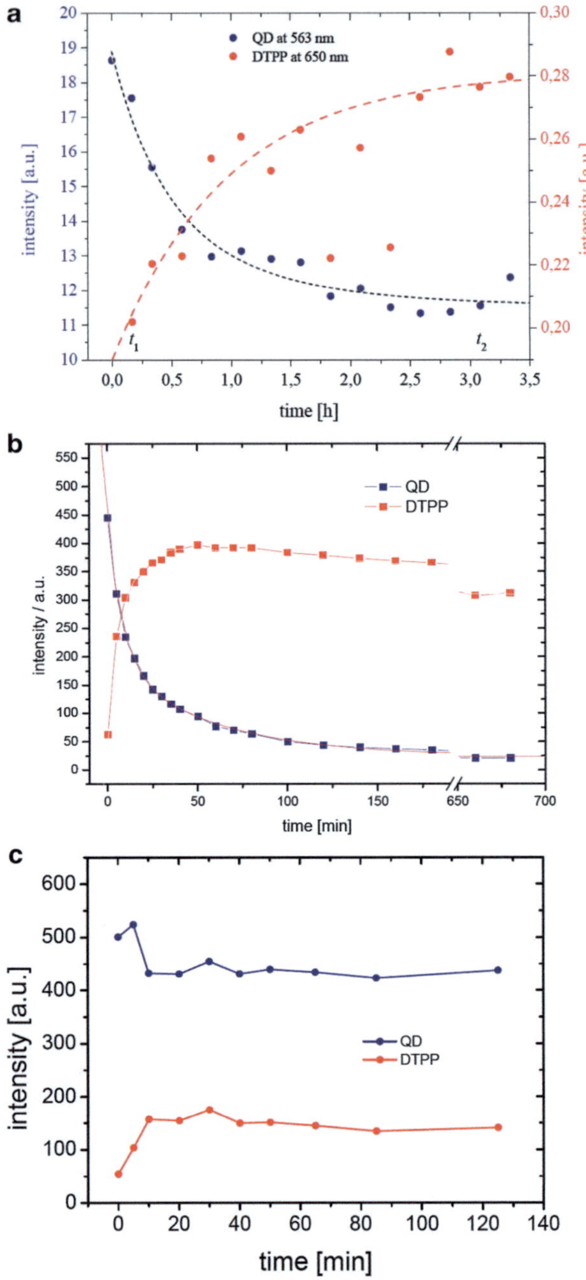

Fig. 4.31 Comparison of time-dependent luminescence intensities of QDs and DTTP as a function of observation time for 3 different solvent ratios R = Toluene:TEHOS. (**a**) $R = 0{:}100$ ($x = 7$); (**b**) $R = 50{:}50$ ($x = 5$); (**c**) $R = 100{:}0$ ($x = 5$). The first spectra have been taken about 1 min after titration, $\lambda_{ex} = 412$ nm

the PL quenching (and DTTP enhancement) main time component is shorter than 1 min, the time needed to register the spectra. On the other hand, the maximum of PL quenching for $R = 0{:}100$ (only TEHOS) is about 0.5 which is very similar as for $R = 100{:}0$, but with the difference that the related FRET is very weak. This can be interpreted as following. At $R = 0{:}100$ it takes a long time to form "QD–Dye" nanoassemblies, but FRET is very effective. At the other end of the relative concentration range ($R = 100{:}0$) nanoassemblies are much more readily formed, but non-FRET processes are more effective than FRET as has also been reported earlier [74, 162].

A qualitative model can be set up following recent arguments [64, 163] (discussed in Sect. 4.4.1), namely that non-FRET quenching can be assigned to a competition of dye-induced ligand detachment with subsequent assembly formation, which gives rise both to FRET and to formation of specific surface states (related to remaining dangling bonds). These surface states give rise to non-FRET quenching due to specific sites, probably localizing charges and thus quenching the PL. Decreasing of polarity will destabilize charge localization [106]. On the other hand, (TOPO) ligand detachment is less favorable in non-polar media, since the polar ligands have a low solubility. For this reason there is not enough space to attach large dye molecules to the QD surface and formation of assemblies might take a very long time. That is why we observe FRET only after a long waiting time but non-FRET quenching is relatively weak. For $R = 100{:}1$ we observe an effective assembly formation within less than 1 min, since ligands are readily detached from the surface. At the same time non-FRET effective surface states are created. Obviously the optimum for effective FRET on a short time scale is close to $R = 50{:}50$. Even at $R = 70{:}30$ most of the PL quenching results in FRET. This observation points to the possibility that in mixed solvents not merely the number of detached ligands has to be considered but also the specific structure of the solvent shell enclosing the QD. Since TEHOS has long aliphatic chains structure presented in Fig. 4.18, these chains probably intercalate with the aliphatic TOPO chains forming a stable TEHOS shell intercalated with the ligands. This shell is only broken up at sufficiently high toluene concentrations $>70\%$. Only then the TEHOS shell is nearly absent and toluene may easily penetrate to the QD surface and stabilizes the non-FRET quenching states.

These experiments in solvent mixtures demonstrate that the kinetics of assembly formation is complex but nevertheless provide detailed information not only on dynamic processes but also on the incorporation of QDs or nanoassemblies into complex and heterogeneous self-organizing environments. Complementing information has been reported via temperature-dependent experiments on similar assemblies in organic glasses [75, 164]. Though most of the properties are inherent to the QDs themselves, dye molecules act as reporters or even initiate reorganization processes especially of QD surface states or within the ligand shell. Such well-defined experiments provide insights, why colloidal QD chemistry is often critically dependent on small amounts of "impurities" during wet-chemical synthesis.

4.5 Influence of Single Dye Molecules on Temperature and Time-Dependent Optical Properties of QDs in Nanoassemblies

In previous sections we presented a microscopic description of QD PL quenching upon formation of nanoassemblies at ambient temperature which takes place in competition with capping ligand (surfactant) dynamics and is accompanied by the reorganization of the capping ligand shell. As we mentioned above PL quantum yield and PL dynamics of QDs and "QD–Dye" nanoassemblies depend drastically on the dielectric properties of the embedding environment [62, 64, 67, 106, 140]. In addition, the properties of these systems are sensitive to changes in temperature [2, 31, 78–80, 82, 83]. As a part of the detailed analysis of self-assembly processes in "QD–Dye" nanoassemblies, this section is devoted to the analysis of the influence of variations of the surface structure such as a ligand-induced phase transitions at low temperatures as well as photodegradiation of QD optical properties and how they are related to nanoassembly formation. Our results are related to temperature (77–300 K)-dependent optical properties of "QD–porphyrin" nanoassemblies as well as to long-term temporal evolution of PL spectra (photodegradation) of QDs assembled with PDI molecules. In case of "QD–porphyrin" nanoassemblies we will report on the temperature-dependent formation of CdSe/ZnS trap states in competition to QD near band edge states, while the temporal variation of trap and near band edge states will be presented for "QD–PDI" nanoassemblies.

4.5.1 Ligand-Related Phase Transitions in QD–Dye Nanoassemblies

Experiments on alone of CdSe/ZnS QDs ($d_{CdSe} = 3.0$ nm, $n_{ZnS} = 2$) and "QD–porphyrin" nanoassemblies with participation of (m-Pyr)$_4$-H_2P molecules have been performed in a methylcyclohexane/toluene (6:1) mixture in a temperature range from 77 to 300 K. These experiments have shown [65, 75] that upon temperature lowering, the band-edge absorption (maximum of the first excitonic transition) for alone QDs is blue shifting which is typical for these QDs [165]. The temperature dependence of the PL peak energy behaves similar as the band-edge absorption, that is a blue shift upon temperature lowering is observed. The analysis of QD PL spectra dependence on the excitation wavelength at 295 K reveals that in addition to size-dependent site selection (observed for QD PL [118]) the presence of at least two luminescent states is characteristic to the QDs under study. Upon temperature lowering PL spectra are not only shifted to the blue but also the line width (FWHM) becomes essentially narrower, while the PL intensity becomes higher. Based on considerations presented recently [75] it may be concluded the QD PL quantum efficiency increases upon temperature lowering and PL emission

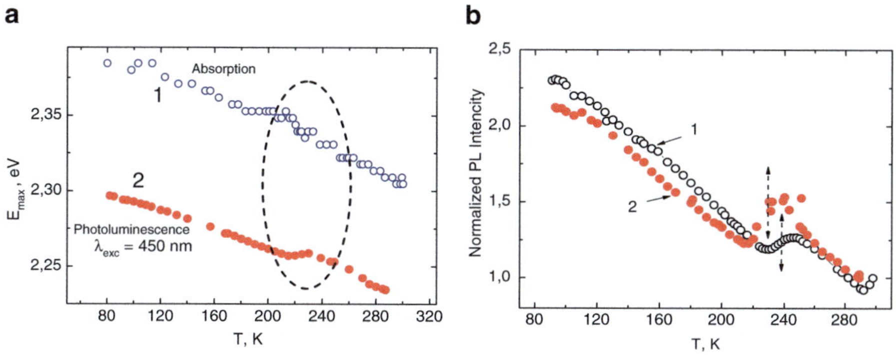

Fig. 4.32 Temperature dependence of absorption and PL properties for alone QDs and "QD–Dye" nanocomposites in methylcyclohexane/toluene (6:1) mixture. (**a**) Temperature dependence of the first absorption peak (*1*, mean weighted data for two Gaussians lines) and PL maximum (*2*, $\lambda_{exc} = 450$ nm) for TOPO capped CdSe/ZnS QDs ($d_{CdSe} = 3.0$ nm, $n_{ZnS} = 2$). The *dashed ellipse* indicates the temperature range at which optical properties change abruptly. (**b**) Temperature dependence of the normalized PL maximum intensity ($\lambda_{exc} = 450$ nm) for CdSe/ZnS-AM QDs ($d_{CdSe} = 3.0$ nm, 2 ZnS monolayers) capped by long-chain amines (*1*) and for nanoassemblies based on CdSe/ZnS-AM QDs and DPP molecules at molar ratio $x = [C_{DPP}]/[C_{QD}] = 1$ (*2*). *Thick dashed arrows* on the curves 1 and 2 indicate temperatures at which optical properties change abruptly. Adapted from [75]

takes place presumably from only lower lying states belonging to surface traps presumably.

Fitting the lowest absorption and PL band positions by Gaussian components shows that besides the above-mentioned blue shift upon temperature lowering a non-monotonous temperature dependence of the respective peak energies becomes evident (Fig. 4.32).

It is seen in Fig. 4.32a that a non-monotonous behavior (which we name for simplicity "kink") is observed for TOPO-capped CdSe/ZnS QDs between 200 and 240 K, far from the glass transition temperatures for methylcyclohexane (146.7 K) and toluene (180 K), respectively, as well as for a methylcyclohexane/toluene (6:1) mixture (151.6 K) [166]. Upon lowering T the "kink" occurs at slightly different temperatures for the first excitonic absorption band (~218 K) and the PL band (~223 K), respectively. Interestingly, that a similar "kink" is observed in the temperature dependence of the PL intensity for AM-capped CdSe/ZnS QDs (Fig. 4.32b, curve 1). In the latter case, the "kink" in PL intensity is more pronounced and is shifted to ~237 K as compared to ~223 K for TOPO-capped CdSe/ZnS QDs. Notably, this non-monotonous behavior becomes stronger for amine-capped CdSe/ZnS QDs with attached DPP molecules at molar ratio $x = [C_{DPP}]/[C_{QD}] = 1$, while no "kink" is observed when following the luminescence of perylene diimide or porphyrin molecules under the same conditions (will be presented and discussed below). From these observations, we therefore exclude that the morphology of the solvent matrix is changing.

Typically, the PL blue shift and intensity rise upon temperature lowering is a commonly observed phenomenon for QDs in colloidal suspensions or in solvent-free systems (polymeric matrices or QD solids). Generally it is explained by reduction of thermally activated carrier trapping [167, 168] and/or a decrease of electron–phonon coupling [10, 80, 165, 168, 169]. Unusual "luminescence anti-quenching" at a well-defined temperature of $T \sim 250$ K (manifested in our experiments as "kink") has been observed for CdTe or CdSe/ZnS QDs in toluene, and has been related to a phase transition in the surfactant capping layer [80]. This effect was connected with surface relaxation and/or surface reconstruction which was strongly dependent on the type of capping ligands (TOPO or aminoethanethiol) and, in turn, may cause a spatially-energetic reordering of trap states. One should note that the existence of the "kink" has up to now only been reported with respect to the PL properties of QDs while we detected it also in absorption spectra at temperatures close to ~218 K [75].

When discussing the appearance of the "kink," one has to take into account that for TOPO- or amine-capped CdSe/ZnS QDs, a phase transition of the capping ligand layer at low T may cause some strain-induced deformation of the ZnS shell. Correspondingly, due to different thermal expansion coefficients for the semi-rigid capping layer and the ZnS shell as well as possible strain effects, slight changes of electron delocalization [148, 171] may take place in the range of the phase transition. In this respect, we believe that slightly different "kink" temperatures for the first excitonic absorption band (~218 K) and the PL band (~223 K) for TOPO-capped CdSe/ZnS QDs may be explained by the different nature of the corresponding transitions in absorption (allowed excitonic ones [10–12]) and in emission (trap dominated ones at low T).

Thus, if trap state emission dominates at low temperatures, these trap states or their distribution might be influenced by dye molecules attached to the QD surface due to replacing some of the surfactant (TOPO or AM) molecules. Figure 4.33 shows the results of the corresponding temperature experiments for "QD–porphyrin" nanoassemblies based on TOPO-capped CdSe/ZnS QD with fixed size and various porphyrin molecules at a relative molar ratio of $x = [C_{\text{porphyrin}}]/[C_{\text{QD}}] = 1$.

It is seen from Fig. 4.33a that in the case of (m-Pyr)$_4$-CuP being attached to QD, upon temperature lowering a long wavelength emission band at $\lambda_{\max} = 770$–780 nm appears. This emission belongs to the phosphorescence of (m-Pyr)$_4$-CuP [172 and references herein]. The intensity of this phosphorescence is continuously increasing upon temperature lowering (Fig. 4.33a, inset) which is typical for Cu-containing porphyrins [172, 173]. Thus, (m-Pyr)$_4$-CuP molecules may be considered as an inner standard showing that the surrounding solvent matrix does not change upon Temperature dependent PL intensities temperature lowering. Under the same conditions, the PL intensity of QDs in "QD–CuP" nanoassemblies shows again a "kink" (Fig. 4.33b, curve 3), which, however, is now much more pronounced (−23%) than that for alone QDs (Fig. 4.33b, curve 1). From Fig. 4.33b it is evident that dye attachment enhances the PL decrease at the phase transition temperature, and the amplification is strongest for (m-Pyr)$_4$-CuP molecules. We suggest that

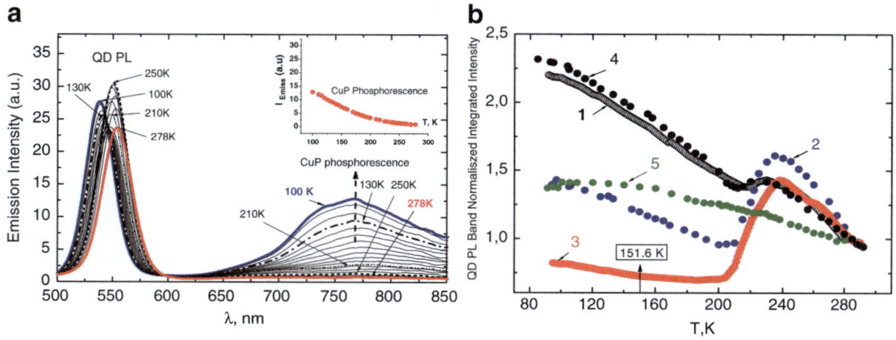

Fig. 4.33 Temperature dependence of emission spectra (**a**) and PL intensities (**b**) for nanocomposites based on TOPO-capped CdSe/ZnS QDs ($d_{CdSe} = 3.0$ nm, $n_{ZnS} = 2$) and various porphyrin molecules at the molar ratio $x = [C_{porphyrin}]/[C_{QD}] = 1$ in a methylcyclohexane/toluene (6:1) mixture ($\lambda_{exc} = 450$ nm). (**a**) The short wavelength band belongs to QD PL and the with temperature strongly increasing long wavelength band belongs to (m-Pyr)$_4$-CuP phosphorescence. The *inset* shows the temperature dependence for CuP phosphorescence bands between 77 and 285 K. (**b**) are shown for the same type of CdSe/ZnS QD ($d_{CdSe} = 3.0$ nm, 2 ZnS monolayers) but various attached porphyrins: (*1*) alone QDs; (*2*) QD + (m-Pyr)$_4$-H_2P; (*3*) QD + (m-Pyr)$_4$-CuP; (*4*) QD+tetraphenylporphyrin; (*5*) QD in dried PMMA film on a quartz plate. The structure of (m-Pyr)$_4$-H_2P molecule is shown in Fig. 4.2. In a (m-Pyr)$_4$-CuP molecule two central protons are replaced by the Cu ion, in tetraphenylporphyrin molecule four *meso*-pyridyl rings are replaced by four *meso*-phenyl rings. The phase transition temperature of the solvent mixture is indicated by an *arrow*. All curves are normalzed to the integrated band intensity of QD PL at 285 K. Adapted from [75]

CuP exhibits the largest effect, since either internal molecular charge transfer states influence the energy distribution of CdSe/ZnS QD trap states considerably or TOPO molecules may be strongly coordinated at low temperatures with the central Cu ion [174, 175], which might increase the disorder of the surfactant capping shell. In contrast, tetraphenylporphyrin without *meso*-pyridyl anchoring groups does not enlarge the QD PL decrease at the "kink" (Fig. 4.33b, curve 4) because of the absence of self-assembly in this case. The "kink" is absent for QDs in a dried rigid PMMA film on a quartz plate (Fig. 4.33b, curve 5). In the latter case, the film rigidity suppresses the reorganization of the TOPO capping layer [80], thus weakening a spatially-energeticreordering of QD trap states.

Figure 4.34 shows band-edge PL decays at various detection wavelengths within the QD PL band for TOPO-capped CdSe/ZnS QDs ($d_{CdSe} = 3.0$ nm, $n_{ZnS} = 2$) in comparison with those for "QD + (m-Pyr)$_4$-CuP" nanoassemblies ($x = 1$) at 293 and 77 K. At 293 K PL decay for alone QDs is multi-exponential at various detection wavelengths, and measured $\langle \tau \rangle$ values do not change much upon variation of detection wavelength (Fig. 4.34a, curve 1). At 77 K the QD PL decay is non-exponential in the short wavelength region, but at the PL maximum and in the long wavelength region the PL decay is mono-exponential exclusively. In the latter case, the mean value $\langle \tau \rangle$ is monotonically rising upon increasing the wavelength of detection. It is evidently seen also that in the long wavelength region the $\langle \tau \rangle$ values at

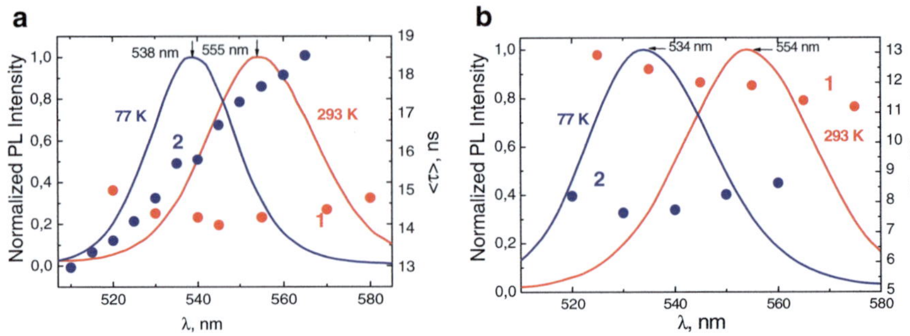

Fig. 4.34 Dependence of the photoluminescence mean decay $\langle\tau\rangle$ on registration wavelength (λ_{reg}) for TOPO-capped CdSe/ZnS QD ($d_{CdSe} = 3.0$ nm, $n_{ZnS} = 2$) at 293 K (*1*) and 77 K (*2*) for QDs (**a**) and "QD + (m-Pyr)$_4$-CuP" nanoassemblies (at molar ratio $x = 1$) (**b**) in methylcyclohexane/toluene (6:1) mixture. QD PL non-exponential decays $\langle\tau\rangle$ have been fitted by $I(t) = \sum A_i \times \exp(-t/\tau_i)$ upon excitation at $\lambda_{exc} = 410$ nm, and the mean lifetime values have been calculated as $\langle\tau\rangle = \sum(A_i \times \tau_i^2)/\sum(A_i \times \tau_i)$

77 K are noticeably longer relative to those measured at 293 K. In addition, at 295 K QD instantaneous spectra are the same practically at various delay times, while at 77 K the red shift of these spectra is observed upon increase of a registration delay time. Taken together, all these PL decay findings for alone QDs in a temperature range of 300–77 K cannot be described by a thermal equilibrium between dark and bright excitonic states as far as such effects manifest themselves at $T < \sim 50$ K [80, 176]. The realistic mechanisms of exciton relaxation in this case may be dominated by trapping dynamics, since the trap depths vary across the QD ensemble and also fluctuate in time for individual QDs [132].

In the case of "QD + (m-Pyr)$_4$-CuP" nanoassemblies, QD PL decays are multi-exponential throughout the total PL band both at room and low temperature (Fig. 4.34b). $\langle\tau\rangle$ values change only slightly upon variation of the detection wavelength. At 293 K, the shortening from $\langle\tau\rangle \approx 14.5$ ns for alone QDs down to $\langle\tau\rangle \approx 11.9$ ns for nanoassemblies is a direct evidence of QD PL quenching in nanoassemblies due to both non-FRET and FRET processes. At 77 K, PL spectra of QDs in nanoassemblies are blue shifted with respect to those for individual QD. This may be considered as an additional support of the influence of attached CuP on the QD ligand-induced phase transition. Data presented in Fig. 4.34 show also that in contrast to alone QDs, the temperature lowering down to 77 K leads to a strong shortening of mean $\langle\tau\rangle$ values even for the PL long wavelength region (from $\sim$18 ns down to $\sim$8 ns). The interpretation of these results is also connected with the TOPO capping layer phase transition perturbed by attached porphyrin molecules. The existence of an additional PL quencher (on average one attached CuP molecule) may influence indirectly the relative position of excitonic and trap states of the QD itself upon temperature lowering as well as from the competitive non-radiative channels connected with (m-Pyr)$_4$-CuP molecules.

To consider the differences in the temperature dependence between QDs and QDs in nanoassemblies, one has to take into account that nanoassemblies at an initial molar ratio $x = [C_{CuP}]/[C_{QD}] = 1$ were prepared in one step at ambient temperature. At this temperature, the formation of such self-aggregated assemblies is thermodynamically characterized by the corresponding complexation constant [101, 150]. Estimations according to a Poisson distribution show that at $x = 1$, the related probabilities P_i of the number of dye molecules per QD are $P_0 = 0.3$, $P_1 = 0.4$, $P_2 = 0.3$, respectively. This implies that some QDs are "porphyrin-free" without PL quenching at ambient temperature. Decrease of temperature might result in a complete complexation of QDs because of the rise of the complexation constant at low temperatures [177]. Correspondingly, this should show up in more effective QD PL quenching. However, this is experimentally not reflected in an increase of PL intensity up to the phase transition. We exclude that at the phase transition more nanoassemblies are formed, since dye molecules become "frozen" in the now rigid surfactant layer.

Summing up this part we conclude that CdSe/ZnS QDs undergo a phase transition in the sense that the "freezing" ligand shell exerts strain on the ZnS shell, thus creating trap states with reduced PL quantum yields. Moreover, this phase transition changes also the CdSe/ZnS absorption, which might be explained by a modification of the CdSe core or ZnS layer structure. Notably, it has been shown recently [84] that the solvent environment and ligands help not only in limiting the size of the QDs in solution but also to stabilize the crystal structure of QDs by minimizing the surface free energy. In solution, the solvent molecules adsorbed onto the QD surface can reduce the total surface free energy, and the surface can undergo reconstruction when the solvent molecules leave the QD surface. In turn, dye attachment creates new and/or more trap states, which obviously quench the PL very effectively. Phase transition temperature and the influence of dye-induced trap states depend critically on the type of ligand and are more pronounced for TOPO as compared to amine ligands. This might be explained by the pyramidal structure of TOPO, which prohibits for steric reasons a complete passivation of the QD surface [64, 178]. Thus, it follows from the above results and considerations that already a few attached dye molecules noticeably influence the optical propertiesof QDs at ligand controlled phase transitions.

4.5.2 Photostability of QD–Dye Nanoassemblies

As we have shown in Sects. 4.2.2.1, 4.2.2.2, and 4.3.1 QD PL quenching caused by both non-FRET and FRET is clearly visible in single "QD–Dye" nanoassembly experiments. Here, we highlight that single functionalized dye molecules can be considered not only as extremely sensitive probes for the complex interface time-dependent physics and dynamics of colloidal semiconductor QDs but also partly control reactions at surfaces.

Before discussing the influence of PDI molecules on the PL for AM-capped CdSe/ZnS QDs in nanoassemblies, we first describe the bulk PL behavior for an ensemble of alone AM-capped CdSe/ZnS QDs in toluene (Fig. 4.35). Similar

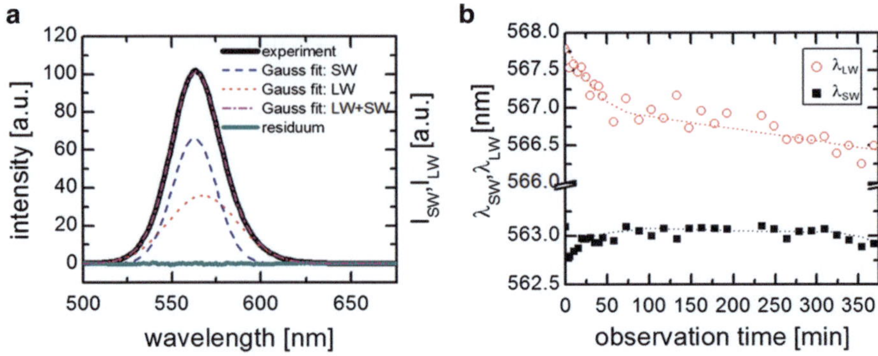

Fig. 4.35 (**a**) PL of AM-capped CdSe/ZnS QDs in toluene at ambient temperature. The PL band has been fitted by two Gaussian components SW and LW shown by *broken lines*. (**b**) Variation of PL emission wavelength for SW and LW as function of observation time. Adapted from [75]

to QDs with TOPO ligands the PL spectrum has been fitted with two Gaussian components, namely "short" (SW) and "long" (LW) wavelength components, respectively (Fig. 4.35a). After sample preparation both the PL intensity and band shape/width change as a function of waiting time. The intensities decrease by about a factor 4 within 10 min and stay nearly constant thereafter. During the total time of several hours the integrated intensity ratio remains constant within experimental error. However, the effective PL wavelengths of the two Gaussian bands change differently as can be seen in Fig. 4.35b. While the LW band shifts systematically to the blue on a time scale of several hours, the SW band remains nearly constant. We tentatively assign the SW band to the near-band edge PL (excitonic by nature), while we assign the LW band to a broad distribution of trap states. Obviously the center position of the trap state distribution is shifted to the blue during waiting time. Specifically, this implies that traps with the lowest PL energy are quenched or the traps are eliminated more effectively than the excitonic-type states.

While for alone QDs the spectral shifts for SW and LW components are extremely small (1–2 nm) during observation time, they are enlarged following titration by PDI-type molecules PP (see the dye structure in Fig. 4.3). It is seen from Fig. 4.36 that for λ_{SW} the shift is at most 4 nm but increases up to 12 nm for λ_{LW}. In both cases the shift becomes larger with increasing PP concentration. At molar ratio $x = 5$ the experimental error becomes large since QD PL is already considerably quenched. The time scale of the respective shift becomes systematically shorter with increasing x. These effects reflect the local influence of attached dye molecules on excitonic and trap states and can be understood in view of recent findings of assembly formation as a function of concentration [64, 74]. These observations indicate that assembly formation goes in line with increased bleaching (oxidation) of the long-wavelength component LW. Basically, also titration experiments show that trap states emitting in the red are effectively bleached during observation time [74].

As was described in Sect. 4.2.2.2 the detection of the spectral properties of single QDs and "QD–Dye" nanoassemblies shows also blue shifts of the PL bands during observation time (see time traces of the PL intensity in the bottom of Fig. 4.13a,

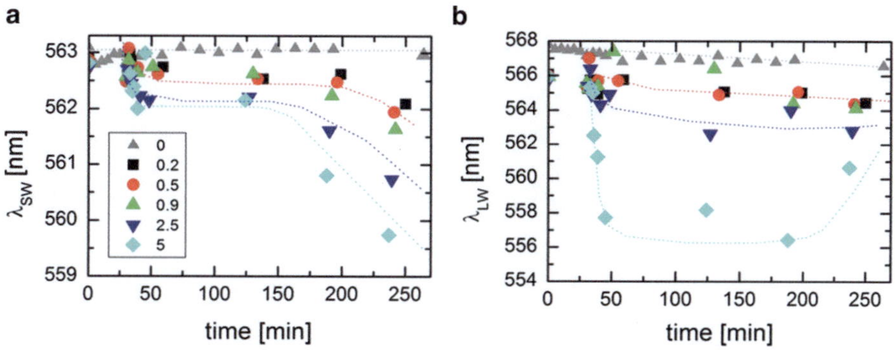

Fig. 4.36 PL maximum positions of the SW (**a**) and LW (**b**) components for AM-capped CdSe/ZnS QDs as a function of observation time and molar ratio $x = [C_{PP}]/[C_{QD}]$ (x: 0–5). *Lines* are only guide lines. Adapted from [75]

b). Like in ensembles experiments this shift is enhanced by even one attached PP molecule [75, 94] and may be caused by the photooxidation of the QD core material [129].

The overall conclusion with respect to "QD–Dye" nanoassemblies is that there is a noticeable difference in the photophysical behavior of QD PL upon assembly formation. In addition, our ensemble experiments can be more convincingly interpreted in a way that predominantly one part of the PL (presumably related to trap states) is effectively quenched. Since this quenching is enhanced following nanoassembly formation, we assign the quenched trap states to surface-related states, since they are expected to react most effectively with attached dye molecules. Single assembly experiments support this conclusion, but additionally show that also photooxidation of the CdSe core takes place since the PL is shifting to the blue by up to 20 nm, which cannot only be explained by modification of surface states. This different behavior with respect to ensemble experiments might be related to the fact that during spin coating the surfactant layer is partly (or even completely) lost and the QD surface can be freely accessed by oxygen, which is often involved in photodegradation of QDs. The enhancement of the photodegradation process for QD by attached dye molecule may be tentatively assigned to two reasons. (1) Upon attachment of dye molecules a large number of ligands are removed, which results in a more unprotected and thus more reactive surface, which allows for a stronger photodegradation either of surface states or even of the QD core. (2) Dye-induced trap states of QDs are more sensitive to photobleaching resulting in apparent spectral shifts and reduced integrated PL. This latter assumption is supported by experiments on "QD–Porphyrin" nanoassemblies discussed earlier in this chapter.

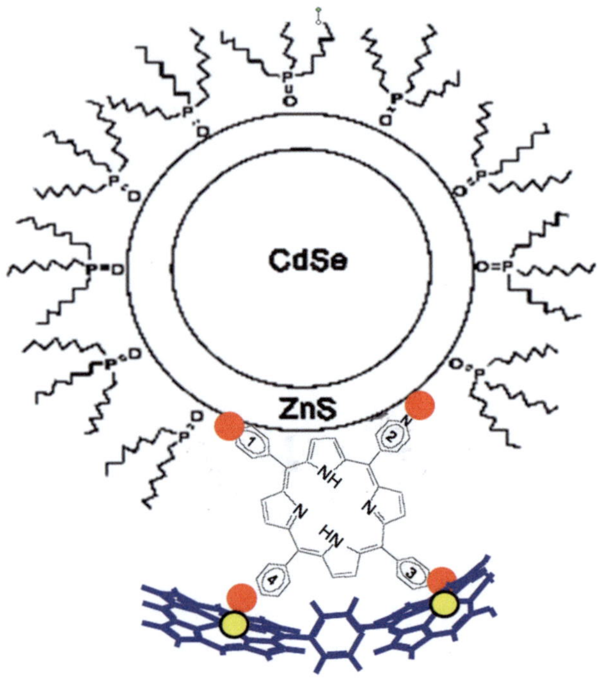

Scheme 4.1 Schematic presentation of a complex nanoassembly with twofold coordination of fourfold functionalized *meso*-pyridyl substituted porphyrin, (m-Pyr)$_4$-H_2P, to the QD and a Zn-porphyrin chemical dimer

4.6 Conclusions

4.6.1 Bottom-Up Approach for "QD–Dye" Nanoassemblies

We highlight that "bottom-up" approach based on self-assembly principles defines a strategy of the formation of organic–inorganic nanoassemblies containing colloidal semiconductor QDs of different types and various heterocyclic molecules with and without functionalized anchoring side substituents (like *meso*-pyridyl substituted porphyrins and perylene diimides in our case). On the basis of a combination of ensemble and single molecule spectroscopy of nanoassemblies, we have shown that single functionalized molecules can be considered as extremely sensitive probes for studying the complex interface physics and exciton relaxation processes in QDs. We have shown also that this approach is the base for the development and study of even more complex assembles containing QDs and porphyrin triads [150] such as in Scheme 4.1.

It has been quantitatively laid out that the major part of the observed QD PL quenching in nanoassemblies, namely non-FRET can be understood, on the one

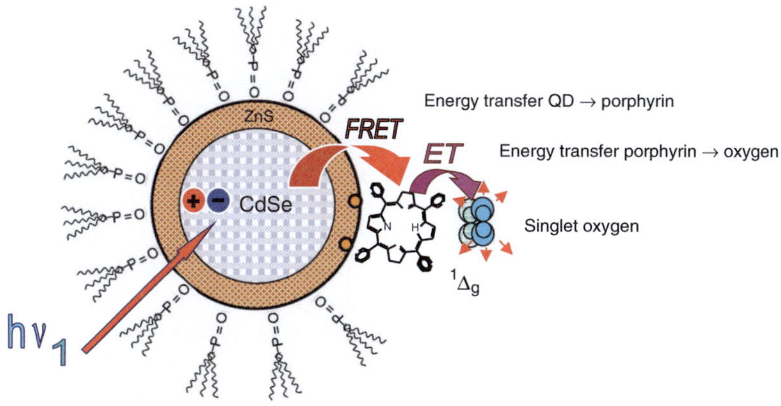

Scheme 4.2 Schematic presentation of a complex nanoassembly with a twofold energy transfer processes from the QD to H_2P (QD → H_2P) and to oxygen ($H_2P \rightarrow O_2$) followed by generation of singlet oxygen

hand, in terms of electron tunneling across the ZnS shell even under the condition of quantum confinement, and, on the other hand, by the influence of ligand dynamic. FRET is—depending on the environment—often only a small part of the PL quenching. The competition between FRET and non-FRET quenching processes drastically depends on the solvent and ligand properties. "FRET quenching" in "QD–Dye" nanoassemblies is effectively suppressed already in slightly polar solvents which is often overlooked in literature data.

In addition based on the quantitative analysis of FRET efficiency in "QD–porphyrin" nanoassemblies as well as on the direct comparative measurements of near-IR emission of singlet oxygen $^1O_2$ it has been proven that namely FRET process QD → porphyrin is a reason of singlet oxygen generation by these nanocomposites in solutions at ambient temperature [114] (Scheme 4.2).

4.6.2 Non-FRET Quenching in QD–Dye Nanoassemblies

Photoinduced FRET and/or charge transfer (CT) has been reported in literature for many QD–Dye assemblies. An unambiguous identification of such processes, however, can only be identified by optical means if the luminescence of both constituents is followed quantitatively. The observation of PL quenching and PL lifetime shortening is not sufficient to identify FRET or CT. If the dynamics observed for both constituents are not in agreement with each other, but PL quenching is observed, non-FRET quenching has to be postulated [62–65, 74, 75]. How can such a process be modeled? In case that dangling bonds at the surface of a QD are not saturated by appropriate ligands new electron–hole-related states are generated in the band gap. Similar trap states might be formed due to surface or interface defects. In case that these states are subject to non-radiative deactivation channels or decreased radiative rates, the PL will be reduced. The magnitude of the quenching will depend on (1) the (strongly distance-dependent) amplitude of the excitonic wave function at the trap site, (2) the number of trap sites, (3) the microscopic nature of the site, which might be additionally influenced by the environment (solvent), and (4) the crystal facets related to the crystal structure. Any change of the conditions (1)–(4) will result in a change of PL properties. In this approach non-FRET quenching of QD PL induced by attaching dye molecules may be caused by (1) a reduced number (dependent on the facette properties) of ligands and thus an increase of intra-band states (traps), since the dye is generally larger than the ligand molecules and (2) a change in the electronic properties at the attachment site since the "anchor" of the dye is different from the ones of the (replaced) ligands. It follows from (2) that the electronic influence of the dye itself is less important as compared to the one of the anchor groups. The general features are shown in Scheme 4.3 with respect to the electronic properties of the QD.

Summariing, non-FRET quenching of QD PL induced by dye attachment (or incorporation in the ligand shell) can be explained by the presence of a limited number of empty or easily accessible sites on the QD surface followed by ligand competitive exchange (TOPO, amines, and dyes) on various time scales. Due to the larger volume of dyes as compared to ligands this results in the creation of more and/or different intra-band states than in the presence of ligands only. We would like to emphasize that also at very long time scales a still increasing attachment of, e.g. $(m\text{-Pyr})_4\text{-}H_2P$ molecules as identified by further increasing FRET is observed. Our present findings of inhomogeneous surface dynamics (of ligands) for semiconductor QDs are also a prerequisite for the establishment of the non-FRET mechanisms caused by dye molecules. We argue that non-FRET quenching is related to depletion of capping ligands by the respective dye molecules followed up by creating more and even new quenching states according to the amplitude of the tunneling ("leaking") excitonic wave function.

Temperature variation and related changes in QD absorption and emission reveal drastic changes of the ligand shell structure in a narrow temperature range for organic (TOPO and amine) ligands (phase transition). The effects on QD PL at this transition become considerably pronounced upon attachment of only a few porphyrin (or perylene diimide) molecules to the QD surface. We conclude from temperature experiments that the ligand phase transition has impact even on the QD core structure and exciton–phonon coupling. This investigation elaborates the importance of (switchable) surface states for the characterization of the PL of QDs. In this respect it should be mentioned that the environment may even induce a phase transition of the CdSe core. Recently, in situ microdiffraction data have shown that the CdSe QDs capped with trioctylphosphine oxide (TOPO) or hexadecylamine (HDA) in toluene show predominantly wurtzite crystal structure, which undergoes a phase transition to a zinc blende crystal structure upon drop-casting onto Si. Furthermore, decreasing the size of the CdSe quantum dots enhances this phase transition [84]. These effects may also lead to a redistribution of trap states which may change the QD PL efficiency.

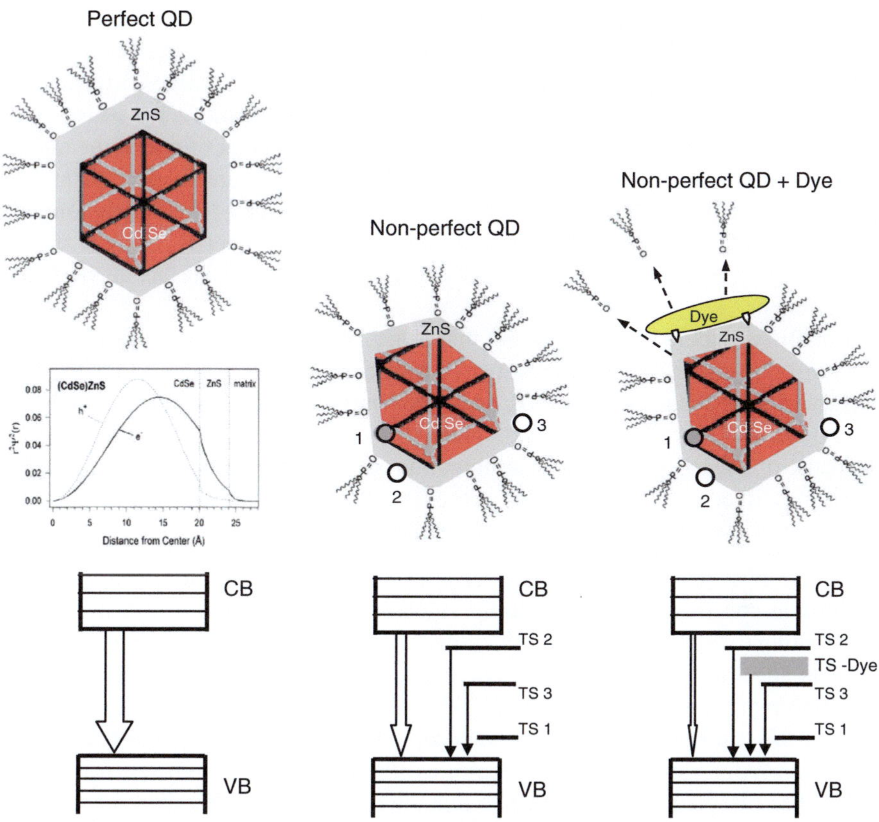

Scheme 4.3 Schematic presentation of a perfect TOPO-capped CdSe/ZnS QD, a non-perfect QD and a "QD–Dye" nanoassembly as well as arbitrary excitonic and trap states (TS) and main relaxation pathways. For perfect CdSe/ZnS QDs, calculated probability density functions $\Psi^2(r)r^2$ for the electron e^- and hole h^+ are also presented. For the nanoassembly an induced ligand detachment is shown by *broken arrows*. *1* is a hole trap, *2* and *3* are electron traps

An alternative but qualitatively similar approach of dye-induced non-FRET can be related to a model recently proposed by Salman et al. [178]. In this paper the authors suggest that for almost all QDs observed non-exponential PL decay might be related to the presence of a permanent electric dipole moment in the electronic ground state. Such a dipole moment will mix the upper two allowed states within the electronic fine structure of the lowest excited QD state. They argue that such a mixing will result in an increase of non-radiative transitions. It is very evident that the surface attachment of a dye molecule will increase or at least change the dipole moment and thus the mixing of excitonic states followed by an increase of non-radiative transitions and will thus give rise to PL quenching. Clearly, the magnitude of such an electric dipole moment will strongly depend on dielectric properties of the solvent as we have observed experimentally. Some of our experimental findings support the above discussed model [178], namely that we observe (see Fig. 4.34) a variation of the PL decay rates across the PL emission wavelength but without a corresponding built-up component for PL (of, e.g., traps) with a time corresponding to the PL decay time of energetically higher states [118]. More PL wavelength-related experiments will be needed to definitely decide in favor of one of the models or a combination of them. Nevertheless, both approaches are in agreement with most of our findings related to various aspects of non-FRET quenching not only for QD–Dye nanoassemblies but in general with related PL properties of QDs such as blinking, non-exponential PL decay, influence of temperature, surfactant shell, and environment.

Finally we would like to point out that properties of "QD–Dye" nanoassemblies are not only interesting in themselves but also provide a valuable tool to study surface-related phenomena in QDs on an extremely low level of surface modification. Instead of investigating ligand dynamics directly, e.g. by NMR, we suggest to make use of QD PL quenching in combination with appropriately functionalized dye molecules. In contrast to NMR, this approach allows to investigate ligand dynamics at extremely low modifications of surface properties.

Note added in proof: In addition, recently, based on the combination of X-ray absorption spectroscopy and ab initio calculations it has been demonstrated that overcoating CdSe QDs with ligands of distinct end group functionalities manifests itself in specific changes in the unoccupied electronic structure, particularly near the bottom of the conduction band [179]. It should be noted that the ligands being used in this study primarily affect unoccupied electronic states >3 eV above the absorption edge rather than in the gap, as one might have predicted for surface states. The calculated spectra indicate that the differences in absorbance associated with changes in surface geometry and chemistry arise primarily due to variations in the unoccupied surface states with *d* character.

Acknowledgments This work was supported by Volkswagen Foundation (Priority Program "Physics, Chemistry and Biology with Single Molecules"), DFG GRK 829/3 ("Accumulation of single molecules to nanostructures"), German Academic Exchange Service (DAAD, grant № A/08/08573, EZ), Belarussian Foundation for Basic Research (grant № Ф10CO-005), Belarussian State Program for Scientific Research "Convergence 3.2.08—Photophysics of Bioconjugates, Semiconductor and Metallic Nanostructures and Supramolecular Complexes and Their Biomedical Applications."

Prof. E. Zenkevich thanks B.I. Stepanov Institute of Physics NAS, Minsk, Belarus and the DFG FOR877 ("From local constraints to macroscopic transport") for financial support.

The first experiments in 2002 have only been possible by providing CdSe/ZnS QDs by Dr. A. Rogach (Ludwig-Maximilians-University Munich, Germany) and Dr. D. Talapin (University of Hamburg, Germany) which are gratefully acknowledged. We also thank Dr. A. Shulga (B.I. Stepanov Institute of Physics NAS, Minsk, Belarus) for the synthesis of all tetrapyrrole compounds including monomers and chemical dimers and Prof. F. Wuerthner (Wuerzburg University) for supplying all perylene diimide dyes.

We thank

Dr. Habil. E. Sagun, Dr. A. Stupak, Dr. V. Knyukshto, Dr. V. Galievsky, A. Stasheuski, and A. Yarovoi (B.I. Stepanov Institute of Physics NAS, Minsk, Belarus), who have performed many experiments in Minsk or in Chemnitz,

Dr. T. Blaudeck (now at ENAS, Chemnitz), Dr. D. Kowerko (now at University of Zürich), S. Krause, Prof. M. Abdel-Mottaleb (now at Nile University, Centre for Nanotechnology, Cairo), K. Szwaykowska (now at California Institute of Technology), for performing mutual experiments, theoretical analysis, and fruitful discussions in Chemnitz.

Titration and FRET experiments in TEHOS have been performed by F. Gerlach (now at Fibotec oberoptics GmbH, Meiningen, Germany).

Dr. D. Kilin (now at the University of South Dakota, USA) has performed helpful calculations on QD–Dye nanoassemblies.

We thank Prof. Dr. S.V. Gaponenko (B.I. Stepanov Institute of Physics, National Academy of Sciences, Minsk, Belarus) for fruitful discussion. We also thank Prof. F. Cichos (now at Molecular Nanophotonics, University of Leipzig, Germany) for stimulating and guiding many of the experiments.

References

1. Moyano DF, Goldsmith M, Solfiell DJ, Landesman-Milo D, Miranda OR, Peer D, Rotello VM (2012) J Am Chem Soc 134:3965
2. De Mello Donega C (2011) Chem Soc Rev 40:1512
3. Mocatta D, Cohen G, Schattner J, Millo O, Rabani E, Banin U (2011) Science 332:77
4. Liang G-X, Li L-L, Liu H-Y, Zhang J-R, Burda C, Zhu J-J (2010) Chem Commun 46:2974
5. Cheng H-M (2011) Chem Commun 47:6763
6. Mansoori GA (2005) Principles of Nanotechnology. Molecular-Based Study of Condensed Matter in Small Systems. University of Illinois at Chicago, Chicago
7. Nicolini C.: Nanobiotechnology and Nanobiosciences. Pan Stanford Series on Nanobiotechnology, vol. 1 (2009) PANStanford Publishing, Singapore
8. Whitesides GM, Grzybowski B (2002) Science 295:2418
9. Lehn J-M (1990) Angew Chem Int Ed Engl 29:1304
10. Klimov V (2000) In: Nalwa HS (ed) Handbook of Nanostructured Materials and Nanotechnology, vol 4. Academic Press, San Diego, Orlando, pp 451–527
11. Woggon U (2006) Optical Properties of Semiconductor Quantum Dots. Springer, Berlin
12. Rogach AL (ed) (2008) Semiconductor Nanocrystal Quantum Dots: Synthesis, Assembly, Spectroscopy and Applications. Springer, Wien
13. Gaponenko SV (2010) Introduction to Nanophotonics. Cambridge University Press, Cambridge
14. Talapin DV, Lee J-S, Kovalenko MV, Shevchenko EV (2010) Chem Rev 110:389
15. Coe-Sullivan S, Woo W-K, Steckel JS, Bawendi M, Bulovic V (2003) Electronics 4:123
16. Wang M, Moon S-J, Xu M, Chittibabu K, Wang P, Cevey-Ha N-L, Humphry-Baker R, Zakeeruddin SM, Graetzel M (2010) Small 6:319
17. Medintz IL, Stewart MH, Trammell SA, Susumi K, Delahanty JB, Mey BC, Melinger JS, Blanco-Canosa JB, Dawson FE, Mattoussi H (2010) Nat Mater 9:676
18. Frasco MF, Chaniotakis N (2009) Sensors 9:7266
19. Hetsch F, Xu X, Wang H, Kershaw SV, Rogach AL (2011) J Phys Chem Lett 2:1879
20. McDowell M, Wright AE, Hammer NI (2010) Materials 3:614
21. Schmitt F-J, Maksimov EG, Suedmeyer H, Jeyasangar V, Theiss C, Paschenko VZ, Eichler HJ, Renger G (2011) Photonics Nanostruct: Fundam Appl 9:190
22. Rakovich A, Savateeva D, Rakovich T, Donegan JF, Rakovich YP, Kelly V, Lesnyak V, Eychmueller A (2010) Nanoscale Res Lett 5:753
23. Knowles KE, Frederick MT, Tice DB, Morris-Cohen AJ, Weiss EA (2012) J Phys Chem Lett 3:18
24. Amelia M, Credi A (2012) Inorg Chim Acta 381:247

25. Michler P (ed) (2009) Single Semiconductor Quantum Dots. Springer, Berlin
26. Efros AL, Rosen M (2000) Annu Rev Mater Sci 30:475
27. Reiss P, Protiere M, Li L (2009) Small 5:154
28. Morris-Cohen AJ, Donakowski MD, Knowles KE, Weiss EA (2010) J Phys Chem C 114:897
29. Evans CM, Evans ME, Krauss TD (2010) J Am Chem Soc 132:10973
30. Ji X, Copenhaver D, Sichmeller C, Peng X (2008) J Am Chem Soc 130:5726
31. Koole R, Schapotschnikow P, De Mello Donega C, Vlugt TJH, Meijerink A (2008) ACS Nano 2:1703
32. Kaluzhny G, Murray RW (2005) J Phys Chem B 109:7012
33. Munro AM, Ginger DS (2008) Nano Lett 8:2585
34. Munro AM, Jen-La Plante I, Ng MS, Ginger DS (2007) J Phys Chem C 111:6220
35. Querner C, Reiss P, Bleuse J, Pron A (2004) J Am Chem Soc 126:11574
36. Dayal S, Lou Y, Samia ACS, Berlin JC, Kenney ME, Burda C (2006) J Am Chem Soc 128:13974
37. Lim TC, Bailey VJ, Ho Y-P, Wang T-H (2008) Nanotechnology 19:75701
38. Ren T, Mandal PK, Erker W, Liu Z, Avlasevich Y, Puhl L, Müllen K, Basché T (2008) J Am Chem Soc 130:17242
39. Bullen C, Mulvaney P (2006) Langmuir 22:3007
40. Knowles KE, Tice DB, McArthur EA, Solomon GC, Weiss EA (2010) J Am Chem Soc 132:1041
41. Jin R (2008) Angew Chem Int Ed 47:6750
42. Gur I, Fromer NA, Chen C-P, Kanaras AG, Alivisatos AP (2007) Nano Lett 7:402
43. Mansur HS, Mansur AAP (2011) Mater Chem Phys 125:709
44. von Holt B, Kudera S, Weiss A, Schrader TE, Manna L, Parak WJ, Braun M (2008) J Mater Chem 18:2728
45. Schmelz O, Mews A, Basché T, Herrmann A, Müllen K (2001) Langmuir 17:2861
46. Williard DM, Carillo LL, Jung J, van Orden A (2001) Nano Lett 1:469
47. Schmitt FJ (2010) J Opt 12:084008
48. Voznyy O (2011) J Phys Chem C 115:15927
49. Smith AM, Duan HW, Rhyner MN, Ruan G, Nie SM (2006) Phys Chem Chem Phys 8:3895
50. Qu LH, Peng XG (2002) J Am Chem Soc 124:2049
51. Xie RG, Kolb U, Li JB, Basche T, Mews A (2005) J Am Chem Soc 127:7480
52. Talapin DV, Rogach AL, Kornowski A, Haase M, Weller H (2001) Nano Lett 1:207
53. Kalyuzhny G, Murray R (2005) J Phys Chem B 109:7012
54. Underwood DF, Kippeny T, Rosenthal SJ (2001) J Phys Chem B 105:436
55. Lee WZ, Shu GW, Wang JS, Shen JL, Lin CA, Chang WH, Ruaan RC, Chou WC, Lu CH, Lee YC (2005) Nanotechnology 16:1517
56. Frenzel J, Joswig J-O, Seifert G (2007) J Phys Chem C 111:10761
57. Kilina S, Ivanov S, Tretiak S (2009) J Am Chem Soc 131:7717
58. Albert V, Ivanov S, Tretiak S, Kilina S (2011) J Phys Chem C 115:15793
59. Anni M, Manna L, Cigolani R, Valerini D, Creti A, Lomascolo M (2004) Appl Phys Lett 85:4169
60. Clapp R, Medintz IL, Mauro JM, Fisher BR, Bawendi MG, Mattoussi H (2004) J Am Chem Soc 126:301
61. Potapova I, Mruk R, Hübner C, Zentel R, Basché T, Mews A (2005) Angew Chem 117:2490
62. Zenkevich E, Cichos F, Shulga A, Petrov E, Blaudeck T, von Borczyskowski C (2005) J Phys Chem B 109:8679
63. Blaudeck T, Zenkevich E, Cichos F, von Borczyskowski C (2008) J Phys Chem C 112:20251
64. Blaudeck T, Zenkevich E, Abdel-Mottaleb M, Szwaykowska K, Kowerko D, Cichos F, von Borczyskowski C (2012) Chemphyschem 13:959
65. Zenkevich EI, Blaudeck T, Kowerko D, Stupak AP, Cichos F, von Borczyskowski C (2012) Macroheterocycles 5:98
66. Califano M, Franceschetti A, Zunger A (2005) Nano Lett 5:2360
67. Issac A, von Borczyskowski C, Cichos F (2005) Phys Rev B 71:161302

68. Klimov V (2000) In: Nalwa HS (ed) Handbook of Nanostructured Materials and Nanotechnology, Chapter 7, vol 4. Academic Press, San Diego, Orlando, pp 451–527
69. Hohng S, Ha T (2004) J Am Chem Soc 126:1324
70. Fomenko V, Nesbitt DJ (2008) Nano Lett 8:287
71. Gomez DE, van Embden J, Jasieniak J, Smith TA, Mulvaney P (2006) Small 2:204
72. Hammer NI, Early KT, Sill K, Odoi MY, Emrick T, Barnes MD (2006) J Phys Chem B 110:14167
73. Odoi MY, Hammer NI, Early KT, McCarthy KD, Tangirala R, Emrick T, Barnes MD (2007) Nano Lett 7:2769
74. Kowerko D, Schuster J, Amecke N, Abdel-Mottaleb M, Dobrawa R, Wuerthner F, von Borczyskowski C (2010) Phys Chem Chem Phys 12:4112
75. Zenkevich EI, Stupak AP, Kowerko D, von Borczyskowski C (2012) Chem Phys 406:21
76. Krebs FC, Senkovskyy V, Kiriy A (2010) IEEE J Sel Top Quantum Electron 16:1821
77. Marjanovic N, Hammerschmidt J, Perelaer J, Farnsworth S, Rawson I, Kus M, Yenel E, Tilki S, Schubert US, Baumann RR (2011) J Mater Chem 21:13634
78. Crooker SA, Barrick T, Hollinthworth JA, Klimov VI (2003) Appl Phys Lett 82:2793
79. Labeau O, Tamarat P, Lounis B (2003) Phys Rev Lett 90:257404
80. De Mello Donega C, Meijerink A (2006) Phys Rev B 74:085320
81. van Driel AF, Allan G, Delerue C, Lodahl P, Vos WL, Vanmaekelbergh D (2005) Phys Rev Lett 95:236804
82. Wuister SF, De Mello Donega C, Bode M, Meijerink A (2004) J Am Chem Soc 126:10397
83. Wuister SF, De Mello Donega C, Bode M, Meijerink A (2004) J Phys Chem B 108:17393
84. Karakoti AS, Sanghavi S, Nachimuthu P, Yang P, Thevuthasan S (2011) J Phys Chem Lett 2:2925
85. Issac A, Jin S, Lian T (2008) J Am Chem Soc 130:11280
86. Ko HC, Yuan CT, Lin SH, Jau T (2011) J Phys Chem C 115:13977
87. Cui S-C, Tachikawa T, Fujitsuka M, Majima T (2011) J Phys Chem C 115:01824
88. Koposov AY, Szymanski P, Cardolaccia T, Meyer TJ, Klimov VI, Sykora M (2011) Adv Funct Mater 21:3159
89. Clapp R, Medintz IL, Mattoussi H (2006) ChemPhysChem 7:47
90. Zenkevich EI, Sagun EI, Yarovoi AA, Shulga AM, Knyukshto VN, Stupak AP, von Borczyskowski C (2007) Opt Spectrosc 103:998
91. Clapp AR, Medintz IL, Fisher BR, Anderson GP, Mattoussi H (2005) J Am Chem Soc 127:1242
92. Halpert JE, Tischler JR, Nair G, Walker BJ, Liu W, Bulovic V, Bawendi MG (2009) J Phys Chem C 113:9986
93. Lee J, Kim H-J, Cheng T, Lee K, Kim K-S, Glotzer SC, Kim J, Kotov NA (2009) J Phys Chem C 113:109
94. Kowerko D, Krause S, Amecke N, Abdel-Mottaleb M, Schuster J, von Borczyskowski C (2009) Int J Mol Sci 10:5239
95. D'Souza S, Antunes E, Litwinski C, Nyokong T (2011) J Photochem Photobiol A Chem 220:011
96. Burda C, Green TC, Link S, El-Sayed MA (1999) J Phys Chem B 103:1783
97. Tsay JM, Trzoss M, Shi L, Kong X, Selke M, Jung ME, Weiss S (2007) J Am Chem Soc 129:6865
98. Rakshit S, Vasudevan S (2009) J Phys Chem C 113:16424
99. Jhonsi MA, Rengnathan R (2010) J Colloid Interface Sci 344:596
100. Pons T, Medintz IL, Wang X, English DS, Mattoussi H (2006) J Am Chem Soc 128:15324
101. Zenkevich E, Blaudeck T, Shulga A, Cichos F, von Borczyskowski C (2007) J Lumin 122–123:784
102. Lim TC, Bailey VJ, Ho Y-P, Wang T-H (2008) Nanotechnology 19:075701
103. Orlova AO, Gubanova MS, Maslov VG, Vinogradova GN, Baranov AV, Fedorov AV, Gounko L (2010) Optika i Spectroscopiya 108:975 (in Russian)

104. Guyot-Sionnest P, Shim M, Matranga C, Hines M (1999) Phys Rev B Condens Matter Mater Phys 60:R2181
105. Klimov VI, Mikhailovsky AA, McBranch DW, Leatherdale CA, Bawendi MG (2000) Phys Rev B 61:13349
106. Schapotschnikow P, Hommersom B, Vlugt TJH (2009) J Phys Chem C 113:12690
107. Chernook AV, Shulga AM, Zenkevich EI, Rempel U, von Borczyskowski C (1996) J Phys Chem 100:1918
108. Chernook AV, Rempel U, von Borczyskowski C, Zenkevich EI, Shulga AM (1996) Chem Phys Lett 254:229
109. Dobrawa, R., Würthner, F.: Chem. Commun. 1878 (2002)
110. Lang E, Würthner F, Köhler J (2005) ChemPhysChem 6:935
111. Zenkevich EI, von Borczyskowski C, Shulga AM, Bachilo SM, Rempel U, Willert A (2002) Chem Phys 275:185
112. Sagun EI, Zenkevich EI, Knyukshto VN, Shulga AM, Starukhin DA, von Borczyskowski C (2002) Chem Phys 275:211
113. Zenkevich EI, von Borczyskowski C (2012) Photoinduced relaxation processes self-assembled nanostructures: multiporphyrin complexes and composites "CdSe/ZnS quantum dot-porphyrin". In: Kim D (ed) Multiporphyrin Arrays: Fundamentals and Applications, Chapter 5. Pan Stanford Publishing Pte. Ltd., Singapore, pp 217–288
114. Zenkevich EI, Sagun EI, Knyukshto VN, Stasheuski AS, Galievsky VA, Stupak AP, Blaudeck T, von Borczyskowski C (2011) J Phys Chem C 115:21535
115. Kilin DS, Tsemekhman K, Prezhdo OV, Zenkevich EI, von Borczyskowski C (2007) J Photochem Photobiol A Chem 190:342
116. Kapitonov AM, Stupak AP, Gaponenko SV, Petrov EP, Rogach AL, Eychmueller A (1999) J Phys Chem B 103:10109
117. Javier A, Magana D, Jennings T, Strouse JF (2003) Appl Phys Lett 83:1423
118. Petrov EP, Cichos F, von Borczyskowski C (2006) J Lumin 119–120:412
119. Pons T, Medintz IL, Sykora M, Mattoussi H (2006) Phys Rev B 73:245302
120. Petrov EP, Cichos F, Zenkevich E, Starukhin D, von Borczyskowski C (2005) Chem Phys Lett 402:233
121. Mattoussi H, Mauro JM, Goldman ER, Anderson JP, Sundar VC, Mikulec FV, Bawendi MG (2000) J Am Chem Soc 122:12142
122. Gouterman M (1961) J Mol Spectrosc 6:138
123. Blaudeck, T.: Self-assembly of functionalized porphyrin molecules on semiconductor nanocrystal surfaces. Dissertation, University of Technology, Chemnitz (2007)
124. Cichos F, von Borczyskowski C, Orrit M (2007) Curr Opin Colloid Interface Sci 12:272
125. Tachiya MJ (1982) Chem Phys Lett 76:340
126. Song N, Zhu H, Jin S, Zhan W, Lian T (2011) ACS Nano 5:613
127. Zenkevich EI, Blaudeck T, Shulga AM, Cichos F, von Borczyskowski C (2007) J Lumin 122–123:784–788
128. Valeur B (2002) Molecular Fluorescence: Principles and Applications. Wiley-VCH, New York
129. van Sark WGJHM, Frederix PLTM, Bol AA, Gerritsen HC, Meijerink A (2002) ChemPhysChem 3:871
130. Kowerko D, Schuster J, von Borczyskowski C (2009) Mol Phys 107:1911
131. Schlegel G, Bohnenberger J, Potapova I, Mews A (2002) Phys Rev Lett 88:137401
132. Fisher BR, Eisler H-J, Stott NE, Bawendi MG (2004) J Phys Chem B 108:143
133. Zhang K, Chang H, Fu A, Alivisatos AP, Yang H (2006) Nano Lett 6:843
134. Fron E, Pilot R, Schweitzer G, Qu J, Herrmann A, Muellen K, Hofkens J, der Auweraer MV, Schryver FCD (2008) Photochem Photobiol Sci 7:597
135. Sirota M, Minkin E, Lifshitz E, Hensel V, Labav M (2001) J Phys Chem B 105:6792
136. Basko D, La Rocca JC, Bassani F, Agranovich VM (1999) Eur Phys J B 8:353
137. Förster T (1965) Delocalized excitation and excitation transfer. In: Sinanoglu O (ed) Modern Quantum Chemistry. Academic, New York, p 93

138. Zenkevich EI, Shulga AM, Chernook AV, Gurinovich GP (1986) J Appl Spectrosc 45:984
139. Lakowicz J (2006) Principles of Fluorescence Spectroscopy. Springer, New York
140. Gerlach F, Täuber D, von Borczyskowski C (2013) Chem Phys Lett 572:90
141. Krause S, Kowerko D, Börner R, Hübner CG, von Borczyskowski C (2011) ChemPhysChem 12:303
142. Reiss P, Carayon S, Bleuse J, Pron A (2003) Synth Met 139:649
143. Kimura J, Uematsu T, Maenosono S, Yamaguchi Y (2004) J Phys Chem B 108:13258–13264
144. Potapova I, Mruk F, Prehl S, Zentel R, Basché T, Mews A (2003) J Am Chem Soc 125:320
145. Querner C, Reiss P, Sadki S, Zagorska M, Pron A (2005) Phys Chem Chem Phys 7:3204
146. Willard DM, Mutschler T, Jung M, Yu J, van Orden A (2006) Anal Bioanal Chem 384:564
147. Leatherdale CA, Bawendi MG (2001) Phys Rev B 63:165315
148. Dabbousi BO, Rodriguez-Viejo J, Mikulec FV, Heine JR, Mattoussi H, Ober R, Jensen KF, Bawendi MG (1997) J Phys Chem B 101:9463
149. Haus JW, Zhou HS, Homma I, Komiyama H (1993) Phys Rev B 47:1359
150. Zenkevich EI, von Borczyskowski C (2009) Macroheterocycles 2:206
151. Nirmal M, Dabbousi BO, Bawendi MG, Macklin JJ, Trautman JK, Harris TD, Brus LE (1996) Nature 383:802
152. Wolf HC, Port H (1976) J Lumin 12–13:33
153. Rempel JY, Trout BL, Bawendi MG, Jensen KF (2005) J Phys Chem B 109:19320
154. Rempel JY, Trout BL, Bawendi MG (2006) J Phys Chem B 110:18007
155. Schuster R, Barth M, Gruber A, Cichos F (2005) Chem Phys Lett 413:280
156. Landes C, Burda C, Braun M, El-Sayed MA (2001) J Phys Chem B 105:2981
157. Mews Z (2007) Phys Chem 221:295
158. Trotzky S, Kolny-Olesiak J, Falke SM, Hoyer T, Lienau C, Tuszynski W, Parisi J (2008) J Phys D Appl Phys 41:102004
159. Pradhan N, Reifsnyder D, Xie R, Aldana J, Peng X (2007) J Am Chem Soc 129:9500
160. Ning Z, Molnár M, Chen Y, Friberg P, Gan L, Ågren H, Fu Y (2011) Phys Chem Chem Phys 13:5848
161. Yu WW, Yang YA, Peng X (2003) Chem Mater 15:4300
162. Kowerko, D.: Dynamic processes in functionalised perylene bisimide molecules, semiconductor nanocrystals and assemblies. Dissertation, University of Technology, Chemnitz (2010)
163. Zenkevich, E.I., Blaudeck, T., Milekhin, A., von Borczyskowski, C.: Int. J. Spectrosc. (2012). doi:10.1155/2012/971791
164. Zenkevich EI, Stupak AP, Kowerko D, von Borczyskowski C (2012) Theoreticheskaya i Experimentalnaya Khimiya 48:18 (in Russian)
165. Liptay TJ, Ram RJ (2006) Appl Phys Lett 89:223132
166. Holzhauer JK, Ziegler WT (1975) J Phys Chem 79:590
167. Dayal S, Burda C (2007) J Am Chem Soc 129:7977
168. Morello G, De Giorgi M, Kudera S, Manna L, Cingolani R, Anni M (2007) J Phys Chem C 111:5846
169. Liptay TJ, Marshall LF, Rao PS, Ram RJ, Bawendi MG (2007) Phys Rev B 76:155314
170. Fernee MJ, Littleton BN, Cooper S, Rubinsztein-Dunlop H, Gomez DE, Mulvaney PJ (2008) Phys Chem C 112:1878
171. Peng X, Schlamp MC, Kadavanich AV, Alivisatos AP (1997) J Am Chem Soc 19:7019
172. Zenkevich EI, von Borczyskowski C (2002) Multiporphyrin self-assembled arrays in solutions and films: thermodynamics, spectroscopy and photochemistry. In: Tripathy SK, Kumar J, Nalwa HS (eds) Handbook of Polyelectrolytes and Their Applications, vol 2. Amer. Sci. Publ, Valencia, pp 301–348
173. Kim D, Holten D, Gouterman M (1984) J Am Chem Soc 106:2793
174. Kruglik SG, Ermolenkov VV, Shvedko AG, Orlovich VA, Galievsky VA, Chirvony VS, Otto C, Turpin PY (1997) Chem Phys Lett 270:293
175. Asano M, Kaizu Y, Kobayashi H (1988) J Chem Phys 89:6567
176. Klimov V (2007) Annu Rev Phys Chem 58:635

177. Meloun M, Havel J, Högfeldt E (1988) Computation of Solution Equilibria. Ellis Horwood, Chichester
178. Al Salman A, Tortschanoff A, van der Zwan G, van Mourik F, Chergui M (2009) Chem Phys 357:96–101
179. Lee JRI, Whitley HD, Meulenberg RW, Wolcott A, Zhang JZ, Prendergast D, Lovingood DD, Strouse GF, Ogitsu T, Schwegler E, Terminello LJ, van Buuren T (2012) Nano Lett 12:2763

Chapter 5
Size-dependent Electronic and Polarization Properties of Multi-Layer InAs Quantum Dot Molecules

Muhammad Usman

Abstract In this chapter, we analyze the polarization response of multi-layer quantum dot molecules (QDMs) containing up to nine vertically stacked quantum dot layers by carrying out a systematic set of multi-million atom simulations. The atomistic modeling and simulations allow us to include correct symmetry properties in the calculations of the electronic and optical spectra: a factor critical to explain the experimental evidence. The values of the degree of polarization (DOP) calculated from our model based on the geometry parameters directly extracted from the experimental TEM images follow the trends of the recently published experimental data. We also present detailed physical insight of the fundamental underlying physics by examining strain profiles, band edges diagrams, and wave function plots. Multi-directional calculations of the DOP reveal a unique property of the InAs QDMs that the TE response in the plane perpendicular to the growth direction is highly anisotropic. Therefore we propose that a single value of the DOP is not sufficient to fully characterize the polarization response. We explain this anisotropy of the TE modes in terms of the orientation of the hole wave functions that align along the $[\bar{1}10]$ direction. Our results provide a new insight that the isotropic polarization response measured in the experimental PL spectra is due to two factors: (i) $TM_{[001]}$ mode increases due to enhanced intermixing of HH and LH bands and (ii) $TE_{[110]}$ mode reduces significantly due to the hole wave function alignments along the $[\bar{1}10]$ direction. This is in contrast to general notion that only an increase in the $TM_{[001]}$ mode is responsible for the isotropic polarization. We also present polarization response as a function of various geometry configurations of the quantum dot layers to provide a guide to experimentalists for the design of optical devices based on multi-layer QDMs.

M. Usman (✉)
Tyndall National Institute, Lee Maltings, Dyke Parade, Cork, Ireland
e-mail: usman@alumni.purdue.edu

J. Wu and Z.M. Wang (eds.), *Quantum Dot Molecules*, Lecture Notes in Nanoscale Science and Technology 14, DOI 10.1007/978-1-4614-8130-0_5,

5.1 Introduction

Deployment of semiconductor quantum dots (QDs) in the active region of optical devices offers unique electronic and optical properties which can be exploited to design several optoelectronic technologies ranging from lasers [1] to semiconductor optical amplifiers (SOAs) [2] or single photon sources [3], where they have successfully overcome critical challenges such as extremely low threshold, high speed response, or entangled photon emission, respectively. However, in these applications, a critical design parameter is the polarization response of QDs, typically characterized in terms of either degree of polarization [DOP = (TE−TM)/(TE+TM)] [4, 5] or TM/TE ratio [6, 7], where TE mode is measured along a direction in the plane of the QD, and TM mode is measured along the growth [001] direction for the GaAs(001) QDs. Engineering of QD nanostructures to achieve isotropic polarization (DOP $\sim$ 0) is critical for the implementation of several optoelectronic devices, for example semiconductor optical amplifier (SOA).

InAs QDs grown by the Stranski–Krastonov (SK) self-assembly growth process typically exhibit very poor polarization response (DOP $\geq$ 0.8) due to the large compressive biaxial strain surrounding the flat shapes of the QDs. The strain-induced splitting between the heavy hole (HH) and the light hole (LH) valence bands leads to a dominant HH character in the few top most valence band states, thus significantly suppressing the TM mode. Therefore, the previous studies of the single InAs QDs have reported very high values of the DOP, typically larger than 0.8 [4, 6, 8, 9].

The polarization response of InAs QDs is influenced by several parameters such as crystal/atomic symmetry, QD shape and aspect ratio (AR = height/base), and composition profile. The atomistic asymmetry of the underlying zincblende crystals implies that the [110] and [$\bar{1}$10] directions are inequivalent. This lowers the overall symmetry of a perfectly circular dome-shaped QD from $C_{\infty v}$ to C_{2v}. As a result, TE mode in the plane of the QD does not remain symmetric and significant in-plane anisotropy may be observed even for an ideal circular-base InAs QD [10]. The anisotropy of the in-plane TE mode can be described as:

$$Pol_{||} = \frac{(TE_{[\bar{1}10]} - TE_{[110]})}{(TE_{[\bar{1}10]} + TE_{[110]})} \tag{5.1}$$

Since $TE_{[\bar{1}10]} \neq TE_{[110]}$, so a single value of the DOP is not sufficient to characterize the polarization response of the QD systems. This, in the past studies [4, 10], has led us to define a direction-dependent value of the DOP,

$$DOP_{[\vec{n}]} = \frac{(TE_{[\vec{n}]} - TM_{[001]})}{(TE_{[\vec{n}]} + TM_{[001]})} \tag{5.2}$$

where the direction, $[\vec{n}]$ = [110] or [$\bar{1}$10], associated with the $DOP_{[\vec{n}]}$ is same as the direction of the $TE_{[\vec{n}]}$-mode in the plane of the QD.

The second factor is composition profile of the quantum dots that impacts their polarization properties. During the self-assembly growth process, In-Ga intermixing and In-segregation effects typically result in a nonuniform composition profile: large In fraction at the center of the QDs surrounded by low In composition close to the QD edges [11]. This nonuniform composition profile results in a decrease in the value of the DOP as recently demonstrated by our atomistic theoretical study [7], where a nonuniform double-layer In composition profile increased the $TM_{[001]}/TE_{[100]}$ ratio from 0.097 (for the pure InAs composition profile) to 0.27, in accordance with the experimental measurement that indicated $TM_{[001]}/TE_{[100]} =$ 0.26. This corresponds to a reduction in the value of the $DOP_{[100]}$ from 0.82 to 0.6 (still considerably higher than the desired value of 0).

The third factor that strongly impacts the polarization response of QDs is their shape. Theoretical studies typically assume perfect circular-base (for dome-shaped QDs) or square-base (for pyramidal-shaped QDs) shapes, whereas the quantum dots grown by the SK self-assembly process have generally ellipsoidal shape, elongated either along the [110] [12–15] direction or along the $[\bar{1}10]$ direction [16–19]. The elliptical shape of the QDs decreases the value of the DOP as recently predicted by our atomistic theoretical calculations [10].

The fourth factor that changes the polarization response of QDs is their aspect ratio. We have recently shown [10] that increasing the QD AR from 0.225 to 0.8 reduces the $DOP_{[110]}$ from 0.97 to 0.83. This decrease is attributed to an enhanced HH–LH intermixing that increases the TM mode component and a stronger confinement of the hole wave functions at the QD interface for the tall QDs that reduces the $TE_{[110]}$ mode. It should also be noted that the decrease in the DOP with AR is accompanied by a significant decrease in the transition strength. By simultaneous increase in the QD AR from 0.225 to 0.8 and changing the QD diameters along the [110] and $[\bar{1}10]$ directions from 20 nm and 20 nm to 14 nm and 26 nm, respectively, a net decrease in the value of the $DOP_{[110]}$ from 0.97 to 0.32 was calculated [10]. However, this theoretical prediction has not been experimentally verified yet.

To conclude this discussion related to the polarization response of the single QD layers, despite various factors significantly impact their DOP and lead to an improvement of the polarization response by reducing the value of the DOP towards zero, still isotropic polarization remains unavailable from these nanostructures.

5.1.1 Multi-Layer Quantum Dot Molecules

The AR of the QD nanostructures can be drastically increased by growing multiple layers of the QDs in the form of vertical stacks, where the adjacent QD layers are closely spaced. In these nanostructures, the long-range strain fields couple the quantum dot layers and therefore result in the formation of molecular electronic states. Due to this characteristic, these multi-layer QD nanostructures are also commonly known as "quantum dot molecules" or "artificial molecules."

It is worth mentioning here that another term quantum dot stack (QDS) is also widely used in the literature to refer to the vertical arrays of QDs. However, the distinction between a QDM and a QDS is that the former specifically refers to the case where the adjacent QD layers are very closely spaced (typically less than 5 nm apart) and therefore exhibit hybridized molecular states. A QDS, on the other hand, is a more general term and the QD layers in a QDS may or may not be strongly coupled. For example, a recent study on the bilayer QDS shows that at 10 nm inter-layer spacing, the two QD layers are weakly coupled and therefore the electronic states remain confined inside the individual QD layers and do not form molecular states. Another theoretical study [20] suggested that the transition from weakly to strongly coupled QDSs roughly occurs at 6 nm inter-layer spacing. This book chapter only focuses on the vertical QDSs with the inter-dot spacing of 4.5 nm (as will be described later in Sect. 5.2.1), so we will refer these strongly coupled QDSs as QDMs in the remainder of this chapter.

A special property of the QDMs is that the electronic states are spread over the whole molecule and therefore provide larger electron–hole wave function overlaps. Another unique characteristic of the QDMs formed as vertical arrays of QDs is that they artificially provide very large ARs. Since the electronic states are hybridized over the whole QDM, so a QDM can be regarded as a single QD nanostructure in which very large height to base ratios can be realized by controlling their constituent number of QD layers. This results in the relaxation of biaxial strain close to the center of QDMs, therefore increasing HH–LH intermixing in the valence band states. As a result, the TM mode response increases, thereby reducing the value of DOP towards the isotropic response. This characteristic of the QDMs makes them very attractive candidates for the design of optical devices where control of polarization response is critical such as semiconductor optical amplifiers.

During the last few years, several experimental [4, 9, 21–25] and theoretical [4, 8, 10] studies have been performed to study polarization-dependent optical spectra from the InAs QDMs. Kita et al. [25] and Saito et al. [8] demonstrated that an isotropic polarization response can be achieved by growing columnar QDs—QDMs with the inter-layer spacing of 0 nm – consisting of nine QD layers.

Recent experiments by Inoue et al. [9, 26, 27] and a subsequent theoretical study by Usman et al. [4] showed that similar tuning of the polarization properties is possible in InAs QDMs where the QD layers are geometrically separated by thin (4.5 nm) GaAs spacers. Such multi-layer QDMs offer twofold advantage over the columnar QDs: (i) a moderately thick GaAs spacer between the QD layers allows a precise control of overall QD shape and size and (ii) a reduced strain accumulation results in isotropic polarization response with fewer number of QD layers in the molecule. The experimental results [9, 26, 27] were provided for the QDMs containing three, six, and nine QD layers. The results indicated that the $DOP_{[110]}$ takes up the values of +0.46 and −0.60 for the samples containing six and nine QD layers, respectively. A change in the sign for the $DOP_{[110]}$ implies that the isotropic polarization response ($DOP_{[110]} \rightarrow 0$) can be achieved by engineering the number of QD layers between six and nine. The theoretical calculations [4] explained this isotropic polarization response in terms of enhancement of the $TM_{[001]}$ mode and a large decrease in the $TE_{[110]}$ mode. This book chapter aims to provide a detailed overview of our recent work [4,10] on the study and design of the multi-layer QDMs for the isotropic polarization properties.

5.1.2 Organization of Chapter

The rest of this chapter is organised as follows: Sect. 5.2.1 documents geometry parameters of the studied quantum dot nanostructures. We study four QDMs with increasing number of QD layers to investigate the impact of the size on the polarization properties. The results for a single QD layer have also been included for reference. Section 5.2.2 outlines a brief overview of our methodologies to compute strain profiles, electronic spectra, optical transition modes, and piezoelectric profiles. In Sect. 5.3, we study electronic structure of the QDMs in detail. First we present an analysis of hydrostatic and biaxial strain profiles, and their impact on the HH–LH intermixing in the valence band states with the aid of band edge diagrams. This is followed by electron and hole wave function plots to show that the strongly coupled QD layers result in molecular states in the conduction band. The hole states in the valence band remain localized in the individual quantum dot layers due to their heavier mass. Section 5.4 provides a detailed investigation of the polarization properties of the QDMs. Our calculations show that the TM mode increases as a function of the size of the QDMs due to the enhanced HH–LH intermixing. We also reveal an interesting property of the QDMs that the in-plane polarization is highly anisotropic ($TE_{[\bar{1}10]} \gg TE_{[110]}$), and therefore only $DOP_{[110]}$ can be designed for the isotropic polarization response. We demonstrate that our calculated values of the $DOP_{[110]}$ and $DOP_{[\bar{1}10]}$ qualitatively follow the trends of the experimentally measured values. In Sect. 5.5 of this book chapter, we provide an analysis of the variations in the geometry parameters on the polarization properties of the QDM consisting of nine QD layers with respect to the single QD layer. In Sect. 5.6, we provide overall summary and main conclusions of our chapter.

5.2 Simulated QD Systems and Methodologies

5.2.1 Simulated QD Systems

In order to investigate the impact of the size of QDMs on their polarization properties, we study four QDMs labelled as QDM-1, QDM-2, QDM-3, and QDM-4 consisting of two, three, six, and nine closely-spaced QD layers, respectively. The schematic diagrams of these QDMs are shown in Fig. 5.1b–e. Each individual QD layer (L_i) consists of a dome-shaped InAs QD with a base diameter of 20 nm and a height of 4.0 nm, lying on top of a 0.5 nm InAs wetting layer. The separation between the adjacent wetting layers in the QDMs is 4.5 nm. This wetting

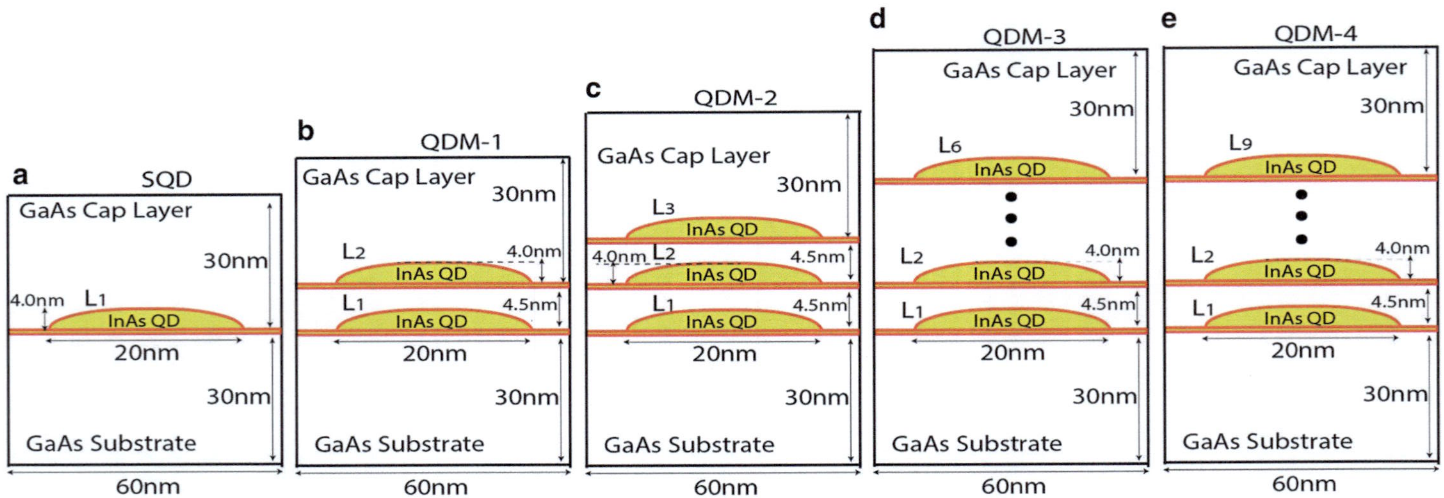

Fig. 5.1 Schematic diagrams of various quantum dot nanostructures are shown: (**a**) SQD: A single InAs QD layer, L_1. (**b**) QDM-1: A bilayer QD molecule, consisting of two identical QD layers, L_1 and L_2. (**c**) QDM-2: A QD molecule comprised of three identical QD layers, L_1, L_2, and L_3. (**d**) QDM-3: A QD molecule containing six identical QD layers, $L_1, L_2, \ldots, L_6$. (**e**) QDM-4: A QD molecule with nine identical QD layers, $L_1, L_2, \ldots, L_9$. Each QD layer (L_i) consists of a dome-shaped InAs QD with base diameter of 20 nm and height of 4 nm, placed on the top of a 0.5 nm InAs wetting layer. For the QDMs shown in (**b**) to (**e**), the adjacent QD layers are separated by 4.5 nm thick GaAs spacer layers

layer-to-layer separation has been experimentally optimized to obtain uniform QD layers from the self-assembly growth process [26]. It should also be noted that the geometrical parameters for the QDM-2, QDM-3, and QDM-4 are taken directly from the recent experimental studies [4, 9, 28] where these QDMs have shown great technological relevance for achieving the isotropic polarization response. For reference purpose, we also simulate a single QD layer labelled as SQD and shown in Fig. 5.1a.

Note that a typical SK growth of a large vertical QDM generally results in an increase in the size of the upper layer QDs [29]; however, no such increase in the QD layer dimensions was reported in the experimental study [9]. Therefore, we keep the size of the QD layers uniform in our study. Also, there is no information available from the experiment on In/Ga intermixing or In-segregation effects during the growth process. Therefore we assume pure InAs composition profiles in our study. It should also be noted that we assume circular-base shape for the QD layers inside the QDMs. However, in Sect. 5.5, we will examine the impact of geometry parameters by simulating ellipsoidal shapes and by varying the base diameters and heights of the QD layers.

The QDMs are embedded inside a sufficiently large GaAs buffer to ensure proper relaxation of atoms and to accurately accommodate the long-range effects of strain. The size of the largest GaAs buffer for the QDM-4 containing nine QD layers is $60 \times 60 \times 106\,\mathrm{nm}^3$, consisting of ≈ 25 million atoms.

5.2.2 Methodologies

The InAs QDMs embedded in the GaAs matrix are simulated using atomistic modeling tool NEMO 3-D [30–32]. The NEMO 3-D simulator has previously been applied to study single [33,34], bilayer [5,35], and multi-layer QD arrays [4,10] and it has demonstrated quantitative agreement with the available experimental data sets.

The atomistic simulations are performed on a large GaAs matrix surrounding the QDs to properly account for the long-range impact of the strain and piezoelectric fields. The size of the GaAs box is selected to be large enough to allow the strain and piezoelectric fields become zero at the edges of the GaAs box. This ensures proper relaxation of atoms and therefore correctly models the impact of strain and piezoelectric potentials in the calculations of electronic structure.

The strain is calculated using atomistic valence force field (VFF) method [36] including anharmonic corrections [37] to the classical Keating potential. Realistic boundary conditions are chosen for the strain domain [38]: the substrate is fixed at the bottom, the GaAs buffer is periodic in the lateral directions, and the capping layer is free to relax from the top. Linear and quadratic piezoelectric potentials are computed following the procedure described in references [35, 39].

The electronic structure calculations are performed by solving empirical tight binding Hamiltonian in which each atomic site is represented by twenty bands in an $sp^3d^5s^*$ model [40] including spin. Electronic domain is chosen to be relatively smaller than the strain domain due to the strong spatial confinement of the electron and hole wave functions inside the QD region. The electronic domain has fixed boundary conditions in all spacial directions. The atoms at the surface are passivated according to our published approach [38].

The inter-band optical transition strengths between the electron and hole energy states are computed using Fermi's golden rule by calculating the squared absolute value of the momentum matrix elements summed over spin degenerate states [33, 35]:

$$T_{E_i - H_i} = | < E_i | [\vec{n}, \mathbf{H}] | H_i > |^2 \tag{5.3}$$

where $\mathbf{H}$ is the single particle tight binding Hamiltonian in the $sp^3d^5s^*$ basis, E_i is an electron energy state, H_i is a hole energy state, and $\vec{n}$ is a selected polarization direction. The polarization-dependent optical modes are calculated by rotating the polarization vector $\vec{n} = (\vec{x} + \vec{y})\cos\phi \sin\theta + \vec{z}\cos\theta$ along the appropriate direction in the polar coordinates: for the $TE_{[110]}$: $\theta = 90°$ and $\phi = 45°$, for the $TE_{[\bar{1}10]}$: $\theta = 90°$ and $\phi = 135°$, and for the $TM_{[001]}$: $\theta = 0°$. Here the angles ϕ and θ are measured from the [100] and [001] axis in the polar coordinate system.

5.3 Electronic Properties

5.3.1 Hydrostatic and Biaxial Strains

Figure 5.2 plots the hydrostatic $\in_H = \in_{xx} + \in_{yy} + \in_{zz}$ (dotted lines) and biaxial $\in_B = \in_{xx} + \in_{yy} - 2\in_{zz}$ (solid lines) strain profiles along the [001] direction through the center of the quantum dot systems shown in Fig. 5.1. The hydrostatic strain exhibits a very slight change from the SQD to the QDM-4. The biaxial strain, however, significantly changes as the size of the QDM increases. For the single QD layer, the biaxial strain is highly negative inside the QD region. This is typical for any flat-shaped InAs QD compressively grown on GaAs substrate.

As the vertical size of the QDM is increased by adding QD layers, the biaxial strain at the center of the molecule reduces. For the QDM-4, as shown in the Fig. 5.2e, the biaxial strain at the center of the molecule approaches zero. The reason for such behavior of the biaxial strain is that in general the InAs unit cells inside the QD region tend to fit over the GaAs matrix by an in-plane compression and an elongation along the [001] direction. This results in highly negative biaxial strain as is observed for the SQD system in Fig. 5.2a. However, when the size of the QDM increases, the unit cells of InAs around the center of the molecule feel lesser and lesser compressive force from the surrounding GaAs. As a result, the vertical lattice constant of the InAs starts matching with the GaAs and hence the biaxial strain tends to change its sign around the middle of the QD stack. Similar strain profiles were reported in an earlier study related to the columnar QDs by Saito et al. [8].

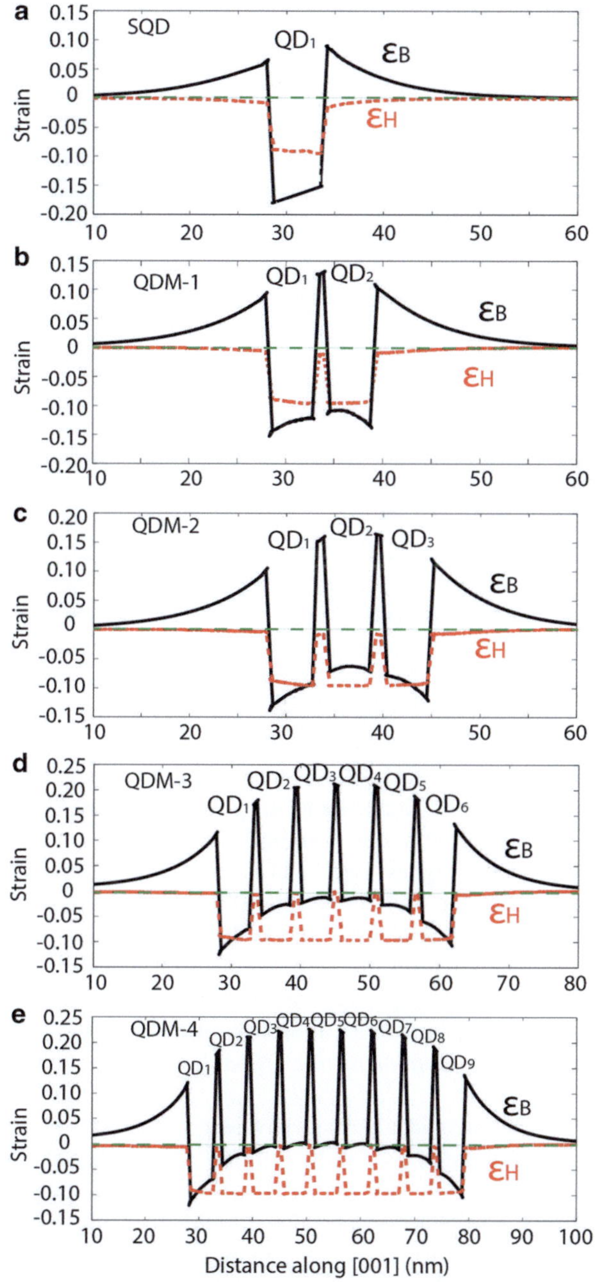

Fig. 5.2 The plots of hydrostatic ($\in_H = \in_{xx} + \in_{yy} + \in_{zz}$) and biaxial ($\in_B = \in_{xx} + \in_{yy} - 2\in_{zz}$) strain components through the center of the quantum dot layers along the [001] direction. The hydrostatic component is dominantly negative inside the QDs indicating strong compression of the InAs and it is almost zero outside the QD regions. As the QDM height increases, the biaxial strain in the QDs evolves from negative to zero with a small increase in positive contributions in the capping layer

5.3.2 HH–LH Intermixing

The negligible change in the magnitude of the hydrostatic strain (as SQD → QDM-4) implies that the lowest conduction band edge will experience very small change as it is only affected by the hydrostatic component [34]. The valence band edges are modified by both the hydrostatic strain and the biaxial strain. The impact of strain on the highest two valence band edges, HH and LH, is analytically expressed as:

$$\delta E_{\mathrm{HH}} = a_v \in_H + \frac{b_v \in_B}{2} \tag{5.4}$$

$$\delta E_{\mathrm{LH}} = a_v \in_H - \frac{b_v \in_B}{2} \tag{5.5}$$

Here a_v and b_v are the valence band deformation potential constants. The values for these constants for InAs systems are $a_v = 1.0$ eV and $b_v = -1.8$ eV, respectively [34]. From the Eqs. (5.4) and (5.5), it is evident that the magnitude of the $\in_B$ determines the HH–LH splitting. For the single QD layer (SQD), due to a large negative value of $\in_B$ around the center of the QD, the HH and LH band edges will be considerably separated inside the QD region. This will induce dominant HH character in the highest few valence band states which will be closer to the HH band edge. As the magnitude of $\in_B$ decreases with the increasing size of the QDMs, the HH–LH splitting reduces, increasing the LH component in the valence band states. For the QDM-4 system, the nearly zero magnitude of the $\in_B$ around the center of the molecule implies that the HH and LH bands will be nearly degenerate around the center of the QDM. The valence band states will therefore be of highly mixed character, consisting of contributions from both the HH and the LH bands.

Figure 5.3a–e plots the highest two local valence band edges, HH and LH, for all of the QD systems under study along the [001] direction through the center of the QDs. Highly negative biaxial strain in the SQD results in $\approx$ 152 meV splitting of the HH and LH bands within the QD region. As the biaxial strain around the center of the QDMs decreases (approaching towards zero for the QDM-4), the HH–LH splitting around the center of the molecule also decreases to $\approx$ 117 meV, $\approx$ 74 meV, $\approx$ 32 meV, and $\leq$ 28 meV for the QDM-2, QDM-3, and QDM-4 systems, respectively.

The HH and LH character of a particular valence band state in the tight binding formulation can be estimated as follows: If the amplitudes of the p_x, p_y, and p_z orbitals at any atomic site are $a_{x,u/d}$, $a_{y,u/d}$, and $a_{z,u/d}$, respectively (where the subscripts u and d refer to up and down spin), then the HH contribution is approximately proportional to $|a_{x,u} - ia_{y,u}|^2 + |a_{x,d} + ia_{y,d}|^2$ summed over all the atomic sites. The LH contribution is approximately proportional to $|a_{z,u}|^2 + |a_{z,d}|^2$ summed over all the atomic sites. By using these expressions, we estimate that the $\frac{HH}{LH}$ ratio for the highest valence band state (H_1) decreases from $\approx$ 108 for the SQD to $\approx$ 15.8, $\approx$ 12.3, and $\approx$ 10.6 for the QDM-2, QDM-3, and QDM-4 systems, respectively. This clearly points towards an increasing LH character in

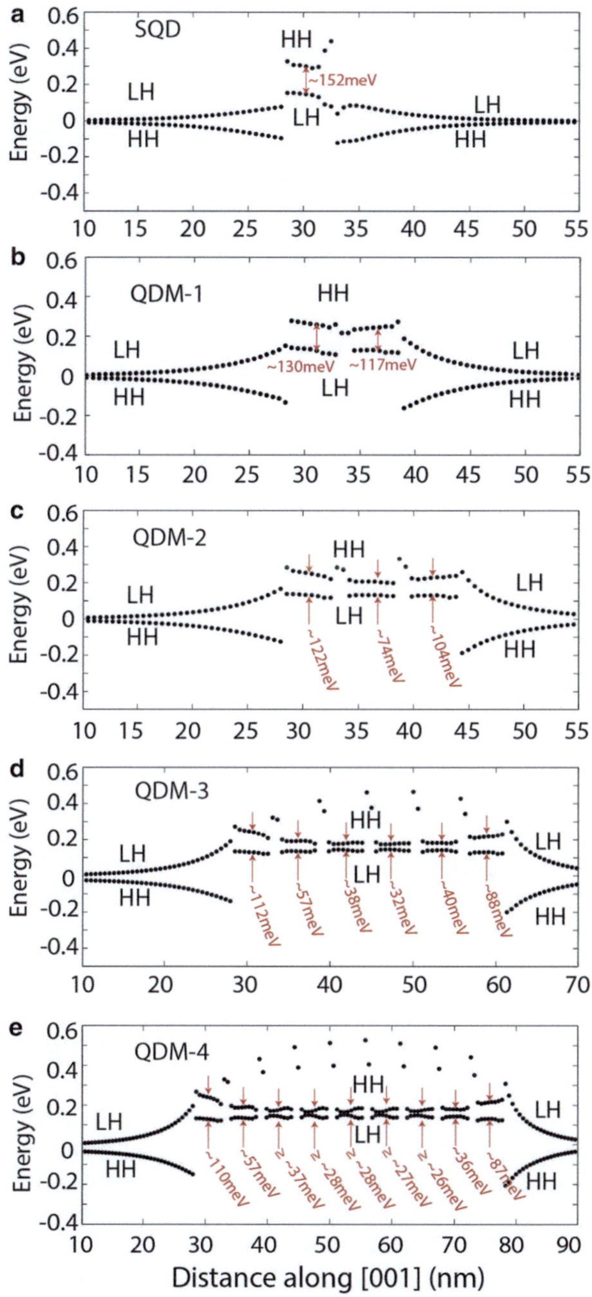

Fig. 5.3 Plots of local band edges for the highest two valence bands (HH and LH) through the center of the QD layers along the [001] direction for the SQD, QDM-1, QDM-2, QDM-3, and QDM-4. The reduction in the magnitude of the biaxial strain results in the enhanced HH/LH intermixing (lesser separation) as the size of the QDM increases

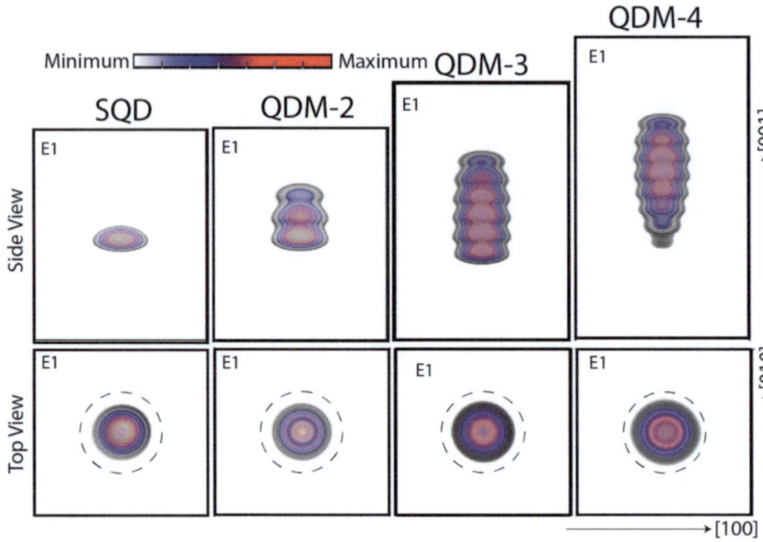

Fig. 5.4 Plots of the lowest conduction band state, E_1, for the quantum dot systems SQD, QDM-2, QDM-3, and QDM-4. *First row*: the side view of the plots is shown. *Second Row*: the top view of the plots is shown. The intensity of the color in the plots indicates the magnitude of the wave function: the *red color* represents the highest magnitude and the *light blue color* represents the lowest magnitude. The *dotted circles* are marked to guide the eye and indicate the boundary of the base of each QD

the valence band states as the size of the QDM increases, in consistent with our earlier discussion regarding the reduced biaxial strain implies enhanced HH–LH intermixing. This also suggests that the TM mode will increase for the QDMs as a function of their size (the number of the QD layers) due to the larger LH character of the valence band states, as will be shown by our calculations later in Sect. 5.4.1.

5.3.3 Electron Wave Functions form Molecular States

Before we move to the study of the polarization properties of the QDMs, let us first analyze the nature of the electron and hole states. Figure 5.4 shows the plots of the lowest conduction band state E_1 for the four QD systems: SQD, QDM-2, QDM-3, and QDM-4. From the top views of the wave functions (second row), it is evident that the lowest electron state is of s-type symmetry. The side views of the wave functions (first row) show that the electron states form hybridized (molecular) states in the QDMs and are spread over all of the QD layers. This is due to the strong coupling between the QD layers at 4.5 nm separation. The presence of the s-like electron wave function in all of the quantum dot layers implies that only the details of the hole wave functions inside the QDM will determine the optical activity of a

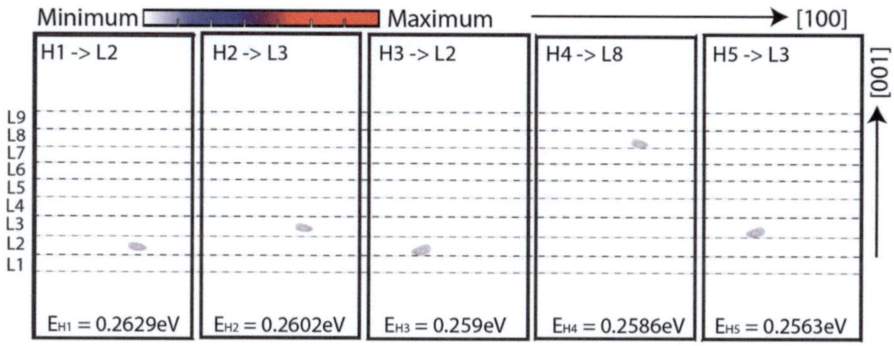

Fig. 5.5 Plots of the highest five valence band state, H_1, H_2, H_3, H_4, and H_5, for the QDM-4. Only the side views of the plots are shown. The *horizontal dotted lines* are marked to guide the eye and indicate the base of the QD layers in the molecule. The intensity of the color in the plots indicates the magnitude of the wave function: the *red color* represents the highest magnitude and the *light blue color* represents the lowest magnitude. The energies of the valence band states are also mentioned

particular quantum dot layer inside the molecule. This is different from the previous study of the bilayer QD stack [5], where the weakly coupled quantum dot layers at 10 nm separation resulted in atomic-like electron wave functions. In that study, the spatial location of both the electron and the hole wave functions determined the optical activity of the QD layers.

Previous studies on the identical bilayer QDM [35] and on the QDM with seven identical QD layers [41] showed that the strain tends to push the electron states towards the lower QDs in these systems. The electron wave functions for the QDM-2 and the QDM-3 systems show similar trend. However we find that this trend is no longer true for the molecule with the nine QD layers, QDM-4, where the electron wave function E_1 vanishes around the edges of the stack due to the larger strain magnitude there (see Fig. 5.4 first row for the QDM-4).

5.3.4 *Hole Wave Functions exhibit Atomic Character*

Although the electrons (lighter-mass particles) are strongly influenced by the inter-dot electronic and strain couplings of the QDs and exhibit tunnelling across the quantum dots forming molecular-like hybridized states, the holes due to their heavier mass remain well confined inside the individual dots and do not show any hybridization. For example, Fig. 5.5 shows the side views of the lowest five valence band wave functions in the QDM-4 system. The horizontal dotted lines are plotted to mark the positions of the base of the QD layers and helps to determine the location of a particular hole wave function inside the molecule. In this system, H_1 and H_3 are inside L_2, H_2 and H_5 are in L_3, and H_4 is in L_8. The location of a hole state inside a

QDM is relatively hard to determine and is strongly influenced by geometry of the QDM, i.e., QD base diameter, QD height, QD layer separation, etc. Ultimately the strain profile that controls the strength of the coupling between the QD layers inside the QDM determines the position of the hole states inside the molecules.

This is also worth mentioning that due to the small inter-layer spacing (4.5 nm) resulting in the strong coupling of the QD layers, the electron wave functions are spread over all of the QD layers (see Fig. 5.4). Therefore, despite the confinement of the hole wave functions is atomic-like (confined in the individual QD layers), the electron–hole wave function overlap remains strong. This is different from the case of the QDMs with weakly coupled QD layers (large inter-layer separations) where both the electron and the hole wave functions will be confined in the individual QD layers and therefore the oscillator strength will be small due to the electron and hole wave functions being in the different QD layers. We suggest that the QDMs with the small inter-layer separations are better candidate for the tuning of the polarization properties while maintaining relatively large oscillator strengths.

5.4 Polarization Properties

5.4.1 Enhanced HH/LH Intermixing Implies $TM_{[001]}$ Mode Increases

In a QD system, the HH states consist of contributions from p_x and p_y orbitals and the LH states consist of contributions from p_x, p_y, and p_z orbitals. These configurations imply that the TM mode (which couples along the z (growth) direction) will only couple to the LH states. The large splitting ($\approx$ 152 meV) of the HH–LH bands (see Fig. 5.3a) resulting in a weak LH contribution in the SQD system will result in very weak $TM_{[001]}$ mode for this system. Thus from Eq. (5.2), the DOP will be close to 1.0 and the polarization response will be highly anisotropic.

As the size of the QDM increases, the larger intermixing of the HH and LH bands (see Fig. 5.3b–e) increases the LH contribution in the valence band states. This will result in an increase of the $TM_{[001]}$ mode of the inter-band optical transitions reducing the anisotropy in the DOP by bringing it closer to 0.

5.4.2 Optical Intensity Functions, $f(\lambda)$

Figure 5.6 plots the optical intensity functions computed from our model as a function of the optical wavelength for the four quantum dot systems under study: SQD, QDM-2, QDM-3, and QDM-4. The calculation of the optical intensity function is done as follows: first we calculate optical transition strengths by using Fermi's golden rule for $TE_{[110]}$, $TE_{[\bar{1}10]}$, and $TM_{[001]}$ modes between the lowest

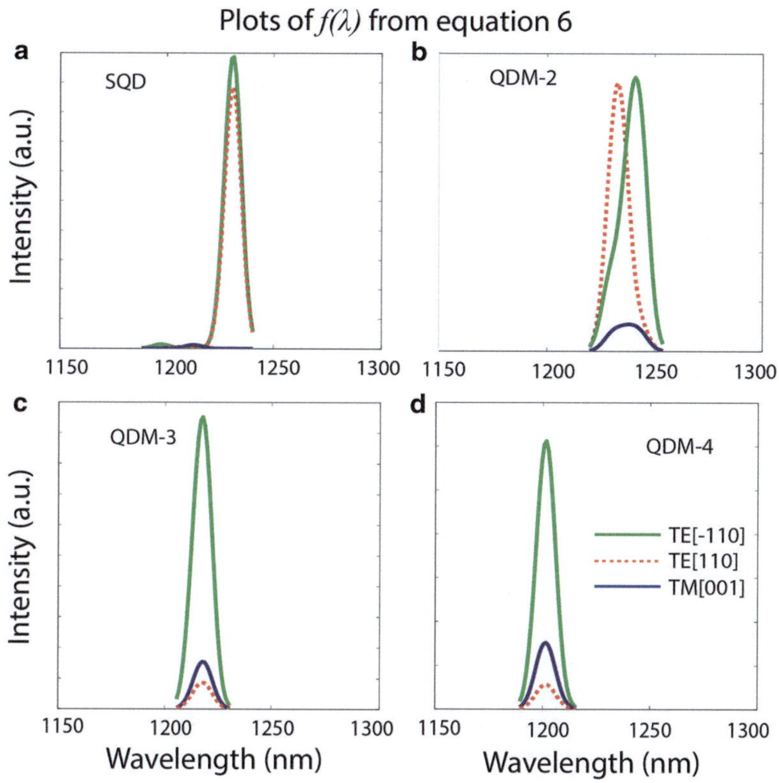

Fig. 5.6 Plots of the optical intensity functions, $f(\lambda)$, are shown for the various QD systems. The optical intensity functions in each case are computed from the Eqs. (5.6) and (5.7)

conduction band state E_1 and the highest five valence band states H_1, H_2, H_3, H_4, and H_5 for a particular QD system. The details of the procedure are described earlier in the Sect. 5.2.2. Next each computed optical transition strength is artificially broadened by multiplication with a Gaussian distribution centered at the wavelength of the transition [42, 43]. Finally we add all of the five Gaussian functions to calculate the total optical intensity function, $f(\lambda)$. The complete expression for the optical intensity function, plotted in Fig. 5.6a–d, is given by Eqs. (5.6) and (5.7) as follows:

$$f(\lambda)_{T^{E_1-H_i}} = \sum_{i=1}^{5} (T^{E_1-H_i}).e^{-\frac{\lambda-\lambda_{E_1-H_i}}{(0.25)^2}} \tag{5.6}$$

where

$$T^{E_1-H_i} = (TE_{[110]}^{E_1-H_i}/TE_{[\bar{1}10]}^{E_1-H_i}/TM_{[001]}^{E_1-H_i}) \tag{5.7}$$

The examination of the optical intensity plots in Fig. 5.6 reveals that the $TM_{[001]}$ mode indeed increases as the number of the QD layers is increased: SQD $\rightarrow$ QDM-4. This is a direct consequence of the change in the biaxial strain component that increases HH and LH intermixing as discussed earlier in Sect. 5.3.2.

5.4.3 *Increase in the $TM_{[001]}$ not Enough for Isotropic Polarization*

Figure 5.6 shows that the increase in the $TM_{[001]}$ mode only partially helps towards an isotropic polarization response. This is in contrast to a general notion where it is described that the increase in the $TM_{[001]}$ mode is mainly responsible for the isotropic polarization. The reason for such understanding is that the previous theoretical [8] or experimental studies [6, 9, 23] of the DOP have assumed only one direction for the TE mode. However, the PL measurements shown in Fig. 5.1 of the [4] indicate that the TE modes along the [110] and $[\bar{1}10]$ have significant anisotropy in the plane of the QDM. Our theoretical model shows that in fact a major contribution to achieve isotropic polarization response in these systems stems from a suppressed $TE_{[110]}$ mode rather than an increased $TM_{[001]}$ mode. Figure 5.6 shows that irrespective of QD geometry, the increase in the $TM_{[001]}$ mode is insufficient to reverse the sign of the $DOP_{[\bar{1}10]}$.

The TE mode is highly anisotropic in the plane of the QD with the magnitudes of the $TE_{[110]}$ and $TE_{[\bar{1}10]}$ modes becoming very different as the QDM size increases. For the SQD system, $TE_{[110]} \sim TE_{[\bar{1}10]}$ and $TM_{[001]}$ is very weak. Hence the measured and calculated DOP is highly anisotropic (close to 1.0), irrespective of the direction for the TE mode. As the QDM size increases, the $TM_{[001]}$ mode increases partially contributing to the reduction in the DOP. However, at the same time, the $TE_{[110]}$ mode decreases drastically such that for the QDM-3 and QDM-4, it becomes smaller than the $TM_{[001]}$ mode. This reverses the sign of the $DOP_{[110]}$. The reason for such a drastic decrease in the $TE_{[110]}$ mode is the orientation of the hole wave functions along the $[\bar{1}10]$ direction for the QDM-3 and QDM-4 systems.

5.4.4 *Hole Wave Function Confinements Reduces the $TE_{[110]}$ Mode*

Figure 5.7 plots the top views of the highest five valence band states H_1, H_2, H_3, H_4, and H_5 for the QD systems SQD, QDM-3, and QDM-4. The five hole wave functions for the SQD system have an almost uniform distribution inside the QD region with nearly symmetric shape. Such symmetry will result in approximately equal magnitude of the TE mode along the [110] and $[\bar{1}10]$ directions, as evident in Fig. 5.6a. For the QDMs with six and nine QD layers, the distribution of the hole

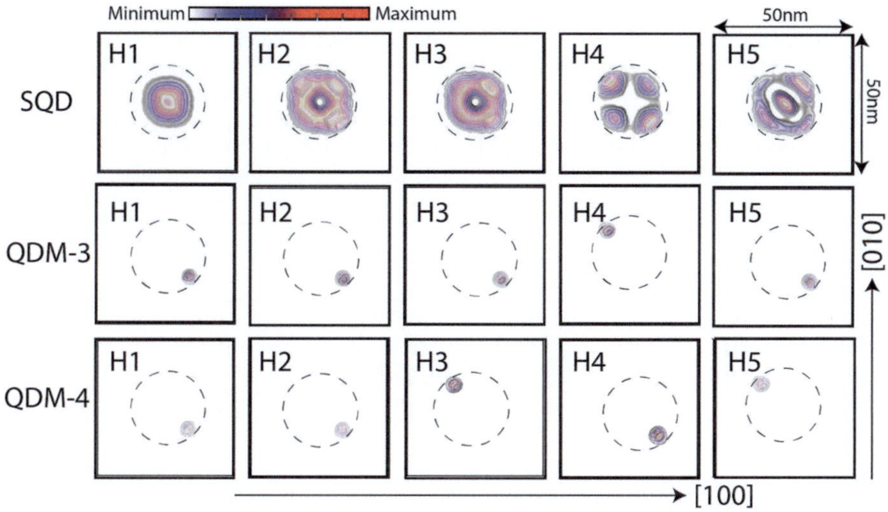

Fig. 5.7 The top views of the plots of the highest five valence band states H_1, H_2, H_3, H_4, and H_5 are shown for the SQD, QDM-3, and QDM-4 systems. The intensity of the colors indicates the magnitude of the hole wave functions: the *red color* indicating the largest magnitude and the *light blue color* indicating the smallest magnitude. The *dotted circles* are marked to guide the eye and indicate the boundary of the QD bases

wave functions is oriented along the $[\bar{1}10]$ directions. This $[\bar{1}10]$-symmetry is mainly due to the strain and piezoelectric potentials that lowers the overall symmetry of the QD system and favors this direction [39]. To verify the impact of the strain and piezoelectricity, if we conduct a numerical experiment and switch off their contributions in the electronic structure calculations, the $TE_{[\bar{1}10]}/TE_{[110]}$ ratio for the QDM-3 system decreases from 10.92 to 0.85 and it decreases from 10.73 to 3.2 for the QDM-4 system. Similar distributions of the hole wave functions are observed for bilayers [5] and a single QD layer with AR ≥ 0.25 [33].

The orientation of the hole wave functions determines the magnitude of the $TE_{[110]}$ and $TE_{[\bar{1}10]}$ modes since the lowest electron wave function (see Fig. 5.4) is symmetrically distributed. All of the highest five hole wave functions are oriented along the $[\bar{1}10]$ direction in the case of the QDM-3 and QDM-4 systems. This results in a strong reduction of the $TE_{[110]}$ mode due to considerably reduced electron–hole wave function overlaps in this direction. The significant reduction in the $TE_{[110]}$ mode makes it to be even smaller than the magnitude of the $TM_{[001]}$ mode as can be seen in Fig. 5.6c,d. This change in the relative magnitudes of the $TE_{[110]}$ and $TM_{[001]}$ modes results in a flip of sign $(+ \rightarrow -)$ for the $DOP_{[110]}$ as indeed measured in the experiments [4, 9, 44]. Hence we conclude that the isotropic polarization response demonstrated by the experiment is a result of two factors: (i) increase in the $TM_{[001]}$ mode due to enhanced HH–LH intermixing and (ii) the reduction of the $TE_{[110]}$ mode due to the orientation of the holes along the $[\bar{1}10]$ direction.

Table 5.1 The comparison of the experimentally measured and theoretically calculated values of the DOP is provided for various in-plane TE mode directions. Column 1: the QD system under study. Column 2: the dimensions of the QD layers in the molecules: B is the base diameter and H is the height of the QD layers. Column 3–6: the values of the DOP calculated from our model. We provide two additional directions for the DOP, [100] and [010], for comparison purpose. Column 7–8: the values of the DOP extracted from the published experimental measurements. The values marked with * are taken from Usman et al. [35] and the values marked with † are taken from Inoue et al. [9]

QDM Geometry		Theoretical Calculations				Experiment	
	(B, H) (nm)	$DOP_{[100]}$	$DOP_{[010]}$	$DOP_{[110]}$	$DOP_{[\bar{1}10]}$	$DOP_{[110]}$	$DOP_{[\bar{1}10]}$
SQD	(20, 4.0)	0.999	0.999	0.999	0.999	0.7^*	0.71^*
QDM-2	(20, 4.0)	0.77	0.77	0.833	0.836	$0.67^\dagger$	
QDM-3	(20, 4.0)	0.548	0.46	−0.244	0.72	$0.46^\dagger$	
QDM-4	(20, 4.0)	0.38	0.371	−0.45	0.603	$-0.6^\dagger$, -0.36^*	0.66^*

Here we want to point out that the relative magnitude of the $TE_{[\bar{1}10]}$ mode does not reduce as the size of the QDM is increased. Even for the largest system under study, QDM-4, the $TE_{[\bar{1}10]}$ mode is much stronger than the $TM_{[001]}$ mode. That means if the experimental measurements are performed for $DOP_{[\bar{1}10]}$, they should still show significant anisotropy. This is verified by the PL measurements on the QDM-4 system as shown in Fig. 5.1 of the Ref. [4]. These PL measurements on QDM-4 system indicate a positive value for the $DOP_{[\bar{1}10]}$ ($TE_{[\bar{1}10]} > TM_{[001]}$) and a negative value for the $DOP_{[110]}$ ($TE_{[110]} < TM_{[001]}$).

5.4.5 Comparison with the Experimental PL Data

Table 5.1 summarizes the calculated values of the DOP from our model along different directions and compares it with the recently published experimental PL measurements. Here we also provide theoretically calculated values of the DOP along the [100] and [010] directions for comparison purpose because some recent experimental studies [6, 8, 45] have chosen these directions for investigation of the DOP and characterization of the polarization response of the QD systems.

The hole state symmetries shown in Fig. 5.7 show strong alignment in the $[\bar{1}10]$ direction indicating that symmetries to the [100] and [010] directions will be almost equivalent. Therefore the values of the DOP are nearly equal for the [100] and [010] directions as mentioned in Table 5.1. The PL measurements along these two directions will not exhibit isotropic polarization even for the QDM-4 system. Similar results were found in an earlier experimental study by Ridha et al. [45]. In their study of the polarization properties of multi-layer stacks based on $TE_{[010]}$ mode, they conclude that such systems cannot provide isotropic polarization and therefore columnar QDs should be studied for this purpose. However, our results show that the QDM-4 exhibits isotropic polarization if the TE mode is measured along the high symmetry [110] axis.

For the SQD system containing only a single QD layer, our calculated DOP values in all the directions are very close to 1.0, exhibiting a very strong polarization anisotropy (TE mode $\gg$ TM mode). This is similar to other theoretical studies based on k·p method [8, 46] where the values of DOP close to 1.0 have been reported for the single QD layers. However the experimentally measured values for the single QD layer are relatively low ($\sim$ 0.7). Since the values for both $DOP_{[110]}$ and $DOP_{[\bar{1}10]}$ are very close so we believe that the reason for this discrepancy between the theory and experiment is not related to the shape of the QD. This is because any elongation in the QD shape will result in considerably enhanced in-plane anisotropy ($DOP_{[110]}$ and $DOP_{[\bar{1}10]}$ being different) [10]. We speculate that one possible reason for this could be due to In-Ga intermixing and In-segregation effects as recently suggested [7]. However, further information on the composition profiles of the experimental samples [9] is required to confirm this understanding.

Our calculated value of the $DOP_{[110]}$ for the QDM-3 is also lower than the experimental measurement, but as we will show later in Sect. 5.5.1 that the polarization response of the QDM-3 is highly sensitive to the height of the QD layers and even a small variation such as 0.5 nm results in a significant increase in the value of the DOP. Therefore, we believe that this discrepancy may be due to the assumed height of the QD layers (4.0 nm) or circular-base shape for the QDM-3. However, our theoretical results exhibit excellent agreement with the experimental measurements for the QDM-4 and therefore confirm that the isotropic polarization response can be achieved by engineering the size of the QDMs.

5.5 Effect of Geometry Parameters

In the last two sections, we have studied the electronic and polarization properties of the multi-layer QDMs assuming an ideal dome-shape for the QD layers with circular-base. However in reality, the actual shape of the SK self-assembled QDs is far from being perfectly circular or square. Several experimental investigations have suggested that the actual shape of the QDs significantly deviates from the ideal circular-base (for dome or lens) or square-base (for pyramid), and usually tends to elongate either along the [110] [12–15] or along the [$\bar{1}$10] [16–19] directions. Therefore it is essential to investigate the impact of the elliptical shapes on the polarization properties of the QDMs.

Furthermore, we have simulated the QDM geometries with the base diameter and height of the QD layers directly extracted from the recent experimental TEM images [4, 9], which suggest that the height and the base diameter of the QD layers are roughly 4.0 nm and 20 nm, respectively. However, as the experimental TEM images are not very clear, so it is worthwhile to analyze the impact of small variations in the base diameter and the height of the QD layers on the calculated polarization properties. Therefore in this section, we provide a quantitative description of the impact of the variations in the geometry parameters on the calculated values of the DOP.

Table 5.2 The comparison of the theoretically calculated values of the DOP for various QD geometry configurations. Column 1: the multi-layer QD system under study. Column 2: the dimensions of the QD layers in the molecules: B is the base diameter and H is the height of the QD. Column 3–6: the values of the DOP calculated from our model

QD Geometry		Theoretical calculations			
	(B, H) (nm)	$DOP_{[100]}$	$DOP_{[010]}$	$DOP_{[110]}$	$DOP_{[\bar{1}10]}$
SQD	(20, 3.5)	0.995	0.996	0.996	0.996
	(20, 4.0)	0.999	0.999	0.999	0.999
QDM-2	(20, 3.5)	0.922	0.884	0.9046	0.933
	(20, 4.0)	0.77	0.77	0.833	0.836
QDM-3	(20, 3.5)	0.594	0.594	0	0.747
	(20, 4.0)	0.548	0.46	−0.244	0.72
QDM-4	(20, 3.5)	0.6	0.6	0.445	0.652
	(20, 4.0)	0.38	0.371	−0.45	0.603
	(20, 4.5)	0.45	0.446	−0.47	0.43

5.5.1 *Height of the QD Layers*

Since the height of the QD layers (L_i) inside the QDMs is not very clear from the TEM images (see Fig. 5.1a in Ref. [4]), so we simulate different height configurations of the QDMs and provide the values of the DOP in Table 5.2. This data serve as a measure of the sensitivity of the DOP with respect to the QDM height parameter and provide a guide to the experimentalists to explore the design space of such complex multi-million atom systems. From Table 5.2, as the QDM height increases in the SQD $\rightarrow$ QDM-2 $\rightarrow$ QDM-3 $\rightarrow$ QDM-4, the values of the degree of polarization reduce. The reduction in the value of the DOP is larger for the QDMs with $H = 4.0$ nm as compared to the QDMs with $H = 3.5$ nm. This is due to the fact that the larger height of the QD layers in the QDM results in stronger coupling between the QD layers. This implies a stronger HH–LH intermixing resulting in larger magnitude of the $TM_{[001]}$ mode.

The dependence of the DOP on the height (H) of the QD layers inside the molecules is generally an unknown factor. The calculated values of the DOP in Table 5.2 show that the DOP becomes very sensitive to the height of the QDs inside the molecule as the size of the molecule grows larger. For the systems SQD and QDM-2, the increase in the height (H) from 3.5 to 4.0 nm results in a small decrease in the values of the DOP. However, for a same change in the value of H, the DOP_{110} significantly decreases from 0 to -0.244 and from 0.445 to -0.45 for the QDM-3 and QDM-4 systems, respectively. This implies that an isotropic polarization response (DOP $\sim$ 0) can either be achieved from the QDM-3 with $H = 3.5$ nm, or from a molecule with fewer number of QD layers and $H = 4.0$ nm. We therefore propose that the polarization response of the QDMs can be tuned by not only increasing the number of QD layers (a parameter typically tuned in the past experimental studies [8, 9, 23, 24, 45]), but also by controlling the height (H) of the individual QD layers inside the molecules.

Another interesting case is when the height of the QD layers is 4.5 nm which is equal to the wetting-layer to wetting-layer separation. In this case, the adjacent QD layers will be connected to each other and the QDM becomes a columnar QD. We investigate such structure for the QDM-4 and find that the values of DOP are relatively higher for the columnar QD when compared to the QDM-4 with 4.5 nm thick GaAs spacer layer. This indicates that to achieve isotropic polarization, overall a fewer number of QD layers are required in a QDM with finite GaAs spacer when compared to a similar diameter columnar QD.

Finally, based on the comparison of the calculated values of the DOP for various QD height configurations and the measured values provided in Table 5.1, we estimate that the dimensions of the QD layers inside the QDM-4 are approximately $B = 20$ nm and $H = 4$ nm.

5.5.2 Nonidentical QD Layers

Another design parameter is the base diameter of the QD layers inside the QDM. As suggested by the experimental studies [5,29] that the base diameter of the QD layers during the SK growth process increases as the size of the QDM/QDS increases. This is due to the fact that the presence of the strain from the lower QD layers tend to reduce the In out diffusion and therefore results in relatively larger QDs in the upper layers. In order to investigate its effect on the polarization properties, we simulate QDM-2v and QDM-3v, where all the geometry parameters are same as for the QDM-2 and QDM-3, respectively, except that the base diameter of the QD layers is increased in steps of 1 nm and the height of the QD layers is increased in steps of 0.25 nm as the size of the QDM increases in the vertical direction. For example in the QDM-2v, the lowest QD layer L_1 has the base diameter of 20 nm and the height of 4 nm. The next QD layer L_2 has the base diameter of 21 nm and the height of 4.25 nm, and so on. Our calculations indicate that such QDMs with the nonidentical QD layers will exhibit relatively higher values of the DOP. The values of the $DOP_{[110]}$ and $DOP_{[\bar{1}10]}$ are calculated to increase from 0.833 and 0.836 (for the QDM-2) to 0.92 and 0.94 (for the QDM-2v), respectively. Similarly, the values of the $DOP_{[110]}$ and $DOP_{[\bar{1}10]}$ are calculated to increase from −0.244 and 0.72 (for the QDM-3) to 0.73 and 0.76 (for the QDM-3v), respectively. Hence we deduce that the QDMs with nonidentical QD layers are not suitable for achieving the isotropic polarization response. No such increase in the size of the QD layers was reported from the TEM images [9].

5.5.3 Elliptical Shape of the QD Layers

The value of the $DOP_{[\vec{n}]}$ strongly depends on the shape of the QD layers inside the QDM. Despite significant experimental evidence for the elongation of the QD

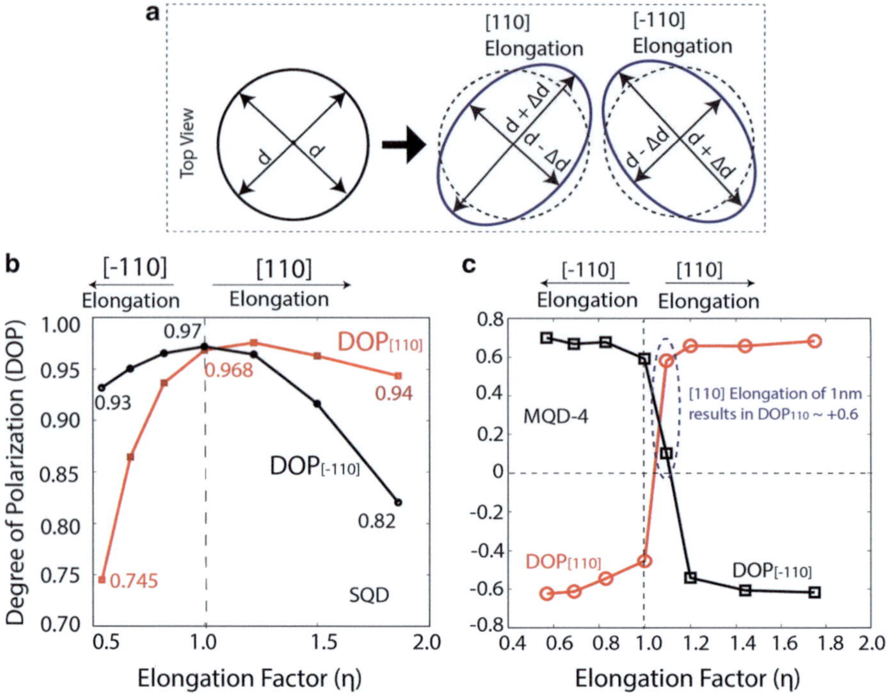

Fig. 5.8 (**a**) Schematic diagram of the top view is shown illustrating the base elongations for the QD layers. The elliptical shape of the QD is formed by simultaneously decreasing (increasing) the base diameter along the [110] direction and increasing (decreasing) the base diameter along the [−110] direction. The elongation factor η is defined as ratio of the diameter along the [110] direction and the diameter along the $[\bar{1}10]$ direction. (**b**) The values of the $DOP_{[110]}$ and $DOP_{[\bar{1}10]}$ are plotted as a function of the elongation factor η for the SQD. The elliptical shape of the SQD reduces the values of the DOP along the both [110] and $[\bar{1}10]$ directions. (**c**) The values of the $DOP_{[110]}$ and $DOP_{[\bar{1}10]}$ are plotted as a function of η for the QDM-4. The large in-plane polarization anisotropy results in $DOP_{[110]} \gg DOP_{[\bar{1}10]}$ for $\eta > 1.0$, and $DOP_{[110]} \ll DOP_{[\bar{1}10]}$ for $\eta < 1.0$. A nearly step-like dependent of the DOP on the values of η is calculated

shapes and its potential strong impact on the polarization properties, the impact of the base elongations on the value of the $DOP_{[\vec{n}]}$ is only barely known. Recently, we have provided, in Ref. [10], a detailed investigation of the impact of elliptical QD shapes on the polarization properties of the single QD layers and vertically stacked QDMs. In this subsection, we provide a brief overview of these results.

Figure 5.8a shows the schematic diagram of the top view of a QD layer to demonstrate the simulated ellipsoidal shape of the QDs. We elongate the QD along the [110] and $[\bar{1}10]$ directions by equally increasing/decreasing the diameters along these directions as illustrated in the figure. We define an elongation factor η as the ratio of the QD diameters along the [110] and $[\bar{1}10]$ directions. Therefore, $\eta > 1.0$ indicates an ellipsoidal QD with its major axis aligned along the [110] direction and $\eta < 1.0$ indicates an ellipsoidal QD with its major axis aligned along the $[\bar{1}10]$ direction. For an ideal circular base QD, $\eta = 1.0$. Note that we do not alter the height of the QD layers which is kept fixed at 4.0 nm, so the ellipsoidal-base QDs will have slightly smaller volume when compared to the circular-base QD (see Fig. 5.1f in Ref. [10]).

Figure 5.8b plots the values of the $DOP_{[110]}$ and $DOP_{[\bar{1}10]}$ as a function of the elongation factor η. For $\eta > 1.0$, [110] elongation of the SQD increases the $TE_{[110]}$ mode and decreases the $TE_{[\bar{1}10]}$ mode. This can be understood as follows: the few top most valence band states have dominant heavy hole (HH) character due to the strain-induced large splitting between the HH and LH bands. These heavy hole states are mainly comprised of $|X\rangle$ and $|Y\rangle$ symmetry wave functions, where X and Y are selected along the high symmetry [110] and $[\bar{1}10]$ directions, respectively. The lowest electron state (E_1) is mainly symmetric $|S\rangle$ type wave function (as a good approximation). The elongation of QD along, for example, the X-direction will have negligible impact on the $|S\rangle$ type wave function, but it will increase (decrease) $|X\rangle$ ($|Y\rangle$) component of the valence band states. Therefore, $TE_X \propto |\langle X|S\rangle|^2$ component of the electron–holes transition will increase and $TE_Y \propto |\langle Y|S\rangle|^2$ component will decrease. The opposite is true for $\eta < 1.0$, where the $[\bar{1}10]$ (Y) elongation of the SQD increases $TE_{[\bar{1}10]}$ mode ($|\langle Y|S\rangle|^2$) and it decreases the $TE_{[110]}$ mode ($|\langle X|S\rangle|^2$).

The analysis of the calculated $TM_{[001]}$ component reveals that it also increases for the elliptical QDs. The reason for this increase is twofold [10]: (i) the LH component in the valence band states increases for the elliptical shapes of the QD and (ii) the elongated shape of the QD layers modify the electron and hole wave functions, thereby increasing their spatial overlap along the z-direction. The increase in the TM mode and the decrease in the TE mode along the minor axis lead to a drastic decrease in the values of the $DOP_{[\vec{n}]}$ along the minor axis of the ellipsoidal SQD as shown in Fig. 5.8b. More interestingly, the values of the DOP along the major axis of the ellipsoidal SQD also slightly decreases. This implies that the elliptical-shape of the SQD, in general, improves the polarization response compared to the circular-base. The largest decrease (≈23 %) in the value of the $DOP_{[110]}$ (from 97 to 74.5 %) is calculated for the $[\bar{1}10]$ elongation ($\eta = 0.54$).

In the previous section for the circular-base QDM-4, we have shown that $TM_{[001]} > TE_{[110]}$ leading to $DOP_{[110]} < 0$. However, a significant anisotropy in the in-plane TE mode results in $TE_{[\bar{1}10]} > TM_{[001]}$ and $DOP_{[\bar{1}10]} > 0$. Similar anisotropies in the $DOP_{[\vec{n}]}$ have been recently reported by Alonso-Alvarez et al. [47] and Humlicek et al. [48]. Our multi-million atom simulations explain that this large in-plane anisotropy ($DOP_{[110]} \neq DOP_{[\bar{1}10]}$) is due to a strong confinement of the hole wave functions at the interfaces of the QDM-4 which tend to align along the $[\bar{1}10]$-direction for the QDM-4 (see Fig. 5.7), and thus significantly reduce the $TE_{[110]}$ mode. The $TE_{[\bar{1}10]}$ mode, on the other hand, does not observe any such decrease. The small increase in the $TM_{[001]}$ mode due to the relaxation of the biaxial strain, in particular around the center of the QDM-4, is also not sufficient to overcome the $TE_{[\bar{1}10]}$ mode and therefore the $DOP_{[\bar{1}10]}$ remains considerably larger than zero.

The good qualitative agreement of our theoretical results with the experimental PL measurements in Table 5.1 even for the assumed ideal circular-base for the QD layers in the QDM-4 leads to a fundamental question that how much is the contribution from a realistic shape such as elongated along the [110] or [$\bar{1}$10] directions. In order to investigate this effect, we systematically elongate the QD layers inside the QDM-4 according to the schematic diagram shown in Fig. 5.8a and analyze its impact on the values of the DOP in Fig. 5.8c. In contrast to the SQD, where the QD base elongations result in a significant tuning of the $DOP_{[\vec{n}]}$, the magnitude of the $DOP_{[\vec{n}]}$ for the QDM-4 is relatively insensitive to the value of η. However, the sign of the $DOP_{[\vec{n}]}$ is a strong function of the orientation of the elongation, and even a very small elongation (0.5–1.0 nm) is sufficient to flip the sign of the $DOP_{[\vec{n}]}$. This is highlighted by using an oval in Fig. 5.8, where even for a 1 nm [110]-elongation, the $DOP_{[110]}$ drastically changes its sign from -0.45 to $+0.6$. This implies that the elliptical shape of the QDM-4 cannot be exploited to simultaneously engineer both $DOP_{[110]}$ and $DOP_{[\bar{1}10]}$ below zero.

By comparing Fig. 5.8b and c, we conclude that although a tuning of the $DOP_{[\vec{n}]}$ over a wide range of values is possible by elongation of the SQD, the values of $DOP_{[\vec{n}]}$ remain relatively insensitive with respect to the magnitude of η for the QDM-4. Therefore, we expect that the elongation of the QDM-4 would not offer much improvement in its polarization response. This also explains the reason for the good agreement between the theory and experiment observed in Sect. 5.4.5 even for the ideal circular-base of QD layers in the QDM-4.

5.5.4 Prediction of Geometry Parameters for the QDM-4 from DOP

The above discussion related to the strong dependence of the polarization properties on the [110]-elongations for the QDM-4 allows us to theoretically probe the geometrical shape of the QDM-4 as grown by Inoue et al. [9]. They reported that the QDM-4 do not have isotropic shape and the TEM images suggested very little anisotropy in the lateral extent [28], possibly a [$\bar{1}$10]-elongation [24]. Our multi-million-atom calculations show that the polarization response is very sensitive to the elongation factor (η) and even a 0.5–1.0 nm [110]-elongation increases $DOP_{[110]}$ above zero. Therefore according to our model results, the experimentally measured $DOP_{[110]} = -0.6$ implies that the shape of the QDM-4 can only have [$\bar{1}$10] elongated base which also agrees with the findings from the TEM images. It should also be noted that as the QDM-4 studied here has pure InAs QD layers, so our finding does not contradict with the conclusions of Mlinar et al. [49], where they report that for the alloyed InGaAs QDs, the alloy random configurations may significantly impact the polarization properties and make the correlation between the measured polarization response and the QD geometry unreliable.

We want to conclude this section with this remark that the recent advancements in the growth techniques have allowed to precisely control the shape of QDs, leading to the fabrication of strongly elongated QD-like nano-structures [50]. These offer an enhanced exciton oscillator strength and allow the realization of single exciton-single photon coupling to build the fundamental blocks for the solid state quantum information [19]. Since our calculations indicate a strong dependence of DOP on the elongation of QDs, so this parameter can be exploited to achieve tailored polarization response for a desired operation.

5.6 Conclusions

In summary, this book chapter presents a detailed analysis of the electronic properties and polarization response of the multi-layer QD molecules (QDMs) by performing a set of systematic multi-million atom tight binding calculations. Our theoretical results follow the trends of the experimental measurements on the QDMs containing three, six, and nine QD layers. We provide in-depth physical insight of the underlying fundamental physics by analyzing the strain profiles, the band edge diagrams, and the wave function plots as a function of the QDM size. The polarization properties of the QDMs indicate that the value of the DOP decreases with the size of the QDM, and therefore an isotropic polarization response can be achieved by engineering the number of QD layers inside a QDM. Furthermore, the large in-plane anisotropy ($TE_{\bar{1}10} \gg TE_{110}$), consistent with the experimental evidence, is explained in terms of hole wave function confinements that occur inside HH pockets at the QD interfaces along the $[\bar{1}10]$ direction. We suggest that the isotropic polarization response ($DOP_{110} \sim 0$) from the multi-layer QDMs is due to two factors: (i) the reduction of the TE_{110}-mode due to the hole wave function confining along the $[\bar{1}10]$ direction and (ii) the increase in the TM_{001}-mode due to enhanced LH–HH intermixing. Our results presented in this chapter for various geometry configurations serve as a guidance for the experimentalists to design future QD-based optical devices. A flip in the sign of the DOP as the size of QDM increases indicates significant potential to achieve polarization insensitive response from the multi-layer QDM systems.

Acknowledgments I am indebted to many colleagues with whom I have had the pleasure to work with, and in particular I wish to gratefully acknowledge Gerhard Klimeck (Purdue University USA), Takashi Kita (Kobe University Japan), Timothy B. Boykin (University of Alabama in Huntsville USA), Eoin P. O'Reilly (Tyndall National Institute Ireland), Stefan Schulz (Tyndall National Institute Ireland), and Shaikh S. Ahmed (Southern Illinois University USA). The use of computational resources from the National Science Foundation (NSF) funded Network for Computational Nanotechnology (NCN) through https://nanohub.org is acknowledged. The NEMO 3-D software package is developed by several researchers at Jet Propulsion Labs (JPL) and Purdue University under supervision of Prof. Gerhard Klimeck whom work has been cited in the corresponding references. The open source tools based on NEMO 3-D simulator are available at https://nanohub.org/groups/nemo_3d_distribution.

References

1. Salhi, A., Raino, G., Fortunato, L., Tasco, V., Visimberga, G., Martiradonna, L., Todaro, M., Giorgi, M.D., Cingolani, R., Trampert, A., Vittorio, M.D., Passaseo, A.: IEEE J. Sel. Top. Quant. Electron. **14**, 1188 (2008)
2. Akiyama, T., Sugawara, M., Arakawa, Y.: Proc. IEEE **95**, 1757 (2007)
3. Dousse, A., Suffczyński, J., Beveratos, A., Krebs, O., Lemaître, A., Sagnes, I., Bloch, J., Voisin, P., Senellart, P.: Nature **466**, 217 (2010)
4. Usman, M., Inoue, T., Harda, Y., Klimeck, G., Kita, T.: Phys. Rev. B **84**, 115321 (2011)
5. Usman, M., Heck, S., Clarke, E., Ryu, H., Murray, R., Klimeck, G.: J. Appl. Phys. **109**, 104510 (2011)
6. Fortunato, L., Todaro, M., Tasco, V., Giorgi, M.D., Vittorio, M.D., Cingolani, R., Passaseo, A.: Superlattice Microst. **47**, 72 (2010)
7. Usman, M., Tasco, V., Todaro, M., Georgi, M.D., O'Reilly, E.P., Klimeck, G., Passaseo, A.: IOP Nanotechnology **23**, 165202 (2012)
8. Saito, T., Ebe, H., Arakawa, Y., Kakitsuka, T., Sugawara, M.: Phys. Rev. B **77**, 195318 (2008)
9. Inoue, T., Asada, M., Yasuoka, N., Kojima, O., Kita, T., Wada, O.: Appl. Phys. Lett. **96**, 211906 (2010)
10. Usman, M.: Phys. Rev. B **86**, 155444 (2012)
11. Biasoil, G., Heun, S.: Phys. Rep. **500**, 117 (2011)
12. Stevenson, R., Thompson, R.M., Shields, A.J., Farrer, I., Kardynal, B.E., Ritchie, D.A., Pepper, M.: Phys. Rev. B **66**, 081302 (2002)
13. Plumhof, J.D., Křápek, V., Ding, F., Jöns, K.D., Hafenbrak, R., Klenovský, P., Herklotz, A., Dörr, K., Michler, P., Rastelli, A., Schmidt, O.G.: Phys. Rev. B **83**, 121302(R) (2011)
14. Pryor, C., Flatte, M.: Phys. Rev. Lett. **91**, 257901 (2003)
15. Fricke, M., Lorke, A., Kotthaus, J.P., Medeiros-Ribeiro, G., Petroff, P.M.: Europhys. Lett. **36**(3), 197 (1996)
16. Křápek, V., Kuldová, K., Oswald, J., Hospodková, A., Hulicius, E., Humlıcek, J.: Appl. Phys. Lett. **89**, 153108 (2006)
17. Hospodkova, A., řá, V., Mates, T., Kuldová, K., Pangrac, J., Hulicius, E., Oswald, J., Melichar, K., Humlicek, J., Simecek, T.: J. Cryst. Growth **298**, 570 (2007)
18. Songmuang, R., Kiravittaya, S., Schmidt, O.G.: J. Cryst. Growth **249**, 416 (2003)
19. Favero, I., Cassabois, G., Jankovic, A., Ferreira, R., Darson, D., Voisin, C., Delalande, C., Roussignol, P., Badolato, A., Petroff, P.M., Gérard, J.M.: Appl. Phys. Lett. **86**, 041904 (2005)
20. Usman, M., Ahmed, S., Klimeck, G.: In: Proceedings of the 8th IEEE Conference on Nanotechnology (IEEE-Nano), Arlington, TX, pp. 541–544 (2008)
21. Kojima, O., Nakatani, H., Kita, T., Wada, O., Akahane, K., Tsuchiya, M.: J. Appl. Phys. **103**, 113504 (2008)
22. Kojima, O., Nakatani, H., Kita, T., Wada, O., Akahane, K.: J. Appl. Phys. **107**, 073506 (2010)
23. Ridha, P., Li, H., Mexis, M., Smowton, P.M., Andrzejewski, J., Sek, G., Misiewicz, J., O'Reilly, E.P., Patriarchi, G., Fiore, A.: IEEE J. Quant. Electron. **46**, 197 (2010)
24. Kita, T., Wada, O., Ebe, H., Nakata, Y., Sugawara, M.: Jpn. J. Appl. Phys. **41**, L1143 (2002)
25. Kita, T., Tamura, N., Wada, O., Sugawara, M., Nakata, Y., Ebe, H., Arakawa, Y.: Appl. Phys. Lett. **88**, 211106 (2006)
26. Inoue, T., Asada, M., Kojima, O., Kita, T., Wada, O.: Proc. SPIE **7597**, 75971J (2010)
27. Inoue, T., Asada, M., Yasuoka, N., Kita, T., Wada, O.: J. Phys. Conf. Ser. **245**, 0120761 (2010)
28. Ikeuchi, Y., Inoue, T., Asada, M., Harada, Y., Kita, T., Taguchi, E., Yasuda, H.: Appl. Phys. Express **4**, 062001 (2011)
29. Xie, Q., Madhukar, A., Chen, P., Kobayashi, N.P.: Phys. Rev. Lett. **75**, 2542 (1995)
30. Klimeck, G., Ahmed, S., Hansang, B., Kharche, N., Clark, S., Haley, B., Lee, S., Naumov, M., Ryu, H., Saied, F., Prada, M., Korkusinski, M., Boykin, T., Rahman, R.: IEEE Trans. Elect. Dev. **54**, 2079 (2007)

31. Klimeck, G., Ahmed, S., Kharche, N., Korkusinski, M., Usman, M., Prada, M., Boykin, T.: IEEE Trans. Elect. Dev. **54**, 2090 (2007)
32. Klimeck, G., Oyafuso, F., Boykin, T.B., Bowen, R.C., Allmen, P.V.: Comput. Model. Eng. Sci. **3**(5), 601 (2002)
33. Usman, M.: J. Appl. Phys. **110**, 094512 (2011)
34. Usman, M., Ryu, H., Woo, I., Ebert, D., Klimeck, G.: IEEE Trans. Nanotechnology **8**(3), 330 (2009)
35. Usman, M., Tan, Y., Ryu, H., Ahmed, S., Krenner, H., Boykin, T., Klimeck, G.: IOP Nanotechnology **22**, 315709 (2011)
36. Keating, P.N.: Phys. Rev. **145**, 637 (1966)
37. Lazarenkova, O.L., von Allmen, P., Oyafuso, F., Lee, S., Klimeck, G.: Appl. Phys. Lett. **85**, 4193 (2004)
38. Lee, S., Lazarenkova, O.L., von Allmen, P., Oyafuso, F., Klimeck, G.: Phys. Rev. B **70**, 125307 (2004)
39. Bester, G., Zunger, A.: Phys. Rev. B **72**, 165334 (2005)
40. Boykin, T., Klimeck, G., Bowen, R.C., Oyafuso, F.: Phys. Rev. B **66**, 125207 (2002)
41. Korkusinski, M., Klimeck, G.: J. Phys. Conf. Ser. **38**, 75 (2006)
42. Stier, O., Grundmann, M., Bimberg, D.: Phys. Rev. B **59**, 5688 (1999 II)
43. Lin, Y.Y., Singh, J.: J. Appl. Phys. **92**, 6205 (2002)
44. Ikeuchi, Y., Inoue, T., Asada, M., Harada, Y., Kita, T., Taguchi, E., Yasuda, H.: Appl. Phys. Express **4**, 062001 (2011)
45. Ridha, P., Li, L., Rossetti, M., Patriarche, G., Fiore, A.: Opt. Quant. Electron. **40**, 239 (2008)
46. Sheng, W., Xu, S.: Phys. Rev. B **77**, 113305 (2008)
47. Alonso-Álvarez, D., Ripalda, J.M., Alén, B., Llorens, J.M., Rivera, A., Briones, F.: Adv. Mat. **23**, 5256 (2011)
48. Humlicek, J., Munzar, D., Navratil, K., Lorenc, M., Oswald, J., Pangrac, J., Hulicius, E.: Phys. E **13**, 229 (2002)
49. Mlinar, V., Zunger, A.: Phys. Rev. B **79**, 115416 (2009)
50. Dusanowski, L., Sek, G., Musiał, A., Podemski, P., Misiewicz, J., Löffler, A., Höfling, S., Reitzenstein, S., Forche, A.: J. Appl. Phys. **111**, 063522 (2012)

Chapter 6
Analysis of Reduced Built-In Polarization Fields and Electronic Structure of InGaN/GaN Quantum Dot Molecules

Stefan Schulz and Eoin P. O'Reilly

Abstract Nitride-based semiconductor materials InN, GaN, AlN, and their alloys are attracting great attention due to their promising applications in optoelectronic devices. However, the emission efficiency of *c*-plane InGaN/GaN quantum wells (QWs) drops significantly when going to longer wavelengths due to the strong electrostatic built-in fields in such heterostructures. We present a surface integral method to show and explain why the polarization potential in an InGaN/GaN quantum dot (QD) grown along the [0001]-direction is strongly reduced compared to that in a QW of the same height. We show that the sign of the shear strain piezoelectric coefficient e_{15} strongly affects the built-in field and therefore the electronic structure both of an isolated QD and also of a system of stacked *c*-plane InGaN/GaN dots. Based on two different approaches we conclude $e_{15} < 0$, in agreement with recent independent studies. We then use a tight-binding model and include strain and polarization fields, to study the electronic structure of InGaN/GaN quantum dot molecules grown along the *c*-axis. This analysis is carried out as a function of the barrier thickness firstly between identical and then between two non-identical dots. Our results show that the built-in field can be further reduced in systems of coupled nitride quantum dots, leading to an increased spatial overlap of electron and hole wave functions compared to an isolated dot. This finding is in agreement with experimental data reported in the literature and is directly related to the behavior of the built-in potential outside an isolated dot.

S. Schulz (✉)
Tyndall National Institute, Lee Maltings, Cork, Ireland
e-mail: stefan.schulz@tyndall.ie

E.P. O'Reilly
Tyndall National Institute, Lee Maltings, Cork, Ireland

Department of Physics, University College Cork, Cork, Ireland
e-mail: eoin.oreilly@tyndall.ie

J. Wu and Z.M. Wang (eds.), *Quantum Dot Molecules*, Lecture Notes in Nanoscale Science and Technology 14, DOI 10.1007/978-1-4614-8130-0_6,

6.1 Introduction

The semiconductor materials InN, GaN, and AlN and their alloys have attracted considerable attention due to their promising applications both in quantum information processing applications, where nitride-based quantum dots (QDs) open new spectral regions for single-photon sources [1], and also for more conventional optoelectronic devices such as lasers and light-emitting diodes (LEDs) [2]. These systems are in principle able to cover a wide wavelength range from ultra-violet to infrared [2]. For energy-efficient solid state lighting, which combines output from blue, green, and red LEDs, InGaN alloys are promising candidates, since the assistance of phosphor is theoretically not required for a white light source [2]. While nitride-based heterostructures have already been utilized in blue LEDs [3] and lasers [4], the emission efficiency of c-plane InGaN/GaN quantum wells (QWs) drops significantly when going to longer wavelengths through use of higher indium composition or thicker QWs [2]. This behavior is known in the literature as the "green gap" problem [2, 5]. One of the main reasons for this reduction in efficiency is the strong electrostatic built-in field in nitride-based heterostructures grown along the polar c-axis [6]. The electrostatic built-in field in wurtzite nitride-based heterostructures is of the order of MV/cm, which is at least an order of magnitude larger than that in more conventional zinc blende InGaAs nanostructures [6–8]. Therefore, these built-in fields significantly affect the electronic and optical properties of nitride-based optoelectronic devices.

Several different approaches have been discussed in the literature to reduce or even eliminate these built-in fields [9–16]. One of the strategies to reduce the built-in field in nitride-based optoelectronic devices is to replace QWs by QDs, since the built-in potential in a QD compared to a QW of the same height and composition is significantly reduced [15, 16]. The indium composition can therefore be increased in a QD compared to a QW, enabling efficient recombination to longer wavelengths. We review this approach in more detail, presenting an analysis that gives insight into the key factors that determine the magnitude of the built-in potential in an isolated nitride-based heterostructure. This analysis can then be used to understand and predict how the built-in potential behaves in a stack of QDs. This is of particular interest for a number of reasons. Firstly, different authors [14, 17, 18] have recently demonstrated that InGaN QD-based LEDs and lasers, operating in the amber and green spectral region, show superior performance compared to their QW-based counterparts. In the active region of laser structures, stacks of InGaN QDs have been used, since the application of vertically stacked InGaAs QD structures has been shown to be beneficial for laser applications [19]. For these applications a small spacing between the QD layers is sometimes used to achieve an electronic coupling between the individual dots along the column [20]. Secondly, experimental data indicate that, compared to a single nitride-based QD, a vertical stacking of nitride-based dots also leads to enhanced photoluminescence (PL) efficiency and efficient emission at room temperature [21]. Again, small barrier thicknesses ($D \approx 2$ nm) have been chosen to achieve a stronger coupling between the QDs along the c-axis [21]. In the experimental study in [22], the influence of the distance D between the dots in a stack of InGaN QDs was analyzed, showing that with decreasing D the PL blue-shifts and the PL lifetime decreases. This indicates a decrease in the magnitude of the built-in field with decreasing D.

Thus, understanding the mechanisms of inter-dot coupling is important not only for fundamental properties of coupled nitride-based QDs but also for designing nitride-based optoelectronic devices. However, knowledge about the electronic structure in quantum dot molecules (QDMs) based on more conventional III–V materials, e.g. InAs/GaAs systems, cannot be simply transferred to a nitride-based system. Amongst other reasons, the electrostatic built-in field in wurtzite nitride-based heterostructures is very different from its III–V zinc blende counterpart. Not only is the magnitude of the piezoelectric coefficients in group-III nitrides up to ten times larger than that in conventional III–V and II–VI compounds [23] but also the built-in potential orientation and symmetry in wurtzite nitride-based heterostructures is very different to that in (001)-oriented zinc blende III–V heterostructures. While there is no net potential drop along the growth direction in conventional (001)-zinc blende InAs/GaAs systems, nitride-based heterostructures show a very pronounced potential drop across the heterostructure. Therefore, when stacking wurtzite nitride-based QDs along the polar c-direction, a complicated interplay can be expected between the strain fields, built-in potential effects and electronic coupling.

In this work we present a detailed analysis of the polarization potential in isolated and coupled $In_xGa_{1-x}N/GaN$ QDs and study its impact on the electronic structure and consequently the optical properties. In a first step, before turning to the stacked systems, we discuss the built-in potential of an isolated dot and analyze the different contributions, spontaneous and piezoelectric polarization, separately and in detail. We show how the built-in potential evolves when going from a QW to an isolated QD of the same height and composition. This analysis reveals that the built-in potential of an isolated QD with a realistic aspect ratio is already strongly reduced compared to that in a QW. Following this general introduction to the behavior of the built-in potential in nitride-based heterostructures, we then focus our attention on the details of the built-in potential. For example, since positive as well as negative values for the shear strain-related piezoelectric coefficient e_{15} can be found in the literature [24], the impact of e_{15} on the results is also studied. This analysis reveals that the sign of e_{15} significantly affects the potential outside an isolated dot, and consequently this will also affect the built-in potential in a stacked QD system. After this detailed analysis of the influence of e_{15} on an isolated $In_xGa_{1-x}N/GaN$ QD, we consider the built-in potential in stacked dots. Here, we focus on the impact of the barrier thickness on the potential and the electronic structure. We show that by engineering the distance between two non-identical $In_xGa_{1-x}N/GaN$ QDs, the built-in potential can be further reduced compared to the potential in an isolated dot, which is already significantly lower than in a QW of the same height and composition. Consequently, the use of coupled $In_xGa_{1-x}N/GaN$ QDs should provide a very beneficial reduction in built-in field for optoelectronic devices operating in the green and yellow spectral regions.

This chapter is organized as follows: we begin in Sect. 6.2 by providing an overview of the theoretical framework used here to study the electronic structure of c-plane InGaN/GaN QDMs, including discussion of our approach to the calculation of strain and built-in fields in such systems. Section 6.3 reviews more general features of the electrostatic built-in field of isolated nitride-based QDs, where we show why the built-in electrostatic field should be significantly reduced in an isolated QD compared to the field in a QW of the same composition and height. In Sect. 6.4 we focus on the experimental findings regarding the shape, the size, and the composition of isolated and stacked InGaN dots. We also discuss uncertainties in one of the key material parameters that determines the built-in fields in nitride-based heterostructures, namely the sign of the piezoelectric coefficient e_{15} and show how its value affects the built-in potential both within and outside an isolated InGaN/GaN QD with realistic dimensions and indium content. Section 6.5 considers the electrostatic built-in field in stacked nitride-based QDs, showing how this can be further reduced compared to the field in an isolated dot. Section 6.6 then discusses the electronic structure of InGaN QDMs. We first deal in Sect. 6.6.1 with the idealized situation of a QDM made up of two identical dots, while Sect. 6.6.2 then focuses on the realistic situation of an InGaN/GaN QDM made up of two non-identical dots. Finally, in Sect. 6.7 we summarize our results.

6.2 Theory: Electrostatic Built-In Fields and Electronic Structure Calculations

In this section we discuss the theoretical framework used to study the electronic structure of InGaN/GaN QDMs. For a realistic description, strain and built-in fields have to be taken into account. Therefore, in a first step, Sect. 6.2.1 describes the determination of the strain field and the built-in potentials in nitride-based QDs using a surface integral method. We use this surface integral approach because it can give very useful insight into trends in the behavior of the strain and built-in potential as a function of dot shape and aspect ratio. In a second step, Sect. 6.2.2 describes the method we use for the electronic structure calculations of InGaN/GaN QDMs. These calculations are carried out in the framework of an empirical tight-binding (TB) model, taking both strain and built-in fields into account.

6.2.1 Strain and Built-In Potentials

The growth of semiconductor heterostructures involving alloys with different lattice constants leads inevitably to the introduction of strain into the system. This strain field couples to charge carriers through the electronic (band) structure and the piezoelectric effect. Therefore, to perform realistic electronic structure calculations, strain and electrostatic built-in fields have to be taken into account. Different techniques are available to calculate the strain and the electrostatic built-in fields. These approaches range from density functional theory-based calculations for small systems (few thousands atoms) [25, 26], through to atomistic descriptions based on semi-empirical force field models [27–30], which can treat up to several millions of atoms, and on to continuum-based descriptions [27, 31, 32], which in turn can deal with even larger systems. Using a continuum-based description, analytic results can be derived by assuming isotropic and homogenous elastic properties. Despite the simplifications involved, these techniques provide considerable insight into the strain field of a variety of different nanostructures [33–36].

On the same footing, assuming isotropic and homogenous material parameters, Davies [37] and Williams et al. [38] derived real space surface integral methods to describe the electrostatic built-in fields in cubic and wurtzite-based semiconductor nanostructures. We summarize here how these real space surface integral methods can be used to calculate and understand the behavior of the strain and built-in potentials in nitride-based heterostructures.

6.2.1.1 Calculation of the Strain Field

To calculate the strain field in a QD system, we apply the surface integral method discussed in detail in Refs. [35, 37]. This approach is based on Eshelby's theory of inclusion [39] to express the stress and strain for a QD embedded in an infinite medium as an integral over the nanostructure under consideration. The strain tensor components $\varepsilon_{ij}(\mathbf{r})$ for isotropic and homogenous elastic constants can be written as integrals over the surface of the QD [35, 37]:

$$\varepsilon_{ij}(\mathbf{r}) = \delta_{ij}\varepsilon_0\chi_{\mathrm{QD}} + \frac{\varepsilon_0 A}{4\pi}\int_{\mathrm{QD}} \frac{(x_i - x_i')}{|\mathbf{r}-\mathbf{r}'|^3}\hat{\mathbf{n}}_j \cdot d\mathbf{S}', \tag{6.1}$$

where the primed quantities denote points on the surface of the dot, $(x_1,x_2,x_3) \equiv (x,y,z)$, ε_0 is the initial misfit and χ_{QD} is defined as the dot characteristic function, which is equal to 1 inside the QD and zero outside. The constant $A = \frac{1+\nu}{1-\nu}$, where ν is the Poisson ratio [38]. The unit vector along the j-direction is denoted by $\hat{\mathbf{n}}_j$. The strain field of a QDM can be calculated by a superposition of the strain fields obtained from Eq. (6.1) for two independent dots centered at different positions. We note also for a cuboid QD that the strain components ε_{ii} at a given point are directly related to the solid angles subtended by the two cuboid surfaces perpendicular to $\hat{\mathbf{n}}_i$. This simple relation can be very useful for the design and analysis of different QD structures [36, 40].

6.2.1.2 Calculation of the Built-In Potential

To treat strain and built-in potential on the same theoretical footing we use the surface integral method developed by Williams et al. [38] to achieve a realistic description of the electrostatic built-in potentials in nitride-based QDs and QDMs.

Again, the benefit of using the surface integral method is that this approach admits analytic solutions for certain QD geometries, and thereby provides useful insights into the parameters that influence the magnitude and the shape of the built-in potential. More details about this method are given in Refs. [38, 41].

In nitride-based heterostructures, we have two contributions to the total built-in potential. The first one is known as the spontaneous polarization $\mathbf{P}^{\text{sp}}$, arising from the lack of inversion symmetry along the c-axis of the wurtzite crystal structure [42]. The second contribution, referred to as the piezoelectric polarization $\mathbf{P}^{\text{pz}}$, is related to the local strain in the system [42]. The total polarization vector of a wurtzite nitride system, treating piezoelectric response to first order, is given by [38]:

$$\begin{aligned}\mathbf{P}^{\text{tot}} &= \mathbf{P}^{\text{sp}} + \mathbf{P}^{\text{pz}} \\ &= \begin{pmatrix} 0 \\ 0 \\ P_{\text{sp}} \end{pmatrix} + \begin{pmatrix} 2e_{15}\varepsilon_{13} \\ 2e_{15}\varepsilon_{23} \\ e_{31}(\varepsilon_{11}+\varepsilon_{22}) + e_{33}\varepsilon_{33} \end{pmatrix},\end{aligned} \tag{6.2}$$

where P_{sp} is the spontaneous polarization, e_{ij} are the piezoelectric coefficients, and ε_{ij} denotes the strain tensor components.

It is simplest when introducing the surface integral technique to first consider the contribution ϕ_{sp} to the total potential arising from the spontaneous polarization, before then extending the analysis to the piezoelectric contribution ϕ_{pz}.

Spontaneous Polarization

When deriving the potential originating from the spontaneous polarization, we make use of the fact that $\mathbf{P}^{\text{sp}}$ is a constant vector in the nanostructure region and the surrounding barrier material, respectively, and points in both regions along the c-axis. Using a standard result from electromagnetism theory [43], the spontaneous polarization potential ϕ_{sp} can be written as [38]:

$$\phi_{\text{sp}} = \frac{1}{4\pi\varepsilon_r\varepsilon_0}\left[\int_{\text{QD}} \frac{\mathbf{P}^{\text{QD}}_{\text{sp}} \cdot d\mathbf{S}'}{|\mathbf{r}-\mathbf{r}'|} + \int_{\text{B}} \frac{\mathbf{P}^{\text{B}}_{\text{sp}} \cdot d\mathbf{S}_B}{|\mathbf{r}-\mathbf{r}_B|}\right] \tag{6.3}$$

$$\approx \frac{P^{\text{QD}}_{\text{sp}} - P^{\text{B}}_{\text{sp}}}{4\pi\varepsilon_r\varepsilon_0}\int_{\text{QD}} \frac{1}{|\mathbf{r}-\mathbf{r}'|}\hat{\mathbf{n}}_3 \cdot d\mathbf{S}', \tag{6.4}$$

where $\mathbf{P}^{\text{QD}}_{\text{sp}} = P^{\text{QD}}_{\text{sp}}\hat{\mathbf{n}}_3$ is the spontaneous polarization of the QD material (e.g., InN) and $\mathbf{P}^{\text{B}}_{\text{sp}} = P^{\text{B}}_{\text{sp}}\hat{\mathbf{n}}_3$ is that of the barrier material (e.g., GaN). In Eq. (6.3), the first integral is taken over the surface of the QD while the second is taken over all the surfaces of the surrounding barrier material. Here we assume that the QD is buried within an infinite matrix. The quantity $\mathbf{P} \cdot d\mathbf{S}$ is equivalent to a fictitious charge $\mathbf{P} \cdot \hat{\mathbf{n}}$ on a surface element dS, where $\hat{\mathbf{n}}$ is the unit vector normal to the surface [38]. When going from Eq. (6.3) to Eq. (6.4), we have neglected the electrostatic field and potential arising from the outer surface of the surrounding matrix. We assume that in an actual structure external charges will accumulate on this outer surface and counteract this potential, so that it is negligible compared to the field arising from the interface between the QD and the barrier material. Consequently, the electrostatic potential ϕ_{sp} in Eq. (6.4), arising from the spontaneous polarization is determined by the difference $\Delta P_{\text{sp}} = P_{\text{sp}}^{\text{QD}} - P_{\text{sp}}^{\text{B}}$ between the spontaneous polarization of the QD and the surrounding barrier material and is therefore equivalent to the potential due to a charge density $\Delta P_{\text{sp}}\hat{\mathbf{n}}_3 \cdot \mathbf{n}$ distributed over the QD surface. Having established the form of the built-in potential due to the spontaneous polarization, we turn now to consider the piezoelectric potential.

Piezoelectric Polarization

In contrast to the spontaneous polarization, the piezoelectric polarization vector can have nonzero components in all three spatial directions in a QD or a QDM. In addition, $\mathbf{P}^{\text{pz}}$ is not a constant vector within the QD, since the strain field $\varepsilon_{ij}(\mathbf{r})$ is position dependent. Using the integral expression of the strain field, Eq. (6.1), in combination with Maxwell's equations, one can derive equations to calculate the position-dependent piezoelectric potential $\phi_{\text{pz}}(\mathbf{r})$ by evaluating two 2-D integrals over the QD surface [38]:

$$\phi_{\text{pz}}(\mathbf{r}) = J \int_{\text{QD}} \frac{(x_i - x_i')^2}{|\mathbf{r}-\mathbf{r}'|^3} \hat{\mathbf{n}}_3 \cdot d\mathbf{S}' + K \int_{\text{QD}} \frac{1}{|\mathbf{r}-\mathbf{r}'|} \hat{\mathbf{n}}_3 \cdot d\mathbf{S}', \tag{6.5}$$

where the constants J and K are functions of the piezoelectric coefficient e_{ij}, the Poisson ratio ν and the initial misfit ε_0. When calculating strain and polarization potentials we apply a linear interpolation for all parameters except for the spontaneous polarization, where a quadratic interpolation is applied [44].

We will discuss in Sect. 6.3 the general features of the built-in potential arising from spontaneous and piezoelectric polarization in an isolated nitride-based QD. However, before discussing this, we give an overview of the theoretical framework which we use for the electronic structure calculations, including how the effects of strain and built-in fields are incorporated in the model.

6.2.2 Electronic Structure Calculations

To study the electronic properties of isolated QDs and QDMs different approaches have been used in the literature, ranging from continuum-based models, e.g. single-band effective mass [45, 46] and multi-band $\mathbf{k} \cdot \mathbf{p}$ [47–49] theory, to atomistic models such as pseudo-potential [29, 50–52] and tight-binding (TB) calculations [28, 53–55]. Here, to analyze the electronic structure of InGaN/GaN QDMs we apply an empirical sp^3 TB model [55, 56]. The TB matrix elements are treated as parameters and are determined by fitting the TB band structures for InN and GaN to ab initio band structures from the literature to reproduce key properties such as the band gaps. The TB parameters for $In_xGa_{1-x}N$ are obtained from a modified virtual crystal approximation [57], which allows us to take the band gap bowing into account. The band gap bowing parameter used for InGaN has been taken from [44]. Such an approach has been often used to calculate the electronic structure of alloyed semiconductor materials [57–59].

Starting from the bulk TB parameters, nanostructures can then be modeled at an atomistic level. To this end one chooses the matrix elements at each lattice site according to the occupying atom. In general, the resulting ith TB wave function of the nanostructure $|\psi_i\rangle$ is expressed in terms of the localized orbitals $|\alpha, \omega, \sigma, \mathbf{R}\rangle$:

$$|\psi_i\rangle = \sum_{\alpha,\omega,\sigma,\mathbf{R}} c^i_{\alpha,\omega,\sigma,\mathbf{R}} |\alpha, \omega, \sigma, \mathbf{R}\rangle. \tag{6.6}$$

Here, $\mathbf{R}$ denotes the unit cell, α the orbital type, σ the spin and ω labels the anions and cations in the given unit cell. The Schrödinger equation $H|\psi_i\rangle = E^i|\psi_i\rangle$ can then be expressed as the following finite matrix eigenvalue problem:

$$\sum_{\alpha,\omega,\sigma,\mathbf{R}} \langle\alpha', \omega', \sigma', \mathbf{R}'|H|\alpha, \omega, \sigma, \mathbf{R}\rangle c^i_{\alpha,\omega,\sigma,\mathbf{R}} - E^i c^i_{\alpha',\omega',\sigma',\mathbf{R}'} = 0, \tag{6.7}$$

where E^i is the corresponding eigenvalue. In the following we use the abbreviation $\langle\alpha', \omega', \sigma', \mathbf{R}'|H|\alpha, \omega, \sigma, \mathbf{R}\rangle = H_{l\mathbf{R}',m\mathbf{R}}$ for the matrix elements with $l = \alpha', \omega', \sigma'$ and $m = \alpha, \omega, \sigma$.

In setting up the Hamiltonian, one has to include the local strain $\varepsilon_{ij}(\mathbf{r})$ and the total built-in potential $\phi_{\text{tot}} = \phi_{\text{pz}} + \phi_{\text{sp}}$ to ensure an accurate description of the electronic properties of QDs and QDMs. Several authors have shown that this can be done by introducing on-site corrections ($\mathbf{R} = \mathbf{R}', \omega = \omega', \sigma = \sigma'$) to the TB matrix elements $H_{l\mathbf{R}',m\mathbf{R}}$ [60, 61]. Therefore, we proceed here in the following way. The strain dependence of the TB matrix elements is included via the Pikus–Bir Hamiltonian [47, 62] as a site-diagonal correction:

$$H^{\text{str}}_{l\mathbf{R},m\mathbf{R}} = \begin{pmatrix} S_s & 0 & 0 & 0 \\ 0 & S_x & S_{xy} & S_{xz} \\ 0 & S_{xy} & S_y & S_{yz} \\ 0 & S_{xz} & S_{yz} & S_z \end{pmatrix}, \tag{6.8}$$

with

$$\begin{aligned}
S_s &= a_{ct}(\varepsilon_{11} + \varepsilon_{22}) + a_{cp}\varepsilon_{zz}, \\
S_x &= (D_2 + D_4)(\varepsilon_{11} + \varepsilon_{22}) + D_5(\varepsilon_{11} - \varepsilon_{22}) + (D_1 + D_3)\varepsilon_{33}, \\
S_y &= (D_2 + D_4)(\varepsilon_{11} + \varepsilon_{22}) - D_5(\varepsilon_{11} - \varepsilon_{22}) + (D_1 + D_3)\varepsilon_{33}, \\
S_z &= D_2(\varepsilon_{11} + \varepsilon_{11}),
\end{aligned}$$

$$S_{xy} = 2D_5\varepsilon_{12},$$
$$S_{xz} = \sqrt{2}D_6\varepsilon_{13},$$
$$S_{yz} = \sqrt{2}D_6\varepsilon_{23},$$

where the D_i's denote the valence-band deformation potentials, while a_{cp} and a_{ct} are the conduction-band deformation potentials.[1] With this approach, the relevant deformation potentials for the highest valence and lowest conduction band states are included directly without any fitting procedure. In the work described below, the deformation potentials for InN and GaN are taken from hybrid-functional density functional theory calculations [63], and a linear interpolation is used to obtain the parameters for $In_xGa_{1-x}N$. Our approach is similar to that used for the strain dependence in an 8-band $\mathbf{k}\cdot\mathbf{p}$ model [47], but has the benefit that the TB Hamiltonian still takes the correct C_{3v} symmetry of the system into account, including explicitly the detail of the dot-barrier interface region.

The built-in potential ϕ_{tot}, arising from both spontaneous and piezoelectric polarization is likewise included as a site-diagonal contribution in the TB Hamiltonian. This also is a widely used approach [54, 64–66].

To perform a realistic calculation of the electronic structure of InGaN/GaN QDs and QDMs, knowledge about the dot size, shape and composition is required, as well as a consistent set of input material parameters, such as band offsets, elastic constants, piezoelectric coefficients, etc. Even though a lot of progress has been made in recent years on the growth of high-quality nitride samples, there is still a large degree of uncertainty in some of the key material parameters, such as the piezoelectric coefficients e_{ij} [13]. We briefly review these uncertainties in the piezoelectric coefficients in Sect. 6.5 and show how these uncertainties affect the calculated electrostatic built-in fields in isolated QDs and QDMs. However, before turning to such details, we focus our attention in the next section on more general aspects of the built-in potential in an isolated dot. This will establish the foundations for understanding the more complex behavior of the built-in field and of the electronic structure in stacked InGaN/GaN QDs.

6.3 Built-In Fields in Nitride-Based QDs: General Aspects and Simple Analytic Solutions

Having outlined the general aspects of the theoretical framework used here to study isolated and stacked InGaN/GaN QDs with realistic dimensions and indium contents, it is useful to next consider the general shape and form of the electrostatic built-in potential of a single InGaN/GaN dot. In the following section we briefly

[1] Note that the quantities a_2 and a_1 given by Vurgaftman and Meyer in their 2003 review article [44] *are not* the conduction-band deformation potentials a_{cp} and a_{ct}, respectively. The quantities denoted by a_1 and a_2 in [44] are the *band gap* deformation potentials, e.g. $a_1 = a_{cp} - D_1$ and $a_2 = a_{ct} - D_2$.

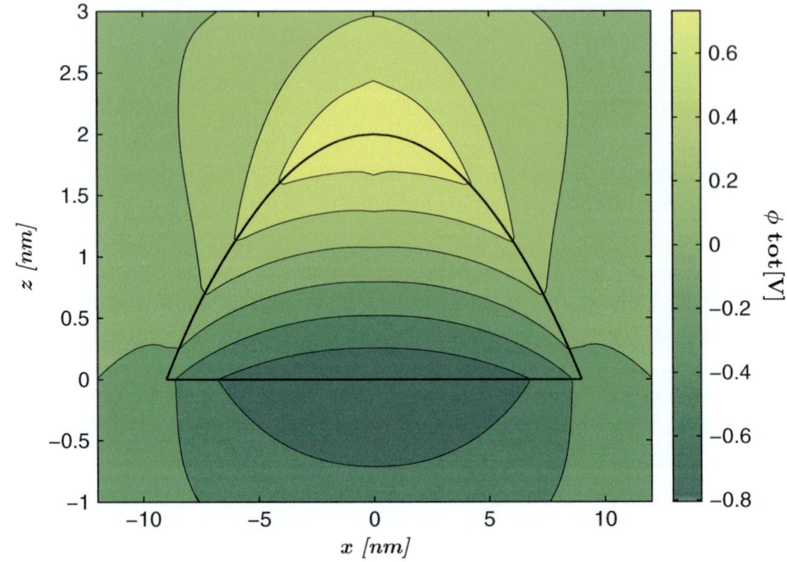

Fig. 6.1 Contour plot of the total (spontaneous + piezoelectric) built-in potential ϕ_{tot} for a c-plane lens-shaped InN/GaN QD. The potential is shown for a slice through the QD center. The z-axis is parallel to the [0001]-direction. [From [67]]

discuss the overall variation of the built-in potential and how it affects the electronic structure and consequently the optical properties. In a second step we then analyze how the aspect ratio of a c-plane nitride-based heterostructure affects the built-in potential drop across such a system. This analysis reveals that the built-in potential across a c-plane nitride-based QD is significantly reduced compared to that across a QW of the same height and with the same indium composition. This opens the possibility to use QDs with higher indium composition to achieve more efficient radiative carrier recombination in optoelectronic devices operating in the green and yellow spectral region.

6.3.1 Built-In Potential in an Isolated QD

Figure 6.1 shows a contour plot of the total built-in potential ϕ_{tot} calculated for a model c-plane InN/GaN QD with a diameter of $d = 18$ nm and a height of $h = 2$ nm, chosen to illustrate the typical properties of the polarization potential in nitride-based QDs. First of all, one observes a large potential difference, $\Delta\phi_{tot} \approx 1.5$ V, between the top and the bottom of the dot. This potential difference causes a spatial separation of electron and hole wave functions, with the negatively charged electrons moving towards the top of the dot while the positively charged holes move towards the bottom of the QD. This behavior has two main consequences. Firstly, the spatial separation of electron and hole wave functions leads to a decrease in the oscillator strength and therefore to an increase in the radiative lifetimes. Secondly, the built-in potential gives rise to a red-shift of the exciton emission energies with respect to the band gap energy. This field-induced separation and red shift is referred to as the quantum confined Stark effect (QCSE).

We note that the electrostatic built-in potential also leads to an additional lateral confinement for both electrons and holes. This ensures that the carriers are well confined within the QD, close to its central axis. Even though our analysis has been presented for an InN/GaN QD, the same conclusions also hold for GaN/AlN systems [7, 38, 41] and for c-plane nanostructures with ternary or quaternary III-N alloys in the dot and/or barrier regions.

6.3.2 Comparison Between the Built-In Potential in a QD and a QW

Having established some of the general features of the electrostatic built-in field in nitride-based QDs, we turn now to discuss in more detail how the built-in potential changes when the QD geometry and also its aspect ratio is changed. Before looking at more realistic QD geometries such as lens-shaped or ellipsoidal structures, we start with the analysis of a cuboid-shaped dot. This simple system is of interest because a complete analytic solution can be obtained for the surface integrals in Eqs. (6.4) and (6.5) for a cuboidal QD [38], thereby permitting a straightforward understanding of how the built-in potential evolves in going from a c-plane QW to a c-plane QD of the same height.

We start by considering the spontaneous polarization potential ϕ_{sp} between the top and bottom center of a cuboid-shaped QD of height $2h$ and base length $2B$, with the dot aspect ratio then given by $F = h/B$, where $F = 0$ corresponds to the QW limit. From the analytic solution for the spontaneous polarization ϕ_{sp} given in [38], we obtain that the potential difference $\Delta\phi_{\text{sp}}$ varies between the center of the top and bottom surfaces of the cuboid for small F as [16]:

$$\Delta\phi_{\text{sp}} \approx \left(\frac{2(P^{\text{sp}}_{\text{QD}} - P^{\text{sp}}_{\text{B}})}{\varepsilon_0\varepsilon_r}\right) h \left(1 - \frac{2\sqrt{2}}{\pi}F\right) = C_{\text{sp}}h\left(1 - \frac{2\sqrt{2}}{\pi}F\right). \tag{6.9}$$

This result shows that $\Delta\phi_{\text{sp}}$ initially decreases linearly with F in a QD compared to a QW structure of the same height, as shown in Fig. 6.2a. This is due to the reduction of the $[0001]$ (c-plane) surface area of the QD and therefore to a reduction of the surface charge compared to that in the corresponding QW system.

The second contribution to the total built-in potential, the piezoelectric part ϕ_{pz}, is modified by two factors when going from a QW to a QD of the same height. Firstly, as in the case of the spontaneous polarization potential ϕ_{sp} the finite size of the QD affects the piezoelectric contribution. Secondly, there is also a strain redistribution in

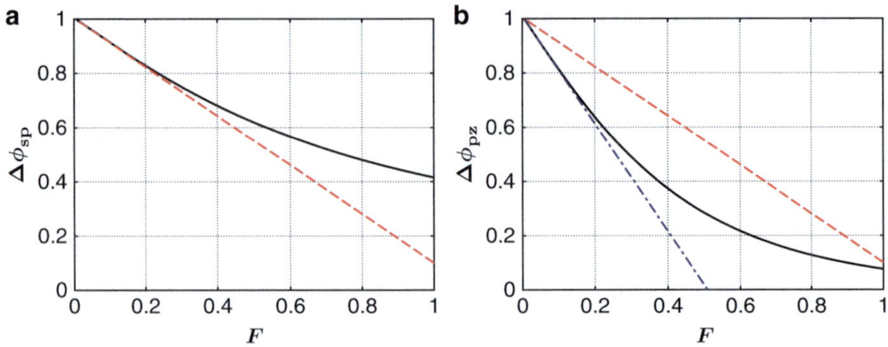

Fig. 6.2 (**a**) Normalized variation of the potential drop in the spontaneous potential $\Delta\phi_{sp}$ across a cuboid-shaped QD as a function of the height to base length ratio $F = h/B$. *Solid line* shows $\Delta\phi_{sp}$ using the exact solution from [38]; *dashed (red) line* is calculated using Eq. (6.9). (**b**) *Solid line*: normalized variation of the strain-related piezoelectric potential $\Delta\phi_{pz}$ using exact solution from [38]. *Dashed (red) and dashed-dotted (blue) lines*: results from Eq. (6.10) assuming, respectively, constant strain only (first term), and allowing also for first-order changes in the strain field (both terms). [From [16]]

the QD compared to a QW. By using the equations given in [38], it can be shown that the potential difference $\Delta\phi_{pz}$ between the center of the top and the bottom surfaces of the QD varies for small F as [16]:

$$\Delta\phi_{pz} \approx C_1 h \left(1 - \frac{2\sqrt{2}}{\pi} F\right) - C_2 h \frac{2\sqrt{2}}{\pi} F, \tag{6.10}$$

with

$$C_1 = \frac{2\varepsilon_0(2e_{31} + (1-A)e_{33})}{\varepsilon_r\varepsilon_0}, \tag{6.11}$$

and

$$C_2 = \frac{\varepsilon_0 A\left[2e_{15} - e_{33} + e_{31}\right]}{\varepsilon_r\varepsilon_0}. \tag{6.12}$$

Here, ε_0 is the isotropic misfit strain, ε_r is the dielectric constant, and $A = \frac{1+\nu}{1-\nu}$, with ν being the Poisson ratio. The first term on the right-hand side (RHS) of Eq. (6.10) is of the same form as the result for the potential difference $\Delta\phi_{sp}$ arising from the spontaneous part given by Eq. (6.9). The second (C_2) term originates from the strain redistribution in the QD, including contributions from the shear strain part of the piezoelectric polarization (e_{15}) as well as axial terms related to e_{31} and e_{33}, cf. Eq. (6.2). In the case of a QW we have a biaxial compressive strain in the c-plane and a tensile strain along the $[0001]$-direction. Due to the finite dot size, a QD structure is able to relax both in the x-y-plane (c-plane) and also along the [0001]-direction, so that the magnitude of each of the axial strain components is then reduced in the QD compared to a QW of the same composition ($|\varepsilon_{ii}^{\mathrm{QD}}| \leq |\varepsilon_{ii}^{\mathrm{QW}}|$). The strain-related piezoelectric potential drop $\Delta\phi_{\mathrm{pz}}$ across the nanostructure is shown as a function of F in Fig. 6.2b. The full calculation, based on the results given in [38], is shown by the solid line. The (red) dashed line shows the contribution from the first term on the RHS of Eq. (6.10), while the (blue) dashed-dotted line includes both terms on the RHS of Eq. (6.10). From Fig. 6.2b we see that the piezoelectric potential drop $\Delta\phi_{\mathrm{pz}}$ is reduced by approximately 60% for $F = 0.4$, with approximately equal contributions to this reduction from the strain redistribution in the QD system and the reduction of the QD surface area compared to a QW.

A similar analysis was presented for a cylindrical QD in [41], which again showed a significant reduction in the potential drop across such a QD compared to a QW of the same composition and height. It should be noted when comparing Fig. 6.2 with Fig. 5 in [41] that the potential drop in the cylindrical shaped QD analyzed there was presented in terms of $f = h/R$, where R is the radius and h the *total* height.

The same analysis can also be carried out for more realistic QD geometries such as ellipsoidal or lens-shaped InGaN QDs [68–70]. We will discuss the experimental findings on QD geometries, dimensions, and composition range in more detail in Sect. 6.4. Here, only the overall geometry is important. Using the results given in [38] for an ellipsoidal dot centered at the origin, with semi-major radius a and semi-minor radius b, we calculate the potential drop across the nanostructure as a function of $F = b/a$.[2] The results for the calculated drop in the spontaneous potential $\Delta\phi_{\mathrm{sp}}$ across the QD are shown in Fig. 6.3a, while those for $\Delta\phi_{\mathrm{pz}}$ are shown in Fig. 6.3b, with the approximate solutions for small F in both cases including terms up to F^3. Comparing the results to those for the cuboid-shaped QD (cf. Fig. 6.2a), we observe that the drop in the spontaneous potential difference $\Delta\phi_{\mathrm{sp}}$ (cf. Fig. 6.3a) is clearly increased in the case of the ellipsoid-shaped dot. The same is true for the drop in the piezoelectric potential difference $\Delta\phi_{\mathrm{pz}}$, as shown in Figs. 6.2b and 6.3b, respectively. This further reduction is directly due to the curved shape of the QD side walls. In a cuboid QD, each point in the dot is below all points on the full [0001]-oriented top surface, so that all surface points then give contributions of the same sign to the total potential. With a curved top surface, points near the top of the dot experience contributions of opposite sign from points on the upper surface of the dot which are below and above the given point, leading to an overall reduction in the total potential change.

A similar effect can be seen in Fig. 6.4, where we plot the normalized variation of the drop in the potential difference when going from a QW to a lens-shaped QD as a function of $F = h/D$, where h is the height of the lens and D the base diameter. By comparing the results of the lens-shaped system, the ellipsoidal QD and the cylindrical shaped dot [41] for the same aspect ratio (e.g., $F = 0.5; f = 1$),

---

[2]Note that the term $\mp 2\pi \mathrm{sgn}(z \pm b)(z \pm b)\frac{a^2}{c^2}\left(\frac{3b^2}{c^2} + \frac{c^2 - b^2 + z(z \pm b)}{c^2 + z^2}\right)$ is missing in Eq. (B13) for I_1 given in the appendix of Ref. [38] for an ellipsoid-shaped system.

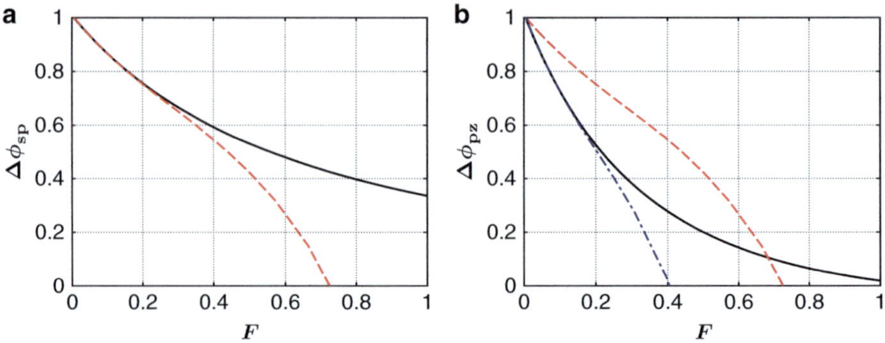

Fig. 6.3 (**a**) Normalized variation of $\Delta\phi_{sp}$ across an ellipsoid shaped QD as a function of the semi-minor to semi-major axis ratio $F = b/a$. *Solid black line*: exact solution from [38]; *Dashed (red) line*: cubic approximation for small F. (**b**) Normalized variation of $\Delta\phi_{pz}$; *solid line*: exact solution from [38]; *dashed (red) line*: results obtained by assuming constant strain; *dashed-dotted (blue) line* accounts for the finite QD size and changes in the strain field. [From [16]]

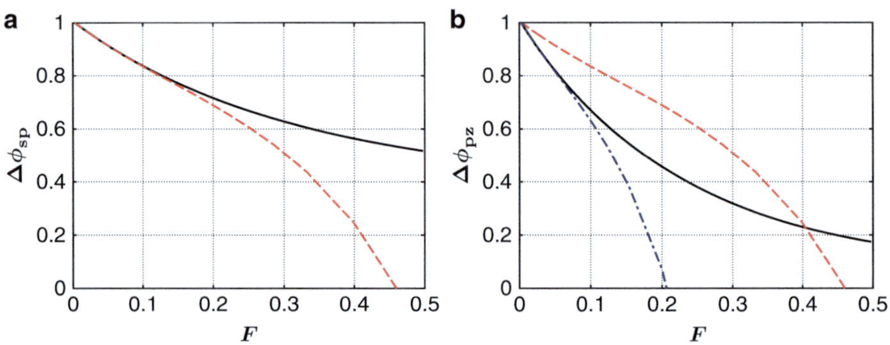

Fig. 6.4 (**a**) Normalized variation of $\Delta\phi_{sp}$ across a lens-shaped QD of height h as a function of the height to diameter ratio $F = h/D$, where D is the diameter of the lens. *Solid black line*: exact solution from [38]; *dashed (red) line*: cubic approximation for small F. (**b**) Normalized variation of $\Delta\phi_{pz}$. *Solid line*: exact solution from [38]; *dashed (red) line*: results obtained assuming constant strain; *dashed-dotted (blue) line* accounts both for the finite QD size and also for changes in the strain field. [From [16]]

we conclude that $\Delta\phi_{tot}$ is almost the same in the first two systems, while the calculated value of $\Delta\phi_{tot}$ is considerably larger for the cylindrical QD. Again, this difference arises because of the vertical QD side walls in the cylindrical dot. Overall we conclude that the combination of three effects can lead to a significant reduction in the potential drop across a QD compared to a QW of the same composition and height, namely (i) the reduced surface area of the QD, (ii) strain relaxation in the QD, and (iii) the sloped side walls which are generally found in a QD structure. Such a reduction in built-in field allows a considerably higher radiative recombination rate in a QD compared to an equivalent QW structure, which is of significant benefit for optoelectronic applications.

6.4 InGaN QDs and Material Parameters

The basic considerations of the previous section allowed us to establish the overall behavior of the electrostatic built-in field in isolated nitride-based QDs and how this affects the electronic and optical properties in general. This analysis also gave insight into the changes in the built-in potential when both the QD geometry and the aspect ratio are changed. However, to perform realistic electronic structure calculations, more details about the QD geometry, such as the size and the composition are required. Here input from growth simulations and/or experimental data is required.

For InGaN/GaN QDs the shape is often approximated by truncated pyramids with a hexagonal base, or by using ellipsoidal or lens-shaped dot geometries [11, 15, 47, 64, 66, 71], as indicated by experimental studies [68–70, 72]. Looking at the geometrical features in more detail, experiments also reveal that the average diameter of InGaN QDs scatters around 15–25 nm while the average height is approximately 2–6 nm [22, 70, 72]. When looking at the indium compositions of InGaN QD systems, structures with 15–35% indium have been reported in the literature [72, 73]. There has been no detailed measurement on the variation of the indium content in stacked InGaN QDs. However, the analysis of coupled InGaAs QDs shows that, due to strain relaxation in the structure, the indium composition of the upper QD is typically higher than in the lower one [74].

When looking at the strain fields, the electrostatic built-in fields and the electronic structure of stacked InGaN/GaN QDs, the relative orientation of the different QDs with respect to each other is also important. Based on the experimental data in [75] and the discussions in [22], we assume a vertical stacking of the QDs in our analysis. This assumption is also consistent with the experimental findings on stacked InGaAs/GaAs QDs [74].

Having outlined the experimental observations on the shape, size, and indium content of InGaN/GaN QDs and QDMs, the theoretical description of these systems also needs as input the values of material parameters such as elastic constants, valence band and conduction band offsets, piezoelectric coefficients, etc. Detailed studies of the influence of different parameter sets on the electrostatic built-in fields and the electronic structure of QWs and QDs can be found, for example, in [13, 76]. Here, we focus our attention on one quantity, namely the shear strain-related piezoelectric coefficient e_{15} which turns out to affect the electrostatic built-in potential and consequently the electronic structure of stacked InGaN/GaN QDs significantly. When looking at the values of the piezoelectric coefficients e_{31} and e_{33} given in the literature, these values scatter mainly in magnitude but not in sign. Turning to the shear strain component, there is not only uncertainty in the magnitude of e_{15}, but there is also conflicting evidence as to its sign [24]. The initial and often used values of Bernardini et al. [77] and Vurgaftman et al. [44] are positive, while more recent results recommend a negative value of e_{15} [78–81]. For example, bias-dependent photoluminescence measurements on semi-polar InGaN QWs have recently been used to determine the sign of e_{15}, concluding that $e_{15} < 0$ [80].

To shed more light on the sign of e_{15}, we have used two different approaches in two different systems to gain further insight into the sign of e_{15}. In the following two subsections we briefly outline these approaches and summarize their outcomes, which further support $e_{15} < 0$. Section 6.4.3 then analyzes the impact of the sign of e_{15} on the total built-in potential in an isolated lens-shaped InGaN/GaN QD with realistic dimensions and indium content.

6.4.1 Sign of e_{15} from Analysis of the Built-In Field in Non-polar Nitride-Based QDs

As discussed above the strong electrostatic built-in fields in c-plane InGaN/GaN heterostructures limit their applicability for optoelectronic devices. To circumvent the problems arising from growth along the polar c-axis ([0001]-direction), there has been a rapid increase in studies of non-polar growth of III-nitride structures, where the [0001]-direction lies within the growth plane [10,82–84]. Growth of QWs along a non-polar direction can eliminate polarization-induced fields, and hence give improved radiative recombination rates [9], since there is no discontinuity in the polarization vector along the [0001]-direction.

In contrast to a non-polar QW, QDs grown along a non-polar direction still have [0001]-oriented side facets. These interfaces give rise to a discontinuity in the polarization vector, Eq. (6.2), leading then to a net polarization potential across the QD. However, optical measurements suggest that the polarization potential is strongly reduced in non-polar (In)GaN QDs compared to their c-plane counterparts [10,83,85].

Recently, we have presented a detailed analysis of the different contributions to the electrostatic potential of non-polar wurtzite GaN/AlN and InGaN/GaN QDs [24,41,71]. Special attention was paid to the sign of the piezoelectric constant e_{15}. Since the contribution from the spontaneous polarization in InN/GaN QDs is negligible compared to the strain-induced component, a significant electrostatic built-in field might be expected even in non-polar *pure InN* systems, independent of the sign of e_{15}. However, our results for non-polar InGaN/GaN QDs with a more realistic indium content (15%) [71] reveal that the built-in fields are negligible in these systems. This finding has recently been confirmed experimentally [85].

While the total potential for InN/GaN QDs is almost entirely due to the piezoelectric contribution, spontaneous and piezoelectric potentials contribute almost equally in GaN/AlN QDs [24, 38]. Our investigations [24] reveal that only a negative value of e_{15} leads to a strong reduction of the built-in field in non-polar GaN/AlN QDs compared to their polar counterparts, as observed experimentally. Furthermore, we have recently shown [11] that when using $e_{15}<0$ and assuming realistic QD geometries for these non-polar structures an increased spatial overlap of the electron and hole ground state wave functions is observed, in agreement with optical measurements discussed in the literature [10,83]. Therefore, we conclude that e_{15} is most likely to be negative in wurtzite nitrides, a conclusion further supported by the results presented in the next section.

6.4.2 Sign of e_{15} from the Piezoelectric Tensor in (111)-Oriented Zinc Blende Structures

We have recently derived expressions for the stiffness tensor and the elastic energy in (111)-zinc blende systems along with equations for the piezoelectric polarization vectors in these structures [86]. These equations allow for an efficient calculation of strain and piezoelectric fields in (111)-oriented zinc blende nanostructures, such as site-controlled (111)-oriented InGaAs/GaAs QDs, which are most promising candidates for sources of entangled photon pairs [87, 88].

Moreover, since (111)-oriented zinc blende and *c*-plane wurtzite systems have several common features in their crystal structure [86, 89], comparison between these two systems can prove very beneficial. In addition to providing insight into the key factors that modify strain and built-in fields in (111)-oriented zinc blende heterostructures, our approach offers furthermore the opportunity to gain insights into material parameters for the wurtzite AlN, InN and GaN systems, including, for example, the sign of the shear strain-related piezoelectric constant e_{15}. From our analysis of the first-order piezoelectric tensor in (111)-zinc blende systems we obtain, that the piezoelectric coefficients e_{31} and e_{15} should be equal and negative, $e_{15}/e_{31} = 1$, and also that the ratio $e_{33}/e_{31} = -2$. When comparing these ratios with ratios derived from literature piezoelectric coefficients, we find very good agreement for GaN and InN and good agreement for AlN, as long as the literature value of e_{15} is negative [86]. From this we conclude that e_{15} should be negative in wurtzite nitrides. More details about the method and in-depth discussions are given in [86].

6.4.3 Impact of the Sign of e_{15} on the Built-In Potential in a Single InGaN/GaN QD

In this subsection we study the impact of the sign of e_{15} on the potential in an isolated QD. Here, the piezoelectric coefficients e_{31} and e_{33} are taken from [78]. In the following we apply the often used *positive* e_{15} values of Vurgaftman et al. [44] and the *negative* values of Shimada [78] for InN and GaN, respectively. We note that the ratios of the ab initio e_{ij} values calculated by Shimada [78] are in good agreement with the ratios predicted from the comparison between (111)-zinc blende and (0001)-wurtzite structures discussed in the previous section.

To investigate the influence of the sign of e_{15} on the total potential ϕ_{tot}, we have calculated and discussed the different contributions to ϕ_{tot} in an isolated lens-shaped InGaN/GaN QD separately in [90]. Here, we summarize the overall result how e_{15} affects the total built-in potential ϕ_{tot}. Figure 6.5 shows the total built-in potential ϕ_{tot} for a line-scan along the *z*-(*c*-) axis through the center of a lens-shaped $In_{0.25}Ga_{0.75}N$/GaN QD with diameter $d = 19.2$ nm and height $h = 3.1$ nm for $e_{15} > 0$ and $e_{15} < 0$, respectively. One can infer from Fig. 6.5 that the sign of e_{15} has mainly two effects on the total built-in potential ϕ_{tot}. Firstly, with $e_{15} < 0$, the potential

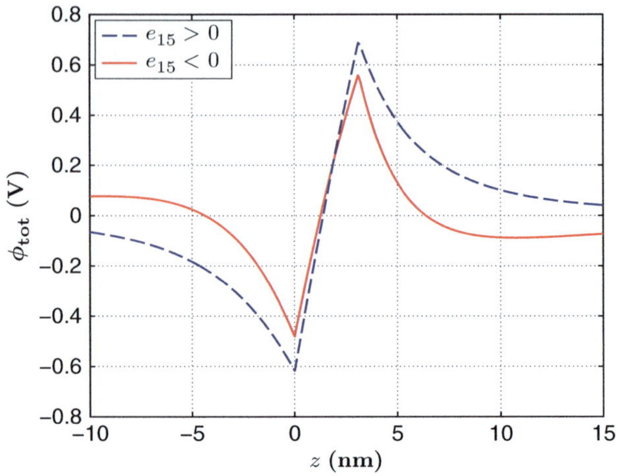

Fig. 6.5 Total built-in potential ϕ_{tot} for a lens-shaped $In_{0.25}Ga_{0.75}N/GaN$ c-plane QD (height $h =$ 3.1 nm, diameter $d = 19.2$ nm) for a line-scan through the center of the dot along the c-direction (z-axis). The results are shown for positive and negative values of the shear strain-related piezoelectric coefficient e_{15}

drop between the top and the bottom surface is reduced compared to the result with $e_{15} > 0$. Secondly, the sign of e_{15} affects the potential profile outside the dot. With $e_{15} < 0$, ϕ_{tot} returns towards zero and changes sign a few nanometers away from the QD. This is to be contrasted with the result obtained for $e_{15} > 0$, where the potential does not change sign outside the QD. The behavior of ϕ_{tot} outside a single QD will also affect ϕ_{tot} in a stacked QD system. Therefore, we turn our attention in the following section to ϕ_{tot} of a system of two InGaN/GaN QDs stacked along the c-axis.

6.5 Built-In Fields in InGaN QDMs

In this section we discuss the electrostatic built-in fields in stacked InGaN/GaN QDs first before turning to the analysis of the electronic structure of InGaN/GaN QDMs in Sect. 6.6. To keep the analysis simple and transparent, we study here the built-in potential ϕ_{tot} for two identical lens-shaped $In_{0.25}Ga_{0.75}N/GaN$ QDs which are stacked along the c-axis. Later, in Sect. 6.6.2, we deal with the realistic situation of non-identical QDs. The QDs considered in this section have a base diameter of $d = 19.2$ nm and a height of $h = 3.1$ nm following the discussions in Sect. 6.4. Based on our findings in the previous section, we use the piezoelectric coefficients from Shimada [78], which predict $e_{15} < 0$ for GaN and InN, respectively. Here, we study the total built-in potential for two different spacer layer thicknesses D, namely $D \approx 1$ and 4.1 nm. In Sect. 6.6, we will extend this discuss to even larger spacer

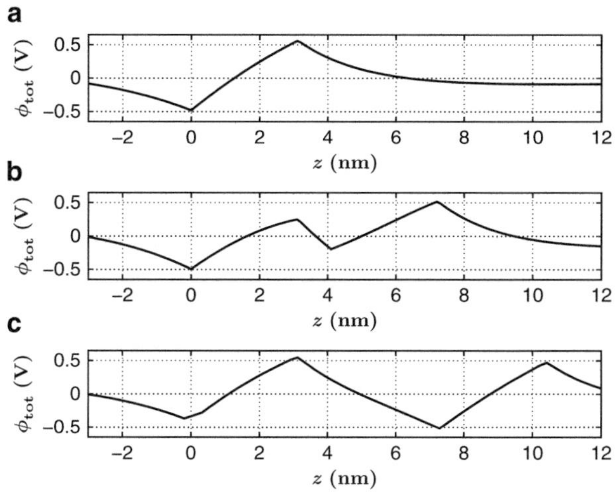

Fig. 6.6 Total built-in potential ϕ_{tot} shown for a line-scan through the center of the lens-shaped $In_{0.25}Ga_{0.75}N$/GaN QDs. (**a**) displays ϕ_{tot} in an isolated QD; (**b**) shows ϕ_{tot} in a QDM with a spacer layer thickness of $D \approx 1$ nm. The total built-in potential of the QDM with a spacer layer thickness of $D \approx 4.1$ nm is shown in (**c**). [From [91]]

layer thicknesses. However, the spacer layer thicknesses $D \approx 1$ and 4.1 nm used here already give a representative picture of the behavior of the built-in potential for InGaN/GaN QDMs made up of two identical QDs with varying spacer layer thicknesses.

Figure 6.6a shows the total built-in potential ϕ_{tot} for a line-scan through the center of an isolated lens-shaped $In_{0.25}Ga_{0.75}N$/GaN QD along the c-axis (z-direction). Outside the QD ϕ_{tot} returns to zero quickly and changes sign a few nanometers away from the dot, in line with the results from Sect. 6.4.3. Figures 6.6b and c shows the same line-scan as in (a), but this time for the $In_{0.25}Ga_{0.75}N$ QDM with $D \approx 1$ nm and $D \approx 4.1$ nm, respectively. The behavior of ϕ_{tot} in (b) and (c) can be understood when superimposing built-in potentials of two isolated, identical lens-shaped $In_{0.25}Ga_{0.75}N$/GaN QDs with their base centered at $z = 0$ and $z \approx 4.1$ nm (b) or $z = 0$ and $z \approx 7.2$ nm (c). In the case of the smaller spacer layer thickness (b), the potential drop across each dot is reduced compared to the potential drop across an isolated dot. We see from Fig. 6.6b that the magnitude of ϕ_{tot} in the upper (lower) QD is reduced at the bottom (top) only, and almost unchanged at the top (bottom) compared to an isolated QD. Therefore, following our discussion in Sect. 6.3, the ground state electron wave function could be expected to be localized at the top of the upper QD while the hole ground state wave function is expected to be localized at the bottom of the lower QD.

For $D > 2$ nm, the change in sign in ϕ_{tot} in an isolated QD becomes important, as can be seen from Fig. 6.6c. Here, ϕ_{tot} is slightly reduced in magnitude at the top (bottom) of the upper (lower) QD while almost unchanged at the bottom (top) of the upper (lower) QD. In this case the electron ground state wave function is expected to be localized at the top of the lower QD while the hole ground state wave function is expected to be localized at the bottom of the upper QD, with the built-in potential change across each dot being very close to that for an isolated QD.

Three key results emerge from this analysis. Firstly, even though we have assumed a QDM made up of two identical dots, the built-in potential breaks the symmetry between the two dots. Therefore, one could expect that the electron and hole ground state wave functions are localized on different QDs in a QDM made up of two identical dots. Secondly, there can be a reduced potential drop across each dot for the case of closely stacked dots. Finally, the built-in potential profile outside an isolated InGaN/GaN QD can lead to electron and hole ground state switching in the here studied QDMs. Having established these general features, we turn now to look in more detail at the electronic structure of QDMs.

6.6 Electronic Structure of InGaN/GaN QDMs

Equipped with an understanding of the electrostatic built-in potential in isolated and stacked InGaN/GaN QDs, we focus now on the electronic structure of InGaN/GaN QDMs. To build on the results of the previous section, and to keep the analysis simple and transparent, we first consider molecules made up of two vertically stacked QDs of identical shape, size and composition with varying inter-dot distance D. We assume here values of $D \approx 1, 2, 4.1, 6.2$, and 8.3 nm, following the experimental findings on stacked InGaN QDs in [22].

In a second step, we extend this analysis in Sect. 6.6.2 to investigate InGaN QDMs made up of two non-identical dots. All other parameters, such as the distance between the dots, are kept as in the case of the two identical dots. We find that the results for the non-identical dots are broadly similar to those for identical dots, except that the electron and hole ground state levels can be found in the same dot for the case of non-identical dots.

6.6.1 Identical QDs: Influence of Strain and Built-In Potentials

We consider here the electronic properties of $In_{0.25}Ga_{0.75}N$/GaN QDMs. We assume for the QD geometry a lens-shaped structure, as suggested by the experimental data in [70]. The molecules considered here are made up of two vertically stacked QDs of identical shape, size, and composition, but with varying inter-dot distance D.

To analyze the electronic structure of $In_{0.25}Ga_{0.75}N$/GaN QDMs in detail we proceed in the following way. To study the impact of strain and built-in fields separately, we switch these contributions off in a first step, Sect. 6.6.1.1. Subsequently, in Sect. 6.6.1.2, we investigate the influence of the strain field on the electronic

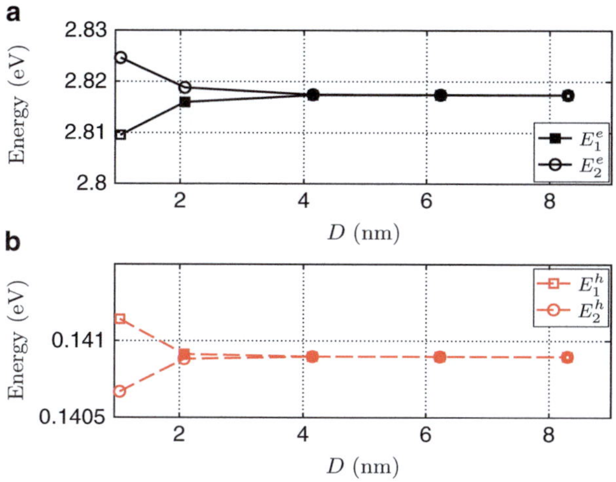

Fig. 6.7 TB single-particle energies of the first two bound (**a**) electron and (**b**) hole states in the absence in the absence of strain and built-in fields as a function of the spacer layer thickness D. The reference energy for our results is set to the unstrained valence band maximum of GaN. [From [91]]

structure on its own. In a third step, Sect. 6.6.1.3, we then add the electrostatic built-in potential, so that the separate impact of each component to the electronic structure can then be clearly seen.

6.6.1.1 Electronic Structure Without Strain and Built-In Field

In this section we analyze the electronic structure of InGaN/GaN QDMs when we artificially switch off strain and built-in fields. The energies $E_{1,2}^{e,h}$ of the first two bound electron (ψ_1^e, ψ_2^e) and hole (ψ_1^h, ψ_2^h) single-particle states are shown in Fig. 6.7 a and b, respectively. Here, electron and hole levels form bonding and anti-bonding molecular orbitals. Therefore, the single-particle energies split symmetrically around the electron and hole ground state energy of the isolated $In_{0.25}Ga_{0.75}N$/GaN QD. This behavior is similar to a real diatomic molecule (e.g., H_2). However, one can infer from Fig. 6.7 that electronic coupling between the dots is only possible for small D values ($D < 4$ nm for electrons and $D < 2$ nm for holes). This weak coupling arises for two main reasons. Firstly, due to the high effective hole masses in the nitride system, ψ_1^h and ψ_2^h are strongly confined in the QD. Therefore very small spacer layer thicknesses D are required to achieve coupling between these valence states. Secondly, even though the electron effective mass is much lower, the large conduction band offset prevents an effective inter-dot coupling between the electron ground state wave functions until small spacer layer values.

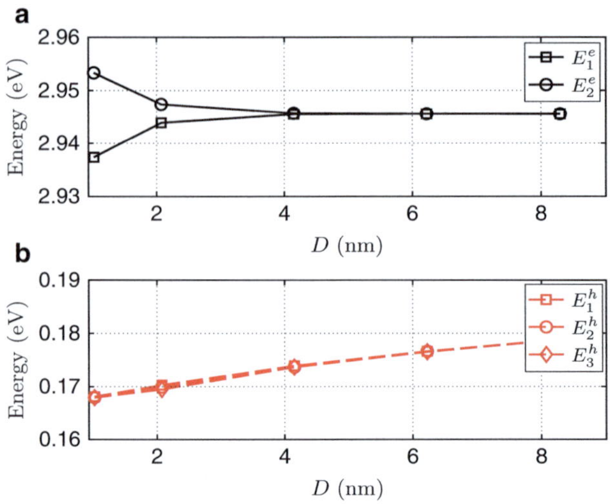

Fig. 6.8 Same as in Fig. 6.7 but this time for the (**a**) first two bound electron and (**b**) first three bound hole states in the presence of strain but in the absence of the built-in field. [From [91]]

6.6.1.2 Influence of Strain on the Electronic Structure

Figure 6.8 shows the calculated single-particle energies of the first two bound electron and the first three bound hole states in the $In_{0.25}Ga_{0.75}N/GaN$ QDM as a function of the spacer layer thickness D, when including in the Hamiltonian the effects of the strain-induced shifts in the bands, as given earlier by Eq. (6.8).

Looking at the electron single-particle energies in the case of large D, we find that ψ_1^e and ψ_2^e are again degenerate. For closely spaced QDs ($D \approx$ 1–2 nm), the states ψ_1^e and ψ_2^e then form bonding and anti-bonding states, similar to the situation in the absence of the strain field. The main difference to the results in the previous section is that ψ_1^e and ψ_2^e are shifted to higher energies due to hydrostatic strain in the system.

The situation is completely different for the hole states. Here, the first two bound hole states ψ_1^h and ψ_2^h do not form bonding and anti-bonding states. When looking at the probability densities of the hole wave functions (not shown) one finds, that for closely spaced QDs ψ_1^h, ψ_2^h and ψ_3^h are localized on the lower QD. This behavior is therefore similar to the behavior in a heteronuclear diatomic molecule (e.g., HF), and occurs primarily due to the following two reasons. Firstly, the strain between the QDs modifies the valence band offsets, leading to an increased effective barrier between the dots. Therefore, tunneling and the ability to form bonding and anti-bonding states is suppressed for the hole states. In addition to this effect, the QDM lacks a center of inversion. Therefore, the strain state of the upper dot is different to the strain state of the lower QD. The localization of the hole states on the lower QD can be understood when looking at the strain field due to an isolated QD. In an isolated lens-shaped QD, the region at the base of the dot experiences a strong

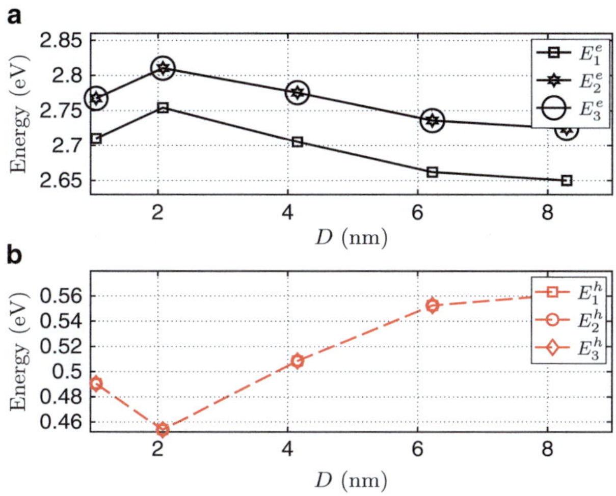

Fig. 6.9 The same as in Fig. 6.7 but this time for the first three bound (**a**) electron and (**b**) hole states when taking strain and built-in field into account. [From [91]]

biaxial compressive strain, with a much smaller biaxial strain at the top of the dot. The barrier material above the QD experiences a biaxial strain of opposite sign to that at the dot base. Therefore, we deduce by applying linear elasticity theory that for closely spaced QDs, the bottom of the upper dot should experience a reduced biaxial strain compared to the bottom of the lower dot. Because these biaxial compressively strained regions are more favorable for (heavy-) hole states, one could expect that the hole states are localized at the bottom of the lower QD. This is exactly the situation we find here for the first few bound hole states.

6.6.1.3 Influence of Built-In Potentials on the Electronic Structure

Having analyzed the impact of the strain field in the previous subsection, we include now also the total built-in polarization potential ϕ_{tot} in the description of the electronic structure of the $\text{In}_{0.25}\text{Ga}_{0.75}\text{N/GaN}$ QDMs.

Figure 6.9a shows, as a function of D, the single-particle energies $E^{\text{e}}_{1,2,3}$ for the first three bound electron states ψ^{e}_1, ψ^{e}_2 and ψ^{e}_3 while (b) depicts the energies $E^{\text{h}}_{1,2,3}$ for the first three bound hole states, ψ^{h}_1, ψ^{h}_2 and ψ^{h}_3. As expected from the QCSE, the electron states are shifted to lower energies while the hole states are shifted to higher energies, compared to the situation without the built-in field (cf. Fig. 6.8). Moreover, when looking at Fig. 6.9, we find that the molecular-like description of the single-particle states now breaks down for both electrons and holes. Additionally, an interesting feature is observed in the energy spectrum for electrons and holes around $D \approx 2\,\text{nm}$, where we observe a kink in the spectrum. Even when ϕ_{tot} breaks the symmetry between the two QDs, as discussed in Sect. 6.5, one might expect that

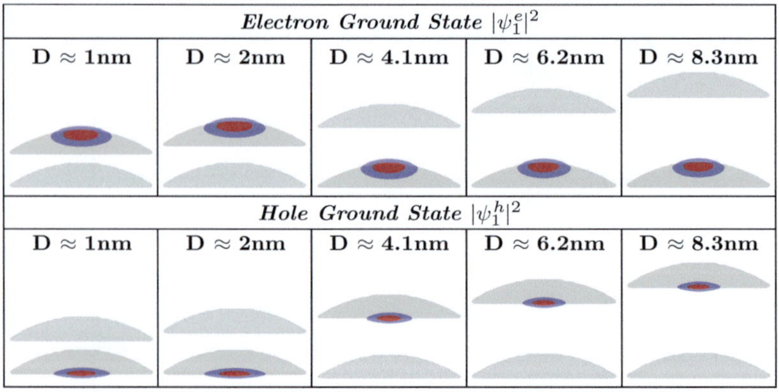

Fig. 6.10 Probability densities of the electron and hole single-particle wave functions ψ_1^e and ψ_1^h, respectively, for different spacer layer thicknesses D. The QD geometry is shown in *light gray* and the *blue* and *red* probability density isosurfaces correspond to 10% and 50% of the maximum value. [From [91]]

the ground state energies for electrons and holes approach the ground state energies of an isolated QD in a continuous way from above or below the respective energy. However, as we discussed in Sect. 6.5, the profile of the built-in potential outside an isolated InGaN/GaN QD can lead to ground state switching for electrons and holes. We saw in Sect. 6.5 that for small D, the ground state electron wave function could be expected to be localized at the top of the upper QD while the hole ground state wave function is expected to be localized at the bottom of the lower QD. For larger D, the change in sign in ϕ_{tot} in an isolated QD becomes important, and in this case the electron ground state wave function is expected to be localized at the top of the lower QD while the hole ground state wave function is expected to be localized at the bottom of the upper QD.

To confirm this behavior, Fig. 6.10 shows isosurfaces of the probability densities of the electron (ψ_1^e) and hole (ψ_1^h) ground states as a function of D. The blue and red isosurfaces correspond to 10% and 50% of the maximum value, respectively. As expected from the single-particle energies, ϕ_{tot} breaks the symmetry between the two QDs, and bonding and anti-bonding states are not formed. Furthermore, Fig. 6.10 shows that the built-potential indeed leads to ground state switching for both electrons and holes. For small spacer layer thicknesses, ψ_1^e is localized at the top of the upper dot, while ψ_1^h is localized at the bottom of the lower QD. This behavior is therefore similar to the behavior one would expect in two coupled QWs. The situation changes for $D > 2\,\text{nm}$. In this case ψ_1^e is localized at the top of the lower QD while ψ_1^h is localized at the bottom of the upper dot. This ground state switching explains therefore the kink in the energy spectrum shown in Fig. 6.9. Again, the reason for this ground state switching is the profile of the total built-in potential ϕ_{tot} above and below an isolated InGaN/GaN QD and is a direct consequence of $e_{15} < 0$.

From this analysis one might expect an even further reduction of the oscillator strength in a QDM compared to an isolated QD, since the wave functions are localized is localized on different QDs, leading to a large spatial separation of the electron and hole wave functions. However, this need not be the case if we assume that electrons and holes can become trapped in both dots. In addition, we have made the simplifying assumption above of two identical QDs. As discussed before, during the growth of a stack of QDs, the geometrical properties of the dots are likely to change from layer to layer. Furthermore, the indium content is probably also different in the different QDs, leading therefore to different confinement energies as well as strain fields. This will then also modify the built-in potential. Consequently, further studies are required to analyze in detail how the change in composition and in geometry affects the built-in field in QDMs and how this impacts the electronic and optical properties of such systems. This will be the topic of the next section.

6.6.2 Non-identical QDs

Having discussed the built-in potential and the electronic structure of InGaN/GaN QDMs made up of two identical dots in the previous section, we turn now and present a detailed analysis of the electronic structure of an InGaN/GaN QDM made up of two *non-identical* dots. Following the analysis of the previous section, we study the electronic structure of the InGaN/GaN QDM as a function of the barrier thickness D. We use values of $D \approx 1, 2, 4.1, 6.2$, and 8.3 nm, which are based on the experimental data in [22]. Here, we consider QDs identical in size and shape but different in their composition. The difference in indium composition mimics therefore already the effects (changes in confinement energy, strain and built-in fields) which can also arise from a change in the QD geometry. Since the average diameter of InGaN QDs scatters around 15–25 nm while the average height is approximately 2–6 nm [22, 70, 72], we assume a diameter $d \approx 19.2$ nm and a height $h \approx 3.1$ nm for both QDs. As discussed before in Sect. 6.4, there is no detailed measurement on the variation of the indium content in stacked InGaN QDs. However, the analysis of coupled InGaAs QDs shows that, due to strain relaxation in the structure, the indium composition of the upper QD is higher than in the lower one [74]. We assume an indium content of 20% in the lower dot and 25% in the upper dot. Based on the experimental data in [75] and the discussions in [22], we assume a vertical stacking of the two QDs. Again, we use $e_{15} < 0$ for this study, according to our recommendation in Sect. 6.4.

In a first step we consider the single-particle electron (ψ_1^{e}) and hole (ψ_1^{h}) ground state wave functions. Figure 6.11 shows the charge densities for ψ_1^{e} and ψ_1^{h} as a function of D. The blue and red isosurfaces correspond to 10% and 50% of the maximum values, respectively. When looking at the results in detail, we find again a ground state switching for the holes, but this time not for the electrons. For $D \approx 1$ nm ψ_1^{h} is localized in the lower QD ($In_{0.2}Ga_{0.8}N$ QD) while ψ_1^{e} is localized in the upper

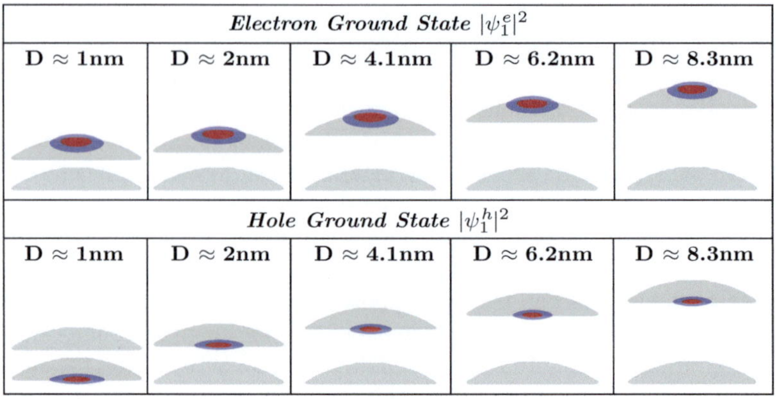

Fig. 6.11 Probability densities of the electron and hole ground state wave functions ψ_1^e and ψ_1^h, respectively, for different barrier thicknesses D. The QD geometry is shown in *light gray* and the *blue* (*red*) probability density isosurface correspond to 10% (50%) of the maximum value. Upper QD: $In_{0.25}Ga_{0.75}N$; Lower QD: $In_{0.2}Ga_{0.8}N$. [From [92]]

dot ($In_{0.25}Ga_{0.75}N$ QD). For larger values of D both ψ_1^h and ψ_1^e are localized on the upper dot. The reason for this switching originates again from the behavior of the built-in potential ϕ_{tot} above and below an isolated QD. Outside the QD, ϕ_{tot} returns to zero quickly along the c-axis and changes sign a few nanometers away from the dot along the c-axis, affecting therefore ϕ_{tot} in a QDM. Line-scans through the center of the QDM along the c-axis for different D are shown in Fig. 6.12. Again, the behavior of ϕ_{tot} can be understood by superimposing built-in potentials for two isolated QDs with their bases centered at $z = 0$ and $z = h + D$, respectively, where h is the height of the lower QD.

The results for the isolated QDs are given by the dashed-dotted line and the dashed line, respectively. In the case of $D \approx 1$ nm (a), the magnitude of ϕ_{tot} in the upper (lower) QD is reduced at the bottom (top), and almost unchanged at the top (bottom) compared to an isolated QD. Therefore, the electron wave functions could be expected to be localized at the top of the upper QD while the hole states are expected to be localized at the bottom of the lower QD. This is exactly the result we obtain from our TB analysis [cf. Fig. 6.11]. For $D \geq 2$ nm, the change in sign in ϕ_{tot} outside a single QD becomes important. For $D \geq 4.1$ nm, ϕ_{tot} is slightly reduced in magnitude at the top (bottom) of the upper (lower) QD while increased in magnitude at the bottom (top) of the upper (lower) QD [cf. Fig. 6.12c and d]. In the case of $D \approx 2$ nm, ϕ_{tot} is decreased in magnitude at all four interfaces [cf. Fig. 6.12b]. From Fig. 6.12b–d one could expect that ψ_1^e is localized at the top of the upper QD while ψ_1^h is expected to be localized at the bottom of the upper QD, in full accordance with our TB results [cf. Fig. 6.11].

The behavior of ϕ_{tot} affects also the ground state transition energies $E_g^{QDM} = E_1^e - E_1^h$, where E_1^e and E_1^h are ground state energies for electrons and holes, respectively. The blue dashed line in Fig. 6.13 shows the calculated value

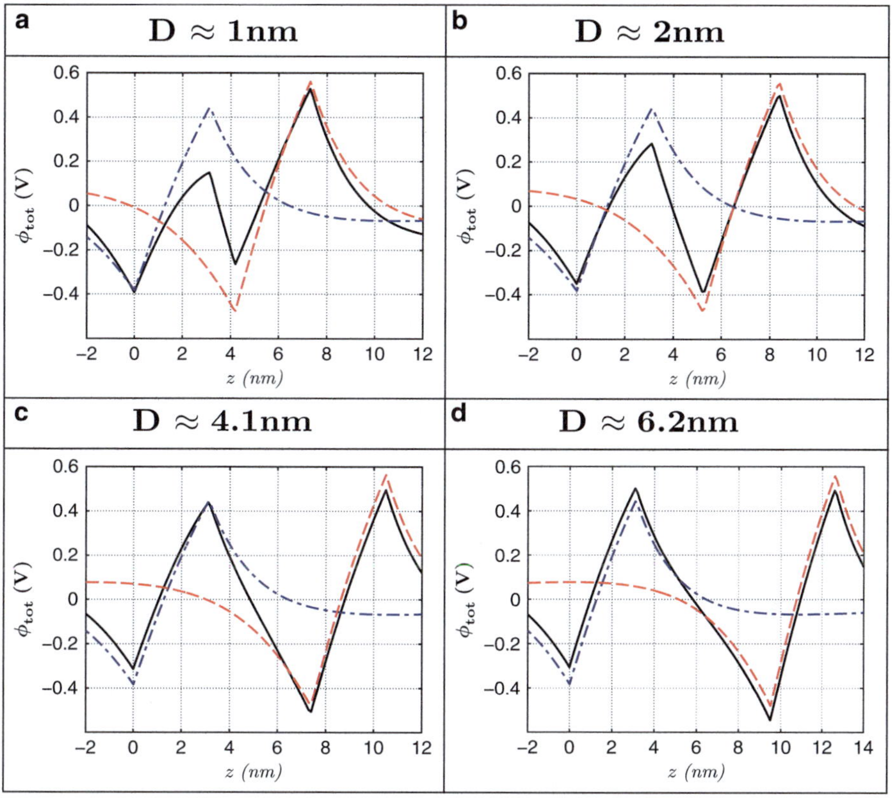

Fig. 6.12 Built-in potential ϕ_{tot} (*solid black line*) in a *c*-plane QDM of two non-identical QDs (lower QD: $In_{0.2}Ga_{0.8}N$; upper QD: $In_{0.25}Ga_{0.75}N$) for a line-scan through the center of the QDs along the *c*-axis and for different barrier thickness *D* between the QDs. The (*blue*) *dashed-dotted line* and the (*red*) *dashed line* indicate the results for isolated QDs. [From [92]]

of E_g^{QDM} as a function of D (dashed line). This is compared to the transition energy E_g^{QD} of a single $In_{0.25}Ga_{0.75}N$ QD (dashed-dotted line). Figure 6.13 also displays the normalized squared dipole matrix element $|\tilde{d}_{11}|^2$, defined by $|\tilde{d}_{11}|^2 = |d_{11}^{\text{QDM}}|^2/|d_{11}^{\text{QD}}|^2$, with $d_{11}^{\alpha} = \mathbf{e} \cdot \langle \psi_1^{\text{e}} | e_0 \mathbf{r} | \psi_1^{\text{h}} \rangle$, where $\mathbf{e} = 1/\sqrt{2}(1,1,0)$ is the light polarization vector and $e_0\mathbf{r}$ the dipole operator, with e_0 being the electron charge. The dipole matrix elements of a single $In_{0.25}Ga_{0.75}N$ QD and the QDM are denoted by d_{11}^{QD} and d_{11}^{QDM}, respectively. The calculation of the dipole matrix elements is performed following the method in [55, 93]

It can be seen from Fig. 6.13 that the energies E_g^{QDM} are blue shifted with respect to E_g^{QD}. This behavior is due in part to the reduction of the biaxial compressive strain in the upper dot compared to an isolated QD. Additionally, the effective reduction of ϕ_{tot} in the QDM compared to a single dot also contributes to the blue-shift in E_g^{QDM}. This effective reduction of ϕ_{tot} also increases the spatial overlap of electron and hole wave functions so that $|\tilde{d}_{11}|^2 > 1$. Neglecting the results for $D \approx 1$ nm, where ψ_1^{e}

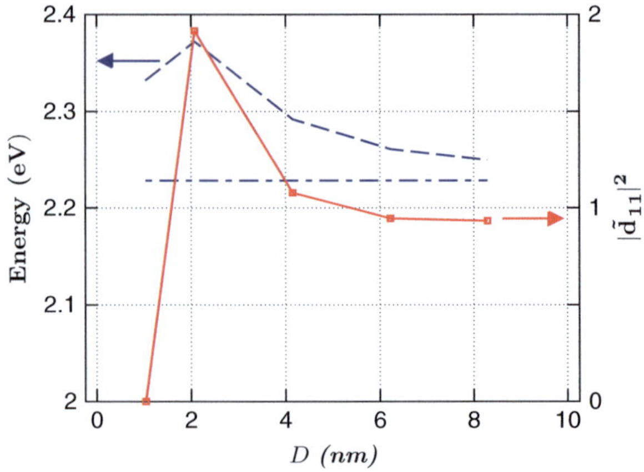

Fig. 6.13 Single-particle energy gap for the InGaN QDM (*dashed line*) and an isolated $In_{0.25}Ga_{0.75}N$ QD (*dashed-dotted line*) as function of the barrier thickness D. *Solid line*: Normalized dipole matrix element squared $|\tilde{d}_{11}|^2$. [From [92]]

and ψ_1^h are localized on different QDs [cf. Fig. 6.11], we find that for $D \approx 2$ nm and $D \approx 4.1$ nm, $|\tilde{d}_{11}|^2 > 1$, indicating an effective reduction of ϕ_{tot} and increase in the radiative recombination rate for carriers in the upper dot. For example, for $D \approx 2$ nm, $|d_{11}^{\text{QDM}}|^2$ is increased by a factor of order two compared to $|d_{11}^{\text{QD}}|^2$, reflecting the change in the slope of ϕ_{tot} in the upper dot of the QDM compared to an isolated QD [cf. Fig. 6.12b]. Note that ϕ_{tot} in an isolated QD is already significantly reduced compared to a QW structure of the same composition and height, as discussed in detail in Sect. 6.3.2 and in [16]. Therefore, the increase of $|\tilde{d}_{11}|^2$ for small D further emphasizes the benefit of using QDs instead of QWs in optoelectronic devices. It should be noted that $|\tilde{d}_{11}|^2 > 1$ for the ground state transition is again a consequence of $e_{15} < 0$. With $e_{15} > 0$, ϕ_{tot} in a QDM would be similar to the $D \approx 1$ nm case [Fig. 6.12a], and as discussed in Sect. 6.5 in detail. Since $|\tilde{d}_{11}|^2 > 1$ is in qualitative agreement with the experiment [21, 22], this further supports our earlier conclusion of $e_{15} < 0$ [24, 86]. For larger values of D ($D > 5$ nm), $|\tilde{d}_{11}|^2$ drops below unity. Looking at $D \approx 6.2$ nm, the slope of ϕ_{tot} inside the upper dot is almost identical to the slope inside a single QD [cf. Fig. 6.12d]. However, the magnitude of ϕ_{tot} at the bottom (top) of the upper dot is slightly increased (decreased) compared to an isolated QD. This change in ϕ_{tot} leads is accompanied by an increased (decreased) lateral confinement for ψ_1^h (ψ_1^e), resulting in $|\tilde{d}_{11}|^2 < 1$ for $D > 5$ nm.

6.7 Conclusions

In summary, we have presented a detailed analysis showing that the built-in electric field in an isolated QD is significantly reduced compared to that in a QW of the same height and the same indium content, with a further reduction in field achievable in a QDM made up of closely spaced stacked dots. This reduction provides a potential route to circumvent the "green gap problem" present in current nitride-based optoelectronic devices.

In contrast to a wurtzite c-plane QW, where the built-in field depends on two piezoelectric coefficients, e_{33} and e_{31}, the potential drop across a QD structure depends also on the shear coefficient, e_{15}. Considerably less attention has been paid to this third coefficient, with disagreement in the literature even as to the sign of e_{15}. We have investigated in detail the impact of e_{15}, showing that the behavior of the potential both inside and outside an isolated QD strongly depends on its sign. In the case of $e_{15} < 0$, the potential outside the dot returns to zero and can even change sign. This behavior has to be contrasted with the result using $e_{15} > 0$, where the potential does not change sign outside the dot, and where there is also a larger potential drop between the top and bottom of the dot. The sign of e_{15} then strongly affects the built-in potential both in an isolated dot and also in a stack of QDs.

An accurate analysis of InGaN QDs therefore requires knowledge of the sign of e_{15}. We have undertaken a number of studies to this end. Firstly we have analyzed the built-in fields in non-polar GaN/AlN QDs. Secondly the first-order piezoelectric tensor in (111)-oriented zinc blende systems has been compared to the first-order piezoelectric tensor in c-plane wurtzite. Both investigations support $e_{15} < 0$. These findings are in agreement both with recent experimental studies and with the outcome of recent ab initio calculations of the piezoelectric coefficients in InN, GaN and AlN. We therefore recommend $e_{15} < 0$ for studies of wurtzite nitride-based heterostructures.

Following this detailed discussion of the built-in potential in InGaN QDs and the sign of the piezoelectric coefficient e_{15} we have investigated the impact of strain and the built-in fields on the electronic structure of InGaN/GaN QDMs. These studies revealed that the molecular-like description of bonding and anti-bonding states breaks down for both electrons and holes, even when assuming that the QDM is made up of two identical QDs. Due to the high effective hole mass and the lack of inversion symmetry along the growth direction, the strain field on its own is already sufficient to prevent the formation of bonding and anti-bonding hole states. In addition to this effect, the electron and hole levels are significantly modified by the presence of the built-in field. When taking the built-in field into account, we find a ground state switching for electrons and holes. For very small spacer layer thicknesses ($D \leq 2$ nm) the electron ground state is localized at the top of the upper QD while the hole ground state is localized near the bottom of the lower QD. However, when going to larger barrier thicknesses ($D > 2$ nm), the electron ground state wave function is localized near the top of the lower QD while the hole ground state wave function is localized near the bottom of the upper QD. This ground state switching follows from the behavior of the built-in field in an isolated QD, where the sign of the potential along the growth direction changes a few nanometers away from the QD, directly as a consequence that $e_{15} < 0$.

Following this analysis of the impact of strain and built-in fields on the electronic structure of an idealized InGaN/GaN QDM system made up of two identical QDs, we then turned to consider the more realistic case of a c-plane InGaN QDM made up of two *non-identical* QDs. Our study revealed that the built-in field in a QDM can be effectively reduced compared to a single QD, leading to a further increase in the spatial overlap of electron and hole wave functions. These results are in qualitative agreement with experimental data [21,22] for stacked nitride QDs, where PL measurements show reduced recombination lifetimes, indicative of an increased spatial overlap of electron and hole wave functions. Furthermore, due to the behavior of the built-in field in a system of stacked QDs, the distance between the two QDs can be used to engineer the optical recombination rate. Overall, we conclude that stacked nitride QDs therefore have significant potential for use in high efficiency light emitters with high indium content, of strong benefit for future optoelectronic devices and their applications.

Acknowledgments The authors acknowledge the financial support from Science Foundation Ireland. S. Schulz was further supported by the Irish Research Council for Science, Engineering and Technology (IRCSET).

References

1. Jarjour, A.F., Oliver, R.A., Taylor, R.A.: Phys. Status Solid A **206**, 2510 (2009)
2. Humphreys, C.J.: MRS Bull. **33**, 459 (2008)
3. Nakamura, S., Mukai, T., Senoh, M.: Appl. Phys. Lett. **64**, 1687 (1994)
4. Nakamura, S.: J. Sel. Top. Quant. Electron. **3**, 712 (1997)
5. Yang, W., Lee, H., Johnson, T.J., Sercel, P.C., Norman, A.G.: Phys. Rev. B **61**, 2784 (2000)
6. Simon, J., Pelekanos, N.T., Adelmann, C., Martinez-Guerrero, E., Andre, R., Daudin, B., Dang, L.S., Mariette, H.: Phys. Rev. B **68**, 035312 (2003)
7. Andreev, A.D., O'Reilly, E.P.: Phys. Rev. B **62**, 15851 (2000)
8. Schulz, S., Caro, M.A., O'Reilly, E.P., Marquardt, O.: Phys. Status Solid B **249**, 521 (2012)
9. Waltereit, P., Brandt, O., Trampert, A., Grahn, H.T., Menniger, J., Ramsteiner, M., Ploog, K.H.: Nature **406**, 865 (2000)
10. Founta, S., Rol, F., Bellet-Amalric, E., Bleuse, J., Daudin, B., Gayral, B., Mariette, H., Moisson, C.: Appl. Phys. Lett. **86**, 171901 (2005)
11. Schulz, S., Caro, M.A., O'Reilly, E.P.: Appl. Phys. Lett. **101**, 113107 (2012)
12. Kim, K.C., Schmidt, M.C., Sato, H., Wu, F., Fellows, N., Saito, M., Fujito, K., Speck, J.S., Nakamura, S., DenBaars, S.P.: Phys. Stat. Sol. (RRL) **1**, 125 (2007)
13. Caro, M.A., Schulz, S., Healy, S.B., O'Reilly, E.P.: J. Appl. Phys. **109**, 084110 (2011)
14. Zhang, M., Banerjee, A., Lee, C.S., Hinckley, J.M., Bhattacharya, P.: Appl. Phys. Lett. **98**, 221104 (2011)
15. Wu, Y.R., Lin, Y.Y., Huang, H.H., Singh, J.: J. Appl. Phys. **105**, 013117 (2009)
16. Schulz, S., O'Reilly, E.P.: Phys. Rev. B **82**, 033411 (2010)
17. Zhang, M., Bhattacharya, P., Guo, W.: Appl. Phys. Lett. **97**, 011103 (2010)
18. Soh, C.B., Liu, W., Chua, S.J., Ang, S.S., Tan, R.J.N., Chow, S.Y.: J. Appl. Phys. **108**, 093501 (2010)
19. Schmidt, O.G., Kirstaedter, N., Ledentsov, N.N., Mao, M.H., Bimberg, D., Ustinov, V.M., Egorov, A.Y., Zhukov, A.E., Maximov, M., Kop'ev, P.S., Alferov, Z.I.: Electron. Lett. **32**, 1302 (1996)
20. Ledentsov, N.N., Shchukin, V.A., Grundmann, M., Kirstaedter, K., Bohrer, J., Schmidt, O., Bimberg, D., Ustinov, V.M., Egorov, A.Y., Zhukov, A.E., Kop'ev, P.S., Zaitsev, S.V., Gordeev, N.Y., Alferov, Z.I., Borovkov, A.I., Kosogov, A.O., Ruvimov, S.S., Werner, P., Gösele, U., Heydenreich, J.: Phys. Rev. B **54**, 8743 (1996)

21. Neogi, A., Morkoc, H., Kuroda, T., Tackeuchi, A., Kawazoe, T., Ohtsu, M.: Nano Lett. **5**, 213 (2004)
22. Davies, S.C., Mowbray, D.J., Ranalli, F., Wang, T.: Appl. Phys. Lett. **96**, 251904 (2010)
23. Bernardini, F., Fiorentini, V., Vanderbilt, D.: Phys. Rev. B **56**, R10024 (1997)
24. Schulz, S., Berube, A., O'Reilly, E.P.: Phys. Rev. B **79**, 081401(R) (2009)
25. Skylaris, C.K., Haynes, P.D., Mostofi, A.A., Payne, M.C.: J. Chem. Phys. **122**, 084119 (2005)
26. Hine, N.D.M., Robinson, M., Haynes, P.D., Skylaris, C.K., Payne, M.C., Mostofi, A.A.: Phys. Rev. B **83**, 195102 (2011)
27. Pryor, C., Kim, J., Wang, L.W., Williamson, A.J., Zunger, A.: J. Appl. Phys. **83**, 2548 (1998)
28. Klimeck, G., Ahmed, S.S., Bae, H., Kharche, N., Rahman, R., Clark, S., Haley, B., Lee, S., Naumov, M., Ryu, H., Saied, F., Prada, M., Korkusinski, M., Boykin, T.B.: IEEE Trans. Electron Dev. **54**, 2079 (2007)
29. Bester, G.: J. Phys. Condens. Matter **21**, 023202 (2009)
30. Caro, M.A., Schulz, S., O'Reilly, E.P.: Phys. Status Solid B **249**, 526 (2012)
31. Schliwa, A., Winkelnkemper, M., Bimberg, D.: Phys. Rev. B **76**, 205324 (2007)
32. Marquardt, O., Schulz, S., Freysoldt, C., Boeck, S., Hickel, T., O'Reilly, E.P., Neugebauer, J.: Opt. Quant. Electron. **44**, 183 (2012)
33. Faux, D.A., Downes, J.R., O'Reilly, E.P.: J. Appl. Phys. **80**, 2515 (1996)
34. Faux, D.A., Downes, J.R., O'Reilly, E.P.: J. Appl. Phys. **82**, 3754 (1997)
35. Downes, J.R., Faux, D.A., O'Reilly, E.P.: J. Appl. Phys. **81**, 6700 (1997)
36. Andreev, A.D., Downes, J., Faux, D.A., O'Reilly, E.P.: J. Appl. Phys. **86**, 297 (1999)
37. Davies, J.H.: J. Appl. Phys. **84**, 1358 (1998)
38. Williams, D.P., Andreev, A.D., O'Reilly, E.P., Faux, D.A.: Phys. Rev. B **72**, 235318 (2005)
39. Eshelby, J.D.: Proc. R. Soc. Lond. Ser. A **241**, 376 (1957)
40. Ridha, P., Li, L.H., Mexis, M., Smowton, P.M., Andrzejewski, J., Sek, G., Misiewicz, J., O'Reilly, E.P., Patriarche, G., Fiore, A.: IEEE J. Quant. Electron. **46**, 197 (2010)
41. Williams, D.P., Schulz, S., Andreev, A.D., O'Reilly, E.P.: J. Sel. Top. Quant. Electron. **15**, 1092 (2009)
42. Ambacher, O.: J. Phys. D Appl. Phys. **31**, 2653 (1998)
43. Kittel, C.: Introduction to Solid State Physics. Wiley, New York (1995)
44. Vurgaftman, I., Meyer, J.R.: J. Appl. Phys. **94**, 3675 (2003)
45. Baer, N., Gartner, P., Jahnke, F.: Eur. Phys. J. B **42**, 231 (2004)
46. Wojs, A., Hawrylak, P., Fafard, S., Jacak, L.: Phys. Rev. B **54**, 5604 (1996)
47. Winkelnkemper, M., Schliwa, A., Bimberg, D.: Phys. Rev. B **74**, 155322 (2006)
48. Marquardt, O., Mourad, D., Schulz, S., Hickel, T., Czycholl, G., Neugebauer, J.: Phys. Rev. B **78**, 235302 (2008)
49. Marquardt, O., Hickel, T., Neugebauer, J.: J. Appl. Phys. **106**, 083707 (2009)
50. Bester, G., Zunger, A.: Phys. Rev. B **71**, 045318 (2005)
51. Bester, G., Zunger, A., Shumway, J.: Phys. Rev. B **71**, 075325 (2005)
52. Singh, R., Bester, G.: Phys. Rev. Lett. **103**, 063601 (2009)
53. Santoprete, R., Koiller, B., Capaz, R.B., Kratzer, P., Liu, Q.K.K., Scheffler, M.: Phys. Rev. B **68**, 235311 (2003)
54. Ranjan, V., Allan, G., Priester, C., Delerue, C.: Phys. Rev. B **68**, 115305 (2003)
55. Schulz, S., Schumacher, S., Czycholl, G.: Phys. Rev. B **73**, 245327 (2006)
56. Schulz, S., Schumacher, S., Czycholl, G.: Eur. Phys. J. B **64**, 51 (2008)
57. Lee, S.J., Chung, H.S., Nahm, K., Kim, C.K.: Phys. Rev. B **42**, 1452 (1990)
58. Bouhafs, B., Aourag, A., Ferhat, M., Zaoui, A., Certier, M.: J. Appl. Phys. **82**, 4923 (1997)
59. Fredj, A.B., Debbichi, M., Said, M.: Microelectron. J. **38**, 860 (2007)
60. Jancu, J.M., Scholz, R., Beltram, F., Bassani, F.: Phys. Rev. B **57**, 6493 (1998)
61. Boykin, T.B., Klimeck, G., Bowen, R.C., Oyafuso, F.: Phys. Rev. B **66**, 125207 (2002)
62. Schulz, S., Badcock, T.J., Moram, M.A., Dawson, P., Kappers, M.J., Humphreys, C.J., O'Reilly, E.P.: Phys. Rev. B **82**, 125318 (2010)
63. Yan, Q., Rinke, P., Scheffler, M., Van de Walle, C.G.: Appl. Phys. Lett. **95**, 121111 (2009)
64. Saito, T., Arakawa, Y.: Phys. E (Amsterdam) **15**, 169 (2002)

65. Zielinski, M., Jaskolski, W., Aizpurua, J., Bryant, G.W.: Acta Phys. Pol. A **108**, 929 (2005)
66. Schuh, K., Barthel, S., Marquardt, O., Hickel, T., Neugebauer, J., Czycholl, G., Jahnke, F.: Appl. Phys. Lett. **100**, 092103 (2012)
67. Schulz, S., O'Reilly, E.P.: Phys. Stat. Sol. (c) **7**, 1900 (2010)
68. Strittmatter, A., Reißmann, L., Seguin, R., Rodt, S., Brostowski, A., Pohl, U., Bimberg, D., Hahn, E., Gerthsen, D.: J. Cryst. Growth **272**, 415 (2004)
69. Gerthsen, D., Hahn, E., Neubauer, B., Potin, V., Rosenauer, A., Schowalter, M.: Phys. Stat. Sol. (c) **0**, 1668 (2003)
70. Senes, M., Smith, K.L., Smeeton, T.M., Hooper, S.E., Heffernan, J.: Phys. Rev. B **75**, 045314 (2007)
71. Schulz, S., O'Reilly, E.P.: Phys. Status Solid C **7**, 80 (2010)
72. Moriwaki, O., Someya, T., Tachibana, K., Ishida, S., Arakawa, Y.: Appl. Phys. Lett. **76**, 2361 (2000)
73. Robinson, J.W., Rice, J.H., Lee, K.H., Na, J.H., Taylor, R.A., Hasko, D.G., Oliver, R.A., Kappers, M.J., Humphreys, C.J., Briggs, G.A.D.: Appl. Phys. Lett. **86**, 213103 (2005)
74. Zhang, Q., Zhu, J., Ren, X., Li, H., Wang, T.: Appl. Phys. Lett. **78**, 3830 (2001)
75. Neogi, A., Gorman, B.P., Morkoc, H., Kawazoe, T., Ohtsu, M.: Appl. Phys. Lett. **86**, 043103 (2005)
76. Christmas, U.M.E., Andreev, A.D., Faux, D.A.: J. Appl. Phys. **98**, 073522 (2005)
77. Bernardini, F., Fiorentini, V.: Appl. Phys. Lett. **80**, 4145 (2002)
78. Shimada, K.: Jpn. J. Appl. Phys. **45**, L358 (2006)
79. Romanov, A.E., Baker, T.J., Nakamura, S., Speck, J.S.: J. Appl. Phys. **100**, 023522 (2006)
80. Shen, H., Wraback, M., Zhong, H., Tyagi, A., DenBaars, S.P., Nakamura, S., Speck, J.S.: Appl. Phys. Lett. **95**, 033503 (2009)
81. Pal, J., Tse, G., Haxha, V., Migliorato, M.A., Tomic, S.: Phys. Rev. B **84**, 085211 (2011)
82. Garro, N., Cros, A., Budagosky, J.A., Cantarero, A., Vinattieri, A., Gurioli, M., Founta, S., Mariette, H., Daudin, B.: Appl. Phys. Lett. **87**, 011101 (2005)
83. Founta, S., Rol, F., Bellet-Amalric, E., Sarigiannidou, E., Gayral, B., Moisson, C., Mariette, H., Daudin, B.: Phys. Stat. Sol. (b) **243**, 3968 (2006)
84. Founta, S., Coraux, J., Jalabert, D., Bougerol, C., Rol, F., Mariette, H., Renevier, H., Daudin, B., Oliver, R.A., Humphreys, C.J., Noakes, T.C.Q., Bailey, P.: J. Appl. Phys. **101**, 063541 (2007)
85. Yang, X., Arita, M., Kako, S., Arakawa, Y.: Appl. Phys. Lett. **99**, 061914 (2011)
86. Schulz, S., Caro, M.A., O'Reilly, E.P., Marquardt, O.: Phys. Rev. B **84**, 125312 (2011)
87. Dimastrodonato, V., Mereni, L.O., Juska, G., Pelucchi, E.: Appl. Phys. Lett. **97**, 072115 (2010)
88. Mohan, A., Felici, M., Gallo, P., Dwir, B., Rudra, A., Faist, J., Kapon, E.: Nat. Photon. **4**, 302 (2010)
89. Yeh, C.Y., Lu, Z.W., Froyen, S., Zunger, A.: Phys. Rev. B **46**, 10086 (1992)
90. Schulz, S., O'Reilly, E.P.: Phys. Status Solid A **208**, 1551 (2011)
91. Schulz, S., O'Reilly, E.P.: Phys. Status Solid B **249**, 516 (2012)
92. Schulz, S., O'Reilly, E.P.: Appl. Phys. Lett. **99**, 223106 (2011)
93. Schulz, S., Mourad, D., Schumacher, S., Czycholl, G.: Phys. Status Solid B **248**, 1853 (2011)

Chapter 7
Interference Single Electron Transistors Based on Quantum Dot Molecules

Andrea Donarini and Milena Grifoni

Abstract We consider nanojunctions in the single electron tunnelling regime which, due to a high degree of spatial symmetry, have a degenerate many-body spectrum. They comprise single molecule quantum dots as well as artificial quantum dot molecules. As a consequence, interference phenomena which cause a current blocking can occur at specific values of the bias and gate voltage. We present here a general formalism providing necessary and sufficient conditions for interference blockade also in the presence of spin-polarized leads. As examples we analyze a triple quantum dot as well as a benzene molecule single electron transistor.

7.1 Introduction

Single particle interference is one of the most genuine quantum mechanical effects. Since the original double-slit experiment [1], it has been observed with electrons in vacuum [2, 3] and even with the more massive C_{60} molecules [4]. Mesoscopic rings threaded by a magnetic flux provided the solid-state analogous [5, 6]. Intra-molecular interference has been recently discussed in molecular junctions for the case of strong [7–18] and weak [19–21] molecule-lead coupling. What unifies these realizations of quantum interference is that the travelling particle has two (or more) *spatially equivalent* paths at disposal to go from one point to another of the interferometer.

Interference, though is hindered by decoherence. Generally, for junctions in the strong coupling regime decoherence can be neglected due to the short time of flight of the particle within the interferometer. In the weak coupling case, instead, the dwelling time is long. It is the regime of the single electron tunnelling devices

A. Donarini • M. Grifoni
Theoretische Physik, Universität Regensburg, 93040 Regensburg, Germany
e-mail: andrea.donarini@physik.uni-r.de; milena.grifoni@physik.uni-r.de

J. Wu and Z.M. Wang (eds.), *Quantum Dot Molecules*, Lecture Notes in Nanoscale Science and Technology 14, DOI 10.1007/978-1-4614-8130-0_7,

(SETs) in which, usually, the decoherence introduced by the leads dominates the picture and the dynamics essentially consists of sequential tunnelling events connecting the many-body eigenstates of the isolated system. Yet, interference is achieved whenever two *energetically equivalent* paths involving degenerate states contribute to the dynamics (see Fig. 7.1) [22]. The associated fingerprints in the transport characteristics are a strong negative differential conductance (NDC) and eventually a current blocking in the case of fully destructive interference.

In the simplest case, NDC and current blocking triggered by interference take place any time a SET presents an N-particle non-degenerate state and two degenerate $N+1$-particle states such that the ratio between the transition *amplitudes* $\gamma_{i\chi}$ ($i=1,2, \quad \chi=S,D$) between those N- and $N+1$-particle states is different for tunnelling at the source (S) and at the drain (D) lead:

$$\frac{\gamma_{1S}}{\gamma_{2S}} \neq \frac{\gamma_{1D}}{\gamma_{2D}}. \tag{7.1}$$

Notice that no asymmetry in the tunnelling *rates*, which are proportional to $|\gamma_{i\chi}|^2$, is implied by Eq. (7.1). This fact excludes the interpretation of the physics of the interference SET in terms of standard NDC with asymmetric couplings. Instead, due to condition (7.1) there exist linear combinations of the degenerate $N+1$-particle states which are connected to the N-particle state via a tunnelling event to one of the leads but *not* to the other. The state which is decoupled from the drain lead (i.e., the lead with the lower chemical potential) represents a blocking state which prevents the current to flow since electrons can populate this state by tunnelling from the source but cannot tunnel out towards the drain.

It should be noticed that several blocking states can be associated with the same system. Let us consider again the example associated with (7.1) and analyze an inversion of the bias polarity which interchanges the source and the drain lead. If the state decoupled from the right lead blocks the current $L \to R$, vice versa the state decoupled from the left lead is a blocking state for the current $R \to L$. Typically these two different blocking states are not orthogonal and cannot form together a valid basis set of the $N+1$ particle space. The basis set that diagonalizes the stationary density matrix (what we call in the manuscript the *physical basis*) contains at large positive biases the $L \to R$ blocking state and is thus different from the physical basis at large negative biases which necessarily contains the $R \to L$ blocking state. More generally, the physical basis depends *continuously* on the bias. Thus only a treatment that includes also coherences and not only populations of the density matrix can capture the full picture at all biases.

It could be argued about the fragility of an effect which relies on the presence of degeneracies in the many-body spectrum. Interference effects are instead rather robust. The exact degeneracy condition can be in fact relaxed and interference survives as far as the splitting between the many-body levels is smaller that the tunnelling rate to the leads. In this limit, the system still does not distinguish between the two energetically equivalent paths sketched in Fig. 7.1. Summarizing, despite the

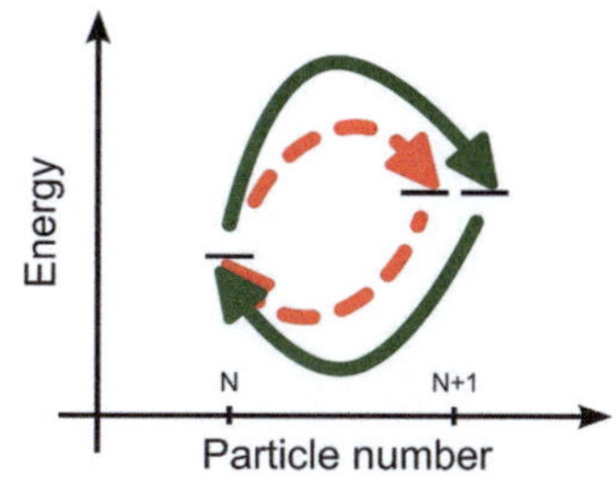

Fig. 7.1 Interference in a single electron transistor (SET). The dynamics is governed by equivalent paths that involve two (or more) degenerate states in the many-body spectrum. From [22]

decoherence introduced by the leads, in such devices, that we called interference single electron transistors [20] (I-SET), interference effects show up even in the Coulomb blockade regime.

In the present chapter we develop a general theory of interference blockade. We give in fact an *a priori* algorithm for the detection of the interference blocking states of a generic I-SET. As concrete examples, we analyze the triple dot and the benzene I-SET. The first is chosen as the simplest structure exhibiting interference blockade and in the second we emphasize the crucial role of the coupling geometry in the interference phenomena. In both cases we further analyze the blockade that involves orbitally and spin degenerate states and we show how to realize all electrical preparation of specific spin states. Thus we obtain an interference mediated control of the electron spin in quantum dots, a highly desirable property for spintronics [23–25] and spin-qubit applications [26–30]. Similar blocking effects have been found also in multiple quantum dot systems in dc [31] and ac [32] magnetic fields. The method of choice for the study of the dynamics in those systems is the generalized master equation approach for the reduced density matrix (RDM), where coherences between degenerate states are retained [19–21,33–43]. Such coherences give rise to precession effects and ultimately cause interference blockade.

The chapter is organized as follows: in Sect. 7.2 we introduce a generic model of I-SET. In Sect. 7.3 we set the necessary and sufficient conditions which define the interference blocking states and a generic algorithm to detect them. In Sects. 7.4 and 7.5 we apply the theory to the benzene and to the triple dot molecule I-SET. Section 7.6 is dedicated to the implications on spin transport of the interference effects in presence of ferromagnetic leads. In Sect. 7.7 we analyze the robustness of the interference phenomena upon relaxation of the exact orbital degeneracy condition. Section 7.8 closes the chapter with a summary of the results and conclusive remarks.

7.2 Generic Model of I-SET

Let us consider the interference single electron transistor (I-SET) described by the Hamiltonian:

$$H = H_{\text{sys}} + H_{\text{leads}} + H_{\text{tun}}, \tag{7.2}$$

where H_{sys} represents the central system and also contains the energy shift operated by a capacitively coupled gate electrode at the potential V_g. The Hamiltonian H_{sys} is invariant with respect to a set of point symmetry operations that defines the symmetry group of the device. This fact ensures the existence of *degenerate* states. To fix the ideas, the system Hamiltonian describing the triple dot and the benzene molecule considered later in this chapter (see Figs. 7.3 and 7.9) is of the Pariser–Parr–Pople form [44–46]:

$$H_{\text{sys}} = \xi \sum_{i\sigma} d_{i\sigma}^{\dagger} d_{i\sigma} + b \sum_{i\sigma} \left(d_{i\sigma}^{\dagger} d_{i+1\sigma} + d_{i+1\sigma}^{\dagger} d_{i\sigma} \right) + U \sum_{i} \left(n_{i\uparrow} - \frac{1}{2} \right) \left(n_{i\downarrow} - \frac{1}{2} \right)$$
$$+ V \sum_{i} \left(n_{i\uparrow} + n_{i\downarrow} - 1 \right) \left(n_{i+1\uparrow} + n_{i+1\downarrow} - 1 \right), \qquad (7.3)$$

where $d_{i\sigma}^{\dagger}$ creates an electron of spin σ in the p_z orbital of site i or in the ground state of the quantum dot i and $i = 1, \ldots, 6(3)$ runs over the six carbon atoms (three quantum dots) of the system. Moreover, $n_{i\sigma} = d_{i\sigma}^{\dagger} d_{i\sigma}$ counts the number of electrons of spin σ on site i. The effect of the gate is included as a renormalization of the on-site energy $\xi = \xi_0 - eV_g$ with V_g being the gate voltage. The parameters U and V describe the Coulomb interaction between electrons, respectively, on the same and on neighboring sites.

We leave a detailed analysis of the many-body spectrum of (7.3) to the Sects. 7.4 and 7.5. Here we just mention that, for these planar structures belonging to the D_n group, the (non-accidental) orbital degeneracy is at maximum twofold and can be resolved using the eigenvalues ℓ of the projection of the angular momentum along the principal axis of rotation. A generic eigenstate is then represented by the ket $|N\ell\sigma E\rangle$ where N is the number of electrons on the system, σ is the spin, and E the energy of the state. The size of the Fock space can make the exact diagonalization of H_{sys} a numerical challenge in its own. We will not treat here this problem and concentrate instead on the transport characteristics. H_{leads} describes two reservoirs of non-interacting electrons with a difference eV_b between their electrochemical potentials. Finally, H_{tun} accounts for the weak tunnelling coupling between the system and the leads, characteristic of SETs:

$$H_{\text{tun}} = \sum_{\chi k i \sigma} t_{ik}^{\chi} c_{\chi k \sigma}^{\dagger} d_{i\sigma} + h.c., \qquad (7.4)$$

where $c_{\chi k\sigma}^{\dagger}$ creates an electron with spin σ and momentum k in lead $\chi = L, R$ and t_{ik}^{χ} is the bare tunnelling amplitude of a k electron in the lead χ to the site i. We assume it for simplicity independent of the spin σ. Naturally, $|t_{ik}^{\chi}|$ is highest for the atom (quantum dot) closest to the lead χ, due to the exponential decay on the atomic scale of the tunnelling probability with the distance between the system and the lead. Moreover, in the case of atomically localized coupling where the tunnelling from the lead is most probable only to a small part of the system it is also reasonable to assume a very weak k-dependence of the tunnelling amplitude. We will simply neglect it in Sect. 7.3 when discussing the general criteria for the identification of blocking states.

In the weak coupling regime, the dynamics essentially consists of sequential tunnelling events at the source and drain lead inducing a flow of probability between the many-body eigenstates of the system. The coupling between the system and the leads, though, also contributes to an internal dynamics of the system that leaves unchanged its particle number. In fact the equation of motion for the reduced density matrix ρ of the system can be cast, to lowest non-vanishing order in the coupling to the leads, in the form [21, 33, 35]:

$$\dot{\rho} \equiv \mathscr{L}\rho = -\frac{i}{\hbar}[H_{\text{sys}}, \rho] - \frac{i}{\hbar}[H_{\text{eff}}, \rho] + \mathscr{L}_{\text{tun}}\rho . \tag{7.5}$$

The commutator with H_{sys} in Eq. (7.5) represents the coherent evolution of the system in absence of the leads. The operator $\mathscr{L}_{\text{tun}}$ describes the sequential tunnelling processes and is defined in terms of the transition amplitudes between the different many-body states. The commutator with H_{eff} is responsible instead for the effective internal dynamics associated with the presence of the leads. It is convenient to analyze the different terms in greater detail. In particular:

$$\begin{aligned}
(\mathscr{L}_{\text{tun}}\rho)^{NE} = & -\frac{1}{2}\sum_{\chi\sigma}\sum_{ij}\Big\{\mathscr{P}_{NE}\Big[d^{\dagger}_{i\sigma}\Gamma^{\chi\sigma}_{ij}(E-H_{\text{sys}})f^{-}_{\chi}(E-H_{\text{sys}})d_{j\sigma} \\
& + d_{j\sigma}\Gamma^{\chi\sigma}_{ij}(H_{\text{sys}}-E)f^{+}_{\chi}(H_{\text{sys}}-E)d^{\dagger}_{i\sigma}\Big]\rho^{NE} + h.c.\Big\} \\
& + \sum_{\chi\sigma}\sum_{ijE'}\mathscr{P}_{NE}\Big[d^{\dagger}_{i\sigma}\Gamma^{\chi\sigma}_{ij}(E-E')\rho^{N-1E'}f^{+}_{\chi}(E-E')d_{j\sigma} \\
& + d_{j\sigma}\Gamma^{\chi\sigma}_{ij}(E'-E)\rho^{N+1E'}f^{-}_{\chi}(E'-E)d^{\dagger}_{i\sigma}\Big]\mathscr{P}_{NE}
\end{aligned} \tag{7.6}$$

where $\rho^{NE} := \mathscr{P}_{NE}\rho\mathscr{P}_{NE}$, being $\mathscr{P}_{NE} := \sum_{\ell\sigma}|N\ell\sigma E\rangle\langle N\ell\sigma E|$ the projection operator on the subspace of N particles and energy E. Moreover, $f^{+}_{\chi}(x)$ is the Fermi function for the lead χ, $f^{+}_{\chi}(x) := f(x-\mu_{\chi})$, and $f^{-}_{\chi}(x) := 1 - f^{+}_{\chi}(x)$. The terms proportional to $f^{+}_{\chi}(x)$ describe in (7.6) tunnelling events *to* the system, while the tunnelling *out* of the system is associated with $f^{-}_{\chi}(x)$. Additionally, μ_{χ} stands for the electro-chemical potential of the lead χ, defined via the applied bias voltage as $\mu_L = \mu_0 + (1-c)eV_{\text{b}}$, $\mu_R = \mu_0 - ceV_{\text{b}}$ and consequently $eV_{\text{b}} = \mu_L - \mu_R$, with the electron charge e, the equilibrium potential μ_0 and the coefficient c governing the relative bias drop at the left and right lead. A symmetrical potential drop is obtained for $c = 1/2$, while for $c = 1$ the bias drops completely at the right-lead interface. Finally, $\mu_0 = -\Phi_0$ relates the equilibrium chemical potential to the work function and, in equilibrium, the work functions of the two leads are assumed equal. Beside the Fermi function, the tunnelling rates are characterized by the geometrical component:

$$\Gamma^{\chi\sigma}_{ij}(\Delta E) = \frac{2\pi}{\hbar}\sum_{k}\left(t^{\chi}_{ik}\right)^{*}t^{\chi}_{jk}\,\delta(\varepsilon^{\chi}_{k\sigma} - \Delta E) . \tag{7.7}$$

The argument ΔE of the rate $\Gamma_{ij}^{\chi\sigma}$ is the energy difference $E_{N+1} - E_N$ of the many-body states involved in the tunnelling process, sometimes written in Eq. (7.6) in terms of the operator H_{sys}.

Until now we only concentrated on the sequential tunnelling processes in the system. We still have to discuss the term in Eq. (7.5) which contains the effective Hamiltonian H_{eff}. The latter is defined as:

$$H_{\mathrm{eff}} = \frac{1}{2\pi} \sum_{NE} \sum_{\chi\sigma} \sum_{ij} \mathscr{P}_{NE} \Big[d_{i\sigma}^{\dagger} \Gamma_{ij}^{\chi\sigma} (E - H_{\mathrm{sys}}) p_{\chi}(E - H_{\mathrm{sys}}) d_{j\sigma} + d_{j\sigma} \Gamma_{ij}^{\chi\sigma} (H_{\mathrm{sys}} - E) p_{\chi}(H_{\mathrm{sys}} - E) d_{i\sigma}^{\dagger} \Big] \mathscr{P}_{NE}, \tag{7.8}$$

where the principal part function $p_{\chi}(x) = -\mathrm{Re}\Psi \left[\frac{1}{2} + \frac{1}{2\pi k_{\mathrm{B}} T}(x - \mu_{\chi}) \right]$, has been introduced, with T being the temperature and Ψ the digamma function. Eq. (7.8) shows that the effective Hamiltonian is block diagonal in particle number and energy, exactly as the density matrix in the secular approximation. Consequently, it only influences the dynamics of the system in presence of degenerate states. The effective Hamiltonian depends on the details of the system, yet in all cases it is bias and gate voltage dependent and this property has important consequences on the interference blocking phenomena that we are considering.

A natural expression for the current operators is obtained in terms of the time derivative of the reduced density matrix:

$$\langle I_{\mathrm{S}} + I_{\mathrm{D}} \rangle = \sum_{NE} \mathrm{Tr} \left\{ N \dot{\rho}^{NE} \right\} , \tag{7.9}$$

where $I_{\mathrm{S/D}}$ are the current operators calculated for the source and the drain interfaces. Conventionally we assume the current to be positive when it increases the charge on the molecule. Thus, in the stationary limit, $\langle I_{\mathrm{S}} + I_{\mathrm{D}} \rangle$ is zero. The stationary current is obtained as the average:

$$\langle I_{\mathrm{S}} \rangle = \mathrm{Tr} \left\{ \rho_{\mathrm{stat}} I_{\mathrm{S}} \right\} = -\langle I_{\mathrm{D}} \rangle , \tag{7.10}$$

where $\rho_{\mathrm{stat}} = \lim_{t \to \infty} \rho(t)$ is the stationary density operator that can be found from

$$\dot{\rho}_{\mathrm{stat}} = \mathscr{L} \rho_{\mathrm{stat}} = 0 , \tag{7.11}$$

where $\mathscr{L}$ is the full Liouville operator defined in (7.5). Finally, by following, for example, the procedure described in detail in [20], one finds the explicit expressions for the current operators:

$$I_{\chi} = \sum_{NE\sigma ij} \mathscr{P}_{NE} \Big[d_{j\sigma} \Gamma_{ij}^{\chi\sigma} (H_{\mathrm{sys}} - E) f_{\chi}^{+} (H_{\mathrm{sys}} - E) d_{i\sigma}^{\dagger} + - d_{i\sigma}^{\dagger} \Gamma_{ij}^{\chi\sigma} (E - H_{\mathrm{sys}}) f_{\chi}^{-} (E - H_{\mathrm{sys}}) d_{j\sigma} \Big] \mathscr{P}_{NE} , \tag{7.12}$$

where the energy renormalization terms, present in the generalized master equation (7.5), do not appear.

7.3 Blocking States

The dynamics of SETs is essentially described by sequential tunnelling events at the source and drain lead which connect the many-body eigenstates of the system. It is natural to define, in this picture, a blocking state as a state which the system can enter but from which it cannot escape. When the system occupies a blocking state the particle number cannot change in time and the current vanishes. If degenerate states participate in transport, they can lead to interference since, like the two arms of an electronic interferometer, they are populated simultaneously (see Fig. 7.1). In particular, depending on the external parameters they can form linear superpositions which behave as blocking states. If a blocking state is the linear combination of degenerate states, we call it *interference blocking state*.

We present in this section the general criteria for the identification of a blocking state and more specifically of an interference blocking state. First of all we will proceed to a classification of the tunnelling processes needed for a many-body description of the electron transport through a nanojunction.

7.3.1 *Classification of the Tunnelling Processes*

For the description of the tunnelling dynamics contained in the superoperator $\mathscr{L}_{\text{tun}}$ (see Eqs. (7.5) and (7.6)), it is convenient to classify all possible tunnelling events according to four categories: (i) Creation (Annihilation) tunnelling events that increase (decrease) by one the number of electrons in the system, (ii) Source (Drain) tunnelling that involves the lead with the higher (lower) chemical potential, (iii) $\uparrow$ ($\downarrow$) tunnelling that involves an electron with spin up (down) with respect of the corresponding lead quantization axis, (iv) Gain (Loss) tunnelling that increases (decreases) the energy in the system.

Using categories (i)–(iii) we can efficiently organize the matrix elements of the system component of H_{tun} in the matrices:

$$T^{+}_{N,EE'} = \begin{pmatrix} \gamma^{+}_{\mathrm{S}\uparrow} \\ \gamma^{+}_{\mathrm{S}\downarrow} \\ \gamma^{+}_{\mathrm{D}\uparrow} \\ \gamma^{+}_{\mathrm{D}\downarrow} \end{pmatrix} \qquad T^{-}_{N,EE'} = \begin{pmatrix} \gamma^{-}_{\mathrm{S}\uparrow} \\ \gamma^{-}_{\mathrm{S}\downarrow} \\ \gamma^{-}_{\mathrm{D}\uparrow} \\ \gamma^{-}_{\mathrm{D}\downarrow} \end{pmatrix} \tag{7.13}$$

where S, D means source and drain, respectively, and

$$\gamma^{+}_{\chi\sigma} = \sum_i (t^{\chi}_{i\sigma})^{*} \langle N+1, \{\ell', \tau'\}, E' | d^{\dagger}_{i\sigma} | N, \{\ell, \tau\}, E \rangle \tag{7.14}$$

is a matrix in itself, defined for every creation transition from a state with particle number N and energy E to one with $N+1$ particles and energy E'. We indicate correspondingly in the following transitions involving $\gamma^{+}_{S\sigma}$ and $\gamma^{+}_{D\sigma}$ as source-creation and drain-creation transitions. The compact notation $\{\ell, \tau\}$ indicates all

Table 7.1 Energy conditions for tunnelling transitions between the many-body eigenstates of the system. The quantity $\Delta E = E_f - E_i$ is the difference between the energies of the final and initial many-body states of the system involved in the transition. The bias energy eV_b is assumed to be positive. From [22]

Tunnelling process	Energy condition
Source-Creation	$\Delta E \leq +eV_{\mathrm{b}}/2$
Source-Annihilation	$\Delta E \leq -eV_{\mathrm{b}}/2$
Drain-Creation	$\Delta E \leq -eV_{\mathrm{b}}/2$
Drain-Annihilation	$\Delta E \leq +eV_{\mathrm{b}}/2$

possible combinations of the quantum numbers ℓ and τ. It follows that the size of $\gamma^{+}_{\chi\sigma}$ is $\mathrm{mul}(N+1,E') \times \mathrm{mul}(N,E)$ where the function $\mathrm{mul}(N,E)$ gives the degeneracy of the many-body energy level with N particles and energy E. Analogously

$$\gamma^{-}_{\chi\sigma} = \sum_i t^{\chi}_{i\sigma} \langle N-1, \{\ell',\tau'\}, E' | d_{i\sigma} | N, \{\ell,\tau\}, E \rangle \tag{7.15}$$

accounts for the annihilation transitions.

The fourth category concerns energy and it is intimately related to the first and the second. Not all transitions are in fact allowed: due to the energy conservation and the Pauli exclusion principle holding in the fermionic leads, the energy gain (loss) of the system associated with a gain (loss) transition is governed by the bias voltage. These energy conditions, for the case of equal potential drop at the source and drain lead ($c = 1/2$), are summarized in Table 7.1 and illustrated in Fig. 7.2.

The quantity $\Delta E := E_f - E_i$ is the difference between the energy of the final and initial state of the system and the condition $\leq$ is in reality smoothed due to the thermal broadening of the Fermi distributions. For simplicity we set the zero of the energy at the chemical potential of the unbiased device. In Table 7.1 one reads, for example, that in a source-creation tunnelling event the system can gain at maximum $\frac{eV_{\mathrm{b}}}{2}$ or that in a source-annihilation and drain-creation transition the system loses at least an energy of $\frac{eV_{\mathrm{b}}}{2}$.

From Table 7.1 one also deduces that, from whatever initial state, it is always possible to reach the lowest energy state (the global minimum) via a series of energetically allowed transitions. Vice versa, not all states can be reached starting from the global minimum. Thus, the only *relevant states* for the transport in the stationary regime are the states that can be reached from the global minimum via a finite number of energetically allowed transitions.

7.3.2 Subspace of Decoupled States

In the process of detecting the blocking states we observe first that some states do not participate in the transport and can be excluded a priori from any consideration.

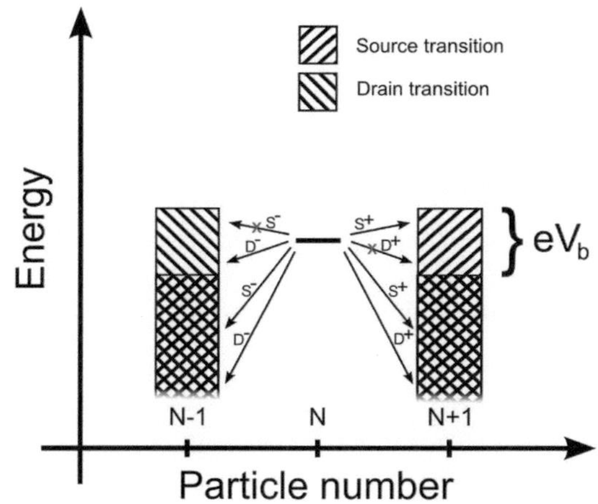

Fig. 7.2 Energetically available transitions *from* an N particle level. The *patterned rectangles* indicate the energy range of energetically available source (S) and drain (D) transitions both to states with $N+1$ and $N-1$ particles. The *arrows* show examples of both allowed and forbidden transitions. From [22]

These are states with zero transition elements to all other *relevant* states. Within the subspace with N particles and energy E the decoupled states span the vector space:

$$\mathscr{D}_{N,E} = \bigcap_{E'} \left[\ker T^{+}_{N,EE'} \cap \ker T^{-}_{N,EE'} \right] \tag{7.16}$$

where E' is the energy of a relevant state with $N+1$ or $N-1$ particles, respectively. The function $\ker M$ returns the null space of the linear application associated with the matrix M.

The decoupled space $\mathscr{D}_{N,E}$ as presented in Eq. (7.16) is constructed as follows. Let us consider a generic many-body state $|\psi_{NE}\rangle$ with N particles and energy E and let $\mathbf{v}$ be the vector of its components in the basis $|N\ell\tau E\rangle$. The vector $T^{+}_{N,EE'}\mathbf{v}$ has thus $4 \times \mathrm{mul}(N+1, E')$ components and consists of all possible transition amplitudes from $|\psi_{NE}\rangle$ to all possible states with $N+1$ particles and energy E'. Consequently $\ker T^{+}_{N,EE'}$ contains the vectors $\mathbf{v}$ associated with states with N particles and energy E which are decoupled from all possible states with $N+1$ particles and energy E'. Analogously holds for the significance of $\ker T^{-}_{N,EE'}$. The intersections in (7.16) and the condition on E' ensure that $\mathscr{D}_{N,E}$ contains only states decoupled at the same time from all other states relevant for transport in the stationary regime. We emphasize that, due to the condition on the energy E', the decoupled space $\mathscr{D}_{N,E}$ is a dynamical concept that depends on the applied gate and bias across the I-SET. The coupled space $\mathscr{C}_{N,E}$ is the orthogonal complement of $\mathscr{D}_{N,E}$ in the Hilbert space with N particles and energy E. The blocking states belong to it.

As a first simple application of the ideas presented so far, let us consider the SET at zero bias. According to Table 7.1 the system can only undergo loss tunnelling events and the global energy minimum is the only blocking state, in accordance with the observation that the system is in equilibrium with the leads and that we measure the energy starting from the equilibrium chemical potential.[1] The potential V_g of the gate electrode defines the particle number of the global minimum and, by sweeping V_g at zero bias, one can change the number of electrons on the system one by one. This situation, the Coulomb blockade, remains unchanged until the bias is high enough to open a gain transition that unblocks the global minimum. Then, the current can flow. Depending on the gate this first unblocking transition can be of the kind source-creation or drain-annihilation. Correspondingly, the current is associated with $N \leftrightarrow N+1$ or $N \leftrightarrow N-1$ oscillations, where N is the particle number of the global minimum.

7.3.3 Blocking Conditions

At finite bias the condition which defines a blocking state becomes more elaborate:

1. The blocking state must be achievable from the global minimum with a finite number of allowed transitions.
2. All matrix elements corresponding to energetically allowed transitions outgoing from the blocking state should vanish: in particular all matrix elements corresponding to $E_f - E_{\text{block}} < -\frac{eV_b}{2}$ and for $|E_f - E_{\text{block}}| < \frac{eV_b}{2}$ only the ones corresponding to the drain-annihilation and source-creation transitions.

The first condition ensures the blocking state to be populated in the stationary regime. The second is a modification of the generic definition of blocking state restricted to energetically allowed transitions and it can be written in terms of the tunnelling matrices $T^{+}_{N,EE'}$ and $T^{-}_{N,EE'}$. For each many-body energy level $|NE\rangle$, the space spanned by the blocking states reads then:

$$\mathscr{B}_{N,E} = \mathscr{B}^{(1)}_{N,E} \cap \mathscr{B}^{(2)}_{N,E} \cap \mathscr{C}_{N,E} \tag{7.17}$$

with

$$\begin{aligned}\mathscr{B}^{(1)}_{N,E} &= \bigcap_{E'} \left\{ \mathscr{P}_{NE} \left[\ker \left(T^{+}_{N,EE'}, T_D \right) \right] \cap \right. \\ &\qquad \left. \mathscr{P}_{NE} \left[\ker \left(T^{-}_{N,EE'}, T_S \right) \right] \right\} \\ \mathscr{B}^{(2)}_{N,E} &= \bigcap_{E''} \left[\ker T^{+}_{N,EE''} \cap \ker T^{-}_{N,EE''} \right].\end{aligned} \tag{7.18}$$

[1] If the equilibrium chemical potential is not set to zero, the many-body energy spectrum should be substituted with the spectrum of the many-body free energy ($H_{\text{sys}} - \mu_0 N$) where μ_0 is the chemical potential of the leads at zero bias. The rest of the argumentation remains unchanged.

In Eq. (7.18) we introduced the matrices $T_D = (\mathbf{0}, \mathbf{1})^T$ and $T_S = (\mathbf{1}, \mathbf{0})^T$ with $\mathbf{1}$ being the identity matrix and $\mathbf{0}$ the zero matrix, both of dimension $2 \times \mathrm{mul}(N+1, E')$ for T_D and $2 \times \mathrm{mul}(N-1, E')$ for T_S. The energies E' and E'' satisfy the inequalities $|E' - E| < \frac{eV_b}{2}$ and $E'' - E < -\frac{eV_b}{2}$, respectively, and $\mathscr{P}_{NE}$ is the projection on the N particle space with energy E.

The first kernel in $\mathscr{B}^{(1)}_{N,E}$ together with the projector $\mathscr{P}_{NE}$ gives all linear combinations of N particle degenerate states which have a finite creation transition involving the drain but not the source lead. This condition can in fact be expressed as a non-homogeneous linear equation for the vector $\mathbf{v}$ of the components in the many-body basis of the generic N particle state with energy E:

$$T^{+}_{N,EE'}\mathbf{v} = \mathbf{b}, \tag{7.19}$$

where $\mathbf{b}$ is a generic vector of length $4 \times \mathrm{mul}(N+1, E')$ whose first $2 \times \mathrm{mul}(N+1, E')$ components (the source transition amplitudes) are set to zero. Due to the form of $\mathbf{b}$, it is convenient to transform Eq. (7.19) into a homogeneous equation for a larger space of dimension $\mathrm{mul}(N, E) + 2 \times \mathrm{mul}(N+1, E')$ which also contains the nonzero elements of $\mathbf{b}$ and finally project the solutions of this equation on the original space. With this procedure we can identify the space of the solutions of (7.19) with:

$$V = \mathscr{P}_{NE}\left[\ker\left(T^{+}_{N,EE'}, T_D\right)\right]. \tag{7.20}$$

The second kernel in $\mathscr{B}^{(1)}_{N,E}$ takes care of the annihilation transitions in a similar way. Notice that V also contains vectors that are decoupled at both leads. This redundance is cured in (7.17) by the intersection with the coupled space $\mathscr{C}_{NE}$.

The conditions (7.18) are the generalization of the conditions over the tunnelling amplitudes that we gave in the introduction (Eq. (7.1)). That very simple condition captures the essence of the effect, but it is only valid under certain conditions: the spin channels should be independent, the relevant energy levels only two and the transition has to be between a non-degenerate and a doubly degenerate level. Equation (7.18), on the contrary, is completely general. At the end of this section we will show explicitly the equivalence of the two approaches in the simple case.

For most particle numbers N and energies E, and sufficiently high bias, $\mathscr{B}_{N,E}$ is empty. Yet, blocking states exist and the dimension of $\mathscr{B}_{N,E}$ can even be larger than one as we will show explicitly in Sect. 7.5 for the triple dot I-SET. Moreover, it is most probable to find interference blocking states among ground states due to the small number of intersections appearing in (7.18) in this situation. Nevertheless also excited states can block the current as we will show in the next section.

The case of spin polarized leads is already included in the formalism in both the parallel and nonparallel configuration. In the parallel case one quantization axis is naturally defined on the whole structure and σ in Eqs. (7.14) and (7.15) is defined along this axis. In the case of nonparallel polarized leads instead it is enough to consider $d^{\dagger}_{i\sigma}$ and $d_{i\sigma}$ in Eqs. (7.14) and (7.15), respectively, with σ along the quantization axis of the lead χ. It is interesting to note that in that case, no blocking states can be found unless the polarization of one of the leads is $P = 1$. The spin channel can in fact be closed only one at the time via linear combination of different spin states.

A last comment on the definition of the blocking conditions is necessary. A blocking state is a stationary solution of the Eq. (7.5) since by definition it does not evolve in time. The density matrix associated with one of the blocking states discussed so far (i) commutes with the system Hamiltonian since it is a state with given particle number and energy; (ii) it is the solution of the equation $\mathscr{L}_{\text{tun}}\rho = 0$ since the probability of tunnelling out from a blocking state vanishes, independent of the final state. Nevertheless, a third condition is needed to satisfy the condition of stationarity:

3. The density matrix ρ_{block} associated with the blocking state should commute with the effective Hamiltonian H_{eff} which renormalizes the coherent dynamics of the system to the lowest non-vanishing order in the coupling to the leads:

$$[\rho_{\text{block}}, H_{\text{eff}}] = 0. \tag{7.21}$$

The specific form of H_{eff} varies with the details of the system. Yet its generic bias and gate voltage dependence implies that, if present, the current blocking occurs only at specific values of the bias for each gate voltage. Further, if an energy level has multiple blocking states and the effective Hamiltonian distinguishes between them, selective current blocking, and correspondingly all electrical preparation of the system in one specific degenerate state, can be achieved. In particular, for spin polarized leads, the system can be prepared in a particular spin state without the application of any external magnetic field as we will show explicitly in Sect. 7.6.

Before continuing with the discussion, in the following sections, we derive here the Eq. (7.1) as a specific example of the general theory presented so far. That equation represents the interference blocking condition for the simplest possible configuration involving only a non-degenerate and a doubly degenerate state. Let us consider for simplicity a spinless[2] system and a gate and bias condition that restricts the set of relevant many-body states to three: one with N particles and two (degenerate) with $N+1$ or $N-1$ particles. The interference blocking state, if it exists, belongs to the $N \pm 1$ level. There is only one interesting tunnelling matrix to be analyzed, namely $T^{\mp}_{N\pm1}$. Let us take for it the generic form:

$$T^{\mp}_{N\pm1} = \begin{pmatrix} \gamma_{S1} & \gamma_{S2} \\ \gamma_{D1} & \gamma_{D2} \end{pmatrix} \tag{7.22}$$

[2]The assumption of a spinless system is not restrictive for parallel polarized leads and transitions between a spin singlet and a doublet since the different spin sectors decouple from each other.

where S and D indicate source and drain, respectively, and 1 and 2 label the two degenerate states with $N \pm 1$ particles. $\gamma_{S(D)i}$ are the elements of the $\gamma^{\mp}_{S(D)}$ matrices introduced in Eqs. (7.14) and (7.15). The decoupled space reads:

$$\mathscr{D}_{N\pm1} = \ker T^{\mp}_{N\pm1}. \tag{7.23}$$

Since the $N \pm 1$ particles relevant Hilbert space has dimension 2 the only possibility to find a blocking state is that $\mathscr{D}_{N\pm1} = \emptyset$. In other terms:

$$\det T^{\mp}_{N\pm1} = \gamma_{S1}\gamma_{D2} - \gamma_{D1}\gamma_{S2} \neq 0 \tag{7.24}$$

This condition is identical to Eq. (7.1). The blocking state can finally be calculated as:

$$\mathscr{B}_{N+1} = \mathscr{P}_{N+1} \ker \begin{pmatrix} \gamma_{S1} & \gamma_{S2} & 1 \\ \gamma_{D1} & \gamma_{D2} & 0 \end{pmatrix} \cap \mathscr{C}_{N+1} \tag{7.25}$$

or

$$\mathscr{B}_{N-1} = \mathscr{P}_{N-1} \ker \begin{pmatrix} \gamma_{S1} & \gamma_{S2} & 0 \\ \gamma_{D1} & \gamma_{D2} & 1 \end{pmatrix} \cap \mathscr{C}_{N-1}, \tag{7.26}$$

where the $\mathscr{C}_{N\pm1}$ is, in the relevant case, the entire space and the projector $\mathscr{P}_{N\pm1}$ simply removes the last component of the vector that defines the one-dimensional kernel.

7.4 The Benzene I-SET

The general ideas on interference blocking presented in the previous section apply to a large class of devices. As a first example of interference SET based on quantum dot molecules we consider a benzene single molecule transistor. We treat the transport through the benzene I-SET in two different setups, the para and the meta configuration, depending on the position of the leads with respect to the benzene molecule (see Fig. 7.3). Similar to [47], we start from an interacting Hamiltonian of isolated benzene where only the localized p_z orbitals are considered and the ions are assumed to have the same spatial symmetry as the relevant electrons. We calculate the $4^6 = 4{,}096$ energy eigenstates of the benzene Hamiltonian numerically. Subsequently, with the help of group theory, we classify the eigenstates according to their different symmetries and thus give a group-theoretical explanation to the large degeneracies occurring between the electronic states. For example, while the six-particle ground state (A_{1g} symmetry) is non-degenerate, there exist four seven-particle ground states due to spin and orbital (E_{2u} symmetry) degeneracy. Fingerprints of these orbital symmetries are clearly visible in the strong differences in the stability diagrams obtained by coupling the benzene I-SET to the leads in the

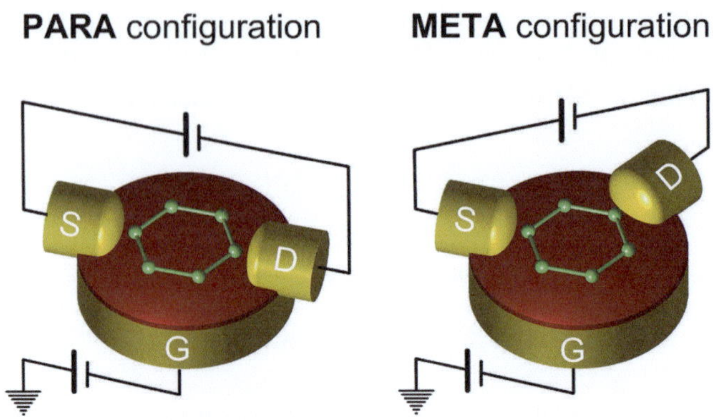

Fig. 7.3 Schematic representation of the two different setups for the benzene I-SET considered in this paper. The molecule, lying on a dielectric substrate, is weakly contacted to source and drain leads as well as capacitively gated. From [20]

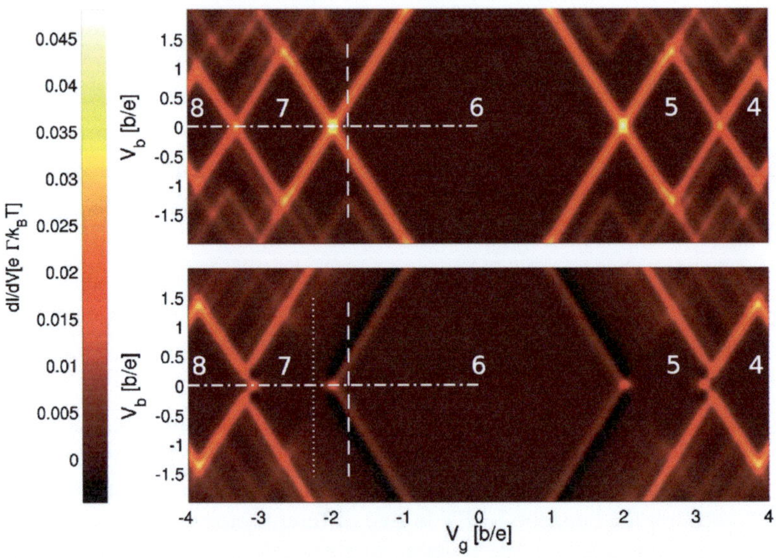

Fig. 7.4 Stability diagram for the benzene I-SET contacted in the para (above) and meta (below) configuration. *Dot-dashed lines* highlight the conductance cuts presented in Fig. 7.5, the *dashed lines* the regions corresponding to the current traces presented in Figs. 7.6 and 7.8, the *dotted line* the region corresponding to the current trace presented in Fig. 7.7. The parameters used are $U = 4|b|, V = 2.4|b|, k_B T = 0.04|b|, \hbar\Gamma_L = \hbar\Gamma_R = 10^{-3}|b|$. From [20]

meta and para configurations (see Fig. 7.4). Striking are the selective reduction of conductance and the appearance of regions of interference-driven current blocking with associated negative differential conductance (NDC) when changing from the para to the meta configuration.

7.4.1 Model Hamiltonian

For the description of the benzene molecule weakly coupled to source and drain leads, we adopt the total Hamiltonian (7.2) introduced in Sect. 7.2 where the first term is now the interacting Hamiltonian for isolated benzene [44–46] that we recall here for clarity:

$$\begin{aligned} H^0_{\text{ben}} = {} & \xi_0 \sum_{i\sigma} d^\dagger_{i\sigma} d_{i\sigma} + b \sum_{i\sigma} \left(d^\dagger_{i\sigma} d_{i+1\sigma} + d^\dagger_{i+1\sigma} d_{i\sigma} \right) \\ & + U \sum_i \left(n_{i\uparrow} - \frac{1}{2} \right) \left(n_{i\downarrow} - \frac{1}{2} \right) \\ & + V \sum_i \left(n_{i\uparrow} + n_{i\downarrow} - 1 \right) \left(n_{i+1\uparrow} + n_{i+1\downarrow} - 1 \right), \end{aligned} \tag{7.27}$$

where $d^\dagger_{i\sigma}$ creates an electron of spin σ in the p_z orbital of carbon i, $i = 1,\ldots,6$ runs over the six carbon atoms of benzene and $n_{i\sigma} = d^\dagger_{i\sigma} d_{i\sigma}$. Only the p_z orbitals (one per carbon atom) are explicitly taken into account, while the core electrons and the nuclei are combined into frozen ions, with the same spatial symmetry as the relevant electrons. They contribute only to the constant terms of the Hamiltonian and enforce particle-hole symmetry. Mechanical oscillations are neglected and all atoms are considered at their equilibrium position.

This Hamiltonian for isolated benzene is respecting the D_{6h} symmetry of the molecule. Since for every site there are 4 different possible configurations ($|0\rangle$, $|\uparrow\rangle$, $|\downarrow\rangle$, $|\uparrow\downarrow\rangle$), the Fock space has the dimension $4^6 = 4{,}096$, which requires a numerical treatment. Though the diagonalization of the Hamiltonian is not a numerical challenge, it turns out to be of benefit for the physical understanding of the transport processes to divide H_{ben} into blocks, according to the number N of p_z electrons (from 0 to 12), the z projection S_z of the total spin and the orbital symmetries of benzene (see Table 7.2).

The parameters b, U, and V for isolated benzene are given in the literature [48] and are chosen to fit optical excitation spectra. The presence, in the molecular I-SET, of metallic electrodes and the dielectric is expected to cause a substantial renormalization of U and V [49–51]. Nevertheless, we do not expect the main results of this work to be affected by this change. We consider the benzene molecule weakly coupled to the leads. Thus, to first approximation, we assume the symmetry of the isolated molecule not to be changed by the screening. Perturbations due to the lead-molecule contacts reduce the symmetry in the molecular junction. They are included in H'_{ben} (see Eqs. (7.64) and (7.65)) and will be treated in Sect. 7.7.

The effect of the gate is included as a renormalization of the on-site energy $\xi = \xi_0 - eV_{\text{g}}$ (V_{g} is the gate voltage) and we conventionally set $V_{\text{g}} = 0$ at the charge neutrality point. Source and drain leads are two reservoirs of non-interacting electrons: $H_{\text{leads}} = \sum_{\chi k \sigma} (\varepsilon_k - \mu_\chi) c^\dagger_{\chi k \sigma} c_{\chi k \sigma}$, where $\chi = L, R$ stands for the left or right lead and the chemical potentials μ_χ of the leads depend on the applied bias voltage $\mu_{\text{L,R}} = \mu_0 \pm \frac{V_{\text{b}}}{2}$. In the following we will measure the energy starting from

Table 7.2 Overview of the 6-particle states of benzene, sorted by S_z and symmetry. Orbitals with A- and B-type of symmetry show no degeneracy, while E-type orbitals are doubly degenerate. From [20]

N	$S_z[\hbar]$	Number of states	States with given symmatry
6	3	1	1 B_{1u}
	2	26	4 A_{1g}
			2 A_{2g}
			2×6 E_{2g}
			4 B_{1u}
			2 B_{2u}
			2×6 E_{1u}
	1	225	16 A_{1g}
			20 A_{2g}
			2×36 E_{2g}
			22 B_{1u}
			17 B_{2u}
			2×39 E_{1u}
	0	400	38 A_{1g}
			30 A_{2g}
			2×66 E_{2g}
			38 B_{1u}
			30 B_{2u}
			2×66 E_{1u}
	−1	225	⋮
	−2	36	
	−3	1	

the equilibrium chemical potential $\mu_0 = 0$. We specialize the tunnelling Hamiltonian (7.4) to the following form

$$H_{\text{tun}} = t \sum_{\chi k \sigma} \left(d^{\dagger}_{\chi\sigma} c_{\chi k \sigma} + c^{\dagger}_{\chi k \sigma} d_{\chi\sigma} \right), \tag{7.28}$$

where we define $d^{\dagger}_{\chi\sigma}$ as the creator of an electron in the benzene carbon atom which is closest to the lead χ. In particular $d^{\dagger}_{\text{R}\sigma} := d^{\dagger}_{4\sigma}, d^{\dagger}_{5\sigma}$, respectively, in the para and meta configuration, while $d^{\dagger}_{\text{L}\sigma} := d^{\dagger}_{1\sigma}$ in both setups (see Fig. 7.6 for the numbering of the benzene carbon atoms).

7.4.2 Symmetry of the Benzene Eigenstates

In this section, we will review the symmetry characteristics of the eigenstates of the interacting Hamiltonian of benzene, focusing on the symmetry operations σ_{v} and C_n which have a major impact on the electronic transport through the molecular I-SET. Benzene belongs to the D_{6h} point group. Depending on their behavior under symmetry operations, one can classify the molecular orbitals by their belonging to a

Table 7.3 Degeneracy, energy, and symmetry of the ground states of the isolated benzene molecule for different particle numbers. We choose the on-site and inter-site Coulomb interactions to be $U = 10\,eV$, $V = 6\,eV$, and the hopping to be $b = -2.5\,eV$. Notice, however, that screening effects from the leads and the dielectric are expected to renormalize the energy of the benzene many-body states. From [20]

N	Degeneracy	Energy[eV] (at $\xi = 0$)	Symmetry	Symmetry behavior under σ_v
0	1	0	A_{1g}	sym
1	2	−22	A_{2u}	sym
2	1	−42.25	A_{1g}	sym
3	4	−57.42	E_{1g}	2 sym, [2 antisym]
4	[3]	[−68.87]	[A_{2g}]	[antisym]
	2	−68.37	E_{2g}	1 sym, [1 antisym]
5	4	−76.675	E_{1g}	2 sym, [2 antisym]
6	1	−81.725	A_{1g}	sym
7	4	−76.675	E_{2u}	2 sym, [2 antisym]
8	[3]	[−68.87]	[A_{2g}]	[antisym]
	2	−68.37	E_{2g}	1 sym, [1 antisym]
9	4	−57.42	E_{2u}	2 sym, [2 antisym]
10	1	−42.25	A_{1g}	sym
11	2	−22	B_{2g}	sym
12	1	0	A_{1g}	sym

certain irreducible representation of the point group. Table 7.2 shows an overview of the states of the neutral molecule (the 6-particle states) sorted by S_z and symmetries. The eigenstates of the interacting benzene molecule have either A-, B- or E-type symmetries. While orbitals having A or B symmetries can only be spin degenerate, states with an E symmetry show an additional twofold orbital degeneracy, essential for the explanation of the transport features occurring in the meta configuration.

Transport at low bias is described in terms of transitions between ground states with different particle number. Table 7.3 shows the symmetries of the ground states (and of some first excited states) of interacting benzene for all possible particle numbers. Ground state transitions occur both between orbitally non-degenerate states (with A and B symmetry), as well as between orbitally degenerate and non-degenerate states (E- to A-type states).

The interacting benzene Hamiltonian commutes with all the symmetry operations of the D_{6h} point group, thus it has a set of common eigenvectors with each operation. The element of D_{6h} of special interest for the *para* configuration is σ_v, i.e., the reflection about the plane through the contact atoms and perpendicular to the molecular plane. The molecular orbitals with A and B symmetry are eigenstates of σ_v with eigenvalue ± 1, i.e., they are either symmetric or antisymmetric with respect to the σ_v operation. The behavior of the E-type orbitals under σ_v is basis dependent, yet one can always choose a basis in which one orbital is symmetric and the other one antisymmetric.

Let us now consider the generic transition amplitude $\langle N|d_{\alpha\tau}|N+1\rangle$, where $d_{\alpha\tau}$ destroys an electron of spin τ on the contact atom closest to the α lead. It is useful to rewrite this amplitude in the form

$$\langle N|d_{\alpha\tau}|N+1\rangle = \langle N|\sigma_{\mathrm{v}}^{\dagger}\sigma_{\mathrm{v}}\, d_{\alpha\tau}\sigma_{\mathrm{v}}^{\dagger}\sigma_{\mathrm{v}}\,|N+1\rangle, \tag{7.29}$$

where we have used the property $\sigma_{\mathrm{v}}^{\dagger}\sigma_{\mathrm{v}} = 1$. Since in the para configuration both contact atoms lie in the mirror plane σ_{v}, it follows $\sigma_{\mathrm{v}}\, d_{\alpha}\sigma_{\mathrm{v}}^{\dagger} = d_{\alpha}$. If the participating states are both symmetric under σ_{v}, Eq. (7.29) becomes

$$\langle N,\mathrm{sym}|\sigma_{\mathrm{v}}^{\dagger} d_{\alpha\tau}\sigma_{\mathrm{v}}\,|N+1,\mathrm{sym}\rangle = \langle N,\mathrm{sym}|d_{\alpha\tau}|N+1,\mathrm{sym}\rangle \tag{7.30}$$

and analogously in the case that both states are antisymmetric. For states with different symmetry it is

$$\langle N,\mathrm{sym}|d_{\alpha\tau}|N+1,\mathrm{antisym}\rangle = -\langle N,\mathrm{sym}|d_{\alpha\tau}|N+1,\mathrm{antisym}\rangle = 0. \tag{7.31}$$

In other terms, there is a selection rule that forbids transitions between symmetric and antisymmetric states. Further, since the ground state of the neutral molecule is symmetric, for the transport calculations in the para configuration we select the effective Hilbert space containing only states symmetric with respect to σ_{v}. Correspondingly, when referring to the N particle ground state we mean the energetically lowest *symmetric* state. For example, in the case of 4- and 8-particle states it is the first excited state to be the *effective* ground state. In the para configuration also the orbital degeneracy of the E-type states is effectively cancelled due to the selection of the symmetric orbital (see Table 7.3).

Small violations of this selection rule, due, e.g., to molecular vibrations or coupling to an electromagnetic bath, result in the weak connection of different metastable electronic subspaces. We suggest this mechanism as a possible explanation for the switching and hysteretic behavior reported in various molecular junctions. This effect is not addressed in this work.

For a simpler analysis of the different transport characteristics it is useful to introduce a unified geometrical description of the two configurations. In both cases, one lead is rotated by an angle ϕ with respect to the position of the other lead. Hence we can write the creator of an electron in the right contact atom $d_{\mathrm{R}\tau}^{\dagger}$ in terms of the creation operator of the left contact atom and the rotation operator:

$$d_{\mathrm{R}\tau}^{\dagger} = \mathscr{R}_{\phi}^{\dagger} d_{\mathrm{L}\tau}^{\dagger} \mathscr{R}_{\phi}, \tag{7.32}$$

where $\mathscr{R}_{\phi}$ is the rotation operator for the anticlockwise rotation of an angle ϕ around the axis perpendicular to the molecular plane and piercing the center of the benzene ring; $\phi = \pi$ for the para and $\phi = (2\pi/3)$ for the meta configuration.

The energy eigenstates of the interacting Hamiltonian of benzene can be classified also in terms their quasi-angular momentum. In particular, the eigenstates of the z-projection of the quasi-angular momentum are the ones that diagonalize all operators $\mathscr{R}_\phi$ with angles multiples of $\pi/3$. The corresponding eigenvalues are phase factors $e^{-i\ell\phi}$ where $\hbar\ell$, the quasi-angular momentum of the state, is an integer multiple of $\hbar$. The discrete rotation operator of an angle $\phi = \pi$ (C_2 symmetry operation) is the one relevant for the para configuration. All orbitals are eigenstates of the C_2 rotation with the eigenvalue ± 1.

The relevant rotation operator for the meta configuration corresponds to an angle $\phi = 2\pi/3$ (C_3 symmetry operation). Orbitals with an A or B symmetry are eigenstates of this operator with the eigenvalue $+1$ (angular momentum $\ell = 0$ or $\ell = 3$). Hence we can already predict that there will be no difference based on rotational symmetry between the para and the meta configuration for transitions between states involving A- and B-type symmetries. Orbitals with E symmetry, however, behave quite differently under the C_3 operation. They are the pairs of states of angular momenta $\ell = \pm 1$ or $\ell = \pm 2$. The diagonal form of the rotation operator on the twofold degenerate subspace of E-symmetry reads:

$$C_3 = \begin{pmatrix} e^{-|\ell|\cdot\frac{2\pi}{3}i} & 0 \\ 0 & e^{|\ell|\cdot\frac{2\pi}{3}i} \end{pmatrix}. \tag{7.33}$$

For the twofold orbitally degenerate 7-particle ground states $|\ell| = 2$. This analysis in terms of the quasi-angular momentum makes easier the calculation of the fundamental interference condition (7.1) given in the introduction. In fact the following relation holds between the transition amplitudes of the 6- and 7-particle ground states:

$$\gamma_{\ell R} \equiv \langle 7_g \ell\tau | d^\dagger_{R\tau} | 6_g \rangle = \langle 7_g \ell\tau | \mathscr{R}^\dagger_\phi d^\dagger_{L\tau} \mathscr{R}_\phi, | 6_g \rangle = e^{-i\ell\phi} \gamma_{\ell L} \tag{7.34}$$

and (7.1) follows directly.

7.4.3 Transport Calculations

With the knowledge of the eigenstates and eigenvalues of the Hamiltonian for the isolated molecule, we implement Eq. (7.5) and look for a stationary solution. The symmetries of the eigenstates are reflected in the transition amplitudes contained in the generalized master equation. We find numerically its stationary solution and calculate the current and the differential conductance of the device. In Fig. 7.4 we present the stability diagram for the benzene I-SET contacted in the para (upper panel) and meta position (lower panel). Bright ground state transition lines delimit diamonds of zero differential conductance typical for the Coulomb blockade regime, while a rich pattern of satellite lines represents the transitions between excited states. Though several differences can be noticed, most striking are the suppression of the linear conductance, the appearance of negative differential conductance (NDC), and the strong suppression of the current at the right(left) border of the 7 (5) particle diamond when passing from the para to the meta configuration. All these features are different manifestations of the interference between orbitally degenerate states and ultimately reveal the specific symmetry of benzene.

7.4.3.1 Linear Conductance

We study the linear transport regime both numerically and analytically. For the analytical calculation of the conductance we consider the low temperature limit where only ground states with N and $N+1$ particles have considerable occupation probabilities, with N fixed by the gate voltage. Therefore only transitions between these states are relevant and we can treat just the terms of (7.5) with N and $N+1$ particles and the ground state energies $E_{g,N}$ and $E_{g,N+1}$, respectively. A closer look at (7.5) reveals that the spin coherences are decoupled from the other elements of the density matrix. Thus we can set them to zero, and write (7.5) in a block diagonal form in the basis of the ground states of N and $N+1$ particles. Additionally, since the total Hamiltonian H is symmetric in spin, the blocks of the GME with the same particle but different spin quantum number τ must be identical. Finally, since around the resonance the only populated states are the N and $N+1$ particle states, the conservation of probability implies that:

$$1 = \sum_n \rho_{nn}^N + \sum_m \rho_{mm}^{N+1}, \tag{7.35}$$

where ρ_{nn}^N is the population of the N-particle ground state and n contains the orbital and spin quantum numbers. With all these observations we can reduce (7.5) to a much smaller set of coupled differential equations that can be treated analytically. The stationary solution of this set of equations can be derived more easily by restricting in (7.5) to the dynamics generated by the sequential tunnelling Liouvillean $\mathscr{L}_{\text{tun}}$. With this simplification we derive an analytical formula for the conductance close to the resonance between N and $N+1$ particle states as the first order coefficient of the Taylor series of the current in the bias:

$$G_{N,N+1}(\Delta E) = 2e^2 \frac{\Gamma_L \Gamma_R}{\Gamma_L + \Gamma_R} \Lambda_{N,N+1} \left[-\frac{S_N S_{N+1} f'(\Delta E)}{(S_{N+1} - S_N) f(\Delta E) + S_N} \right] \tag{7.36}$$

where $\Delta E = E_{g,N} - E_{g,N+1} + eV_g$ is the energy difference between the benzene ground states with N and $N+1$ electrons diminished by a term linear in the gate voltage. Interference effects are contained in the overlap factor $\Lambda_{N,N+1}$:

$$\Lambda_{N,N+1} = \frac{\left| \sum_{nm\tau} \langle N,n | d_{L\tau} | N+1,m \rangle \langle N+1,m | d_{R\tau}^\dagger | N,n \rangle \right|^2}{S_N S_{N+1} \sum_{nm\alpha\tau} \left| \langle N,n | d_{\alpha\tau} | N+1,m \rangle \right|^2}, \tag{7.37}$$

where n and m label the S_N-fold and S_{N+1}-fold degenerate ground states with N and $N+1$ particles, respectively. In order to make the interference effects more visible we remind that $d^{\dagger}_{\mathrm{R}\tau} = \mathscr{R}^{\dagger}_{\phi} d^{\dagger}_{\mathrm{L}\tau} \mathscr{R}_{\phi}$, with $\phi = \pi$ for the para while $\phi = 2\pi/3$ for the meta configuration. Due to the behavior of all eigenstates of H^0_{ben} under discrete rotation operators with angles multiples of $\pi/3$, we can rewrite the overlap factor:

$$\Lambda_{N,N+1} = \frac{\left| \sum\limits_{nm\tau} |\langle N,n|d_{\mathrm{L}\tau}|N+1,m\rangle|^2 e^{i\phi_{nm}} \right|^2}{S_N S_{N+1} 2 \sum\limits_{nm\tau} \left| \langle N,n|d_{\mathrm{L}\tau}|N+1,m\rangle \right|^2}, \tag{7.38}$$

where ϕ_{nm} encloses the phase factors coming from the rotation of the states $|N,n\rangle$ and $|N+1,m\rangle$.

The effective Hamiltonian H_{eff} neglected in (7.36) only influences the dynamics of the coherences between orbitally degenerate states. Thus, Eq. (7.36) provides an exact description of transport for the para configuration, where orbital degeneracy is cancelled. Even if Eq. (7.36) captures the essential mechanism responsible for the conductance suppression, we have derived an exact analytical formula also for the meta configuration which can be found in Appendix B of [20].

In Fig. 7.5 we present an overview of the results of both the para and the meta configuration. A direct comparison of the conductance (including the H_{eff} term of (7.5)) in the two configurations is displayed in the upper panel. The lower panel illustrates the effect of the energy non-conserving terms on the conductance in the meta configuration. The number of p_z electrons on the molecule and the symmetry of the lowest energy states corresponding to the conductance valleys are reported. The symmetries displayed in the upper panel belong to the (effective) ground states in the para configuration, the corresponding symmetries for the meta configuration are shown in the lower panel.

Figure 7.5 shows that the results for the para and the meta configuration coincide for the $10 \leftrightarrow 11$ and $11 \leftrightarrow 12$ transitions. The ground states with $N = 10, 11, 12$ particles have A- or B-type symmetries, they are therefore orbitally non-degenerate, no interference can occur and thus the transitions are invariant under configuration change. For every other transition we see a noticeable difference between the results of the two configurations (Fig. 7.5). In all these transitions one of the participating states is orbitally degenerate. First we notice that the linear conductance peaks for the $7 \leftrightarrow 8$ and $8 \leftrightarrow 9$ transitions in the para configuration are *shifted* with respect to the corresponding peaks in the meta configuration. The selection of an effective symmetric Hilbert space associated with the para configuration reduces the total degeneracy by cancelling the orbital degeneracy. In addition, the ground state energy of the 4- and 8-particle states is different in the two configurations, since in the para configuration the *effective* ground state is in reality the first excited state. The degeneracies S_N, S_{N+1} of the participating states as well as the ground state energy are both entering the degeneracy term of Eq. (7.36)

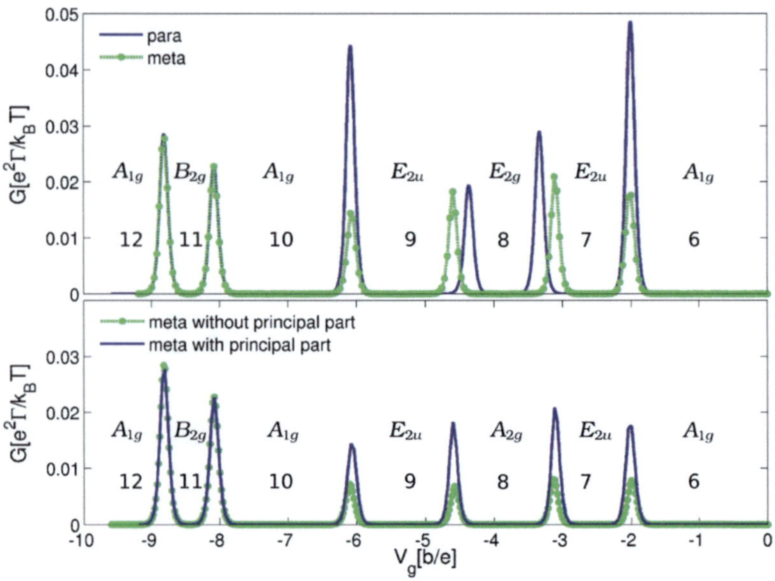

Fig. 7.5 Conductance of the benzene I-SET as a function of the gate voltage. Clearly visible are the peaks corresponding to the transitions between ground states with N and $N+1$ particles. In the low conductance valleys the state of the system has a definite number of particles and symmetry as indicated in the upper panel for the para, in the lower for the meta configuration. Selective conductance suppression when changing from the meta to the para configuration is observed. From [20]

$$\Delta = -\frac{f'(\Delta E)}{(S_{N+1}-S_N)f(\Delta E)+S_N}, \tag{7.39}$$

and determine the shift of the conductance peaks.

Yet, the most striking effect regarding transitions with orbitally degenerate states participating is the *systematic suppression* of the linear conductance when changing from the para to the meta configuration. The suppression is appreciable despite the conductance enhancement due to the principal part contributions to the GME (see Fig. 7.5, lower panel). Thus, we will for simplicity discard them in the following discussion.

The conductance suppression is determined by the combination of two effects: the reduction to the symmetric Hilbert space in the para configuration and the interference effects between degenerate orbitals in the meta configuration. The reduction to the symmetric Hilbert space implies also a lower number of conducting channels (see Table 7.4). One would expect a suppression of transport in the para configuration. As we can see from Table 7.4 on the example of the $6 \leftrightarrow 7$ transition peak, $\Delta_{\max}$ is higher in the para configuration but not enough to fully explain the difference between the two configurations. The second effect determining transport is the interference between the E-type states, which is accounted for in the overlap

Table 7.4 Number of channels participating in transport, overlap factor and resonance value of the degeneracy term in the para and the meta configuration for the $6 \leftrightarrow 7$ transition peak. It is $C = |\langle 6_g|d_{L\tau}|7_g l\tau\rangle|^2$, where τ and ℓ are the spin and the quasi-angular momentum quantum numbers, respectively. The values of Δ_{max} are given for $k_B T = 0.04|b|$. From [20]

Configuration	Number of channels ($S_N S_{N+1}$)	Overlap factor Λ	Degeneracy term Δ_{max} [$1/k_B T$]
Para	2	C	0,17
Meta	4	$\frac{1}{8}C$	0,11

factor Λ. The overlap factor is basis independent, thus we can write the transition probabilities for the $6 \leftrightarrow 7$ transition as $|\langle 6_g|d_{L\tau}|7_g \ell \tau\rangle|^2 = C$, where τ and ℓ are the spin and the quasi-angular momentum quantum number, respectively. The transition probabilities have the same value, since all four 7-particle states are in this basis equivalent (see Appendix C of [20] for a detailed proof). Under the C_2 rotation the symmetric 7-particle ground state does not acquire any phase factor. Under the C_3 rotation however, the two orbitally degenerate states acquire different phase factors, namely $e^{\frac{4\pi}{3}i}$ and $e^{-\frac{4\pi}{3}i}$, respectively. Thus the overlap factors Λ for the $6 \leftrightarrow 7$ transition are:

$$\Lambda_{para} = \frac{1}{2 \cdot 8C} \cdot |4C|^2 = C,$$

$$\Lambda_{meta} = \frac{1}{4 \cdot 8C} \cdot \left|2Ce^{+\frac{4\pi}{3}i} + 2Ce^{-\frac{4\pi}{3}i}\right|^2 = \frac{1}{8}C.$$

The linear conductance is determined by the product between the number of conducting channels, the overlap factor and the degeneracy term. Yet, it is the *destructive interference* between degenerate E-type orbitals, accounted for in the overlap factor Λ, that gives the major contribution to the strong suppression of the conductance in the meta configuration.

7.4.3.2 Negative Differential Conductance and Current Blocking

Interference effects between orbitally degenerate states are also affecting nonlinear transport and producing in the *meta* configuration current blocking and thus NDC at the border of the 6-particle state diamond (Fig. 7.4). The upper panel of Fig. 7.6 shows the current through the benzene I-SET contacted in the meta configuration as a function of the bias voltage. The current is given for parameters corresponding to the white dashed line of Fig. 7.4. In this region only the 6- and 7-particle ground states are populated. At low bias the 6-particle state is mainly occupied. As the bias is raised, transitions $6 \leftrightarrow 7$ occur and current flows. Above a certain bias threshold a blocking state is populated and the current drops. For the understanding of this

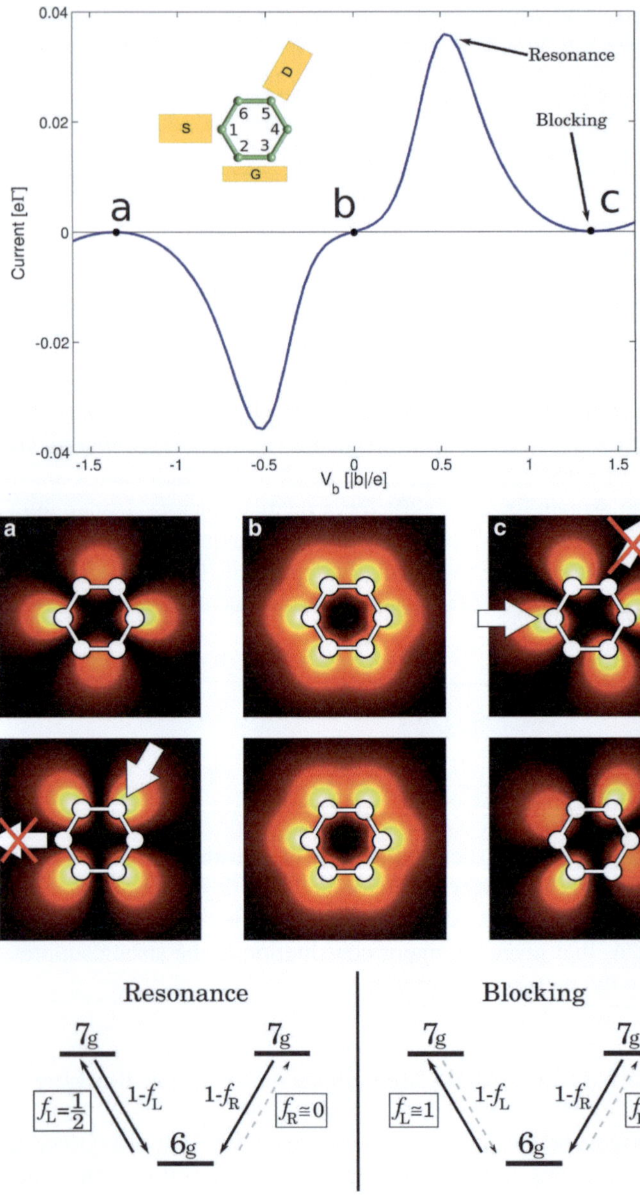

Fig. 7.6 *Upper panel*—Current through the benzene I-SET in the meta configuration calculated at bias and gate voltage conditions indicated by the *dashed line* of Fig. 7.4. A pronounced NDC with current blocking is visible. *Middle panels*—Transition probabilities between the 6-particle and each of the two 7-particle ground states for bias voltage values labelled $a - c$ in the upper panel. The transition to a blocking state is visible in the upper (lower) part of the c (a) panels. *Lower panels*—Sketch of the energetics for the $6 \rightarrow 7$ transition in the meta configuration at bias voltages corresponding to the resonance current peak and current blocking as indicated in the upper panel of this figure. From [20]

nonlinear current characteristics, we have to take into account energy conservation, the Pauli exclusion principle and the interference between participating states. For the visualization of the interference effects, we introduce the transition probability (averaged over the z coordinate and the spin σ):

$$P(x,y;n,\tau) = \lim_{L\to\infty} \sum_{\sigma} \frac{1}{2L} \int_{-L/2}^{L/2} dz |\langle 7_g\, n\, \tau | \psi_{\sigma}^{\dagger}(\mathbf{r}) | 6_g \rangle|^2 \tag{7.40}$$

for the *physical* 7-particle basis, i.e., the 7-particle basis that diagonalizes the stationary density matrix at a fixed bias. Here τ is the spin quantum number, $n = 1,2$ labels the two states of the physical basis which are linear combinations of the orbitally degenerate states $|7_g \ell \tau\rangle$ and can be interpreted as conduction channels. Each of the central panels of Fig. 7.6 are surface plots of (7.40) at the different bias voltages *a*–*c*. The 7-particle ground states can interfere and thus generate nodes in the transition probability at the contact atom close to one or the other lead, but, in the meta configuration, never at both contact atoms at the same time.

Energetic considerations are illustrated in the lower panels of Fig. 7.6 for two key points of the current curve at positive biases. The left panel corresponds to the resonance peak of the current. Due to energy conservation, electrons can enter the molecule only from the left lead. On the contrary the exit is allowed at both leads. The current is suppressed when transitions occur to a state which cannot be depopulated (a blocking state). Since, energetically, transmissions to the 6-particle state are allowed at both leads, each 7-particle state can always be depopulated and no blocking occurs.

The current blocking scenario is depicted in the lower right panel of Fig. 7.6. For large positive bias the transition from a 7-particle ground state to the 6-particle ground state is energetically forbidden at the left lead. Thus, for example, the *c* panel in Fig. 7.6 visualizes the current blocking situation yielding NDC: while for both channels there is a non-vanishing transition probability from the source lead to the molecule, for the upper channel a node prevents an electron from exiting to the drain lead. In the long time limit the blocking state gets fully populated while the non-blocking state is empty. At large negative bias the blocking scenario is depicted in the panel *a* that shows the left–right symmetry obtained by a reflection through a plane perpendicular to the molecule and passing through the carbon atoms 6 and 3. The temperature sets the scale of the large bias condition and, correspondingly, the width of the current peak presented in Fig. 7.6 grows with it. The peak is not symmetric, though, its shape depends also on the energy renormalization introduced by the coupling to the leads [21] and described by the effective Hamiltonian (7.8). In fact the interference blocking is not a threshold effect in the bias. The complete blocking corresponds to a very precise bias which is determined by the form of H_{eff}. We will return to this point in Sect. 7.6, while discussing the spin-dependent transport. Moreover, we remark that only a description that retains coherences between the degenerate 7-particle ground states correctly captures NDC at both positive and negative bias.

In contrast to the $6 \to 7$ transition, one does *not* observe NDC at the border of the 7-particle Coulomb diamond, but rather a strong suppression of the current. The upper panel of Fig. 7.7 shows the current through the benzene I-SET contacted in the meta configuration as a function of the bias voltage corresponding to the white dotted line of Fig. 7.4. The middle panels show the transition probabilities between each of the 7-particle and the 6-particle ground state. The lower panel of Fig. 7.7 shows a sketch of the energetics at positive bias corresponding to the "expected" resonance peak. Here electrons can enter the molecular dot at both leads, while the exit is energetically forbidden at the left lead. Thus, if the system is in the 7-particle state which is blocking the right lead, this state cannot be depopulated, becoming the blocking state. On the other hand, transitions from the 6-particle ground state to both 7-particle ground states are equally probable. Thus the blocking state will surely be populated at some time. The upper plot of the *b* panel in Fig. 7.7 shows the transition probability to the blocking state that accepts electrons from the source lead but cannot release electrons to the drain. As just proved, in this case the current blocking situation occurs already at the resonance bias voltage. For a higher positive bias, the transition probability from the blocking state at the drain lead increases and current can flow. This effect, though, can be captured only by taking into account also the H_{eff} contribution to (7.5).

In the *para* configuration, the current as a function of the bias voltage is shown in Fig. 7.8. The current is given for parameters corresponding to the white dashed line of Fig. 7.4. In this case, no interference effects are visible. We see instead the typical step-like behavior of the current in the regime of single electron tunnelling. The panels on the right are the surface plots of

$$P(x,y;\tau) = \lim_{L\to\infty} \sum_{\sigma} \frac{1}{2L} \int_{-L/2}^{L/2} \mathrm{d}z |\langle 7_{\mathrm{g}}\, \tau; (\mathrm{a})\mathrm{sym}| \psi_{\sigma}^{\dagger}(\mathbf{r}) |6_{\mathrm{g}}\rangle|^2. \tag{7.41}$$

The upper plot shows the transition probability to the symmetric 7-particle state, the lower to the antisymmetric. Remember that in the para configuration only the symmetric states contribute to transport. Evidently the symmetric state is in the para configuration non-blocking. Additionally, since the coherences between orbitally degenerate states and therefore the energy non-conserving terms do not play any role in the transport, the physical basis states are not bias dependent. Thus in the para configuration there are always non-blocking states populated and no NDC can occur.

7.5 The Triple Dot I-SET

As a second example of I-SET we consider an artificial quantum dot molecule: i.e. the triple dot I-SET. The triple dot SET has been recently in the focus of intense theoretical [31, 32, 52–56] and experimental [57–60] investigation due to

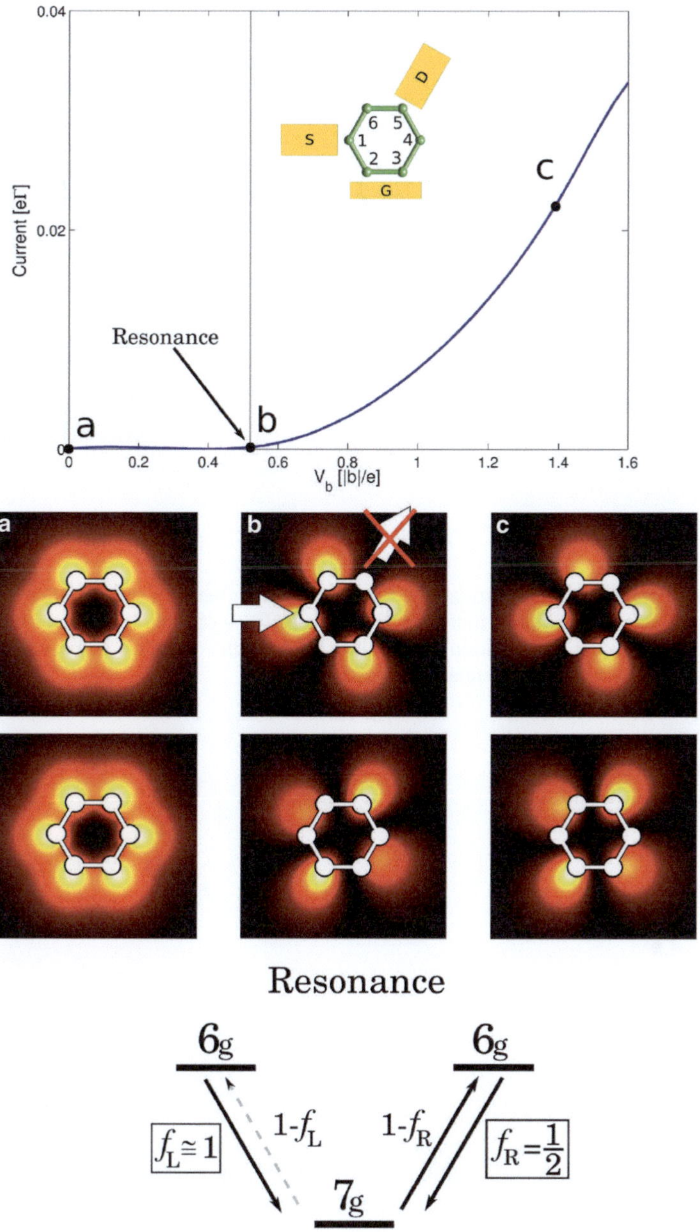

Fig. 7.7 *Upper panel*—Current through the benzene I-SET in the meta configuration calculated at bias and gate voltage conditions indicated by the dotted line of Fig. 7.4. No NDC is visible. *Middle panels*—Transition probabilities between each of the 7-particle and the 6-particle ground state for bias voltage values labelled $a - c$ in the upper panel. *Lower panel*—Sketch of the energetics for the $7 \rightarrow 6$ transition in the meta configuration at bias voltage corresponding to the expected resonance peak. (compare to Fig. 7.6). From [20]

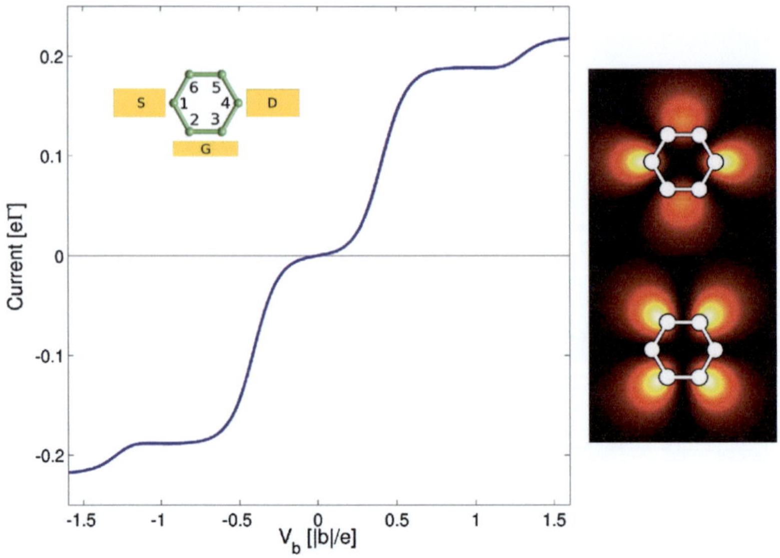

Fig. 7.8 *Left panel*—Current through the benzene I-SET in the para configuration calculated at bias and gate voltage conditions indicated by the *dashed line* of Fig. 7.4. No interference effects are visible. *Right panels*—Transition probabilities between the 6-particle and the symmetric and antisymmetric 7-particle ground states. From [20]

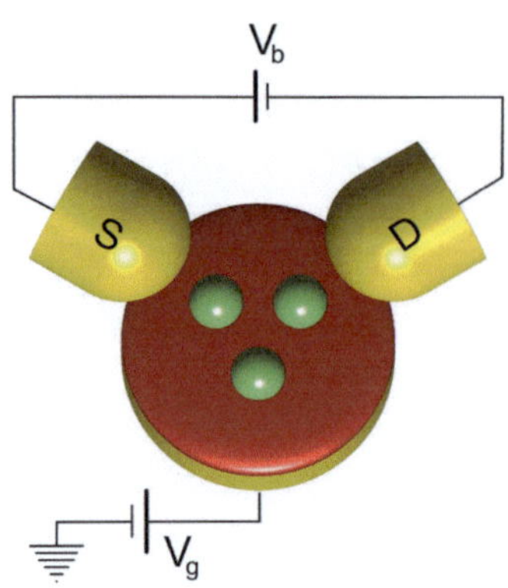

Fig. 7.9 Schematic representation of a triple dot interference single electron transistor (I-SET). From [22]

its capability of combining incoherent transport characteristics and signatures of molecular coherence. The triple dot I-SET that we consider here (Fig. 7.9) is the simplest structure with symmetry protected orbital degeneracy exhibiting interference blockade. Despite its relative simplicity this system displays different kinds of current blocking and it represents for this reason a suitable playground for the ideas presented in Sect. 7.3. In particular we will concentrate on the blockade that involves an *excited* triplet state: a regime which is not achievable in the benzene I-SET.

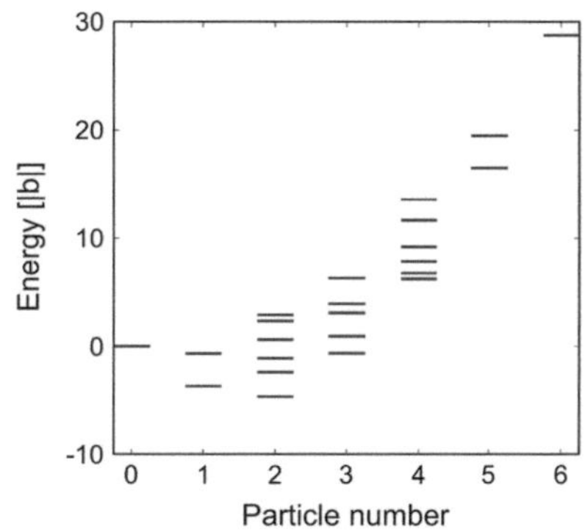

Fig. 7.10 Spectrum of the triple dot system for the specific gate voltage $eV_g = 4.8b$ chosen to favor a configuration with two electrons. The other parameters in the system are $U = 5|b|$ and $V = 2|b|$, where b is the hopping integral between the different dots. From [22]

7.5.1 The Model

The total Hamiltonian of the I-SET is in the generic form (7.2). We describe the system with an Hamiltonian in the extended Hubbard form[3]:

$$H_{3\mathrm{d}} = \xi_0 \sum_{i\sigma} d^\dagger_{i\sigma} d_{i\sigma} + b \sum_{i\sigma} \left(d^\dagger_{i\sigma} d_{i+1\sigma} + d^\dagger_{i+1\sigma} d_{i\sigma} \right)$$

$$+U \sum_i \left(n_{i\uparrow} - \frac{1}{2} \right) \left(n_{i\downarrow} - \frac{1}{2} \right) \tag{7.42}$$

$$+V \sum_i \left(n_{i\uparrow} + n_{i\downarrow} - 1 \right) \left(n_{i+1\uparrow} + n_{i+1\downarrow} - 1 \right), \tag{7.43}$$

where $d^\dagger_{i\sigma}$ creates an electron of spin σ in the ground state of the quantum dot i. Here $i = 1, \ldots, 3$ runs over the three quantum dots of the system and we impose the periodic condition $d_{4\sigma} = d_{1\sigma}$. Moreover $n_{i\sigma} = d^\dagger_{i\sigma} d_{i\sigma}$. The effect of the gate is included as a renormalization of the on-site energy $\xi = \xi_0 - eV_\mathrm{g}$ where V_g is the gate voltage. We measure the energies in units of the modulus of the (negative) hopping integral b. The parameters that we use are $\xi_0 = 0$, $U = 5\,|b|$, $V = 2\,|b|$.

H_leads in (7.2) describes two reservoirs of non-interacting electrons with a difference eV_b between their electrochemical potentials. Finally, H_tun accounts for the weak tunnelling coupling between the system and the leads, characteristic of SETs, and we consider the tunnelling events restricted to the atoms or to the dots closest to the corresponding lead.

The number of electrons considered for the triple dot structure goes from 0 to 6. Thus the entire Fock space of the system contains $4^3 = 64$ states. By exact diagonalization we obtain the many-body eigenstates and the corresponding eigenvalues that we present in Fig. 7.10 for a gate voltage of $V_\mathrm{g} = 4.8\,b/e$. In Table 7.5

[3]This denomination of the Pariser–Parr–Pople Hamiltonian is more common in the solid state community.

Table 7.5 Degeneracy of the triple dot system energy levels as it follows from the underlying D_3 symmetry. A level N_i is the ith excited level with N particles. The total degeneracy of the level is the product of its orbital and spin degeneracies. From [22]

Many-body energy level	Orbital degeneracy	Spin degeneracy
0	1	1
1_0	1	2
2_0	1	1
2_1	2	3
3_0	2	2
4_0	1	3
5_0	2	2
6	1	1

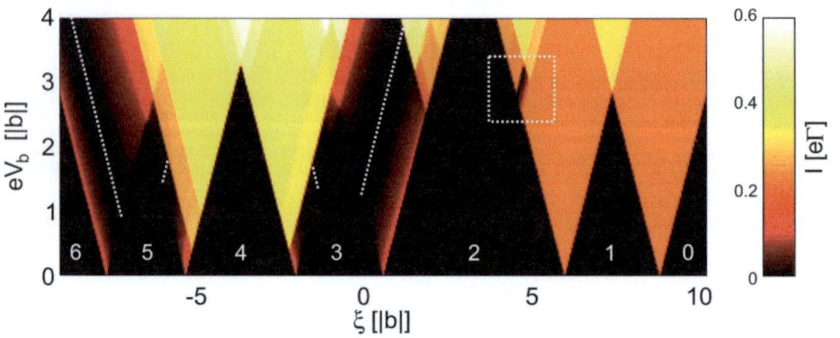

Fig. 7.11 Stationary current for the triple dot I-SET. Coulomb blockade diamonds are visible at low biases. Ground state and excited state interference blockades are also highlighted. The temperature is $k_bT = 0.002|b|$. The other parameters are the same as the ones in Fig. 7.10. From [22]

we also give the degeneracies of all levels relevant for the blocking states analysis which will follow. We distinguish between spin and orbital degeneracy since the latter is the most important for the identification of the blocking states. The total degeneracy of a level is simply the product of the two.

7.5.2 *Excited State Blocking*

In Fig. 7.11 we show the stationary current through the triple dot I-SET as a function of bias and gate voltage. At low bias the current vanishes almost everywhere due to Coulomb blockade. The particle number is fixed within each Coulomb diamond by the gate voltage and the zero particle diamond is the first to the right. The zero current lines running parallel to the borders of the 6-, 4-, and 2-particle diamonds are instead signatures of ground state interference that involves an orbitally non-degenerate ground state (with 2-, 4-, and 6-particle) and an orbitally double-degenerate one (with 3 and 5 particles).

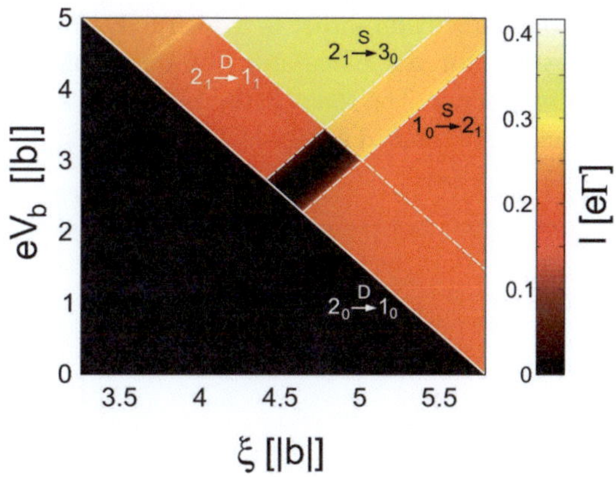

Fig. 7.12 Blow up of the stationary current through the triple dot I-SET around the 2- to 1-particle degeneracy point. The black area sticking out of the 2-particle Coulomb diamond denotes the excited states blocking. From [22]

The striking feature in Fig. 7.11 is the black area of current blocking sticking out of the right side of the two-particle Coulomb diamond. It is the fingerprint of the occupation of an excited interference blocking state. Figure 7.12 is a zoom of the current plot in the vicinity of this excited state blocking. The dashed lines indicate at which bias and gate voltage a specific transition is energetically allowed, with the notation N_i labelling the ith excited many-body level with N particles. These lines are physically recognizable as abrupt changes in the current and run all parallel to two fundamental directions determined by the ground state transitions. For positive bias, positive (negative) slope lines indicate the bias threshold for the opening of source-creation (drain-annihilation) transitions. The higher the bias the more transitions are open, the higher, in general, the current.

The anomalous blockade region is delimited on three sides by transitions lines associated with the first excited two-particle level 2_1. Our group theoretical analysis shows that the two particle first excited state is a twofold orbitally degenerate spin triplet (see Table 7.5). In other terms we can classify its six states with the notation $|2_1,\ell,S\rangle$ with $\ell = \pm\hbar$ being the projection of the angular momentum along the main rotation axis, perpendicular to the plane of the triple dot, and $S_z = -\hbar, 0, \hbar$ the component of the spin along a generic quantization axis. The 1_0 energy level is instead twice spin degenerate and invariant under the symmetry operations of the point group D_3.

In order to identify the 2-particle blocking states we perform the analysis presented in the previous section for the 2_1 energy level with the gate and bias in the blocking region. Firstly, we find that the 2_1 energy level can be reached from 2_0 via the drain-annihilation transition $2_0 \to 1_0$ followed by the source-creation transition $1_0 \to 2_1$. Secondly, the space of the decoupled states $\mathscr{D}_{2_1}$ is empty and the only energetically allowed outgoing transition is the drain-annihilation $2_1 \to 1_0$ transition. Thus the blocking space is given by the expression:

$$\mathscr{B}_{2_1} = \mathscr{P}_{2_1}\left[\ker\left(T^{-}_{2,2_1 1_0}, T_S\right)\right] \tag{7.44}$$

and has dimension three. It is instructive to calculate explicitly the $T^{-}_{2,2_1\,1_0}$ matrix necessary for the calculation of the triplet blocking states and the associated blocking states. The states in the 1_0 doublet and in the two times orbitally degenerate triplet 2_1 are labeled and ordered as follows:

$$1_0 \begin{cases} |1_0, \ell = 0, \uparrow\rangle \\ |1_0, \ell = 0, \downarrow\rangle \end{cases}, \quad 2_1 \begin{cases} |2_1, \ell = +\hbar, S_z = +\hbar\rangle \\ |2_1, \ell = +\hbar, S_z = 0\rangle \\ |2_1, \ell = +\hbar, S_z = -\hbar\rangle \\ |2_1, \ell = -\hbar, S_z = +\hbar\rangle \\ |2_1, \ell = -\hbar, S_z = 0\rangle \\ |2_1, \ell = -\hbar, S_z = -\hbar\rangle \end{cases}. \tag{7.45}$$

The elements of the $\gamma_{\alpha\sigma}$ matrices that compose $T^{-}_{2,2_1\,1_0}$ have thus the general form:

$$\gamma^{-}_{\alpha\sigma}(S_z, S'_z, \ell) = \langle 1_0, \ell' = 0, S'_z | d_{\alpha\sigma} | 2_1, \ell, S_z\rangle. \tag{7.46}$$

By orbital and spin symmetry arguments it is possible to show that

$$\gamma^{-}_{\alpha\sigma}(S_z, S'_z, \ell) = t\, e^{\frac{i}{\hbar}\ell\phi_\alpha}\, \delta_{S'_z, S_z - \sigma}\, (\sqrt{2}\delta_{S'_z,\uparrow} \cdot + \delta_{S'_z,\downarrow}) \tag{7.47}$$

where $t = \langle 1_0, \ell' = 0, \downarrow | d_{M\uparrow} | 2_1, \ell = 1, S_z = 0\rangle$. The subscript M labels a reference dot and ϕ_α is the angle of the rotation that brings the dot α on the dot M. The explicit form of $T^{-}_{2_1,2_1\,1_0}$ reads:

$$T^{-}_{2,2_1\,1_0} = t \begin{pmatrix} \sqrt{2}e^{-i2\pi/3} & 0 & 0 & \sqrt{2}e^{+i2\pi/3} & 0 & 0 \\ 0 & e^{-i2\pi/3} & 0 & 0 & e^{+i2\pi/3} & 0 \\ 0 & e^{-i2\pi/3} & 0 & 0 & e^{+i2\pi/3} & 0 \\ 0 & 0 & \sqrt{2}e^{-i2\pi/3} & 0 & 0 & \sqrt{2}e^{+i2\pi/3} \\ \sqrt{2}e^{+i2\pi/3} & 0 & 0 & \sqrt{2}e^{-i2\pi/3} & 0 & 0 \\ 0 & e^{+i2\pi/3} & 0 & 0 & e^{-i2\pi/3} & 0 \\ 0 & e^{+i2\pi/3} & 0 & 0 & e^{-i2\pi/3} & 0 \\ 0 & 0 & \sqrt{2}e^{+i2\pi/3} & 0 & 0 & \sqrt{2}e^{-i2\pi/3} \end{pmatrix}. \tag{7.48}$$

The rank of this matrix is six since all columns are independent. Thus $\mathscr{C}_{2,2_1}$ coincides with the full Hilbert space of the first excited two-electron energy level. The blocking space $\mathscr{B}_{2,2_1,1_0}$ reads:

$$\mathscr{B}_{2,2_1,1_0} = \mathscr{P}_{2_1} \ker(T^{-}_{2,2_1\,1_0}, T_S) \tag{7.49}$$

where T_S reads

$$T_S = \begin{pmatrix} 1\,0\,0\,0\,0\,0\,0\,0 \\ 0\,1\,0\,0\,0\,0\,0\,0 \\ 0\,0\,1\,0\,0\,0\,0\,0 \\ 0\,0\,0\,1\,0\,0\,0\,0 \end{pmatrix}^T, \tag{7.50}$$

in accordance with its general definition given in Eq. (7.17), and the projector $\mathscr{P}_{2_1}$ removes the last four components from the vectors that span $\ker(T^-_{2,2_1\,1_0}, T_S)$. It is then straightforward to calculate the vectors that span the blocking space $\mathscr{B}_{2,2_1,1_0}$:

$$\mathbf{v}_1 = \begin{pmatrix} \frac{e^{-i\frac{\pi}{6}}}{\sqrt{2}} \\ 0 \\ 0 \\ \frac{e^{+i\frac{\pi}{6}}}{\sqrt{2}} \\ 0 \\ 0 \end{pmatrix}, \quad \mathbf{v}_2 = \begin{pmatrix} 0 \\ \frac{e^{-i\frac{\pi}{6}}}{\sqrt{2}} \\ 0 \\ 0 \\ \frac{e^{+i\frac{\pi}{6}}}{\sqrt{2}} \\ 0 \end{pmatrix}, \quad \mathbf{v}_3 = \begin{pmatrix} 0 \\ 0 \\ \frac{e^{-i\frac{\pi}{6}}}{\sqrt{2}} \\ 0 \\ 0 \\ \frac{e^{+i\frac{\pi}{6}}}{\sqrt{2}} \end{pmatrix}. \tag{7.51}$$

The vectors $\mathbf{v}_1$, $\mathbf{v}_2$, and $\mathbf{v}_3$ are the components of the blocking states written in the 2_1 basis set presented in (7.45). Thus, the three blocking states correspond to the three different projectors of the total spin $S_z = \hbar$, 0, and $-\hbar$, respectively. Essentially, there is a blocking state for each of the three projection of the spin S_z. This result is natural since, for unpolarized or parallel polarized leads, coherences between states of different spin projection along the common lead quantization axis do not survive in the stationary limit.

7.6 Spin-Dependent Transport

In the previous sections we have shown different types of interference blocking, involving both ground and excited many-body states. All of them were essentially described in terms of the sequential tunnelling dynamics generated by $\mathscr{L}_{\text{tun}}$ (see Eq. (7.6)). We neglected H_{eff} in the analysis of the numerical results and correspondingly the role of the third condition ($[H_{\text{eff}}, \rho_{\text{block}}] = 0$) in the definition of a blocking state. Indeed, the consequences of the dynamics generated by H_{eff} on the transport characteristics of the benzene and triple dot I-SETs are marginal for unpolarized leads.

The scenario changes completely for the case of spin polarized leads (Fig. 7.13). Thanks to (7.21) the destructive interference between orbitally degenerate electronic states typical of I-SETs produces current blocking at *specific* bias voltages (see Figs. 7.16 and 7.11). In the presence of parallel polarized ferromagnetic leads the interplay between interference and the exchange coupling on the system generates an effective energy renormalization yielding different blocking biases for majority and minority spins. Hence, by tuning the bias voltage full control over the spin of

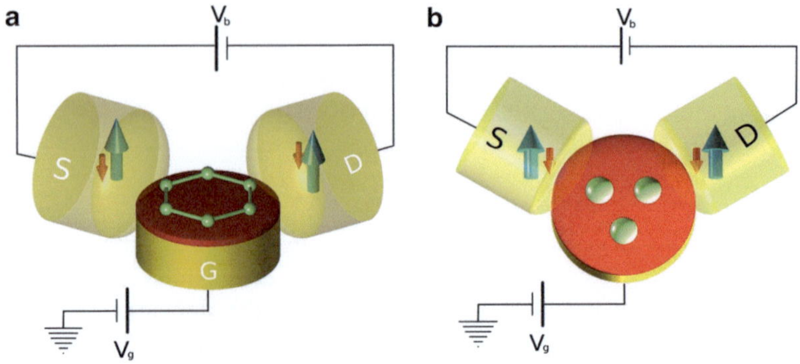

Fig. 7.13 Two examples of interference single electron transistors (I-SETs): a benzene molecular junction contacted in the meta configuration (**a**) and a triple quantum dot artificial molecule (**b**). The source and drain are parallel polarized ferromagnetic leads. From [21]

the trapped electron is achieved. Notice that we assume the leads to be parallel polarized so that no spin torque is active in the device and we can exclude the spin accumulation associated with that [33, 35]. In conclusion, the spin-dependent renormalization of the system dynamics introduced by H_{eff} allows to exploit interference to achieve all-electrical control of a single electron spin in quantum dots, a highly desirable property for spintronics [23–25] and spin-qubit applications [26–30].

7.6.1 Spin Polarized Leads

The lead polarization P_χ with $\chi = L, R$ is defined by means of the density of states $D_{\chi\sigma}$ at the Fermi energy for the different spin states:

$$P_\chi = \frac{D_{\chi\uparrow} - D_{\chi\downarrow}}{D_{\chi\uparrow} + D_{\chi\downarrow}} \tag{7.52}$$

and is taken equal for the two leads $P = P_L = P_R$. The spin polarization influences the dynamics of the system via the tunnelling rate matrices $\Gamma_{ij}^{\chi\sigma}$ which are proportional to the density of states $D_{\chi\sigma}$ and enter both the definition of the tunnelling component of the Liouvillian $\mathscr{L}_{\text{tun}}$ (see Eq. (7.6)) and the effective Hamiltonian H_{eff} (see Eq. (7.8)).

The particular form of the effective Hamiltonian for the benzene and the triple dot I-SET is the key to the understanding of the spin selective interference blockade. For the sake of simplicity we give in the following the explicit form of H_{eff} only for the benzene I-SET and for the ground state transition $6_g \to 7_g$ that is characterized by interference blocking. The argumentation is nevertheless very general and can be repeated for all the systems exhibiting interference due to rotational symmetry.

Let us start from (7.8) and project it on the subspace of the fourfold degenerate 7-particle ground state. One obtains a 4×4 matrix that can be cast into the tensor product of a spin and an orbital component:

$$H_{\mathrm{eff}}|_{7_g} = \sum_{\chi} \begin{pmatrix} \omega_{\chi\uparrow} & 0 \\ 0 & \omega_{\chi\downarrow} \end{pmatrix} \otimes L_{\chi}. \tag{7.53}$$

The spin component has the units of a frequency

$$\omega_{\chi\sigma} = \frac{1}{\pi} \sum_{\sigma'\{E\}} \Gamma^0_{\chi\sigma'} \Big[\langle 7_g \ell\sigma | d_{M\sigma'} | 8\{E\}\rangle\langle 8\{E\}| d^{\dagger}_{M\sigma'} | 7_g \bar{\ell}\sigma\rangle p_{\chi}(E - E_{7_g}) \\ + \langle 7_g \ell\sigma | d^{\dagger}_{M\sigma'} | 6\{E\}\rangle\langle 6\{E\}| d_{M\sigma'} | 7_g \bar{\ell}\sigma\rangle p_{\chi}(E_{7_g} - E) \Big], \tag{7.54}$$

and it weights the energy renormalization given to the states of spin σ by their coupling to the lead χ. In (7.54) we have introduced $d_{\mathrm{M}\sigma}$ which destroys an electron of spin σ in a reference carbon atom M placed in the middle between the two contact atoms, $|7_g\,\ell\,\sigma\rangle$ are the orbitally degenerate 7-particle ground states, $\ell = \pm 2$ is the z projection of the angular momentum in units of $\hbar$ and $\bar{\ell} \equiv -\ell$. The compact notation $|N\{E\}\rangle$ indicates all possible states with particle number N and energy E, $p_{\chi}(x) = -\mathrm{Re}\psi\left[\frac{1}{2} + \frac{i\beta}{2\pi}(x - \mu_{\chi})\right]$ where $\beta = 1/k_{\mathrm{B}}T$, T is the temperature, and ψ is the digamma function. Moreover $\Gamma^0_{\chi\sigma'} = \frac{2\pi}{\hbar}|t|^2 D_{\chi\sigma'}$ is the bare tunnelling rate to the lead χ of an electron of spin σ', where t is the tunnelling amplitude and $D_{\chi\sigma'}$ is the density of states for electrons of spin σ' in the lead χ at the corresponding chemical potential μ_{χ}. Due to the particular choice of the arbitrary phase of the 7-particle ground states, $\omega_{\chi\sigma}$ does not depend on the orbital quantum numbers ℓ. It depends instead on the bias and gate voltage through the energy of the 6, 7-ground, and 8-particle states. The orbital component of (7.53) has the units of an angular momentum and it reads:

$$L_{\chi} = \frac{\hbar}{2} \begin{pmatrix} 1 & e^{i2|\ell|\phi_{\chi}} \\ e^{-i2|\ell|\phi_{\chi}} & 1 \end{pmatrix}. \tag{7.55}$$

where ϕ_{α} is the angle of which we have to rotate the molecule to bring the reference atom M into the position of the contact atom χ. The present choice of the reference atom implies that $\phi_L = -\phi_R = \frac{\pi}{3}$.

It is important to notice for the following discussions that L_{χ} not only has the units of an angular momentum but it *is* also a quasi-angular momentum since it generates the rotations of the 7-particle ground states of the benzene molecule along the horizontal C_2 symmetry axis passing through the contact atom χ. Let us prove the last statement. It is convenient, for the purpose, to choose the arbitrary phases of the states $|7_g\ell\sigma\rangle$ in such a way that the rotation of π around the axis passing through a reference atom M and the center of the molecule transforms $|7_g\ell\sigma\rangle$ into $-|7_g\bar{\ell}\sigma\rangle$. In other terms

$$\exp\left(i\pi\frac{L_M}{\hbar}\right) = -\sigma_x, \tag{7.56}$$

where σ_x is the first Pauli matrix. The relation is in fact an equation for L_M and the solution reads:

$$L_M = \frac{\hbar}{2}\begin{pmatrix} 1 & 1 \\ 1 & 1 \end{pmatrix}. \tag{7.57}$$

Eventually we obtain L_α by rotation of L_M in the molecular plane, namely:

$$L_\alpha = e^{-\frac{i}{\hbar}\phi_\chi L_z}\, L_M\, e^{\frac{i}{\hbar}\phi_\chi L_z} = \frac{\hbar}{2}\begin{pmatrix} 1 & e^{i2|\ell|\phi_\chi} \\ e^{-i2|\ell|\phi_\chi} & 1 \end{pmatrix}, \tag{7.58}$$

where $L_z = 2\hbar\sigma_z$ is the generator of the rotations along the principal rotational axis for the 7-particle ground states of the benzene molecule.

7.6.2 All-Electrical Spin Control

We come now to the phenomenology of the spin-dependent transport through a benzene and a triple dot I-SET. The different panels of Figs. 7.14 and 7.15 show the current through the benzene and triple dot I-SET, respectively, as a function of bias and gate voltage. As in all SETs at low bias the so-called Coulomb diamonds, where transport is energetically forbidden, occur. Within the diamonds the particle number is fixed as indicated in the figures.

The characteristic fingerprint of I-SETs is represented by the *interference blockade* where the current decreases for increasing bias generating negative differential conductance (NDC) and eventually vanishes (see green lines in the panels B and C of Figs. 7.14 and 7.15). Panels B in the same figures indicate moreover that, for a given gate voltage and in absence of polarization in the leads, the current is blocked only at one specific bias voltage. For parallel polarized leads, however, at a given gate voltage, the current is blocked at *two specific* bias voltages, one for each spin configuration (panels C). As demonstrated below, the blocking of the minority electrons occurs for the smaller bias voltages. As such full control of the spin configuration in the I-SET can be electrically achieved. The interference blockade and its spin selectivity is also demonstrated in panels A and B of Fig. 7.16. Along the dotted (dashed) line a majority (minority) spin electron is trapped into the molecule. The molecular spin state can thus be manipulated simply by adjusting the bias across the I-SET. In the following we discuss the physics of the spin-selective interference blocking and present the necessary ingredients for its occurrence.

From the analysis of the negative differential conductance and current blocking associated with interference presented in Sects. 7.4 and 7.5 one would conclude that the interference blocking is a threshold effect appearing when the bias opens transitions to a specific set of degenerate states and surviving until transitions to

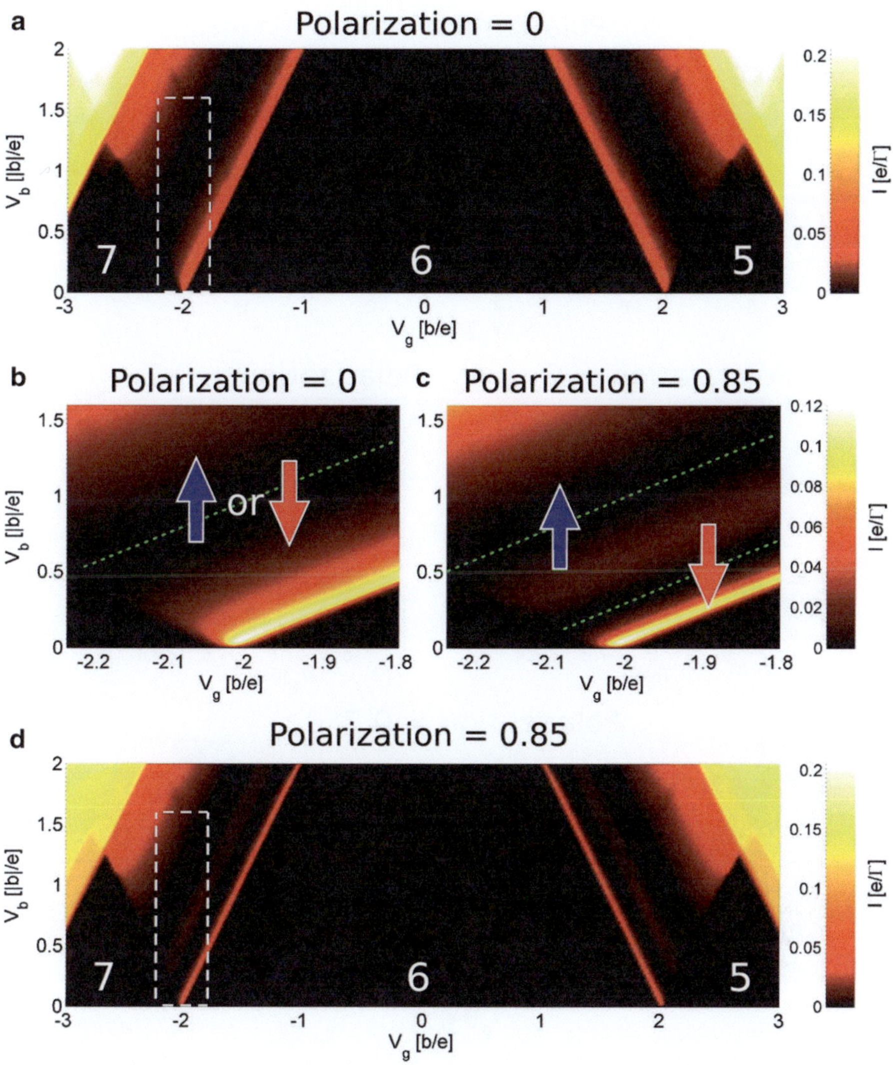

Fig. 7.14 Benzene I-SET: polarized vs. unpolarized configuration. Panel A—Current vs. bias and gate voltage for unpolarized leads. Panel D—Current vs. bias and gate voltage for polarized leads (polarization $P = 0.85$). Panels B and C—Blow up of the $6 \to 7$ particle transition for both configurations. The unpolarized case shows a single current blocking line and the trapped electron has either up or down polarization. The polarized case shows two current blocking lines, corresponding to the different spin of the trapped electron. The current is given in units of e/Γ where Γ is the bare average rate, and the temperature $k_B T = 0.01b$ where b is the hopping parameter. From [21]

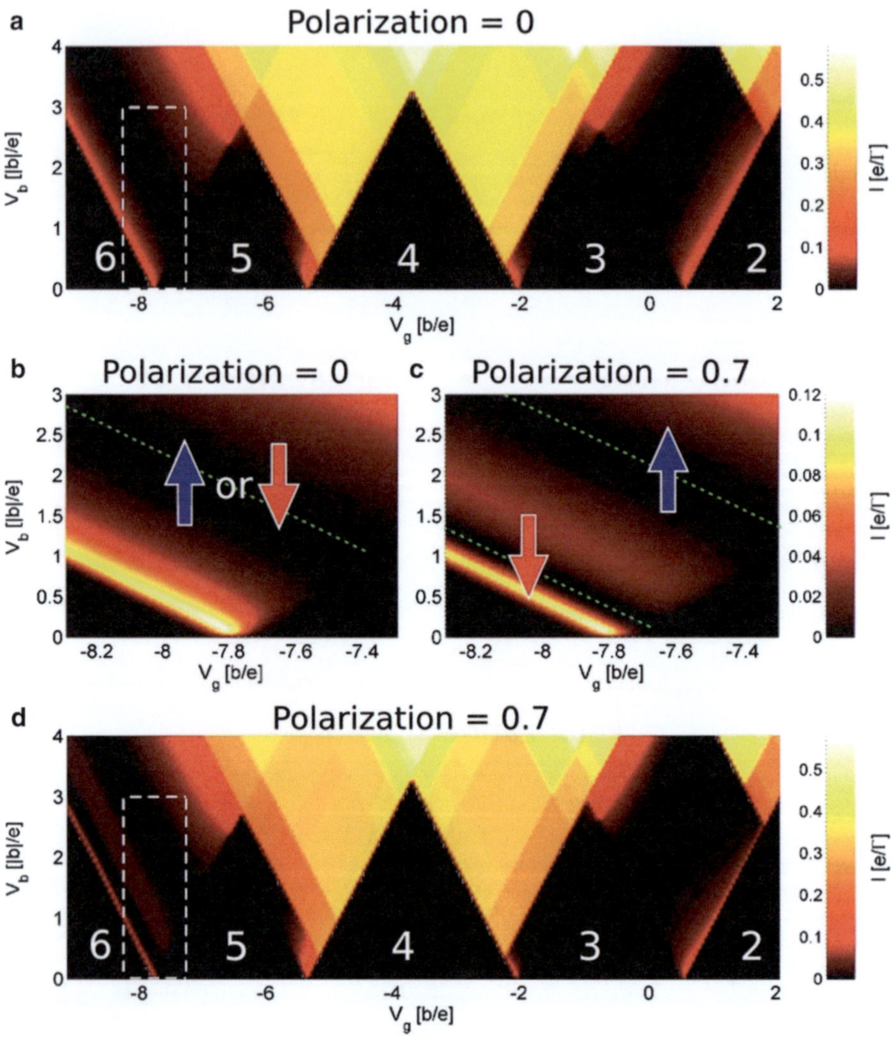

Fig. 7.15 Triple dot I-SET: polarized vs. unpolarized configuration. Panel A—Current vs. bias and gate voltage for unpolarized leads. Panel D—Current vs. bias and gate voltage for polarized leads (polarization $P = 0.7$). Panels B and C—Blow up of the $6 \to 5$ particle transition for both configurations. The selective spin blocking is analogous to the one of the benzene I-SET (Fig. 7.14). From [21]

other states lift the blocking. However, as shown in Figs. 7.14 and 7.15, the current is completely blocked only at specific values of the bias voltage. The explanation of this phenomenon relies on two observations:

(i) The blocking state (Fig. 7.16) must be antisymmetric with respect to the plane perpendicular to the system and passing through its center and the atom

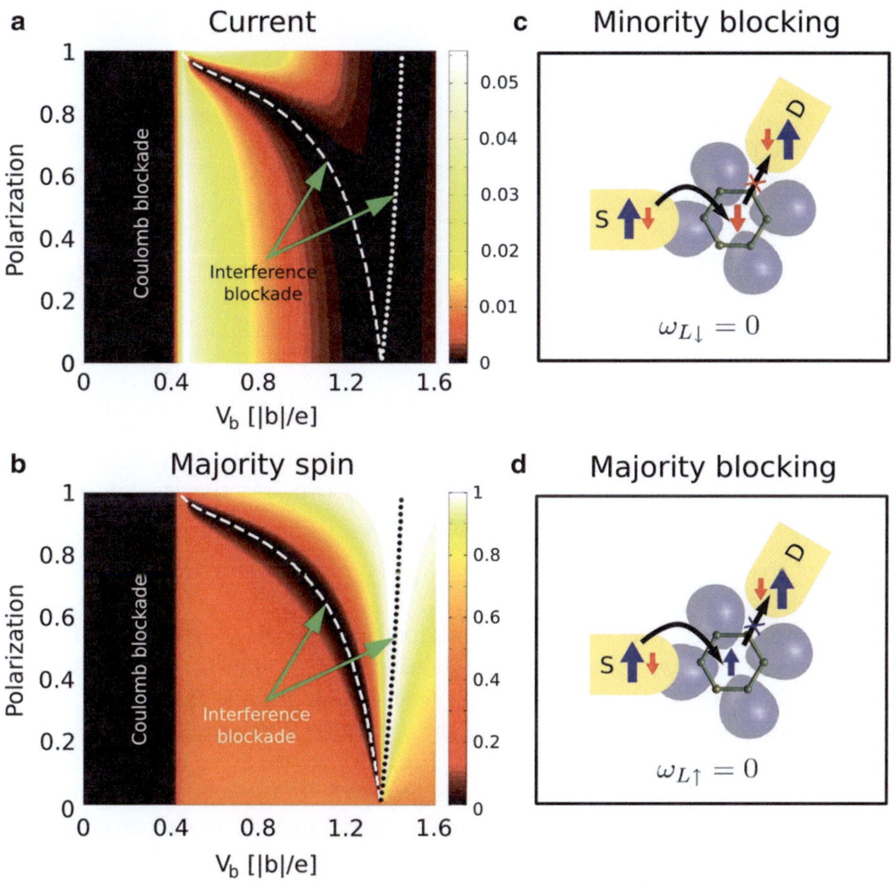

Fig. 7.16 Spin control. Panel (**a**)—Current through the benzene I-SET vs bias and polarization at the $6 \to 7$ electrons transition. Panel (**b**)—Population of the majority spin 7-particle state. The two zero current lines at high bias correspond to the maximum or minimum population of the 7-particle majority spin state and thus identify the spin state of the trapped electron on the molecule. Panels (**c**) and (**d**)—Schematic representation of the spin selective blocking corresponding to the *dashed* (**c**) and *dotted* (**d**) *lines* of the panels A and B. From [21]

(quantum dot) closest to the drain; this state is thus also an eigenstate of the projection of the angular momentum in the direction of the drain lead.[4] At positive (negative) bias voltages we call this state the $R(L)$-antisymmetric state $|\psi_{R(L),a}\rangle$.

(ii) The complete interference blocking is only achieved when $[\rho_{\text{block}}, H_{\text{eff}}] = 0$.

---

[4]The corresponding eigenvalue depends on the symmetry of the atomic (quantum dot) wave function with respect to the molecular (artificial molecule) plane: $\hbar$ or 0 for symmetric or antisymmetric wave functions, respectively.

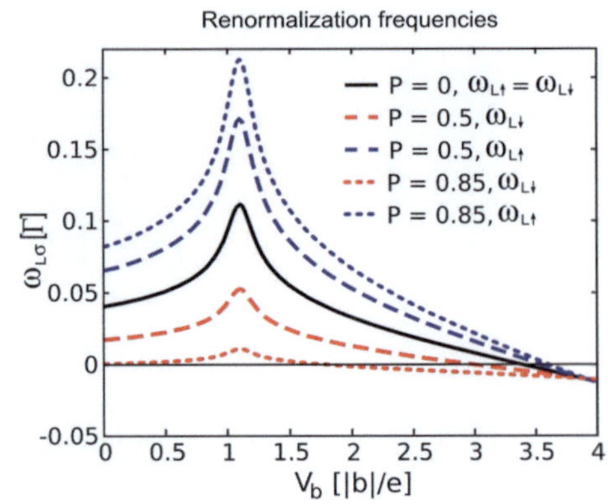

Fig. 7.17 Blocking condition. Renormalization frequencies $\omega_{L\sigma}$ of a benzene I-SET as function of the bias and for different lead polarizations. The current blocking condition $\omega_{L\sigma} = 0$ is fulfilled at different biases for the different spin states. From [21]

In Fig. 7.17 the black curve depicts $\omega_{L\sigma}$ as a function of the bias in absence of polarization: the frequencies corresponding to the two spin species coincide and thus vanish at the same bias. The same condition,

$$\omega_{L\sigma} = 0, \tag{7.59}$$

also determines the bias at which the current is completely blocked. In fact, at that bias the effective Hamiltonian contains only the projection of the angular momentum in the direction of the right lead (the drain) and the density matrix corresponding to the full occupation of the 7-particle R-antisymmetric state ($\rho = |\psi_{R,a}\rangle\langle\psi_{R,a}|$) is the stationary solution of Eq. (7.5). As we leave the blocking bias the effective Hamiltonian contains also the projection of the angular momentum in the direction of the left lead and the R-antisymmetric state is no longer an eigenstate of H_{eff}. The corresponding density matrix is not a stationary solution of (7.5) and current flows through the system. The $L \leftrightarrow R$ symmetry of the system implies, for negative biases, the blocking condition $\omega_{R\sigma} = 0$.

All-electric-spin control is achieved, in an I-SET, only in presence of ferromagnetic leads and with exchange interaction on the system as we prove by analyzing the splitting of the renormalization frequencies $\omega_{\chi\sigma}$. (see Eq. (7.54).) By introducing the average bare rate $\Gamma = \frac{\Gamma^0_{\alpha\uparrow}+\Gamma^0_{\alpha\downarrow}}{2}$, for simplicity equal in both leads, and using the fact that benzene is paramagnetic we get:

$$\begin{aligned}\omega_{\alpha\uparrow} - \omega_{\alpha\downarrow} = 2\bar{\Gamma}^0_\alpha P_\alpha \frac{1}{\pi} \sum_{\{E\}} \Big[&\langle 7_g\ell \uparrow |d_{M\uparrow}|8\{E\}\rangle\langle 8\{E\}|d^\dagger_{M\uparrow}|7_g m \uparrow\rangle p_\alpha(E - E_{7_g}) \\ &+\langle 7_g\ell \uparrow |d^\dagger_{M\uparrow}|6\{E\}\rangle\langle 6\{E\}|d_{M\uparrow}|7_g m \uparrow\rangle p_\alpha(E_{7_g} - E) \\ &-\langle 7_g\ell \uparrow |d_{M\downarrow}|8\{E\}\rangle\langle 8\{E\}|d^\dagger_{M\downarrow}|7_g m \uparrow\rangle p_\alpha(E - E_{7_g}) \\ &-\langle 7_g\ell \uparrow |d^\dagger_{M\downarrow}|6\{E\}\rangle\langle 6\{E\}|d_{M\downarrow}|7_g m \uparrow\rangle p_\alpha(E_{7_g} - E)\Big],\end{aligned} \tag{7.60}$$

where one appreciates the linear dependence of the spin splitting on the lead polarization P_α. The first and the third terms of the sum would cancel each other if the energy of the singlet and triplet 8-particle states would coincide. An analogous condition, but this time on the 6-particle states, concerns the second and the fourth terms. For this reason the exchange interaction on the system is a necessary condition to obtain spin splitting of the renormalization frequencies and thus the full all-electric spin control.

In Fig. 7.17 we show the frequencies $\omega_{L\sigma} = 0$ vs. bias voltage also for a finite values of the polarization P calculated for the benzene I-SET, where exchange splitting is ensured by the strong Coulomb interaction on the system. The interference blocking conditions $\omega_{L\sigma} = 0$ for the $L \to R$ current are satisfied at different biases for the different spin species. The dotted and dashed lines in Fig. 7.16 are the representation of the relations $\omega_{L\uparrow} = 0, \omega_{L\downarrow} = 0$ as a function of the bias and polarization, respectively.

7.7 Robustness

One could argue about the fragility of an effect which relies on the degeneracy of the many-body spectrum. Interference effects are instead rather robust. The exact degeneracy condition can in fact be relaxed and interference survives also for a *quasi-degeneracy* condition: i.e., as far as the splitting between the many-body levels is smaller than the tunnelling rate to the leads. In this limit, the system still does not distinguish between the two energetically equivalent paths sketched in Fig. 7.1.

To quantify the robustness of the effect we will address, in this section, two issues: the first is the modification of the master equation, Eq. (7.5), necessary to capture the interference between quasi-degenerate states, the second is the detailed study of an example of I-SET (the benzene single molecule junction) under several perturbations that lower the symmetry of the system.

7.7.1 GME and Current in the Non-secular Approximation

The bias and the contact perturbations in our model for a benzene I-SET lower the symmetry of the active part of the junction and consequently lift the degeneracy that appeared so crucial for the interference effects. The robustness of the latter relies on the fact that the necessary condition is rather quasi-degeneracy, expressed by the relation $\delta E \ll \hbar\Gamma$.

Nevertheless, if the perfect degeneracy is violated, the secular approximation applied to obtain Eqs. (7.5)–(7.8) does not capture this softer condition. We report here the general expression for the generalized master equation and the associated current operator in the Born–Markov approximation and under the only further condition that coherences between states with different particle number are decoupled from the populations and vanish exactly in the stationary limit:

$$
\begin{aligned}
\dot{\rho}^N_{\mathrm{EE'}} &= -\frac{i}{\hbar}(E-E')\rho^N_{\mathrm{EE'}} \\
&-\sum_{\chi\sigma}\sum_{ij}\sum_{\mathrm{F}}\frac{1}{2}\mathscr{P}_{\mathrm{NE}}\left\{d^\dagger_{i\sigma}\Gamma^{\chi\sigma}_{ij}(F-H_{\mathrm{sys}})\left[-\frac{i}{\pi}p_\chi(F-H_{\mathrm{sys}})+f^-_\chi(F-H_{\mathrm{sys}})\right]d_{j\sigma}\right. \\
&\qquad\left.+d_{j\sigma}\Gamma^{\chi\sigma}_{ij}(H_{\mathrm{sys}}-F)\left[-\frac{i}{\pi}p_\chi(H_{\mathrm{sys}}-F)+f^+_\chi(H_{\mathrm{sys}}-F)\right]d^\dagger_{i\sigma}\right\}\rho^N_{\mathrm{FE'}} \\
&-\sum_{\chi\sigma}\sum_{ij}\sum_{\mathrm{F}}\frac{1}{2}\rho^N_{\mathrm{EF}}\left\{d^\dagger_{i\sigma}\Gamma^{\chi\sigma}_{ij}(F-H_{\mathrm{sys}})\left[+\frac{i}{\pi}p_\chi(F-H_{\mathrm{sys}})+f^-_\chi(F-H_{\mathrm{sys}})\right]d_{j\sigma}\right. \\
&\qquad\left.+d_{j\sigma}\Gamma^{\chi\sigma}_{ij}(H_{\mathrm{sys}}-F)\left[+\frac{i}{\pi}p_\chi(H_{\mathrm{sys}}-F)+f^+_\chi(H_{\mathrm{sys}}-F)\right]d^\dagger_{i\sigma}\right\}\mathscr{P}_{\mathrm{NE'}} \\
&+\sum_{\chi\sigma}\sum_{ij}\sum_{\mathrm{FF'}}\frac{1}{2}\mathscr{P}_{\mathrm{NE}}\left\{d^\dagger_{i\sigma}\rho^{N-1}_{\mathrm{FF'}}d_{j\sigma}\,\Gamma^{\chi\sigma}_{ij}(E'-F')\left[+\frac{i}{\pi}p_\chi(E'-F')+f^+_\chi(E'-F')\right]\right. \\
&\qquad+d^\dagger_{i\sigma}\rho^{N-1}_{\mathrm{FF'}}d_{j\sigma}\Gamma^{\chi\sigma}_{ij}(E-F)\left[-\frac{i}{\pi}p_\chi(E-F)+f^+_\chi(E-F)\right] \\
&\qquad+d_{j\sigma}\rho^{N+1}_{\mathrm{FF'}}d^\dagger_{i\sigma}\Gamma^{\chi\sigma}_{ij}(F'-E')\left[+\frac{i}{\pi}p_\chi(F'-E')+f^-_\chi(F'-E')\right] \\
&\qquad\left.+d_{j\sigma}\rho^{N+1}_{\mathrm{FF'}}d^\dagger_{i\sigma}\Gamma^{\chi\sigma}_{ij}(F-E)+\left[-\frac{i}{\pi}p_\chi(F-E)+f^-_\chi(F-E)\right]\right\}\mathscr{P}_{\mathrm{NE'}}. \quad (7.61)
\end{aligned}
$$

Equation (7.5) represents a special case of Eq. (7.61) in which all energy spacings between states with the same particle number are either zero or much larger than the level broadening $\hbar\Gamma$. The problem of a master equation in presence of quasi-degenerate states in order to study transport through molecules has been addressed in the work of Schultz et al. [43]. The authors claim in their work that the singular coupling limit should be used in order to derive an equation for the density matrix in presence of quasi-degenerate states. Equation (7.61) is derived in the weak coupling limit and bridges all the regimes as illustrated by Figs. 7.18–7.20.

The current operators associated with the master equation just presented read:

$$
\begin{aligned}
I_\chi &= \frac{1}{2}\sum_{\mathrm{NEF}}\sum_{ij}\sum_{\sigma}\mathscr{P}_{\mathrm{NE}} \\
&\left\{d^\dagger_{i\sigma}\Gamma^{\chi\sigma}_{ij}(E-H_{\mathrm{sys}})\left[+\frac{i}{\pi}p_\chi(E-H_{\mathrm{sys}})+f^-_\chi(E-H_{\mathrm{sys}})\right]d_{j\sigma}\right. \\
&+d^\dagger_{i\sigma}\Gamma^{\chi\sigma}_{ij}(F-H_{\mathrm{sys}})\left[-\frac{i}{\pi}p_\chi(F-H_{\mathrm{sys}})+f^-_\chi(F-H_{\mathrm{sys}})\right]d_{j\sigma}
\end{aligned}
$$

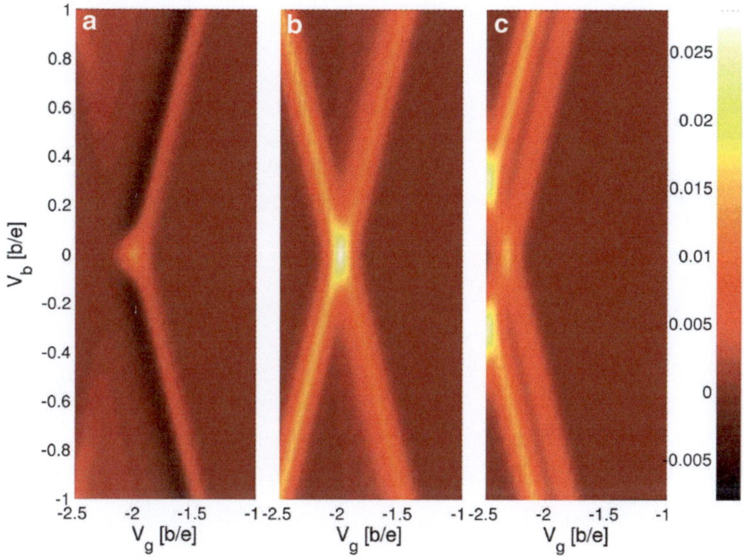

Fig. 7.18 Closeup views of the stability diagram around the $6 \leftrightarrow 7$ resonance for the system under contact perturbation. The perturbation strength grows from left to right. The parameter that describes the contact effect assumes the values $\xi_c = 0.15\hbar\Gamma, 2\hbar\Gamma, 15k_BT$ from left to right, respectively, and $k_BT = 10\hbar\Gamma$. From [20]

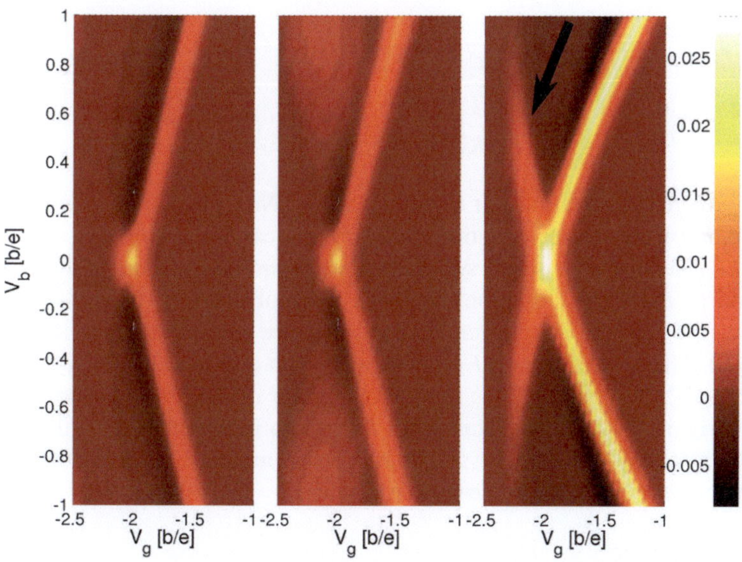

Fig. 7.19 Closeup views of the stability diagram around the $6 \leftrightarrow 7$ resonance for the system under the effect of the bias potential, displayed for different strengths of the electrostatic potential drop over the molecule. The parameter that describes the strength of the electrostatic drop over the molecule assumes the values $a = 25, 12, 0.6$ from left to right, respectively. From [20]

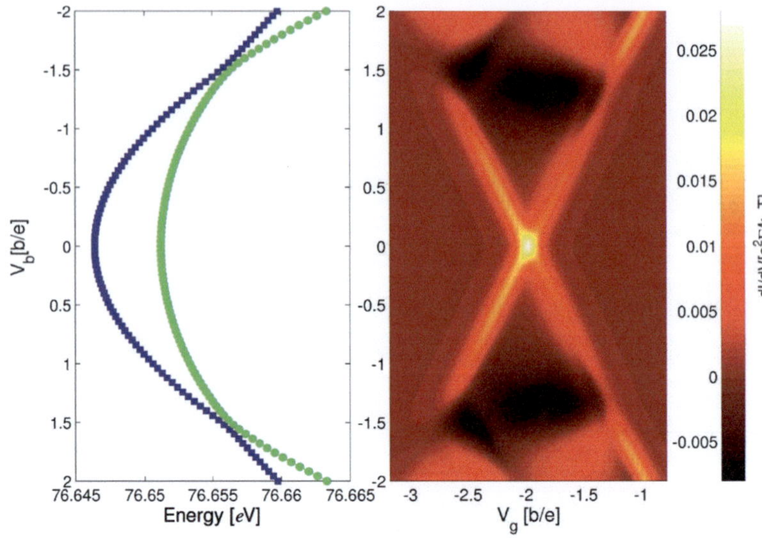

Fig. 7.20 Combination of the bias and contact perturbations. *Left panel*—Energy levels of the 7-particle ground and first excited state as functions of the bias voltage. *Right panel*—Stability diagram around the $6 \leftrightarrow 7$ resonance. The perturbation parameters are in this case $\xi_c = 2\hbar\Gamma$ and $a = 12$. From [20]

$$-d_{j\sigma}\Gamma_{ij}^{\chi\sigma}(H_{\rm sys}-E)\left[+\frac{i}{\pi}p_\chi(H_{\rm sys}-E)+f_\chi^+(H_{\rm sys}-E)\right]d_{i\sigma}^\dagger$$

$$-\,d_{j\sigma}\Gamma_{ij}^{\chi\sigma}(H_{\rm sys}-F)\left[-\frac{i}{\pi}p_\chi(H_{\rm sys}-F)+f_\chi^+(H_{\rm sys}-F)\right]d_{i\sigma}^\dagger\Bigg\}\mathscr{P}_{\rm NF} \quad (7.62)$$

where $\chi = L, R$ indicates the left or right contact. Nevertheless, within the limits of derivation of the master equation, this formula can be simplified. Actually, if $E - F \leq \hbar\Gamma$, then F can be safely substituted with E in the argument of the principal values and of the Fermi functions, with an error of order $\frac{E-F}{k_{\rm B}T} < \frac{\hbar\Gamma}{k_{\rm B}T}$ which is negligible (the generalized master equation that we are considering is valid for $\hbar\Gamma \ll k_{\rm B}T$). The approximation $E \sim F$ breaks down only if $E - F \sim k_{\rm B}T$, but this implies $E - F \gg \hbar\Gamma$ which is the regime of validity of the secular approximation. Consequently, in this regime, terms with $E \neq F$ do not contribute to the average current because they vanish in the stationary density matrix. Ultimately we can thus reduce the current operators to the simpler form:

$$I_\chi = \sum_{NE}\sum_{ij}\sum_{\sigma}\mathscr{P}_{NE}\Big[d_{j\sigma}\Gamma_{ij}^{\chi\sigma}(H_{\rm sys}-E)f_\chi^+(H_{\rm sys}-E)d_{i\sigma}^\dagger$$

$$-d_{i\sigma}^\dagger\Gamma_{ij}^{\chi\sigma}(E-H_{\rm sys})f_\chi^-(E-H_{\rm sys})d_{j\sigma}\Big]\,, \quad (7.63)$$

which is almost equal to the current operator corresponding to the secular approximation. The only difference is here the absence of the second projector operator that allows contributions to the current coming from coherences between different energy eigenstates.

7.7.2 *Interference in a Reduced Symmetry I-SET*

In this section we study the effect of reduced symmetry on the transport characteristics of a benzene I-SET. For this purpose, we generalize the model Hamiltonian by taking into account the perturbations on the molecule due to the contacts and the bias voltage. The contact between molecule and leads is provided by different anchor groups. These linkers are coupled to the contact carbon atoms over a σ bond, thus replacing the corresponding benzene hydrogen atoms. Due to the orthogonality of π and σ orbitals, the anchor groups affect in first approximation only the σ orbitals of benzene. In particular the different electron affinity of the atoms in the linkers imply a redistribution of the density of σ electrons. Assuming that transport is carried by π electrons only, we model the effect of this redistribution as a change in the on-site energy for the p_z orbitals of the contact carbon atoms:

$$H'_{\text{ben}} := H_{\text{contact}} = \xi_{\text{c}} \sum_{\chi\sigma} d^{\dagger}_{\chi\sigma} d_{\chi\sigma}, \quad \chi = L, R \tag{7.64}$$

where $d_{R\sigma} = d_{4\sigma}, d_{5\sigma}$, respectively, in the para and meta configuration, while $d_{L\sigma} = d_{1\sigma}$ in both setups.

We also study the effect of an external bias on the benzene I-SET. In particular we release the strict condition of potential drop all concentrated at the lead-molecule interface. Nevertheless, due to the weak coupling of the molecule to the leads, we assume that only a fraction of the bias potential drops across the molecule. For this residual potential we take the linear approximation $V_{\text{b}}(\mathbf{r}) = -\frac{V_{\text{b}}}{a}(\mathbf{r} \cdot \hat{\mathbf{r}}_{\text{sd}}/a_0)$, where we choose the center of the molecule as the origin and $\hat{\mathbf{r}}_{\text{sd}}$ is the unity vector directed along the source to drain direction. $a_0 = 1.43\,\text{Å}$ is the bond length between two carbon atoms in benzene, a is the coefficient determining the intensity of the potential drop over the molecule. Since the p_z orbitals are strongly localized, we can assume that this potential will not affect the inter-site hopping, but only the on-site term of the Hamiltonian:

$$H'_{\text{ben}} := H_{\text{bias}} = e \sum_{i\sigma} \xi_{\text{b}_i} d^{\dagger}_{i\sigma} d_{i\sigma} \tag{7.65}$$

with $\xi_{\text{b}_i} = \int d\mathbf{r}\; p_z(\mathbf{r} - \mathbf{R}_i) V_{\text{b}}(\mathbf{r}) p_z(\mathbf{r} - \mathbf{R}_i)$.

Under the influence of the contacts or the bias potential, the symmetry of the molecule changes. Table 7.6 shows the point groups to which the molecule

Table 7.6 Point groups to which the molecule belongs under the influence of the contacts and the external bias potential. From [20]

Type of perturbation	Symmetry (Para config.)	Symmetry (Meta config.)
Contact perturbation	D_{2h}	C_{2v}
Bias perturbation	C_{2v}	C_{2v}

belongs in the perturbed setup. This point groups have only A- and B-type reducible representations. Thus the corresponding molecular orbitals do not exhibit orbital degeneracy.

No interference effects influence the transport in the para configuration. Thus we do not expect its transport characteristics to be qualitatively modified by the new setup with the corresponding loss of degeneracies. In the meta configuration, on the other hand, interferences between orbitally degenerate states play a crucial role in the explanation of the occurring transport features. Naively one would therefore expect that neither conductance suppression nor NDC and current blocking occur in a benzene I-SET with reduced symmetry.

Yet we find that, under certain conditions, the mentioned transport features are robust under the lowered symmetry.

The perturbations due to the contacts and the bias lead to an expected level splitting of the former orbitally degenerate states. Very different current-voltage characteristics are obtained depending of the relation between the energy splitting δE and other two important energy scales of the system: the tunnelling rate Γ and the temperature T. In particular, when $\delta E \ll \Gamma \ll T$, interference phenomena persist. In contrast when $\Gamma < \delta E \ll T$ interference phenomena disappear, despite the fact that, due to temperature broadening, the two states still cannot be resolved. In this regime, due to the asymmetry in the tunnelling rates introduced by the perturbation, standard NDC phenomena, see Fig. 7.19, occur.

Figure 7.18 shows from left to right closeup views of the stability diagram for the setup under the influence of increasing *contact perturbation* around the $6 \leftrightarrow 7$ resonance. The orbital degeneracy of the 7-particle states is lifted and the transport behavior for the $6 \leftrightarrow 7$ transition depends on the energy difference between the formerly degenerate 7-particle ground states. In panel *a* the energy difference is so small that the states are quasi-degenerate: $\delta E \ll \hbar\Gamma \ll k_B T$. As expected, we recover NDC at the border of the 6-particle diamond and current suppression at the border of the 7-particle diamond, like in the unperturbed setup. Higher on-site energy-shifts correspond to a larger level spacing. Panel *b* displays the situation in which the latter is of the order of the level broadening, but still smaller than the thermal energy ($\delta E \simeq \hbar\Gamma \ll k_B T$): no interference causing NDC and current blocking can occur. Yet, due to thermal broadening, we cannot resolve the two 7-particle states. Eventually, panel *c* presents the stability diagram for the case $\delta E > k_B T > \hbar\Gamma$: the level spacing between the 7-particle ground and first excited state is now bigger than the thermal energy, thus the two transition lines corresponding to these states are clearly visible at the border of the 6-particle stability diamond.

Figure 7.19 shows closeup views of the stability diagram for the setup under the influence of the *bias perturbation* at the border of the 6- and 7-particle diamonds.

The same region is plotted for different strengths of the external potential over the molecule. In contrast to the contact perturbation, the amount of level splitting of the former degenerate states is here bias dependent. This fact imposes a bias window of interference visibility. The bias must be small enough, for the 7-particle states to be quasi-degenerate and at the same time bigger than the thermal energy, so that the occurring NDC is not obscured by the thermally broadened conductance peak. A strong electrostatic potential perturbation closes the bias window and no interference effect can be detected. Panel *a* of Fig. 7.19 represents the weak perturbation regime with no qualitative differences with the unperturbed case. The typical fingerprints of interference (NDC at the border of the 6-particle diamond and current blocking for the $7 \rightarrow 6$ transition) are still visible for intermediate perturbation strength (panel *b*) but this time only in a limited bias window. Due to the perturbation strength, at some point in the bias, the level splitting is so big that the quasi-degeneracy is lifted and the interference effects destroyed. In panel *c* the quasi-degeneracy is lifted in the entire bias range. There is NDC at the border of the 6-particle diamond, but is not accompanied by current blocking as proved by the excitation line at the border of the 7-particle diamond (see arrow): no interference occurs. The NDC is here associated with the sudden opening of a slow current channel, the one involving the 6-particle ground state and the 7-particle (non-degenerate) excited state (standard NDC).

Figure 7.20 refers to the setup under both the *bias and contact perturbations*. The left panel shows the energy of the lowest 7-particle states as a function of the bias. In the right panel we present the stability diagram around the $6 \leftrightarrow 7$ resonance. NDC and current blocking are clearly visible only in the bias region where, due to the combination of bias and contact perturbation, the two seven-particle states return quasi-degenerate. Also the fine structure in the NDC region is understandable in terms of interference if in the condition of quasi-degeneracy we take into account the renormalization of the level splitting due to the energy non-conserving terms.

Interference effects predicted for the unperturbed benzene I-SET are robust against various sources of symmetry breaking. Quasi-degeneracy, $\delta E \ll \hbar\Gamma \ll k_B T$, is the necessary condition required for the detection of the interference in the stability diagram of the benzene I-SET.

7.8 Conclusions

In this chapter we addressed the interference effects that characterize the electronic transport through a symmetric single electron transistor based on quantum dot molecules. Interestingly, in this class of devices interference effects survive even in the weak tunnelling coupling regime when usually the decoherence introduced by the leads dominates the picture and transport consists of a set of incoherent tunnelling events.

After introducing the concept of interference single electron transistor (I-SET) we formulate a simple interference condition (7.24) for I-SETs in terms of the tunnelling transitions *amplitudes* of degenerate states with respect to the source and drain lead. A generic model of I-SET is then introduced, together with our method of choice to study the dynamics of the molecular I-SET: i.e., the density matrix approach which starts from the Liouville equation for the total density operator which enables the treatment of quasi-degenerate states, so crucial for the description of the interference effects which are the focus of our investigation. As a further step, we derive the most generic conditions for interference blockade and an algorithm for the identification of the interference blocking states as linear combination of degenerate many-body eigenstates of the system. The theory is sufficiently general to be applied to any device consisting of a system with degenerate many-body spectrum weakly coupled to metallic leads, e.g., molecular junctions, graphene or carbon nanotube quantum dots, artificial molecules. In particular, the algebraic formulation of the blocking condition in terms of kernels of the tunnelling matrices $T^{\pm}$, Eq. (7.18), allows a straightforward numerical implementation and makes the algorithm directly applicable to complex junctions with highly degenerate spectrum. For example, we have recently applied the same theory to study the transport through STM junctions of single molecules on thin insulating films [61, 62].

As an application of the theory we study the benzene and the triple dot I-SET.

For the first system, two different setups are considered, the para and the meta configuration, depending on the position of the leads with respect to the molecule. Within an effective p_z orbital model, we diagonalize exactly the Hamiltonian for the molecule. We further apply a group theoretical method to classify the many-body molecular eigenstates according to their symmetry and quasi-angular momentum. With the help of this knowledge we detect the orbital degeneracy and, in the para configuration, we select the states relevant for transport. The application of the simple interference condition (7.24) enables us to predict the existence of interference effects in the meta configuration. The stability diagrams for the two configurations show striking differences. In the linear regime a selective conductance suppression is visible when changing from the para to the meta configuration. Only transitions between ground states with well-defined particle number are affected by the change in the lead configuration. With the help of the group theoretical classification of the states we recognize in this effect a fingerprint of the *destructive interference* between orbitally degenerate states. We derive an analytical formula for the conductance that reproduces exactly the numerical result and supports their interpretation in terms of interference. Other interference effects are also visible in the nonlinear regime where they give rise to NDC and current blocking at the border of the 6-particle Coulomb diamond as well as to current suppression for transitions between 7- and 6-particle states.

Despite its relative simplicity, the triple dot I-SET exhibits different types of interference blocking and it represents an interesting playground of the general theory. Specifically, we concentrated on the interference blockade that involves an excited triplet state, a condition not accessible in the benzene I-SET.

In both cases we further analyze the blockade that involves orbitally and spin degenerate states and we show how to realize all electrical preparation of specific spin states. Thus we obtain an interference-mediated control of the electron spin in quantum dots, a highly desirable property for spintronics [23–25] and spin-qubit applications [26–30]. Similar blocking effects have been found also in multiple quantum dot systems in dc [31] and ac [32] magnetic fields.

Finally, we provide a detailed discussion of the impact of the reduced symmetry due to linking groups between the molecule and the leads or to an electrostatic potential drop over the molecule. We classify different transport regimes and set up the limits within which the discussed transport features are robust against perturbations. We identify in the *quasi-degeneracy* of the many-body states the necessary condition for interference effects.

Acknowledgements We would like to thank Dr. Georg Begemann and Dana Darau for their important contribution to the development of the research presented in this chapter. We also acknowledge the German Research Foundation (DFG) for the financial support through the research programs SPP 1243, SFB 689 and GRK 1570.

References

1. Young, T.: Phil. Trans. Roy. Soc. Lond. **94**, 12 (1804)
2. Jönsson, C.: Z. Physik **161**, 454 (1961)
3. Merli, P.G., Missiroli, G.F., Pozzi, G.: Am. J. Phys. **44**, 306 (1976)
4. Arndt, M., Nairz, O., Vos-Andreae, J., Keller, C., van der Zouw, G., Zeilinger, A.: Nature **401**, 680 (1999)
5. Yacoby, A., Heiblum, M., Mahalu, D., Shtrikman, H.: Phys. Rev. Lett. **74**, 4047 (1995)
6. Gustavsson, S., Leturcq, R., Studer, M., Ihn, T., Ensslin, K.: Nano Lett. **8**, 2547 (2008)
7. Gutiérrez, R., Grossmann, F., Schmidt, R.: ChemPhysChem **4**, 1252 (2003)
8. Cardamone, D.V., Stafford, C.A., Mazumdar, S.: Nano Lett. **6**, 2422 (2006)
9. Ke, S.-H., Yang, W., Baranger, U.: Nano Lett. **8**, 3257 (2008)
10. Quian, Z., Li, R., Zhao, X., Hou, S., Sanvito, S.: Phys. Rev. B **78**, 113301 (2008)
11. Solomon, G.C., Andrews, D.Q., Hansen, T., Goldsmith, R.H., Wasielewski, M.R., Duyne, R.P.V., Ratner, M.A.: J. Chem. Phys. **129**, 054701 (2008)
12. Markussen, T., Stadler, R., Thygesen, K.S.: Nano Lett. **10**, 4260 (2010)
13. Tsuji, Y., Staykov, A., Yoshizawa, K.: J. Am. Chem. Soc. **133**, 5955 (2011)
14. Markussen, T., Stadler, R., Thygesen, K.S.: Phys. Chem. Chem. Phys. **13**, 14311 (2011)
15. Ernzerhof, M.J.: Chem. Phys. **135**, 014104 (2011)
16. Taniguchi, M., Tsutsui, M., Mogi, R., Sugawara, T., Tsuji, Y., Yoshizawa, K., Kawai, T.: J. Am. Chem. Soc. **133**, 11426 (2011)
17. Aradhya, S.V., Meisner, J.S., Krikorian, M., Ahn, S., Parameswaran, R., Steigerwald, M.L., Nuckolls, C., Venkataraman, L.: Nano Lett. **12**, 1643 (2012)
18. Guédon, C.M., Valkenier, H., Markussen, T., Thygesen, K.S., Hummelen, J.C., van der Molen, S.J.: Nature Nanotechnology **7**, 304 (2012)
19. Begemann, G., Darau, D., Donarini, A., Grifoni, M.: Phys. Rev. B **77**, 201406(R) (2008); **78**, 089901(E) (2008)
20. Darau, D., Begemann, G., Donarini, A., Grifoni, M.: Phys. Rev. B **79**, 235404 (2009). Copyright (2009) by the American Physical Society
21. Donarini, A., Begemann, G., Grifoni, M.: Nano Lett. **9**, 2897 (2009). Copyright (2009) by the American Chemical Society
22. Donarini, A., Begemann, G., Grifoni, M.: Phys. Rev. B **82**, 125451 (2010). Copyright (2010) by the American Physical Society

23. Wolf, S.A., Awschalom, D.D., Buhrman, R.A., Daughton, J.M., von Molnár, S., Roukes, M.L., Chtchelkanova, A.Y., Treger, D.M.: Science **294**, 1488 (2001)
24. Awschalom, D.D., Flatté, M.E.: Nat. Phys. **3**, 153 (2007)
25. Ohno, H., Chiba, D., Matsukura, F., Omiya, T., Abe, E., Dietl, T., Ohno, Y., Ohtani, K.: Nature **408**, 944 (2000)
26. Golovach, V.N., Borhani, M., Loss, D.: Phys. Rev. B **74**, 165319 (2006)
27. Levitov, L., Rashba, E.: Phys. Rev. B **67**, 115324 (2003)
28. Debald, S., Emary, C.: Phys. Rev. Lett. **94**, 226803 (2005)
29. Walls, J.: Phys. Rev. B **76**, 195307 (2007)
30. Nowack, K.C., Koppens, F.H.L., Nazarov, Yu.V., Vandersypen, L.M.K.: Science **318**, 1430 (2007)
31. Emary, C.: Phys. Rev. B **76**, 245319 (2007)
32. Busl, M., Sanchez, R., Platero, G.: Phys. Rev. B **81**, 121306(R) (2010)
33. Braig, S., Brouwer, P.W.: Phys. Rev. B **71**, 195324 (2005)
34. Gurvitz, S.A., Prager, Ya.S.: Phys. Rev. B **53**, 15932 (1996)
35. Braun, M., König, J., Martinek, J.: Phys. Rev. B **70**, 195345 (2004)
36. Wunsch, B., Braun, M., König, J., Pfannkuche, D.: Phys. Rev. B **72**, 205319 (2005)
37. Donarini, A., Grifoni, M., Richter, K.: Phys. Rev. Lett. **97**, 166801 (2006)
38. Harbola, U., Esposito, M., Mukamel, S.: Phys. Rev. B **74**, 235309 (2006)
39. Mayrhofer, L., Grifoni, M.: Eur. Phys. J. B **56**, 107 (2007)
40. Koller, S., Mayrhofer, L., Grifoni, M.: New J. Phys. **9**, 348 (2007)
41. Pedersen, J., Lassen, B., Wacker, A., Hettler, M.: Phys. Rev. B **75**, 235314 (2007)
42. Hornberger, R., Koller, S., Begemann, G., Donarini, A., Grifoni, M.: Phys. Rev. B **77** 245313 (2008)
43. Schultz, M.G., von Oppen, F.: Phys. Rev. B **80**, 033302 (2009)
44. Pariser, R., Parr, R.G.: J. Chem. Phys. **21**, 466 (1953)
45. Pople, J.A.: Trans. Faraday Soc. **49**, 1375 (1953)
46. Linderberg, J., Öhrn, Y.: J. Chem. Phys. **49**, 716 (1968)
47. Hettler, M.H., Wenzel, W., Wegewijs, M.R., Schoeller, H.: Phys. Rev. Lett. **90**, 076805 (2003)
48. Barford, W.: Electronic and Optical Properties of Conjugated Polymers. Clarendon Press, Oxford (2005)
49. Kubatkin, S., Danilov, A., Hjort, M., Cornil, J., Brédas, J.-L., Stuhr-Hansen, N., Hedegård, P., Bjørnholm, T.: Nature **425**, 698 (2003)
50. Kaasbjerg, K., Flensberg, K.: Nano Lett. **8**, 3809 (2008)
51. Kaasbjerg, K., Flensberg, K.: Phys. Rev. B **84**, 115457 (2011)
52. Delgado, F., Shim, Y.-P., Korkusinski, M., Gaudreau, L., Studenikin, S.A., Sachrajda, A.S., Hawrylak, P.: Phys. Rev. Lett. **101**, 226810 (2008)
53. Gong, W., Zheng, Y., Lü, T.: Appl. Phys. Lett. **92**, 042104 (2008)
54. Kostyrko, T., Bułka, B.R.: Phys. Rev. B **79**, 075310 (2009)
55. Shim, Y.-P., Delgado, F., Hawrylak, P.: Phys. Rev. B **80**, 115305 (2009)
56. Pöltl, C., Emary, C., Brandes, T.: Phys. Rev. B **80**, 115313 (2009)
57. Gaudreau, L., Studenikin, S.A., Sachrajda, A.S., Zawadzki, P., Kam, A., Lapointe, J., Korkusinski, M., Hawrylak, P.: Phys. Rev. Lett. **97**, 036807 (2006)
58. Rogge, M.C., Haug, R.J.: Phys. Rev. B **78**, 153310 (2008)
59. Gaudreau, L., Sachrajda, A.S., Studenikin, S., Kam, A., Delgado, F., Shim, Y.P., Korkusinski, M., Hawrylak, P.: Phys. Rev. B **80**, 075415 (2009)
60. Austing, G., Payette, C., Yu, G., Gupta, J., Aers, G., Nair, S., Amaha, S., Tarucha, S.: Jpn. J. Appl. Phys. **49**, 04DJ03 (2010)
61. Sobczyk, S., Donarini, A., Grifoni, M.: Phys. Rev. B **85**, 205408 (2012)
62. Donarini, A., Siegert, B., Sobzyk, S., Grifoni, M.: Phys. Rev. B **86**, 155451 (2012)

Chapter 8
Quantum Interference Effects on the Electronic Transmission Through Quantum Dot Molecules

Rodolfo H. Romero, Daniel A. Lovey, Diego Sebastian Acosta Coden, and Sergio S. Gomez

Abstract The fabrication and control of devices at nanometric scale emphasizes the importance of quantum effects on the electronic motion in semiconductor heterostructures. The size of the systems through which the electrons move makes mandatory to take into account their wave-like character. As a consequence, the typical interference effects play an important role. In this chapter the noticeable Fano and Aharonov–Bohm effects in the transmission through a ring of quantum dots threaded by a magnetic flux will be discussed. They arise from the interference of the electron wave function when propagating along several transmission paths. The effects manifest themselves as peaks of high conductance and dips of cancellations of transmission along the ring. The control of the gate potentials applied to quantum dots and the magnetic field threading the ring allow one to tune the energies at which high and low transmissions occur.

8.1 Introduction

The persistent search for miniaturization of electronic devices has led to the development of new physical systems where the charge carriers are tightly confined. In semiconductor heterostructures, the electronic motion can be confined close to its interface, forming a two-dimensional electron gas (2DEG). Further confinement in the lateral dimensions can be achieved by applying bias voltages to metallic gates patterned in-plane or on-top of the interface. In this way, nanometric island of electron charges—quantum dots (QDs)—can be produced and precisely controlled. The discretization of their energy levels, due to the spatial restriction of the

R.H. Romero (✉) • D.A. Lovey • D.S. Acosta Coden • S.S. Gomez
Facultad de Ciencias Exactas y Naturales y Agrimensura, and Instituto de Modelado e Innovación Tecnológica (CONICET-UNNE) Av. Libertad 5500 (3400) Corrientes, Argentina
e-mail: rhromero@exa.unne.edu.ar; lovey.da@gmail.com; diegoacoden@gmail.com; ssgomez@exa.unne.edu.ar

J. Wu and Z.M. Wang (eds.), *Quantum Dot Molecules*, Lecture Notes in Nanoscale Science and Technology 14, DOI 10.1007/978-1-4614-8130-0_8,

electronic motion, has propitiated to name them as "artificial atoms", because of the similarity of their energy spectra with those from the elements of the periodic table. By extension, quantum dot arrays are also termed "artificial molecules" or QD molecules. Remarkably, the most noticeable difference is their typical energies. While the excitation energies of natural atoms and molecules are of the order of eV, those for quantum dots are three order of magnitude smaller. This allows one to take advantage of effects hardly observable in natural molecules because of the required energy scale.

As the dimensions of the region accessible to the electronic motions become smaller, the quantum effects turn more notorious and a wave function description of the electronic motion becomes mandatory. This wave character manifests itself in intrinsically quantum effects, like coherence and interference. The control of such effects is important to provide both a better understanding of the quantum realm and new functionalities to the circuits [1, 2]. Quantum interference allows to enhance or to cancel, total or partially, the response of the system beyond the simple classical additive behavior. Such an effect not only can pose a problem to be avoided but also could provide new capabilities to the device with respect to its classical counterpart [3, 4]. Two remarkable effects arose from this quantum behavior, namely, the Fano and the Aharonov–Bohm effects. In this chapter, we will show a way of including these effects in a particular type of theoretical models and their consequences for nanoscale structures. On the one hand, the ability of tuning the device configuration and parameters allows one to control its response to externally applied fields or voltages. On the other hand, the modification of the charge transport through the nanostructure provides useful information on the material properties as well as to design new applications with unprecedented functionalities [5].

Here, we present the non-equilibrium Green function (NEGF) formalism for quantum transport and apply it to a system in a phase coherent non-interacting regime. In experiments with nanostructures, various electrostatic gate voltages are used to tune the device. The tunability of QD couplings allows one to change the electronic transmission of the system. Under proper connection topologies, the transmission can be completely suppressed at a given energy.

When a magnetic field perpendicular to a ring in the 2DEG is present, such as the one shown in Fig. 8.1, a periodic modulation of the phase of the wave function arises. For this configuration, two or more spatial paths for transmission between the source and drain terminals exist. In such a case, the phase varies with the magnitude of the magnetic flux threading it. It has been experimentally shown that connecting or disconnecting the arms of a ring of QDs by inducing the formation of QD molecules between them produces an abrupt change in the conductance. This result is usually interpreted in terms of the simplest fully-coherent single-mode wave picture. We use the NEGF formalism together with a Hamiltonian for nearest-neighbor interacting QDs placed at fixed discrete sites. This description leads to the appealing interpretation of transmission pathways. They are formed by the chain of sites (QDs) occupied by the electron when successively hopping from one QD to its nearest neighbor. Throughout the chapter, we shall stress the qualitatively modeling of the relevant effects and its interpretation, rather than their

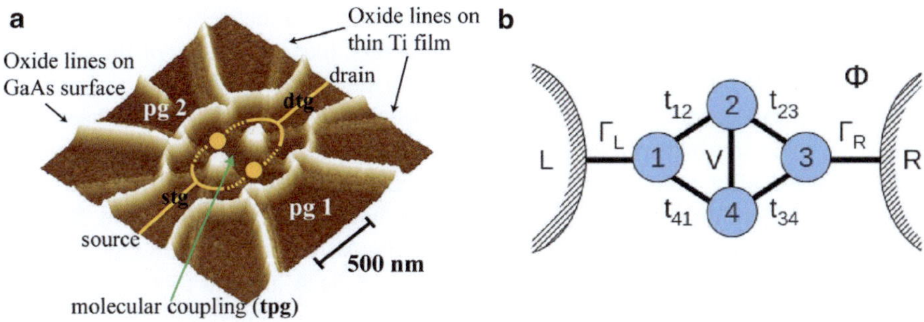

Fig. 8.1 (**a**) SFM image of the sample surface of the experimental setup of [6]. (**b**) Scheme of the theoretical model based on a 4-QDs ring. Leads L and R represent the source and drain terminals, and the QD2-QD4 coupling parameter allows the formation of molecular states

quantitative connection to specific experimental measurements. In line with that, we shall analyze the sensitivity of the conductance upon variation of the various model parameters. In particular, it will be shown that the concept of pathways becomes useful in predicting the energy at which the onset and cancellation of conductance occur.

8.2 Basic Concepts

8.2.1 *Dynamical and Magnetic Phases*

The phase of the wave function describing the electron motion $\psi(\mathbf{r}) = A\exp[i\phi(\mathbf{r})]$ depends on the potential energy $U(\mathbf{r})$ at a given position. In the simplest case of an electron moving in a one-dimensional constant potential U, its phase varies with a wave vector $k = \sqrt{2m(E-U)}/\hbar$. The phase acquired after traveling from x_1 to x_2 is $k(x_2 - x_1) = k\Delta x$. When the potential energy varies with the position, the phase adds according to

$$\phi = \phi_0 + \int_{x_1}^{x_2} k(x)dx, \tag{8.1}$$

where ϕ_0 is an arbitrary initial phase and $k(x) = \sqrt{2m[E-U(x)]}/\hbar$ now depends on the position. The phase accumulated along its propagation, $\phi - \phi_0$, is the analogous of the phase acquired by light traveling along an optical path, with the potential energy playing the role of the optical refraction index. Making optical waves travel successively along a path and its reverse does not cancel both phases, but adds them. In the same way, the quantum phase of an electron traveling along a trajectory and its reverse does not vanishe but evolves increasing it always in the same direction. This phase due to the motion along a trajectory through external potentials is named *dynamical phase*. In QD arrays, such external potentials are experimentally realized by gate terminals which produce tunable potential bias. In theoretical models, those potentials are included through variable parameters, like the energy of the electron when it is localized in a given QD, referred to as the on-site energy.

There is still another way in which the electron wave function can change its phase. In their seminal article, Aharonov and Bohm showed the now well-known effect, named after them, that the magnetic vector potential **A** can influence the motion of a quantum particle even in regions where the magnetic field **B** itself vanishes [7]. In such a case, the phase shift of the wave function evaluated at two different points in space changes by the integral of the **A** along a trajectory joining those two points. The phase accumulated along a closed curve can be related to the flux of the magnetic field enclosed by the curve, invoking the Stoke's theorem. This local change of phase, entirely due to a non-vanishing vector potential, has no classical analog [8].

8.2.2 Fano and Aharonov–Bohm Effects

In the following, we shall restrict ourselves to QDs under low biases, thus disregarding processes leading to energy relaxation. In such a case, the linear conductance of quantum dots is mostly due to elastic transport processes. In particular, we assume a ballistic regime, as opposite to a diffusive one. A ring having QDs embedded into their arms allows one to combine the two causes of interference aforementioned. In order to be observable in the measured currents, the dimensions of the ring must be of the order or smaller than the length after which the interaction of electrons produces lost of coherence. The AB effect has been envisioned for the feasible exploitation of the quantum phase in electronic devices [4, 9–14]. Phase coherent effects in AB rings have been treated theoretically in the literature [15–31]. Due to its origin, the presence of AB effect is used as an experimental probe showing the signature of coherent behavior of the electronic transport in 2DEG nanostructures [32–35], in graphene rings [36–39], or even at the leads [40]; to study the lost of coherence of a system due to inelastic interactions [14, 41], or to determine the dependence of the coherence of the system on its parameters [42]. Interestingly, it has also been used as probe of coherence in experiments not directly related to electronic transport, such as those measuring properties depending on the entanglement between electrons [43].

Closely related to the concept of coherence is the quantum interference between a discrete state with a continuum of states. This phenomenon was firstly studied by Fano in the spectrum of photoionization of atoms and termed Fano effect after him [44], although was found ubiquitous in a wide range of physical phenomena. Interestingly, it has been observed in properly tailored nanoscale systems [6,45–47]. The first tunable experiment showing the characteristic asymmetric Fano profile in the electron transmission was reported in an AB ring with a quantum dot embedded in one of its arms [45]. Various parameters, such as the gate voltage and the magnetic flux through the ring among others, allowed to tune the peak and dip of the profile.

Theoretical models show that the typical asymmetric line profile is composed of a Lorentzian centered at the bonding energy, and a Fano line shape at the energy of the anti-bonding state [18]. In recent experiments a QD molecule has been embedded between the arms of an AB interferometer, that has a number of tunable parameters, allowing the transmission to exhibit a large variety of behaviors as a function of them [6]. It has also been experimentally demonstrated that, within the Coulomb blockade regime and keeping the number of electrons in the dots constant, there is a phase-coherent contribution that can dominate the measured conductance when tunneling times are shorter than dephasing times [9, 10, 14].

8.2.3 Non-Equilibrium Approach to Electronic Transport

One of the approaches for the study of the electron transport is through the use of Green functions. The Green function of a quantum system $G(x,t;x',t')$ propagates a solution of the Schrödinger equation from a position x at the instant t to a position x' at time t' through

$$\psi(x',t') = \int G(x',t';x,t)\psi(x,t)dx. \tag{8.2}$$

$G(x',t';x,t)$ is the resolvent of the Hamiltonian

$$[E - H(x)]G(x',t';x,t) = \delta(x'-x)\delta(t'-t). \tag{8.3}$$

For time-independent Hamiltonians, G depends on t and t' only through their difference $\tau = t' - t$, in which case it is convenient to Fourier transform to the energy space. Then, $G(x',x,E)$ can be written as the matrix element $\langle x'|G(E)|x\rangle$ of the operator $G(E)$, formally defined as

$$G^{r,a}(E) = [(E \pm i0) - H]^{-1}, \tag{8.4}$$

where $\pm i0$ has to be understood as an infinitesimal imaginary part added to E, giving the retarded or advanced Green function, respectively. Inserting the identity resolution, $\sum_k |\psi_k\rangle\langle\psi_k| = 1$, in terms of the eigenstates ψ_k of H, we get the *spectral representation* of the Green function

$$G^{r,a}(E) = \sum_k \frac{|\psi_k\rangle\langle\psi_k|}{E - \varepsilon_k \pm i0}. \tag{8.5}$$

If an interaction term is added to the Hamiltonian, $H \to H + \mathcal{V}$, the Green function of the new Hamiltonian (suppressing, for abbreviation, the indexes labeling the advanced or retarded function)

$$\mathcal{G}=[(E\pm i0)-H-\mathcal{V}]^{-1}=[1-(E\pm i0-H)^{-1}\mathcal{V}]^{-1}(E\pm i0-H)^{-1}=(1-G\mathcal{V})^{-1}G, \tag{8.6}$$

is related to the old one by the Dyson equation

$$\mathcal{G}=G+G\mathcal{V}\mathcal{G}. \tag{8.7}$$

Usually $\mathcal{V}$ will be the interaction that couples the QD array to the terminals. Therefore, G and $\mathcal{G}$ will represent the Green functions of the isolated and connected device, respectively.

Let us consider in the following the electron transport through a quantum dot array under steady-state conditions only. The application of a bias voltage between the leads gives rise to a non-equilibrium situation, with the electrons of both leads being at different chemical potentials μ_L and μ_R. A great variety of effects can occur as the electrons passes throughout the system. They are subjected to interactions with the substrate, losing coherence; with phonons, producing heating and dissipation; and with each other, changing the occupation within the accessible states, even precluding their occupation (Coulomb blockade effect). We shall focus here on the regime of coherent transport where the relevant physics of interference-based devices works.

The current through the device can be calculated with the Landauer equation

$$I=\frac{2e}{h}\int dE\; T(E)\,[f_L(E)-f_R(E)], \tag{8.8}$$

where f_L and f_R are the Fermi distributions at the L and R leads [1]. At low temperatures, the transmission function represents the dimensionless conductance (in units of the quantum $e^2/2h$) and is calculated as

$$T(E)=4\mathrm{Tr}(\mathbf{\Gamma}^L\mathbf{G}^r(E)\mathbf{\Gamma}^R\mathbf{G}^a(E)), \tag{8.9}$$

where $\mathbf{G}^a$ and $\mathbf{G}^r$ are the matrix representation of the advanced and retarded Green functions of the system isolated from the terminals, and $\mathbf{\Gamma}^L$ and $\mathbf{\Gamma}^R$ are proportional to the spectral densities of the leads. The Green functions of the system connected to the leads can be determined by using Dyson equation

$$\mathcal{G}^a(E)=[\mathcal{G}^r(E)]^\dagger=[(E-i0^+)\mathbf{1}-\mathbf{H}-\mathbf{\Sigma}_L^a-\mathbf{\Sigma}_R^a]^{-1}, \tag{8.10}$$

where $\mathbf{H}$ is the matrix representation of the Hamiltonian of the central region connected to the terminals, and $\mathbf{\Sigma}_L$ and $\mathbf{\Sigma}_R$ are the self-energies, with $\mathbf{\Gamma}_{L,R}=\mathrm{Im}\mathbf{\Sigma}_{L,R}$.

In theoretical investigations, typically, two different approaches can be followed. In first-principles calculations, the confinement potential is modeled as accurately as possible, and the various effects are included into the Hamiltonian from fundamental interactions, like the electron–electron Coulomb potential. In contrast, the second approach employs phenomenological models capturing the main physical mechanisms and providing qualitative physical insight of the underlying effects. While the former is more precise and rigorously grounded, the latter gives results of a wider generality, beyond the details or the peculiarities of the system under study. Therefore, both approaches are complementary and a number of methods lie in between these pure extremes. Here, we shall consider the second approach, neglecting the spatial extension of the wave function, and considering a discretization of the Hamiltonian. The electron is assumed to be in states of given sites (QD levels) with a probability of transition between them (hopping term).

8.2.4 A Quantum Dot Molecule in an AB Interferometer

As a model for a quantum dot molecule embedded into a Aharonov–Bohm interferometer, we consider four quantum dots forming a ring and coupled to two leads L and R. Sites 2 and 4 of the ring are connected to each other forming the artificial molecule, as shown in Fig. 8.1b. We consider only one energy level in each dot and both the intradot and interdot electron–electron interactions are neglected. The system shall be described by a Hamiltonian

$$H = H_{\rm ring} + H_{\rm lead} + H_{\rm tunnel}, \tag{8.11}$$

where $H_{\rm ring}$ is the Hamiltonian of the isolated bicyclic ring,

$$H_{\rm ring} = \sum_{i=1}^{4} \varepsilon_i d_i^\dagger d_i + \sum_{i=1}^{4} t_{i,i+1}(d_i^\dagger d_{i+1} e^{i\varphi} + d_{i+1}^\dagger d_i e^{-i\varphi}) + V(d_2^\dagger d_4 + d_4^\dagger d_2) \tag{8.12}$$

$H_{\rm lead}$ is the Hamiltonian of the leads

$$H_{\rm lead} = \sum_{k,\alpha\in L,R} \varepsilon_{k\alpha} c_{k\alpha}^\dagger c_{k\alpha}, \tag{8.13}$$

and $H_{\rm tunnel}$ is the Hamiltonian describing the tunneling between the leads and the ring

$$H_{\rm tunnel} = \sum_k (V_L c_{kL} d_1^\dagger + V_R c_{kR} d_n^\dagger) + \text{H.c.}, \tag{8.14}$$

where ε_i are the on-site energy at the dots, $t_{i,i+1}$ are the nearest-neighbor hopping parameters (where $t_{45} = t_{41}$ should be understood), V is the interdot hopping that couples the upper and lower arms of the interferometer, $\varphi = 2\pi\Phi/4\Phi_0 = \pi\Phi/2\Phi_0$ is the phase $\varphi = \pi\phi/2$ acquired due to interdot hopping in a magnetic field threading the ring with a reduced flux $\phi = \Phi/\Phi_0$ (i.e., in units of the quantum $\Phi_0 = h/e$), and n is the site of contact to the right lead. The left lead is always attached to the dot 1, as shown in Fig. 8.1b; for brevity we refer to it as the connection (1,3). In experimental setup, the quantum dots 2 and 4 have a molecular coupling controlled by a plunger gate; this allows to open or to pinch-off gradually the direct connection between them. In our model, the interdot hopping V parameter is allowed to vary continuously between zero (pinch-off) and a finite value (open).

8.2.5 Properties of the Transmission Function T(E)

Some characteristics of the transmission spectrum can be derived in general terms, from the relation between the Green functions for the isolated system, G, and for the system connected to the terminals, $\mathcal{G}$. We will assume that the ring is connected to each terminal through single sites, l and r, attached to leads L and R, respectively.

We assume the wide band approximation, in which the spectral densities at the leads are energy-independent, and a symmetric coupling to the leads ($\Gamma_l = \Gamma_r = \Gamma$). Using Dyson equation (8.7), the Green function of the connected system, $\mathcal{G}$, can be written in terms of the Green function of the isolated molecule, G, as

$$\mathcal{G}_{lr} = \frac{G_{lr}}{1 - \Gamma^2(G_{ll}G_{rr} - |G_{lr}|^2) - i\Gamma(G_{ll} + G_{rr})}, \tag{8.15}$$

from which the transmission T_{lr} can be obtained:

$$T_{lr} = 4\Gamma^2|\mathcal{G}_{lr}|^2. \tag{8.16}$$

In the present tight-binding treatment of the electronic structure of QD arrays, the charge transport results from a competition between the on-site energy ε that tends to localize the electron in the dot positions, and the hopping energy t that favors the motion from a site to its nearest neighbor. As a result, the spectrum of transmission for a weakly coupled QD molecule ($\Gamma \ll t$) typically consists of narrow peaks of high (eventually perfect) conductance and narrower antiresonances (i.e., complete suppression of transmission) or dips at a discrete set of energy values, on top of a background of a smooth transmission function. The system is characterized by a set of energy eigenvalues, roughly in the range $|E - \varepsilon| \lesssim 2t$, where the Green function behaves approximately as $\mathcal{G}_{lr} \sim 1/2t$, so the transmission becomes of order $T_{lr} = 4\Gamma^2|\mathcal{G}_{lr}|^2 \sim \Gamma^2/t^2$. The expression for the connected Green function, (8.15), and its corresponding transmission, (8.16), shows that the transmission through the connected molecule depends on the electronic structure of the isolated molecule both through the diagonal Green functions (G_{ll} and G_{rr}) at the connecting sites and through the off-diagonal function G_{lr} between them. The choice of the pair (l,r) corresponds to the dependence on the topology of the connection, which we will refer to as the (l,r) connection.

8.2.6 *Resonant Zeros of the Transmission Function*

For the sake of simplicity, consider a nondegenerate energy eigenvalue E_k of the disconnected device with eigenfunction $|\psi_k\rangle$, which can be written as a linear combination of the site orbitals $|i\rangle$, $|\psi_k\rangle = \sum_i c_{ki}|i\rangle$, with $c_{ki} = \langle i|\psi_k\rangle$. If the state $|\psi_k\rangle$ has a non-vanishing weight at the connection sites l and r simultaneously, i.e. $c_{kl} \neq 0 \neq c_{kr}$, then the spectral representation

$$G_{lr}(E) = \sum_k \frac{\langle l|\psi_k\rangle\langle\psi_k|r\rangle}{E - E_k} = \sum_k \frac{c_{kl}c^*_{kr}}{E - E_k} \tag{8.17}$$

shows that E_k is a pole of G_{lr} because the term $c_{kl}c^*_{kr}/(E - E_k)$ is present in the expansion (8.17). Therefore, also the terms

$$\frac{c_{kl}c^*_{kl}}{E - E_k} \quad \text{and} \quad \frac{c_{kr}c^*_{kr}}{E - E_k} \tag{8.18}$$

will be present in the spectral representation of G_{ll} and G_{rr}, respectively, and E_k will also be a pole of them. That is, the poles of G_{lr} also become poles of G_{ll} and G_{rr} and all three Green functions G_{ij} diverge as $G_{ij}(E) \approx R_{ij}/(E - E_k)$, where $R^k_{ij} = c_{ki}c^*_{kj}$ $(i, j = l, r)$ is the residue of G_{ij} at the simple pole E_k. Therefore, the Green function of the connected ring, (8.15), can be approximated as

$$\begin{aligned} \mathcal{G}_{lr}(E) &\approx \frac{R_{lr}(E - E_k)^{-1}}{1 - \Gamma^2(R_{ll}R_{rr} - R^2_{lr})(E - E_k)^{-2} - i\Gamma(R_{ll} + R_{rr})(E - E_k)^{-1}} \\ &= \frac{R_{lr}}{(E - E_k) - \Gamma^2(R_{ll}R_{rr} - R^2_{lr})(E - E_k)^{-1} - i\Gamma(R_{ll} + R_{rr})}. \end{aligned} \tag{8.19}$$

Taking into account that $R_{ll}R_{rr} - R^2_{lr} = |\langle l|\psi_k\rangle|^2|\langle r|\psi_k\rangle|^2 - |\langle l|\psi_k\rangle\langle\psi_k|r\rangle|^2 = 0$, it reduces to

$$\mathcal{G}_{lr} \approx \frac{R_{lr}}{(E - E_k) - i\Gamma(R_{ll} + R_{rr})} \xrightarrow{E \to E_k} \frac{iR_{lr}}{\Gamma(R_{ll} + R_{rr})} \tag{8.20}$$

which shows that $\mathcal{G}$ has a pole at $E = E_k + i\Gamma(R_{ll} + R_{rr})$ and a finite transmission $T_{lr} = 4R^2_{lr}/(R_{ll} + R_{rr})^2$. The pole E_k acquires a finite width proportional to the coupling to the leads Γ. In the particular case where the sites l and r are topologically equivalent because of the symmetry of the system, $R_{ll} = R_{rr} = R_{lr}$ so that perfect transmission occurs. This is the case for the connection (1,3), because the sites $l = 1$ and $r = 3$ are equivalent when their coupling strengths to their respective terminals are the same. It should be noted that is not the case for the (1,2) connection.

On the other hand, if $E = E_k$ is a pole of G_{ll} or G_{rr}, but not of G_{lr}, the numerator $G_{lr}(E_k)$ of (8.15) is finite whilst its denominator diverges; therefore, T_{lr} will show an antiresonance at $E = E_k$. In other words, a finite transmission occurs when the

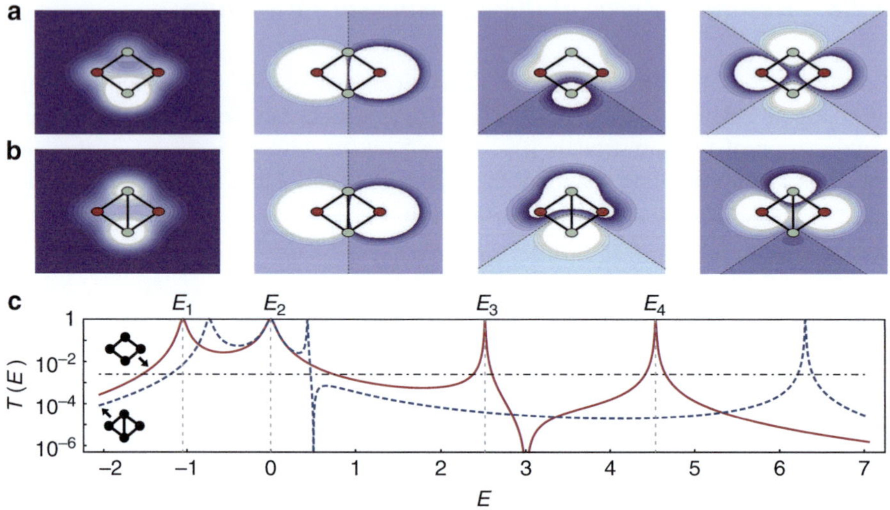

Fig. 8.2 Scheme of molecular orbitals for the 4-QDs ring: (**a**) without interarm coupling and (**b**) with interarm coupling QD2-QD4. (**c**) Transmission functions through the ring without coupling (*solid line*) and with coupling $V = 2.5$ (*dashed line*). The *horizontal dot-dashed line* shows the Γ^2/t^2 background of transmission. The *vertical dashed lines* show the position of the energy eigenvalues E_i $(i = 1,\ldots,4)$ at which perfect transmission occurs. The antiresonance is located at the energy $(\varepsilon_2 + \varepsilon_4)/2 - V$ and is due to coupling of the quantum dot molecule QD2-QD4. QD1 and QD3 are connected to the left and right leads and have on-site energies $\varepsilon_1 = \varepsilon_2 = 0$, while $\varepsilon_2 = 2$ and $\varepsilon_4 = 4$

eigenstate ψ_k of the isolated system have non-vanishing projection on the orbitals $|l\rangle$ and $|r\rangle$: $\langle l|\psi_k\rangle$ and $\langle r|\psi_k\rangle$. Reciprocally, if one of them equals zero, the electron of energy $E = E_k$ has a vanishing probability of being at both sites, and therefore no transmission can occur. Since such a vanishing of transmission occurs at the energy eigenvalues of the system, they have been named as *resonant zeros*.

Figure 8.2 depicts the scheme of the orbitals for the four-QD ring without coupling between the arms (Fig. 8.2a) and with an interarm coupling $V = 2.5$ (Fig. 8.2b). In this representation, the spatial wave function for the orbital k is given by $\sum_i c_{ki}\varphi_i(\mathbf{r})$, where φ_i are trial Gaussian functions centered at the QD locations. The dark and light regions are zones of opposite signs and the dotted curves between them are nodal lines. The transmission spectra are shown in Fig. 8.2c with solid $(V = 0)$ and dashed lines $(V = 2.5)$. As discussed above, the peaks of perfect transmission occurs at the energy eigenvalues of the ring E_i $(i = 1,\ldots, 4)$. The origin of the antiresonance at $E = (\varepsilon_2 + \varepsilon_4)/2 - V$ is the interference of the wave function along the two arms and will be discussed at Sect. 8.2.8. This allows one to change the energy at which cancellation of transmission occurs by tuning the V coupling. The energy of the second orbital is $E = 0$ independently of V because of the QD2-QD4 molecule is symmetrically coupled to terminal dots QD1 and QD3.

Therefore, it suffices to know the pole structure of the Green functions of the disconnected device to predict its transport properties when connected to the terminals. At the energies of the poles of G_{lr} there exists peaks of finite (or even perfect) transmission, while at the energies at which G_{ll} or G_{rr} do have poles (but G_{lr} do not), the electron transmission is forbidden.

8.2.7 Transmission Pathways

Spatially discretized models with hopping terms, allowing transitions from a site to its neighbor, like the present one, suggest the appealing interpretation of an electron successively jumping along a path. This interpretation allows one to recover the image of electrons moving along trajectories through the device. A similar approach has been used in an analytical treatment of quantum interference in a benzene ring [48].

Partitioning of a Hilbert space is usually employed for isolating the effects on the part of interest from the rest of the system [49]. Here, we apply a partitioning technique to define spatial transmission pathways. The 4×4 Hamiltonian can be partitioned in terms of 2×2 matrices as follows:

$$H = \begin{pmatrix} H^P & U \\ U^\dagger & H^Q \end{pmatrix}, \tag{8.21}$$

where $H^P = PHP$ is the part of the Hamiltonian projected on the subspace of orbitals centered on the sites of connection 1 and n, where $P = |1\rangle\langle 1| + |n\rangle\langle n|$, whilst H^Q is the projection of H on the complementary subspace $Q = 1 - P$. Matrices U and $U^\dagger$ contain matrix elements connecting states belonging to P and Q. The Green function can be obtained by block matrix decomposition

$$g = (E - H)^{-1} = \begin{pmatrix} E - H^P & -U \\ -U^\dagger & E - H^Q \end{pmatrix}^{-1}. \tag{8.22}$$

We are interested here in the Green function projected on the subspace of the connection sites, i.e., its P-block. Hence, g^P can be obtained from the inverse of the Schur complement of the Q-block, $E - H^Q$,

$$g^P = [E - H^P - U(E - H^Q)^{-1}U^\dagger]^{-1} = (E - H^P - Ug^QU^\dagger)^{-1}, \tag{8.23}$$

from which an effective Hamiltonian can be defined as

$$H_{\text{eff}} = E - (g^P)^{-1} = H^P + Ug^QU^\dagger = H^P + \Sigma, \tag{8.24}$$

where $\Sigma = Ug^QU^\dagger$ is the self-energy that contains the interactions involving the orbitals not connected to the leads.

The Green function written in terms of the self-energy,

$$g^P(E) = \frac{1}{\Delta}\begin{pmatrix} E-\varepsilon_n-\Sigma_{nn} & t_{1n}+\Sigma_{1n} \\ t_{n1}+\Sigma_{n1} & E-\varepsilon_1-\Sigma_{11} \end{pmatrix}, \tag{8.25}$$

with $\Delta = \det(E-H_{\text{eff}}) = (E-\varepsilon_1-\Sigma_{11})(E-\varepsilon_n-\Sigma_{nn}) - |\Sigma_{1n}|^2$, contains all the matrix elements needed for the calculation of the conductance between the sites 1 and n. Their poles are given by the zeros of the secular determinant $\Delta = 0$, while the antiresonances come from zeros in $g_{1n} = (t_{1n}+\Sigma_{1n})/\Delta$. When $t_{1n} = 0$, as in the (1,3) connection, the antiresonances arise from the zeros of Σ_{1n}. Interestingly, the self-energy $\Sigma_{ij} = \Sigma^A_{ij} + \Sigma^B_{ij} + \Sigma^C_{ij}$ becomes a sum of contributions throughout the paths from above (A), from below (B), and through the interarm coupling (C). The vanishing of the transmission occurs when the contributions from those paths interfere destructively, thus cancelling the element $t_{1n}+\Sigma_{1n}$ of the effective Hamiltonian.

For the (1,3) connection, the contributions become

$$\Sigma^{\text{A}}_{13} = t_{12}g_{22}t_{23}e^{2i\varphi}, \quad \Sigma^{\text{B}}_{13} = t_{41}g_{44}t_{34}e^{-2i\varphi}, \quad \Sigma^{\text{C}}_{13} = g_{24}(t_{41}t_{23}+t_{12}t_{34}), \tag{8.26}$$

where

$$g_{22} = (E-\varepsilon_4)/D, \qquad g_{44} = (E-\varepsilon_2)/D, \qquad g_{24} = V/D, \tag{8.27}$$

and $D = (E-\varepsilon_2)(E-\varepsilon_4) - V^2 = (E-E_a)(E-E_b)$, with E_a and E_b being the energies of the bonding (E_b) and antibonding (E_a) orbitals of the molecule QD2-QD4 described by H^Q.

8.2.8 *Antiresonances from Interfering Pathways*

As we discussed above antiresonances occur at the energy eigenvalues of the molecule. Nevertheless, (8.15) shows that also the roots of G_{lr} can lead to zeros of conductance that are termed *multipath zeros* [48].

The Green function of the isolated system can be written as the resolvent $(E - H_{\text{eff}})^{-1}$ of an effective 2×2 Hamiltonian in the P-space, in terms of the self-energy $\Sigma(E)$ as [48, 50]

$$G(E) = \frac{1}{\Delta}\begin{pmatrix} E-\varepsilon_r-\Sigma_{rr} & t_{lr}+\Sigma_{lr} \\ t_{rl}+\Sigma_{rl} & E-\varepsilon_l-\Sigma_{ll} \end{pmatrix}, \tag{8.28}$$

where $\Delta(E) = \det(E - H_{\text{eff}}) = (E - \varepsilon_r - \Sigma_{rr})(E - \varepsilon_l - \Sigma_{ll}) - |t_{lr} + \Sigma_{lr}|^2$. The energy-dependent diagonal and off-diagonal elements of the self-energy are interpreted here as effective or dressed on-site energies and hoppings between the sites l and r, respectively. Now assume that sites l and r are coupled to the rest of the molecule through sites l', l'', r' and r'' through identical hoppings.

Then,

$$G_{lr} = \frac{t_{lr} + \Sigma_{lr}}{\Delta}, \tag{8.29}$$

where $t_{lr} = t$ if l and r are adjacent to each other, like in the (1,2)-connection, and zero otherwise.

In our four-QD ring with (1,3) connection, the P-space of the connected sites is formed by the states $\{|1\rangle, |3\rangle\}$ centered at QD1 and QD3. The Q-space of the rest of the system is $\{|2\rangle, |4\rangle\}$. Thus, for the (1,3)-connection, $G_{13} = \Sigma_{13}/\Delta$, with

$$\Sigma_{13} = (g_{22} + g_{44} + g_{24} + g_{42})t^2, \tag{8.30}$$

where g refers to the Q-block of the Green function G. This gives Σ_{13} as the sum of pathways defined from above: $\Sigma_{13}^A = g_{22}t^2$, below: $\Sigma_{13}^B = g_{44}t^2$, and through the coupling: $\Sigma_{13}^C = 2g_{24}t^2$, and the Green function has the form

$$g = \begin{pmatrix} g^A & g^C \\ g^{C\dagger} & g^B \end{pmatrix}. \tag{8.31}$$

From (8.27) and (8.30), it can be seen that $\Sigma_{13} = 0$ when $E = (\varepsilon_2 + \varepsilon_4)/2 - V$, thus explaining the antiresonance of Fig. 8.2c.

When the molecule has two paths disconnected from each other, i.e., two disjoint sets of sites, as in a ring, the Green function in the Q-space becomes blocked

$$g = \begin{pmatrix} g^A & 0 \\ 0 & g^B \end{pmatrix}, \tag{8.32}$$

so that, $g_{24} = g_{42} = 0$, and there are only two contributing pathways A (QD1-QD2-QD3) and B (QD1-QD4-QD3):

$$\Sigma_{13} = (g_{22} + g_{44})t^2. \tag{8.33}$$

Then, a zero of transmission coming from $G_{13} = 0$ can be attributed to a cancellation of the two contributions of the dynamical phase solely, one from each path: $g_{22} + g_{44} = 0$.

8.3 Transmission Through Laterally Coupled QDs

Let us analyze in this section some numerical calculations for a four-QD ring with on-site energies $\varepsilon_1 = \varepsilon_3 = 0$, $\varepsilon_2 = 2$ and $\varepsilon_4 = 4$ in arbitrary units. The connecting dots are assumed to be weakly coupled to the leads by $\Gamma = 0.05$. Figure 8.3 shows the dimensionless conductance $T(E)$ for the ring connected to the leads in the configuration (1,3) (solid line), and for the three-site chains forming the upper (dotted line) and lower (dashed line) arms with the side-dot QD4 and QD2, respectively, coupled by V. The transmission with an applied magnetic flux $\Phi = 0.1\Phi_0$ is shown in dot-dashed line. The transmission shows four peaks at the energy eigenvalues of the rings. The ring with disconnected arms, Fig. 8.3a, shows also an antiresonance, not present in the chains because, for $V = 0$, there is no side-coupled dot. This suppression of the transmission in the ring is due to the cancellation of the contributions to the self-energy throughout the upper and lower paths ($\Sigma_{13} = \Sigma_{13}^{A} + \Sigma_{13}^{B} = 0$). In general, (8.26) shows that the self-energy vanishes if

$$\Sigma_{13} = t^2(g_{22}e^{2i\varphi} + g_{44}e^{-2i\varphi} + 2g_{24}) = \frac{t^2}{D}\left[(E-\varepsilon_4)e^{2i\varphi} + (E-\varepsilon_2)e^{-2i\varphi} + 2V\right] = 0 \tag{8.34}$$

Figure 8.3a and b depict the tuning of the antiresonance with $V = 0$ and $V = (\varepsilon_2 - \varepsilon_4)/2$, respectively. Since, as mentioned above, in absence of magnetic flux ($\varphi = 0$), Σ_{13} vanishes at the energy $E = \bar{\varepsilon} - V = (\varepsilon_2 + \varepsilon_4)/2 - V$. When there is a finite magnetic flux, Σ_{13} is complex and its cancellation requires vanishing its real and imaginary parts, i.e., $(2E - \varepsilon_2 - \varepsilon_4)\cos 2\varphi + 2V = 0$, and $(\varepsilon_2 - \varepsilon_4)\sin 2\varphi = 0$. Both equations cannot be satisfied simultaneously, except when $\varepsilon_2 = \varepsilon_4$, which presents an antiresonance at $\varepsilon_2 - V/\cos 2\varphi$. Therefore, the magnetic field eliminates the antiresonance for a ring with different site energies $\varepsilon_2 \neq \varepsilon_4$ for arbitrary V. This suppression of the antiresonance is shown with dot-dashed lines in Fig. 8.3 for a flux $\Phi = 0.1\Phi_0$.

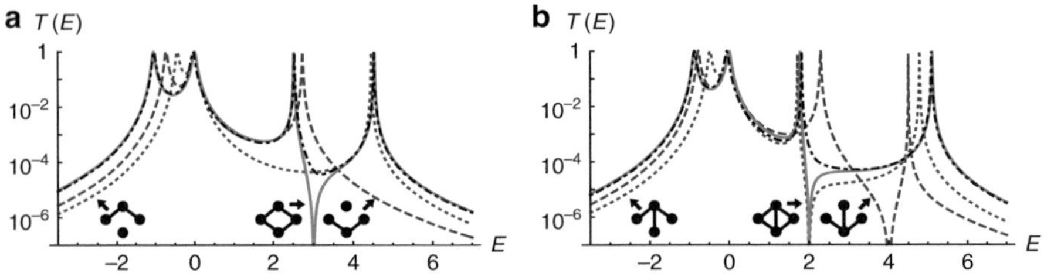

Fig. 8.3 Transmission function of a four-QD ring in the connection (1,3) in absence of magnetic field (*solid line*) and with an applied magnetic flux $\Phi = 0.1\Phi_0$ (*dot-dashed line*). For comparison, transmission for the three-site chains QD1-QD2-QD3 (*upper arm*) and QD1-QD4-QD3 (*lower arm*) is shown in *dotted and dashed lines*, respectively. (**a**) Without interarm coupling and (**b**) with coupling $V = 1$ between the arms through QD2-QD4. The *insets* show the scheme of the chains pointing to the respective curve

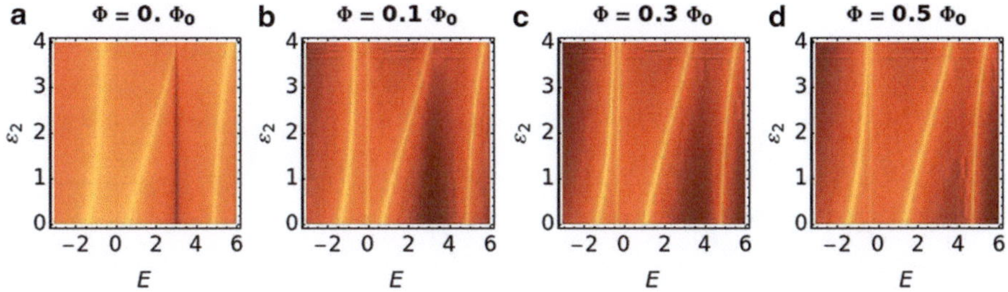

Fig. 8.4 The logarithm of the transmission function of a ring with coupling between its arms in the configuration of connection (1,2), i.e., one of the dots of the molecule (QD2) is connected to the contacting lead R. Electron energy E and on-site energy ε_2 are represented in the horizontal and vertical axes. All interdot hopping parameters are the same, $t_{12} = t_{23} = t_{34} = t_{41} = V = 1$. *Bright lines and dark regions* represent zones of high and low transmission, respectively

Application of gate potentials at dots 2 and 4 allows to tune their on-site energies ε_2 and ε_4. Figure 8.4 shows the effect on the transmission T_{12} of varying ε_2 and the electron energy E with and without magnetic flux for the connection (1,2). Bright lines and dark regions represent zones of high and low transmission, respectively. In absence of magnetic field (Fig. 8.4a) three peaks of conductance are clearly visible as depicted by the bright curves. The central peak has a marked linear dependence on ε_2 while the external ones are weakly dependent on it, as seen from the slope of the curves. Two antiresonances are also visible, namely, a faint vertical dark line at $E = 0$ independent on ε_2, and a vertical dark thicker line at $E = \varepsilon_4 - V$. Figure 8.4b–d shows the effect of switching on and increasing the magnetic flux. As discussed for the connection (1,3), the antiresonances are cancelled out. In particular, it should be noted that the antiresonance at $E = 0$ turned into a peak of transmission. Also the second antiresonance at $E = \varepsilon_4 - V$ becomes weakened at low flux ($\Phi = 0.1\Phi_0$) and the dark sharp straight line turns into a diffuse dark region of low transmission (Fig. 8.4b and c), then increasing at a larger magnetic flux ($\Phi = 0.5\Phi_0$).

8.3.1 Tuning Fano Profiles with a Magnetic Field

Figure 8.5 shows in solid lines the transmission through an asymmetric ring where the upper arm has hoppings much smaller than those of the lower arm ($t_{12} = t_{23} \ll t_{34} = t_{41}$ in Fig. 8.1b). The dotted blue and dashed red lines are the transmissions of the upper and lower arms separately, calculated as three-site chains with a central site having energy ε_2 and ε_4, respectively, and lateral sites having on-site energies $\varepsilon = 0$.

Figure 8.5a–c represents in solid line the transmission function of the four-QD ring without coupling between the arms ($V = 0$) for $\Phi = 0$, $\Phi_0/4$, and $\Phi_0/2$, respectively. The dotted and dashed lines show $T(E)$ for the upper and lower arm, respectively, as sketched in the inset schemes. They are the same for panels (a)–(c)

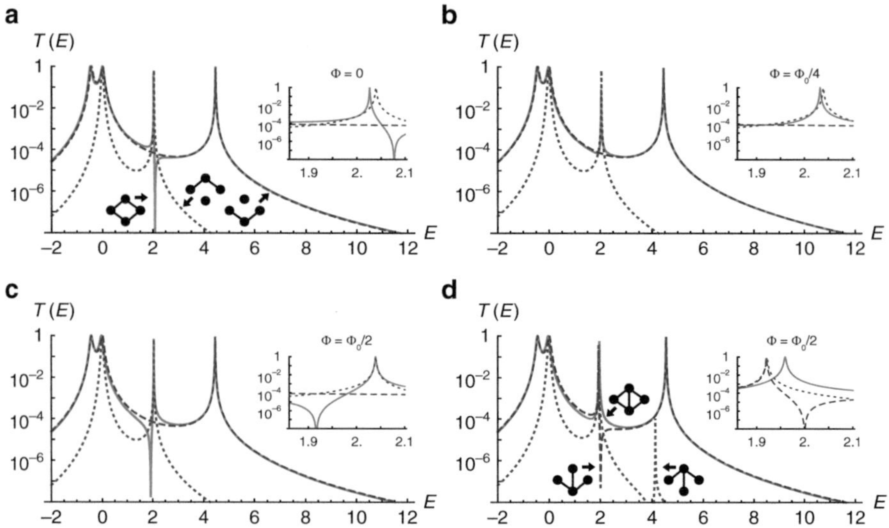

Fig. 8.5 Transmission for the connection (1,3) as a function of the Fermi energy for an asymmetric ring with hopping parameters $t_{12} = t_{23} = 0.2$ (*upper arm*), $t_{34} = t_{41} = 1$ (*lower arm*), and on-site energies $\varepsilon_2 = 2$ and $\varepsilon_2 = 4$ (*solid line*), as compared to the transmissions through the upper arm only (*dotted blue line*), and throughout the lower arm only (*dashed red line*). Figure (**a**)–(**c**) corresponds to a decoupled ring ($V = 0$) (**a**) at zero magnetic field, (**b**) at $\Phi = \Phi_0/4$, and (**c**) magnetic flux $\Phi = \Phi_0/2$. Figure (**d**) has an interarm coupling $V = 0.5$ and magnetic flux $\Phi = \Phi_0/2$. The *insets* show in more detail the behavior of the curves around the resonance

because they do not enclose any magnetic flux and, therefore, are not affected by Φ. In panel (d), the connection between the arms was switched on ($V = 0.5$), which affects the position of the peaks and dips.

In Fig. 8.5, the four peaks of conductance, corresponding to connection (1,3), can be recognized as those from the lower arm along with the on-site energy of site 2. The transmission of the ring nearly coincides in almost the whole range with that of the lower arm, thus showing that the conduction is throughout such a pathway at almost every energy, except for $E \approx \varepsilon_2$. When the incident electron is resonant with the site 2, the transmission is well described by the resonant peak of the upper pathway. Nevertheless, none of the paths by themselves can provide the onset of the antiresonance close to $E \approx \varepsilon_2$. The self-energy $\Sigma_{13} = \Sigma_{13}^{A} + \Sigma_{13}^{B} \approx \Sigma_{13}^{B}$ because $\Sigma_{13}^{A} = t_{\rm up}^2/(E - \varepsilon_2) \ll \Sigma_{13}^{B} = t_{\rm down}^2/(E - \varepsilon_4)$, except for $E \approx \varepsilon_2$ when they can become comparable. Hence, in a neighborhood of $E \approx \varepsilon_2$, the self-energy can be approximated as $\Sigma_{13} = t_{\rm up}^2/(E - \varepsilon_2) + t_{\rm down}^2/(\varepsilon_2 - \varepsilon_4)$ which vanishes for $E = \varepsilon_2 + t_{\rm up}^2(\varepsilon_4 - \varepsilon_2)/t_{\rm down}^2$, that is, slightly to the right of the peak $E = \varepsilon_2$, thus giving the Fano-like profile.

The Fano-like peak shows the signature of the interference between both paths, typical when a localized state interferes with a continuum. Figure 8.5a–c shows the dependence of the Fano profile with the magnetic flux. With no magnetic flux, there is the above discussed Fano resonance due to the path interference; the application of a flux $0 < \Phi < \Phi_0/4$ suppresses the antiresonance leaving only a dip in the transmission which also disappears at $\Phi = \Phi_0/4$ leaving only the resonant peak, as seen in Fig. 8.5b. Further increase of the flux in the range $\Phi_0/4 < \Phi < \Phi_0/2$ produces a new dip at an energy slightly smaller than ε_2 moving the peak to slightly higher energies. At $\Phi = \Phi_0/2$ the dip in the transmission of the ring becomes an antiresonance, with the resonant peak tuned with that of the upper chain as seen in the inset of Fig. 8.5c. Between $\Phi_0/2$ and Φ_0, the behavior of $T(E)$ is reversed, such that a cycle is completed in a period of Φ_0. Figure 8.5d depicts the transmission through the ring with an interarm coupling $V = 0.5$. Now the transmission through the lower (dashed line) and upper (dotted line) arms, including the site laterally coupled by V, shows Fano-like resonances at $\varepsilon = \varepsilon_2$ and at $\varepsilon = \varepsilon_4$, respectively. This can be understood as the electron propagation along a linear chain between macroscopic leads (continuous state) with a laterally coupled dot (discrete state). The transmission through the ring (solid line) still remains close to that of the lower arm with a lateral connection to site 2. Then, the overall picture for the transmission through a ring with different connection strengths along each arm is that of the transmission throughout the stronger pathway (i.e., the one with larger hoppings) at almost every energy, except at the one resonant with the energy of the site connecting the arms where a Fano interference occurs.

In the presence of a magnetic flux, the self-energy is a complex quantity and its modulus should be considered. The self-energy contributions (8.26) show that the flux Φ introduces a phase $\pm 2\varphi$ in the paths throughout the arms while no phase change occurs in the self-energy corresponding to the interarm coupling. Let us call $\Sigma_{13}(\varphi)$ the self-energy with magnetic field. Then, $\Sigma_{13}(\varphi) = \Sigma_{13}^{A} e^{2i\varphi} + \Sigma_{13}^{B} e^{-2i\varphi} + \Sigma_{13}^{C}$, where $\Sigma_{13}^{A,B,C}$ are the real self-energies at zero magnetic field, such that

$$|\Sigma_{13}(\varphi)|^2 = (\Sigma_{13}^{A})^2 + (\Sigma_{13}^{B})^2 + (\Sigma_{13}^{C})^2 + 2\Sigma_{13}^{A}\Sigma_{13}^{B}\cos 4\varphi + 2(\Sigma_{13}^{A} + \Sigma_{13}^{B})\Sigma_{13}^{C}\cos 2\varphi, \tag{8.35}$$

where the first three terms represent the non-interfering transmission along the paths A, B, and C. The last two terms contain the effect of the interference due to the quantum and magnetic phases. Even for $\varphi = 0$ there is an interference between the path contributions to the self-energy. Interestingly, there are two periods in the magnetic phase; a period $\Phi = \Phi_0$ (associated with $\cos 4\varphi$) and a period $\Phi = 2\Phi_0$ (associated with $\cos 2\varphi$). When there is no interarm coupling, $\Sigma_{13}^{C} = 0$, the latter is not present. On the other hand, as soon as a finite V exists, the self-energy acquires the longer period modulated by the shorter one. Such a behavior has been observed in experiments [6] and were termed as Fano resonances of the *big* and *small orbits*.

Figure 8.6a shows the conductance in the (1,3) connection, with the hopping t_{23} set equal to zero, such that there is a single small orbit. By setting $t_{23} = 1$ and $V = 0$, one has a big orbit enclosing twice the flux threading the small one, whose transmission is shown in Fig 8.6b. For a given energy E, the passage from the former to the latter can be realized by decreasing the interarm coupling at the time that the

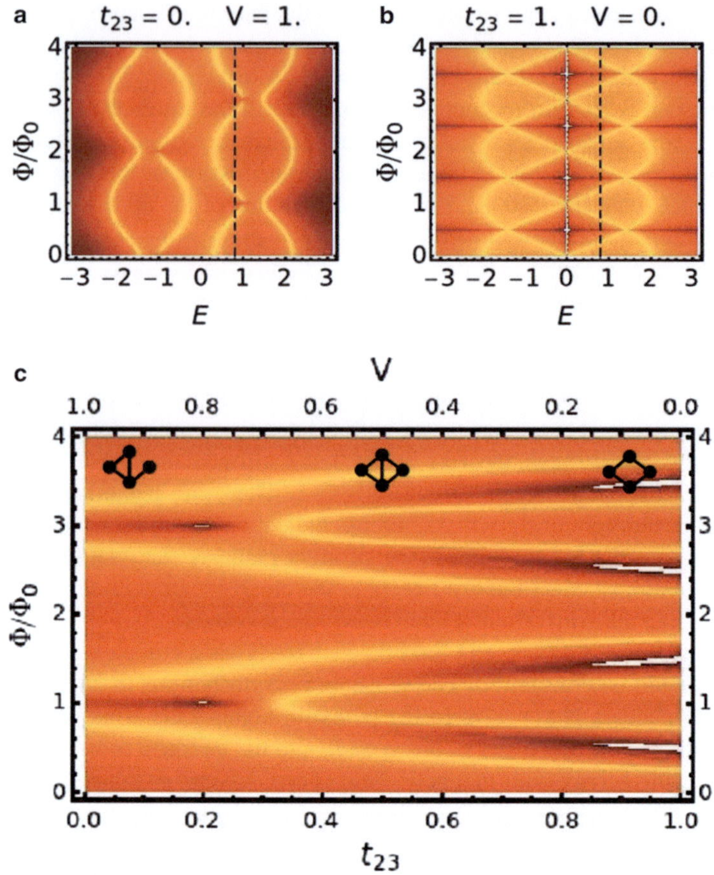

Fig. 8.6 Transmission function $T_{13}(E,\Phi)$ for (**a**) the small orbit with $t_{23}=0$ and $V=1$ and (**b**) the big orbit with $t_{23}=1$ and $V=0$. (**c**) Transmission $T_{13}(V,\Phi)$ along the cut $E=0.8$ is shown by the *vertical dashed lines* in (**a**) and (**b**). The *left and right edges* of (**c**) correspond to the transmission throughout the small and large orbits, respectively, as shown in the *insets*. The transition from one to the other is performed by varying the interarm coupling along $V=1-t_{23}$. *Bright lines and dark regions* represent zones of high and low transmission, respectively. The transmission $T(V,\Phi)$ along the left edge ($V=1$) shows a period of $2\Phi_0$ while along the right edge ($V=0$) has a period of Φ_0

coupling between QD2 and QD3 (t_{23}) increases as $V=1-t_{23}$, shown in Fig. 8.6c. The transmission through the small orbit (Fig. 8.6a), for $V=1$, shows a period $2\Phi_0$ in the flux. In the conductance through the large orbit (Fig. 8.6b), for $V=0$, the period Φ_0 becomes apparent. Finally, the bottom panel (c) shows the variation of the transmission at $E=0.8$ as a function of V and Φ. The smooth transition from Fig. 8.6a to b is realized by interpolating along $V=1-t_{23}$, in such a way that as V increases t_{23} decreases, and reciprocally.

As a final example consider the (1,2) configuration with the lead L attached to QD1 and lead R connected to QD2. There are three interfering paths, namely, the direct path through sites 1 and 2 (t_{12}), the path Σ_{12}^{C} ($1 \to 4 \to 2$) and the path Σ_{12}^{B} ($1 \to 4 \to 3 \to 2$), contributing to the self-energy $\Sigma_{12}(\varphi) = t_{12}e^{i\varphi} + Vt_{14}g_{44}e^{-i\varphi} + t_{14}g_{43}t_{31}e^{-3i\varphi}$. The absolute square of the self-energy is

$$\Sigma_{12} = t_{12}^2 + (\Sigma_{12}^{B})^2 + (\Sigma_{12}^{C})^2 + 2(t_{12} + \Sigma_{12}^{B})\Sigma_{12}^{C}\cos 2\varphi + 2t_{12}\Sigma_{12}^{B}\cos 4\varphi, \quad (8.36)$$

where the phase of the big orbit (4φ) characterizes the interference between the pathways along the arms of the ring, whilst the phase of the small orbit (2φ) corresponds to the interference between them and the molecular bond.

8.4 Concluding Remarks

In this chapter, we have discussed two quantum interference-based phenomena, namely, the Fano and AB effects. It has been argued that they have different origins. The former originates in the external gate potentials used to produce and couple the QDs, while the latter arises from an applied magnetic field perpendicular to the heterostructure when a configuration of two or more alternative electron paths there exists.

The Fano effect produces a typical asymmetric profile in the transmission coefficient as a function of the electron energy. It is characterized by a peak and an antiresonance in the conductance at two close energies. Using the Green function formalism for the electron transport together with a discrete Hamiltonian description, the conductance can be analyzed from the Green function of the system disconnected from the source and drain terminals. The resonances in the transmission occur at the poles of the Green functions between QD sites connected to the source and drain, respectively. In particular, one type of antiresonance is due to the cancellation of contributions to the self-energy along alternative transmission paths. Including QDs in such paths, like in an AB interferometer, offers interesting prospects for technological applications, e.g., by encoding those cancellations of conductance as quantum bits.

The application of a magnetic field precludes the occurring of Fano antiresonances, because the magnetic phase induced by the AB effect, prevents the complete cancellation of the electron transmission. This phase depends on the magnetic flux enclosed by a given loop path and, therefore, on the closed trajectories available to the electrons. Hence, this suppression of the conductance cancellation becomes strongly sensitive to the formation of QDs molecular states due to their couplings.

Acknowledgments This work was partly supported by SGCyT (Universidad Nacional del Nordeste), National Agency ANPCYT and CONICET (Argentina) under grants PI F007/11, PICTO-UNNE 204/07 and PIP 11220090100654/2010.

References

1. Imry, Y.: Introduction to Mesoscopic Physics, 2nd edn. Oxford University Press, Oxford (2002)
2. Di Ventra, M.: Electrical Transport in Nanoscale Systems. Cambridge University Press, Cambridge (2008)
3. Aronov, A.G., Shavin, Yu.V.: Magnetic flux effects in disordered conductors. Rev. Mod. Phys. **59**, 755 (1987)
4. Hod, O., Baer, R., Rabani, E.: Magnetoresistance of nanoscale molecular devices based on Aharonov-Bohm interferometry. J. Phys. Condens. Matter **20**, 383201 (2008)
5. Barnham, K., Vvedensky, D. (eds.): Low-Dimensional Semiconductor Structures. Fundamentals and Device Applications. Cambridge University Press, Cambridge (2001)
6. Ihn, T., Sigrist, M., Ensslin, K., Wegscheider, W., Reinwald, M.: Interference in a quantum dot molecule embedded in a ring interferometer. New J. Phys. **9**, 111 (2007)
7. Aharonov, Y., Bohm, D.: Significance of electromagnetic potentials in the quantum theory. Phys. Rev. **115**, 485 (1959)
8. Nazarov, Y.V., Blanter, Y.M.: Quantum Transport - Introduction to Nanoscience. Cambridge University Press, Cambridge (2009)
9. Yacobi, A., Heiblum, M., Mahalu, D., Shtrikman, H.: Coherence and phase sensitive measurements in a quantum dot. Phys. Rev. Lett. **74**, 4047 (1995)
10. Sigrist, M., Fuhrer. A., Ihn, T., Ensslin, K., Ulloa, S.E., Wegscheider, W., Bichler, M.: Magnetic-field-dependent transmission phase of a double-dot system in a quantum ring. Phys. Rev. Lett. **93**, 066802 (2004)
11. Hackenbroich, G.: Phase coherent transmission through interacting mesoscopic systems. Phys. Rep. **343**, 463 (2001)
12. Hod, O., Baer, R., Rabani, E.: Feasible nanometric magnetoresistance devices. J. Phys. Chem. B **108**, 14807 (2004)
13. Hod, O., Baer, R., Rabani, E.: Inelastic effects in Aharonov-Bohm molecular interferometers. Phys. Rev. Lett. **97**, 266803 (2006)
14. Sigrist, M., Ihn, T., Ensslin, K., Loss, D., Reinwald, M., Wegscheider, W.: Phase coherence in the inelastic cotunneling regime. Phys. Rev. Lett. **96**, 036804 (2006)
15. Büttiker, M., Imry, Y., Azbel, M.Ya.: Quantum oscillations in one-dimensional normal-metal rings. Phys. Rev. A **30**, 1982 (1984)
16. Taniguchi, T., Büttker, M.: Friedel phases and phases of transmission amplitudes in quantum scattering systems. Phys. Rev. B **60**, 13814 (1999)
17. Levy Yeyati, A., Büttiker, M.: Scattering phases in quantum dots: An analysis based on lattice models. Phys. Rev. B **62**, 7307 (2000)
18. Ladrón de Guevara, M.L., Claro, F., Orellana, P.A.: Ghost Fano resonance in a double quantum dot molecule attached to leads. Phys. Rev. **67**, 195335 (2003)
19. Ladrón de Guevara, M.L., Orellana, P.A.: Electronic transport through a parallel-coupled triple quantum dot molecule: Fano resonances and bound states in the continuum. Phys. Rev. **73**, 205303 (2006)
20. Gómez, I., Domínguez-Adame, F., Orellana, P.: Fano-like resonances in three-quantum-dot Aharonov-Bohm rings. J. Phys. Condens. Matter **16**, 1613 (2004)
21. Hedin, E.R., Joe, Y.S., Satanin, A.M.: Resonance and phase shift in an open Aharonov-Bohm ring with an embedded quantum dot. J. Phys. Condens. Matter **21**, 015303 (2009)
22. An, X.-T., Liu, J.-J.: Aharonov-Bohm ring with a side-coupled quantum dot array as a spin switch. Appl. Phys. Lett. **96**, 223508 (2010)
23. Ladrón de Guevara, M.L., Lara, G.A., Orellana, P.A.: Quantum interference effects in two double quantum dots-molecules embedded in an Aharonov-Bohm ring. Phys. E **42**, 1637 (2010)

24. Rai, D., Hod, O., Nitzan, A.: Circular currents in molecular wires. J. Phys. Chem. C **114**, 20583 (2010)
25. Rai, D., Hod, O., Nitzan, A.: Magnetic field control of the current through molecular ring junctions. J. Phys. Chem. Lett. **2**, 2118 (2011)
26. Akera, H.: Aharonov-Bohm effect and electron correlation in quantum dots. Phys. Rev. B **47**, 6835 (1993)
27. Izumida, W., Sakai, O., Shimizu, Y.: Many body effects on electron tunneling through quantum dots in an AB circuit. J. Phys. Soc. Jpn. **66**, 717 (1997)
28. Aharony, A., Entin-Wohlman, O.: Measuring the Kondo effect in the Aharonov-Bohm interferometer. Phys. Rev. B **72**, 073311 (2005)
29. Kang, K., Cho, S.Y.: Tunable molecular resonances of a double quantum dot Aharonov-Bohm interferometer. J. Phys. Condens. Matter **16**, 117 (2004)
30. Recher, P., Trauzettel, B., Rycerz, A., Blanter, Ya.M., Beenakker, C.W.J., Morpurgo, A.F.: Aharonov-Bohm effect and broken valley degeneracy in graphene rings. Phys. Rev. B **76**, 235404 (2007)
31. Rai, D., Hod, O., Nitzan, A.: Magnetic fields effects on the electronic conduction properties of molecular ring structures. Phys. Rev. B **85**, 155440 (2012)
32. Buchholz, S.S., Fischer, S.F., Kunze, U., Reuter, D., Wieck, A.D.: Nonlocal Aharonov-Bohm conductance oscillations in an asymmetric quantum ring. Appl. Phys. Lett. **94**, 022107 (2009)
33. Miyamoto, S., Moutanabbir, O., Ishikawa, T., Eto, M., Haller, E.E., Sawano, K., Shiraki, Y., Itoh, K.M.: Excitonic Aharonov-Bohm effect in isotopically pure ^{70}Ge/Si self-assembled type-II quantum dots. Phys. Rev. B **82**, 073306 (2010)
34. Neder, I., Heiblum, M., Levinson, Y., Mahalu, D., Umansky, V.: Unexpected behavior in a two-path electron interferometer. Phys. Rev. Lett. **96**, 016804 (2006)
35. Holleitner, A.W., Qina, H., Blicka, R.H., Eberlb, K., Kotthausa, J.P.: Aharonov-Bohm oscillations for charge transport through two parallel quantum dots. Phys. E **12**, 774 (2002)
36. Smirnov, D., Schmidt, H., Haug, R.J.: Aharonov-Bohm effect in an electron–hole graphene ring system. Appl. Phys. Lett. **100**, 203114 (2012)
37. Russo, S., Oostinga, J.B., Wehenkel, D., Heersche, H.B., Sobhani, S.S.: Observation of Aharonov-Bohm conductance oscillations in a graphene ring. Phys. Rev. B **77**, 085413 (2008)
38. Huefner, M., Molitor, F., Jacobsen, A., Pioda, A., Stampfer, C., Ensslin, K., Ihn, T.: The Aharonov-Bohm effect in a side-gated graphene ring. New J. Phys. **12**, 043054 (2010)
39. Yoo, J.S., Park, Y.W., Skákalová, V., Roth, S.: Shubnikov-de Haas and Aharonov Bohm effects in a graphene nanoring structure. Appl. Phys. Lett. **96**, 143112 (2010)
40. Hatano, T., Kubo, T., Tokura, Y., Amaha, S., Teraoka, S., Tarucha, S.: Aharonov-Bohm oscillations changed by indirect interdot tunneling via electrodes in parallel-coupled vertical double quantum dots. Phys. Rev. Lett. **106**, 076801 (2011)
41. Sigrist, M., Ihn, T., Ensslina, K., Reinwald, M., Wegscheider, W.: Is inelastic cotunneling phase coherent? J. Appl. Phys. **101**, 081701 (2007)
42. Mühle, A., Wegscheider, W., Haug, R.J.: Quantum dots formed in a GaAs/AlGaAs quantum ring. Appl. Phys. Lett. **92**, 013126 (2008)
43. Neder, I., Ofek, N., Chung, Y., Heiblum, M., Mahalu, D., Umansky, V.: Interference between two indistinguishable electrons from independent sources. Nature **448**, 333 (2007)
44. Fano, U.: Effects of configuration interactions of intensities and phase shifts. Phys. Rev. **124**, 1866 (1961)
45. Kobayashi, K., Aikawa, H., Katsumoto, S., Iye, Y.: Tuning of the Fano effect through a quantum dot in an Aharonov-Bohm interferometer. Phys. Rev. Lett. **88**, 256806 (2002)
46. Fuhrer, A., Brusheim, P., Ihn, T., Sigrist, M., Ensslin, K., Wegscheider, W., Bichler, M.: Fano effect in a quantum-ring-quantum-dot system with tunable coupling. Phys. Rev. B **73**, 205326 (2006)

47. Miroshnichenko, A.E., Flach, S., Kivshar, Y.S.: Fano resonances in nanoscale structures. Rev. Mod. Phys. **82**, 2257 (2010) and references therein
48. Hansen, T., Solomon, G.C., Andrews, D.Q., Ratner, M.A.: Interfering pathways in benzene: An analytical treatment. J. Chem. Phys. **131**, 194704 (2009)
49. Löwdin, P.O.: Studies in perturbation theory. IV. Solution of eigenvalue problem by projection operator formalism. J. Math. Phys. **3**, 969 (1962)
50. Lovey, D.A., Gomez, S.S., Romero, R.H.: Transmission through a quantum dot molecule embedded in an Aharonov-Bohm interferometer. J. Phys. Condens. Matter **23**, 425303 (2011)

Chapter 9
Phonon-Assisted Processes and Spontaneous Emission in Double Quantum Dots

Paweł Karwat, Krzysztof Gawarecki, Katarzyna Roszak, Anna Sitek, and Paweł Machnikowski

Abstract We describe, from a theoretical point of view, phonon-induced dephasing and relaxation processes in double quantum dots, the specific features of the spontaneous emission in these systems, and the interplay of these two classes of physical processes. First, we discuss the phonon-assisted transfer of electron, hole and exciton occupation in coupled dots, the dephasing of spatially delocalized superpositions, and the phonon-induced decay of entanglement. Second, we study the signatures of collective emission in the linear absorption and the four-wave mixing optical response of these systems, as well as in time-resolved luminescence, where collective effects are manifested by long-living tails in the emission due to the vacuum-induced coherence effect. Finally, we analyze the interplay of phonon-related processes and spontaneous emission. We show that occupation transfer between bright and dark states in the ground state doublet of a double-dot structure can lead to a temperature dependence of the spontaneous emission rate, which may result in a non-monotonic temperature dependence of the luminescence decay rate.

9.1 Introduction

Nanostructures composed of two closely spaced semiconductor quantum dots (QDs) have attracted much attention in recent years. This is because they not only open the path for new applications but also provide ways of controlling the states and dynamics of confined carriers. For example, double quantum dot (DQD) systems provide the basis for an implementation of excitonic qubits (quantum bits—the building blocks of quantum information processing) with extended lifetimes [1], allow conditional quantum control of carrier states [2], and have been proposed as a

---

P. Karwat • K. Gawarecki • K. Roszak • A. Sitek • P. Machnikowski (✉)
Institute of Physics, Wrocław University of Technology, 50-370 Wrocław, Poland
e-mail: Pawel.Karwat@pwr.wroc.pl; Krzysztof.Gawarecki@pwr.wroc.pl; Katarzyna.Roszak@pwr.wroc.pl; Anna.Sitek@pwr.wroc.pl; Pawel.Machnikowski@pwr.wroc.pl

J. Wu and Z.M. Wang (eds.), *Quantum Dot Molecules*, Lecture Notes in Nanoscale Science and Technology 14, DOI 10.1007/978-1-4614-8130-0_9,

possible source of entangled photon pairs [3]. Furthermore, carrier transfer in DQD structures can be used for initializing spin states of a dopant Mn atom localized in one of the dots [4].

Pairs of vertically stacked QDs with a spatial separation down to single nanometers can be obtained in a two-layer self-assembled growth process, where the strain distribution favors the nucleation of QDs in the second layer directly on top of the QDs formed in the first layer [5,6]. The development of manufacturing technologies [7, 8] has made it possible to achieve DQD structures made of nearly identical dots, with the splitting of ground state transition energies down to single meV (possibly interaction-limited) [9, 10]. The state space of such a double quantum dot is obviously richer than that of a single QD [11, 12] and allows, e.g., for the formation of entanglement between the dots. Also the recombination and relaxation processes in DQDs show many features which cannot appear in individual QDs. The quantum coherence of carrier states in DQDs is affected by interference and collective effects that appear in the interaction of such systems with their radiative environment (electromagnetic vacuum) and with the surrounding crystal lattice (phonons).

Compared to single QDs, DQDs display more complexity when the decoherence resulting from the charge carrier–phonon coupling is studied. The interaction with the phonon bath leads to pure dephasing in both types of systems, which has been experimentally observed as the decay of the nonlinear optical response in a four-wave mixing experiment with ultrashort pulses [13, 14]. A characteristic feature of the phonon-induced dephasing in QDs is that it is always only partial, i.e., after a few picoseconds of carrier–phonon dynamics, the degree of coherence (i.e., the amplitudes of the off-diagonal elements of the density matrix) reaches a certain finite level, depending on the system geometry on and temperature [15,16]. In DQDs and regular QD arrays, the degree of dephasing may be reduced by encoding the logical qubit values into many-exciton states over a QD array [17]. Phonon-induced dephasing is also detrimental to entanglement in DQDs and larger QD arrays. The impact of the partial pure dephasing on entanglement is very strong, since it is more prone to dephasing than local coherence. It turns out that entanglement may be completely destroyed by the interaction with phonons even though the decoherence is always only partial [18–20]. Moreover, because of the importance of the delicate inter-subsystem coherences, phonon-induced entanglement decay strongly depends on the nature of the interaction with the environment, meaning that an interaction with separate reservoirs disentangles QD subsystems effectively than an interaction with a common reservoir [21].

Optical properties of DQDs may be strongly modified due to the collective interaction of sufficiently closely spaced QDs with the electromagnetic (EM) field. These collective effects have been extensively studied for atomic systems [22] where they manifest themselves by superradiant emission [23]. A signature of superradiant behavior was also observed in ensembles of QDs [24]. On the other hand, the collective interaction leads to the appearance of subradiant states which are decoupled from the environment and, therefore, do not undergo decoherence. It has been proposed to use these states for noiseless encoding of quantum information [25, 26].

Various types of coupling between the dots change the properties of the system even further. Theoretical calculations show that for closely spaced dots, the tunnel coupling, which is proportional to the wave function overlap, between the dots should strongly affect their electronic structure [27–29] leading to a delocalization of the low energy eigenstates of the DQD. Optical spectra of such structures indeed show clear manifestations of electronic coupling [10, 30–33]. The tunnel coupling is analogous to chemical bonding between atoms and, therefore, turns the two dots into one quantum system (also referred to as a quantum dot molecule). On the other hand, the wave function overlap is not the only mechanism of interaction between the QDs. In fact, for QD separations of about 10 nm, the energetically lowest states in the absence of external fields correspond to spatially direct excitons localized in individual QDs [29]. Such states are still bound by the Coulomb interaction. While the static ("direct") dipole coupling preserves the occupations of the individual QDs, the Förster interaction via interband dipole moments [34, 35] (first introduced in the context of molecular systems [36, 37]) enables the transfer the exciton occupation between the dots. These dipole couplings are rather like van der Waals forces between separate entities. Therefore the distance between the dots is a crucial parameter, which can lead to a crossover between regimes of distinguishable (Förster coupled) and indistinguishable (tunnel coupled) entities occupying the DQD.

In a closed system, the (usually very weak) Förster interaction has considerable effects only very close to resonance [34, 35]. However, the carrier–phonon coupling provides the necessary dissipation channel which enables excitation transfer driven by the Förster interaction, even if the energy mismatch between the dots is much larger than the interaction energy. Phonon-assisted excitation transfer between the quantum states of a molecule was in fact observed in many experiments [9, 38–42]. In closely stacked dots this excitation transfer process is mostly due to phonon-assisted tunneling of charge carriers, but for larger separations the tunneling is exponentially suppressed. Therefore, phonon-assisted transitions involving tunneling are very inefficient for a 10 nm separation even though a small energy splitting matches the acoustic phonon energies [42]. In such cases, the transfer is most likely to occur due to the Förster coupling.

In this chapter we present a review of a number of phenomena that appear in DQD systems due to the exciton–phonon and exciton–photon interactions, and the interplay between the two. In Sect. 9.2 we discuss the system under study, introduce the full Hamiltonian, and outline the methods used to find the evolution of the DQD system. Section 9.5 is devoted to phonon-assisted effects, such as phonon-induced relaxation, phonon-induced partial pure dephasing, and phonon-induced entanglement decay. In Sect. 9.4 the effects on the DQD subsystem resulting from the interaction with the radiative environment are discussed. This includes collective luminescence effects, the linear and nonlinear optical responses of DQD ensambles, and vacuum-induced coherence. In Sect. 9.5 we focus on the role of the interplay between the phonon and photon environments when discussing collective luminescence, vacuum-induced coherence, and entanglement decay. Section 9.6 concludes the chapter.

9.2 The Model and the Methodology

9.2.1 Model

The system under study is composed of two vertically stacked semiconductor QDs spatially separated by a distance D of the order of few nanometers (Fig. 9.1).

In the simplest approach, the wave functions of carriers confined in the QDs are modeled by identical anisotropic Gaussians with identical extensions l in the xy plane and l_z along the growth direction for electrons and holes,

$$\Psi(\boldsymbol{r}) \sim \frac{1}{\pi^{3/4} l_z l} \exp\left[-\frac{1}{2}\frac{x^2+y^2}{l^2} - \frac{1}{2}\frac{z^2}{l_z^2}\right]. \tag{9.1}$$

This is an approximation, since the QDs formed in the two layers differ slightly in size and shape, but most of the effects discussed within this chapter do not depend considerably on the carrier wave function shapes.

A more accurate model can be based on Bir-Pikus 8 band $\boldsymbol{k}\cdot\boldsymbol{p}$ Hamiltonian accounting also for the strain distribution in the system. The strain tensor is found by minimization of the elastic energy of the system [43] and using for calculating the position-dependent band edges and the effective masses. The electron and hole wave functions are then found by a combination of Löwdin perturbation theory [44], a generalization of "adiabatic" separation of variables [45] and Ritz variational method [46]. While most of the results presented here, in particular those related to optical properties of the system, rely on the simplified model described above, the $\boldsymbol{k}\cdot\boldsymbol{p}$ theory is indispensable to reveal the actual physics of phonon-assisted tunneling, to be described in Sect. 9.3.1.

In all the theoretical results presented here, the discussion is restricted to the ground states of electrons and heavy holes in both dots. Taking spin into account, this yields 16 electron–hole (exciton) configurations. This superfluous richness is reduced by restricting the model to spin-bright configurations (see [47] for a discussion of optical selection rules) and to "spatially direct" states, i.e., states for which electron–hole pairs forming the excitons reside in the same QD. Due to strong

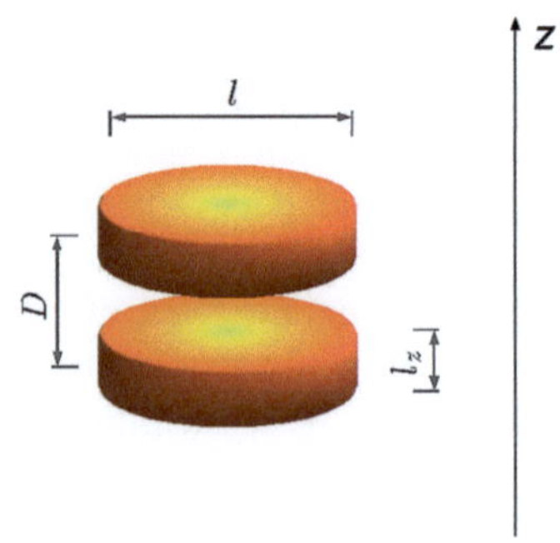

Fig. 9.1 Schematic representation of a pair of vertically stacked QDs

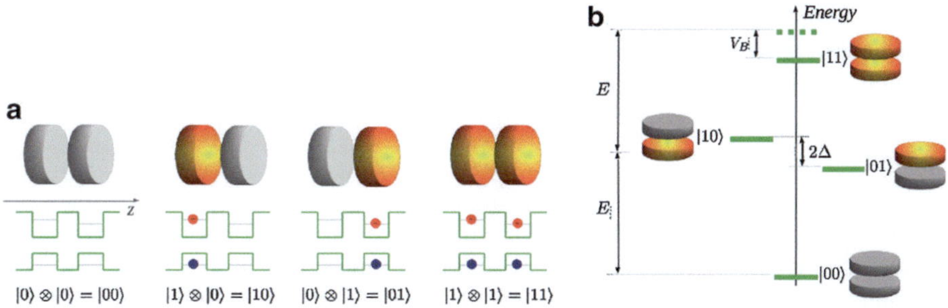

Fig. 9.2 (**a**) The basis states of a pair of QDs. (**b**) The energy levels in a pair of QDs

Coulomb electron–hole attraction the energy of such states is much lower than the energy of "dissociated" states (this can change if external electric fields are applied [11, 12]). In modeling excitonic systems, we also assume that the polarization of the exciting laser pulse is adjusted in such a way that it corresponds to the polarization eigenstate of the excitons. This allows us to include only one out of two bright states in each dot, since in both dots only one fundamental transition is allowed. With these restrictions, we may describe each QD as a two-level system with the basis states $|0\rangle$ and $|1\rangle$ denoting an empty dot and a dot occupied with an exciton, respectively.

Consequently, excitons in a pair of QDs may be modeled as a four-level system with the ground state $|00\rangle$ referring to two empty dots; single-exciton states are $|10\rangle$ and $|01\rangle$ with electron–hole pairs residing in the lower or upper QD, respectively; and the "molecular biexciton" state $|11\rangle$ representing both QDs occupied with single excitons (Fig. 9.2a).

Typically, the QDs have different fundamental excitonic transition energies E_1 and E_2 for the two QDs. It is convenient to use the average transition energy $E = (E_1 + E_2)/2$ and transition energy mismatch $\Delta = (E_1 - E_2)/2$ to express the exciton binding energies of the two QDs,

$$E_1 = E + \Delta \quad \text{and} \quad E_2 = E - \Delta. \tag{9.2}$$

Due to the interaction of static dipole moments of excitons confined in the two QDs, the energy of the biexcitonic state is shifted from the sum of single-exciton transition energies $E_1 + E_2$ by V_B (biexcitonic shit). Assuming that the energy of the ground state is zero, the Hamiltonian of the excitons confined in a pair of QDs in this simple model takes the form

$$\begin{aligned} H_\mathrm{DQD} = {} & (E+\Delta)\,|10\rangle\langle 10| + (E-\Delta)\,|01\rangle\langle 01| + (2E+V_\mathrm{B})\,|11\rangle\langle 11| \\ & + V\big(|10\rangle\langle 01| + |01\rangle\langle 10|\big), \end{aligned} \tag{9.3}$$

where V is an amplitude of the coupling between single-exciton states of the system which may originate either from Coulomb (Förster) interactions or from the tunnel coupling. The schematic diagram of excitonic states in a DQD is presented in Fig. 9.2b.

The analyzed QD system is stacked in a semiconductor material, and thus subjected to lattice vibration of the surrounding material, that is, phonons. The free phonon modes are described by the Hamiltonian

$$H_{\mathrm{ph}} = \sum_{\boldsymbol{q}} \hbar w_{\boldsymbol{q}} b_{\boldsymbol{q}}^{\dagger} b_{\boldsymbol{q}}, \tag{9.4}$$

where $b_{\boldsymbol{q}}^{\dagger}$, $b_{\boldsymbol{q}}$ are bosonic operatorscreating and annihilating phonon modes with wave vector $\boldsymbol{q}$ and $\hbar w_{\boldsymbol{q}}$ are the corresponding energies.

Interaction of carriers confined in the DQD with phonons is modeled by the Hamiltonian

$$H_{\mathrm{DQD-ph}} = \sum_{nm} \sum_{\boldsymbol{q}} F_{nm}(\boldsymbol{q}) |n\rangle\langle m| \left(b_{\boldsymbol{q}} + b_{-\boldsymbol{q}}^{\dagger} \right), \tag{9.5}$$

where $F_{nm}(\boldsymbol{q})$ are the coupling constants specific for a given problem (see, e.g., [48–50] for a review). Depending on the system under consideration, either only deformation potential coupling or both deformation and piezoelectric coupling has to be included.

We also investigate the interaction of the QD system with its radiative surrounding. The Hamiltonian of the photon reservoir is

$$H_{\mathrm{rad}} = \sum_{\boldsymbol{k}} \hbar \omega_{\boldsymbol{k}} c_{\boldsymbol{k}\lambda}^{\dagger} c_{\boldsymbol{q}\lambda},$$

where operators $c_{\boldsymbol{k}\lambda}^{\dagger}$, $c_{\boldsymbol{k}\lambda}$ create and annihilate photon modes with wave vectors $\boldsymbol{k}$ and polarization λ, while $\hbar\omega_{\boldsymbol{k}}$ denote the corresponding energies. In all the cases discussed here, the spatial separation D of the two dots is of the order of a few nanometers, that is two to three orders of magnitude smaller than the wavelength of the radiation with which the QDs interact (for the wide-gap semiconductors investigated here, the fundamental transition energy is of the order of 1 eV which corresponds to the wavelength $\sim 1{,}240$ nm in vacuum). This allows us to describe the coupling between excitons and the photon environment in the Dicke limit [51] where the spatial dependence of an electromagnetic field within the pair of QDs is neglected. The relevant coupling Hamiltonian in the dipole and rotating wave approximation is

$$H_{\mathrm{DQD-rad}} = \Sigma_{-} \sum_{\boldsymbol{k}} \sum_{\lambda=1,2} g_{\boldsymbol{k}\lambda} c_{\boldsymbol{k}\lambda}^{\dagger} + \mathrm{H.c.}, \tag{9.6}$$

where $\Sigma_{-} = \sigma_{-}^{(1)} + \sigma_{-}^{(2)}$ is a collective exciton annihilation operator and

$$g_{\boldsymbol{k}\lambda} = i\sqrt{\frac{\hbar\omega_{\boldsymbol{k}}}{2\varepsilon_0\varepsilon_r v}} \boldsymbol{d} \cdot \hat{e}_{\lambda}(\boldsymbol{k}). \tag{9.7}$$

is the exciton–photon coupling constant. Here $\boldsymbol{d}$ is the inter-band dipole moment (equal for both dots), $\hat{e}_{\lambda}(\boldsymbol{k})$ is the unit polarization vector of the photon mode with polarization λ, ε_0 is the vacuum dielectric constant, ε_r is the relative dielectric constant of the semiconductor, and v is the normalization volume.

9.2.2 *Simulation of the Evolution*

The dynamics of the system has been studied using various versions of the density matrix formalism, ranging from Fermi golden rule description of the occupation dynamics to non-Markovian simulations based on correlation expansion or time-convolutionless (TCL) methods.

9.2.2.1 Non-Markovian Master Equation for Carrier–Phonon Dynamics

The dynamic induced by the lattice vibrations is governed by the time-convolutionless equation in the interaction picture

$$\dot{\rho}(t) = \mathscr{L}_{\text{ph}}[\rho(t)],$$

where $\rho(t)$ is a reduced density operator for the carrier subsystem and

$$\mathscr{L}_{\text{ph}}[\rho(t)] = -\int_0^t d\tau \text{Tr}_{\text{ph}}\left[H_{\text{DQD-ph}}(t), \left[H_{\text{DQD-ph}}(\tau), \rho(t) \otimes \rho_{\text{ph}}\right]\right], \quad (9.8)$$

with the carrier–phonon interaction Hamiltonian in the interaction picture $H_{\text{DQD-ph}}(t)$. Here, ρ_{ph} is the phonon density matrix at thermal equilibrium and Tr_{ph} denotes a partial trace with respect to the phonon degrees of freedom. In the limit of long times and short reservoir memory, this equation reduces to the Markovian Lindblad equation with the transition rate from the initial state n to the final state m given by the Fermi golden rule, $\Gamma_{\text{ph},n\to m} = 2\pi R_{nm}[(E_n - E_m)/\hbar]$, where

$$R_{nm}(\omega) = \frac{1}{\hbar^2}\sum_{\boldsymbol{q}} |F_{nm}(\boldsymbol{q})|^2 \left[\delta(\omega - w_{\boldsymbol{q}})(n_{\boldsymbol{q}} + 1) + \delta(\omega + w_{\boldsymbol{q}})n_{\boldsymbol{q}}\right]$$

is the appropriate spectral density of the phonon reservoir (for a given pair of states). The details are given in [52].

9.2.2.2 The Weyl Operator Method for the Exact Diagonalization of the Electron–Phonon Coupling Term

If the system under study consists only of the DQD without tunneling between the dots and off-diagonal terms in the carrier–phonon interaction (which is typical, since phonon energies are usually too small to enable transitions between the dots without the mediation of other interactions), the Hamiltonian can be diagonalized exactly [49, 53]. The Hamiltonian is now restricted to the DQD Hamiltonian of Eq. (9.3) without the tunnel coupling ($V = 0$), the free phonon Hamiltonian [Eq. (9.4)], and the exciton–phonon interaction given by Eq. (9.5), where $n = m$ and n correspond to states of charge carriers localized in single QDs.

If we further restrict the system to the lowest energy excitonic states, then the unitary transformation that diagonalizes this Hamiltonian is as follows

$$\tilde{H} = \mathbb{W}H\mathbb{W}^{\dagger} = \tilde{H}_{\text{DQD}} + H_{\text{ph}},$$

with the operator

$$\mathbb{W} = \sum_{i=0}^{1}\sum_{j=0}^{1} |ij\rangle\langle ij| W_{ij}. \tag{9.9}$$

W_{ij} are the Weyl shift operators and are given by

$$W_{00} = 0, \qquad W_{11} = \exp[\sum_{q}(g_{q}^{(1)*} - g_{q}^{(2)*})b_{q} - \text{H.c.}]$$

$$W_{01} = \exp[\sum_{q} g_{q}^{(2)*} b_{q} - \text{H.c.}], \qquad W_{10} = \exp[\sum_{q} g_{q}^{(1)*} b_{q} - \text{H.c.}],$$

with $g_{q}^{(i)} = f_{q}^{(i)}/w_{q}$, where $f_{q}^{(i)}$ are phononic coupling constants corresponding to the two dots, $f_{q}^{(1)} = F_{10,10}(\boldsymbol{q})$, $f_{q}^{(2)} = F_{01,01}(\boldsymbol{q})$ (cf. Eq. (9.5)). The DQD Hamiltonian is now equal to

$$\tilde{H}_{\text{DQD}} = E_1(|1\rangle\langle 1| \otimes \mathbb{I}) + E_2(\mathbb{I} \otimes |1\rangle\langle 1|) + \Delta E(|1\rangle\langle 1| \otimes |1\rangle\langle 1|), \tag{9.10}$$

while the free phonon Hamiltonian remains unchanged. The DQD energies are shifted and are equal to $E_i = \varepsilon_i - \sum_q w_q |g_q^{(i)}|^2$, and $\Delta E = \Delta\varepsilon - 2\text{Re}\sum_q w_q g_q^{(1)} g_q^{(2)*}$.

The evolution operator generated by the full Hamiltonian can be written in terms of the Weyl operators and the diagonalized Hamiltonian

$$U_t = \mathbb{W}^{\dagger}\mathbb{W}_t\tilde{U}_t, \tag{9.11}$$

where $\tilde{U}_t = \exp(-i\tilde{H}t)$ and $\mathbb{W}_t = \tilde{U}_t\mathbb{W}\tilde{U}_t^{\dagger}$. Since $\tilde{H}$ is diagonal, the evolution described by $\tilde{U}_t$ is trivial and the evolution of the reduced DQD density matrix can be found in the form

$$\tilde{\rho}(t) = \text{Tr}_{\text{ph}}\left[\mathbb{W}^{\dagger}\mathbb{W}_t e^{-i\tilde{H}t}\sigma_0 e^{i\tilde{H}t}\mathbb{W}_t^{\dagger}\mathbb{W}\right], \tag{9.12}$$

where σ_0 denotes the combined initial state of the DQD and phonon system, and the trace is taken over phononic degrees of freedom.

9.2.2.3 Correlation Expansion Techniques

The correlation expansion technique is a standard method used for the description of quantum kinetics of interacting carriers and phonons in semiconductor systems of any dimensionality [54]. It has been successfully applied to carrier–phonon kinetics in QDs driven by an optical field [55–57], yielding a reliable description beyond the instantaneous excitation limit, where an exact solution is available for decoupled systems [58], and beyond the weak perturbation case which allows a perturbative treatment [59]. Compared to higher-dimensional systems, in QDs coherent and non-equilibrium phonons play a larger role because of the localized polaron effect. Therefore, a reliable description of the carrier phonon-kinetics in these systems requires a high enough degree of the correlation-expansion technique [57].

Various implementations of this technique differ in notation and in the choice of dynamical variables. Here, let us start from the Hamiltonian (9.5) restricted to only diagonal ($n = m$) couplings to the two ground exciton states in the two QDs and to LA phonons. By shifting the phonon modes in $H^{(1)}_{\mathrm{DQD-ph}} + H^{(2)}_{\mathrm{DQD-ph}}$ (where the two terms describe the phonon interaction with each dot) according to $b_{\boldsymbol{q}} \to b_{\boldsymbol{q}} - [F_{01,01}(\boldsymbol{q}) + F_{10,10}(\boldsymbol{q})]/(2\hbar w_{\boldsymbol{q}})$ one gets the coupling Hamiltonian in the form

$$H_{\mathrm{DQD-ph}} = \sum_{\boldsymbol{q}} f_{\boldsymbol{q}} \left(|01\rangle\langle 01| - |10\rangle\langle 10|\right) \left(b_{\boldsymbol{q}} + b^{\dagger}_{-\boldsymbol{q}}\right),$$

where $f_{\boldsymbol{q}} = [F_{01,01}(\boldsymbol{q}) - F_{10,10}(\boldsymbol{q})]/2$. In the first step, one defines three dynamical variables x, y, z describing the carrier state, $x = \langle \sigma_x(t) \rangle, \ldots$, where $\sigma_i(t) = e^{i\tilde{H}t/\hbar} \sigma_i e^{i\tilde{H}t/\hbar}$ are Pauli operators, written in the $|01\rangle, |10\rangle$ basis, in the Heisenberg picture. These three variables are the coordinates of the evolving Bloch vector, uniquely determining the reduced density matrix of the carrier subsystem.

From the Heisenberg equations of motion one finds the dynamical equations for these three variables,

$$\dot{x} = i\langle [H, \sigma_x] \rangle = -\Delta y - 4y \sum_{k} \mathrm{Re}\, B_k - 4y \sum_{k} \mathrm{Re}\, y_k, \tag{9.13}$$

and analogous for y and z. Obviously, this set of equations is not closed, but involves the new phonon variables $B_k = f_k \langle b_k \rangle$, as well as phonon-assisted variables of the form $y_k = f_k \langle\langle \sigma_y b_k \rangle\rangle = f_k(\langle \sigma_y b_k \rangle - \langle \sigma_y \rangle \langle b_k \rangle)$. The double angular brackets, $\langle\langle \ldots \rangle\rangle$, denote the correlated part of a product of operators, obtained by subtracting all possible factorizations of the product.

Next, one writes down the equations of motion for the new variables that appeared in the previous step, for instance,

$$\begin{aligned} \dot{y}_k = i\langle [H, y_k] \rangle &= \Delta x_k - 2V z_k - i w_k y_k + |f_k|^2 (iyz + x) \\ &+ 2\sum_{q} (x_{qk} + \tilde{x}_{qk}) + 4x_k \sum_{q} \mathrm{Re}\, B_q + 2x \sum_{q} (B_{qk} + \tilde{B}_{qk}), \end{aligned} \tag{9.14}$$

where the new two-phonon and two-phonon-assisted variables are defined as $B_{\boldsymbol{qk}} = f_q f_k \langle\langle b_q b_k \rangle\rangle$, $\tilde{B}_{\boldsymbol{qk}} = f_q^* f_k \langle\langle b_q^\dagger b_k \rangle\rangle$, $x_{\boldsymbol{qk}} = f_q f_k \langle\langle \sigma_x b_q b_k \rangle\rangle$, $\tilde{x}_{\boldsymbol{qk}} = f_q^* f_k \langle\langle \sigma_x b_q^\dagger b_k \rangle\rangle$, etc. In the next step, one finds the equation of motion for these new variables, introducing three-phonon variables. It is clear that the resulting hierarchy of equations is infinite and has to be truncated at a certain level. For a QD system, this can be done by setting all the correlated parts of the three-phonon and three-phonon-assisted variables equal to zero. This amounts to neglecting the correlations involving three or more phonons or, physically, to neglecting three-phonon processes (that is, emission or absorption of three or more phonons within the memory time of the phonon reservoir, which is of the order of 1 ps). The motivation for this procedure is that higher order correlations should play a decreasing role in the dynamics. From the equations of motion it is also clear that such higher order correlations develop in higher orders with respect to the coupling constants $f_{\boldsymbol{k}}$. On the other hand, truncation on this level allows one to account for the back-action of non-thermal and coherent phonons, which is important for a confined system [57].

9.2.2.4 Lindblad Master Equation for Carrier–Photon Dynamics

The decoherence effects induced by radiative environments are described in the Markov limit by the Master equation of motion in the Lindblad form

$$\dot{\rho}(t) = \mathscr{L}_{\mathrm{rad}}[\rho(t)],$$

where

$$\mathscr{L}_{\mathrm{rad}}[\rho(t)] = \Gamma \left[\Sigma_-(t)\rho(t)\Sigma_+(t) - \frac{1}{2}\{\Sigma_+(t)\Sigma_-(t), \rho(t)\} \right], \tag{9.15}$$

is the Lindblad dissipator with $\Sigma_-(t) = (\Sigma_+(t))^\dagger$ denoting the operators in the interaction picture and

$$\Gamma = \frac{E^3 |\boldsymbol{d}|^2}{3\pi\varepsilon_0\varepsilon_r \hbar^4} \tag{9.16}$$

is a spontaneous decay rate for a single QD.

9.2.2.5 Joint Influence of Phonon-Induced Dynamics and Spontaneous Emission

The evolution of a pair of QDs may be solved exactly only in special cases: An exact solution is available for an uncoupled system ($V = 0$) interacting only with lattice vibrations in the limit of instantaneous state preparation or ultrafast optical excitation [58, 60] or in the Markov limit for dots coupled only to its radiative surrounding [61]. In order to describe the dynamics of the QD system simultaneously interacting with both, phonon and photon, environments we have proposed a method [62] based on the equation of motion for the reduced density matrix of the exciton subsystem, ρ, in the interaction picture,

$$\dot{\rho}(t) = \mathscr{L}_{\mathrm{ph}}[\rho(t)] + \mathscr{L}_{\mathrm{rad}}[\rho(t)], \tag{9.17}$$

where the dissipators $\mathscr{L}_{\mathrm{ph}}[\rho(t)]$ and $\mathscr{L}_{\mathrm{rad}}[\rho(t)]$ are defined by formulas (9.8) and (9.15), respectively.

The presented method allows us to strictly reproduce the limiting cases mentioned at the beginning of the subsection. Moreover, the results for coupled ($V \neq 0$) QDs interacting with the phonon reservoir are reasonably close to those obtained by the correlation expansion technique [63].

9.3 Phonon-Assisted Processes

In this section, we discuss phonon-induced transitions in a double-dot system. First, we describe phonon-assisted relaxation and tunneling processes. Then we present the theory of phonon-assisted excitation (Förster-like) transfer. Next, pure dephasing effects are briefly discussed. Finally, we discuss the phonon-induced decay of entanglement between the charge states in the two dots.

From the experimental point of view, dissipative carrier transfer in self-assembled structures has been studied with optical spectroscopy methods (time-integrated and time-resolved photoluminescence, and photoluminescence excitation experiments) both in lateral double-dot systems [4, 40, 64] and in stacked quantum dot molecules (QDMs) [9, 38, 39, 41, 42, 65–72] and QD chains (both stacked and lateral) [73, 74]. Various mechanisms have been invoked to account for the observed properties. In most cases, the kinetics is attributed to tunneling [38–40, 42, 65, 70–73]. In some other experiments [9, 66, 69], signatures of radiative (Förster-like) transfer have been observed. Coulomb scattering [41] and thermally activated processes [68, 72] also seem to play an important role, at least in some systems. In a recent experiment [75], the inelastic contribution was extracted from the electron kinetics in a double dot placed in a photodiode structure in which the splitting between the levels could be tuned. The resulting tunneling rate followed the dependence on the energy detuning typical for a phonon-assisted process.

9.3.1 Phonon-Assisted Relaxation and Tunneling

In this section, we discuss phonon-assisted relaxation processes in a tunnel-coupled QDM. We investigate the system of a QDM doped with a single electron [76] as well as the QDM containing a single hole [77]. In both cases, we describe phonon dynamics within the Fermi golden rule approach. We study phonon-assisted transitions between two lowest carrier states. In particular, if these states describe a carrier localized in different dots, the phonon-assisted relaxation leads to a charge transfer in the system.

Phonon-assisted relaxation in the QDM in the case of a single electron has been widely studied in the literature. In [78], the system containing two lateral dots coupled by an inter-dot tunneling barrier was investigated. Two lowest electron states were calculated in a relatively simple model using a perturbation method combined with unitary transformation. The calculation of the phonon-assisted relaxation rate in this paper accounts for coupling by a deformation potential as well as coupling by a piezoelectric field. Furthermore, in [79] a similar system was investigated. Here, the phonon-induced decoherence caused by the piezoelectric field coupling was calculated within Born–Markov approximation. A laterally coupled DQD system has been studied in [80]. In that paper, the confining potential in the z direction was modeled as an infinite quantum well, while the in-plane confinement was described as two parabolic wells separated by an inter-dot distance. Calculations of the phonon-assisted relaxation rate were performed within the Fermi golden rule. In [81] phonon-related processes in the system of stacked QDs were investigated. In a plane, the potential was modeled as an isotropic parabolic confinement, while in the growth direction the system was approximated by two rectangular wells with a finite barrier. The interaction of the electron with LO phonons was described using self-consistent Green's function formalism combined with degenerate perturbation approach. Finally, the model of the phonon-assisted relaxation in the case of two electrons in a lateral DQD was described in [82]. The wave functions were assumed to be a linear combination of Gaussians localized on each dot. In [83] a similar system was investigated. In that work, authors, using non-perturbative quantum kinetics theory, described carrier–phonon kinetics in the presence of both diagonal and off-diagonal couplings. The result shows the importance of off-diagonal terms and indicates the need of using non-Markovian theory. Experimentally, the phonon-assisted relaxation in the system containing a single hole in a vertically stacked QDM was measured [84], but also in this case only a simple model was used to interpret the results.

Here, we present the result of our modeling based on Bir-Pikus 8 band $\boldsymbol{k} \cdot \boldsymbol{p}$ theory [76, 77]. The shape of both dots is modeled as a spherical segment with the base radius of 10 nm and the height 3.7 nm. The dots are grown on the wetting layer with thickness of 0.6 nm. Due to the lattice mismatch between dots and matrix material, the strain appears in the system and affects the band structure. The strain tensor is calculated by minimization of the elastic energy of the system [43]. Using the strain tensor components, the position-dependent band edges and the effective masses are found. In the case of a single electron as well as for a single hole, Löwdin perturbation method is used to separate the electron and hole part from the full Hamiltonian. By generalizing the method of "adiabatic" separation of variables [46], we calculate z-dependent components of the wave functions. The radial parts are then obtained using Ritz variational method [76]. The phonon-assisted transition rates are calculated from the Fermi golden rule, using the carrier wave functions obtained in this way.

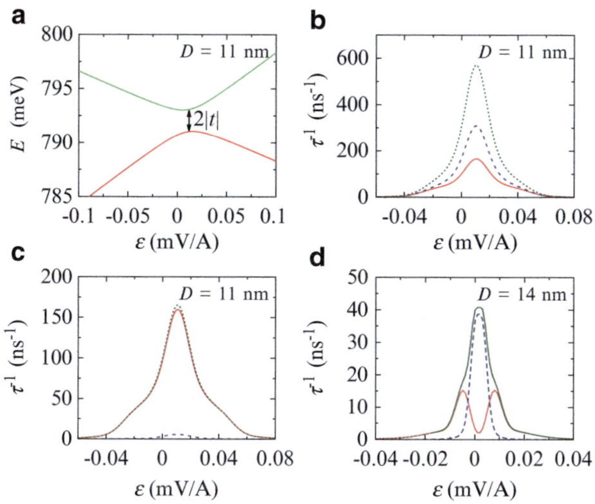

Fig. 9.3 (**a**) The two lowest electron energy levels at $D = 11$ nm as a function of the electric field. Half of the energy splitting at the resonance is equal to the value of the tunnel coupling parameter. (**b**) Relaxation rate between the two lowest one-electron states at $T = 0$ K (*red solid line*), 20 K (*blue dashed line*), and 40 K (*green dotted line*). (**c**,**d**) Contributions to the relaxation rate from the DP coupling (*red solid line*), PE coupling (*blue dashed line*) and the total relaxation rate (*green dotted line*) at $T = 0$ K for two different inter-dot distances

The two lowest energy levels as a function of the electric field for an electron are shown in Fig. 9.3a. The resonance (the point which corresponds to the minimal splitting) describes the situation when the electron in both states is delocalized. In order to describe the tunnel properties of the system, we introduce the tunnel coupling parameter t, defined as a half of the energy splitting at the resonance. In the case of electrons, the value of t decays exponentially with an increasing inter-dot separation D.

The quantity describing phonon-related dynamics is a phonon-assisted relaxation rate γ. We calculated the phonon-assisted relaxation rate between the first excited state and the ground state for two different distances between the dots: at $D = 11$ nm and at $D = 14$ nm. All the material parameters, as well as the exact system geometry are described in [76, 77]. We took into account the carrier–phonon coupling by piezoelectric field (PE) as well as by deformation potential (DP). The results of the total phonon-assisted relaxation rate (defined as a sum of DP coupling and PE coupling contributions) at three different temperatures is presented in Fig. 9.3b. The temperature dependence has been modeled using the Bose–Einstein distribution of phonons. Relaxation rates are large near the energy resonance and become small for the electric fields far from the resonance point. The interpretation of this behavior is based on the Fermi golden rule where the main role is played by spectral density of the phonon reservoir (see Sect. 9.2; Fig. 9.4a,b). The magnitude of the spectral density critically depends on the spatial overlap between the wave functions. On the

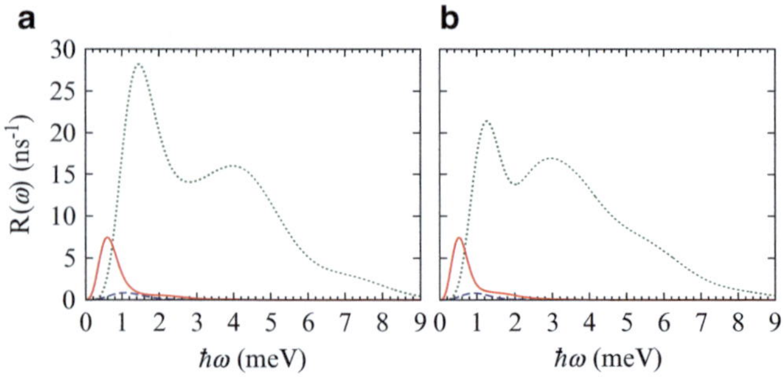

Fig. 9.4 The contributions to the spectral density at $\varepsilon = 0$ from the DP coupling to longitudinal acoustic (LA) phonons (*green dotted line*) and from the piezoelectric coupling to LA phonons (*red solid line*) and transverse acoustic (TA) phonons (*blue dashed line*) at $T = 0$ K for (**a**) $D = 11$ nm and (**b**) $D = 14$ nm

other hand, at the resonance, the wave functions of both states are delocalized (what leads to a large overlap between them). In consequence, as can be seen in Fig. 9.3b, the phonon-assisted relaxation rate has a maximum at the point corresponding to the energy resonance. Furthermore, the spectral density oscillates. The reason is the emission of the short wavelength phonons along the direction with the strongest confinement [85]. Because there are two confinement centers along this direction, an interference takes place and the amplitude of phonon emission has a maximum for each $\omega = (2n+1)\pi c/D$, where n is an integer. The oscillations are clearly reflected in the behavior of the DP contribution to the relaxation rate at $D = 14$ nm (Fig. 9.3d). In the case of $D = 11$ nm (Fig. 9.3c), only one maximum can be seen, because the period of oscillations is greater than for $D = 11$ nm. On the other hand, the envelope of the spectral density drops down for large ω because the high energy (short wavelength) phonons are not effectively coupled to the relatively weakly confined electron states. The contributions to the overall magnitude of spectral density are shown in Fig. 9.4a,b. One can see that for small ω, the PE coupling dominates as a mechanism of electron–phonon coupling. On the other hand, DP coupling is preferred for moderate and high values of frequency. This results from the fact that the DP coupling is isotropic while the PE coupling is anisotropic and is suppressed for emission along the z direction, which is preferred for higher phonon energies [76]. The properties of spectral density are clearly reflected in Fig. 9.3c,d. As can be seen in Fig. 9.3c, the DP coupling dominates for small inter-dot separations (large energy splittings). On the other hand, the PE coupling becomes dominant mechanism for large distances (Fig. 9.3d).

We also performed calculations for the system containing a single hole. The two lowest eigenstates have a very small light hole admixture (from 0.5 to 2% contribution to the total state). The reason is a very weak confinement of light holes in the QDM. In order to analyze the tunnel coupling between the hole states in the system we calculated the tunnel parameter t as a function of the distance D (Fig. 9.5a). The results for holes show qualitatively different behavior compared

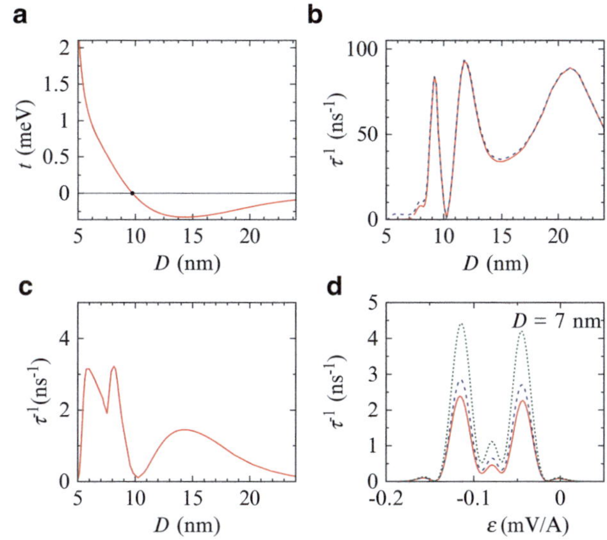

Fig. 9.5 (**a**) The tunnel coupling parameter as a function of the inter-dot separation. At $t = 0$ the bonding–antibonding transition of the ground state takes place. (**b**) The total maximum relaxation rate (*blue dashed line*) and the contribution from the PE coupling (*red solid line*). (**c**) Contribution to the total relaxation rate from the DP coupling. (**d**) Relaxation rate for hole states at $T = 0$ K (*red solid line*), 20 K (*blue dashed line*), and 40 K (*green dotted line*) for $D = 7$ nm

to the electron case. In contrast to the electron case, this dependence for holes is non-exponential. The sign of t is related to the character of the ground state: the positive one means that the ground state is bonding, while the negative sign implies an antibonding ground state. The vanishing of tunnel coupling which takes place about $D = 10.23$ nm in our system is caused by the transition from a bonding to antibonding ground state [86–88].

We calculated the maximum magnitude of the relaxation rate as a function of the distance between the dots. The resulting total magnitude as well as its contributions from the DP coupling and the PE coupling is shown in Fig. 9.5b,c. For small distances ($D \to 0$), both relaxation rates vanish. Similarly like for electrons, the reason is that for small D, the energy splitting is large and there is a low density of phonon states at very high frequencies. In the case of $D \to \infty$ the relaxation rates also drop down because of vanishing overlap between the wave functions. On the other hand, for distances near $D = 10.23$ nm (where $t = 0$) the maximum magnitude of the relaxation rate drops down by two orders of magnitude compared to the highest value. The reason is an extremely narrow energy splitting. However, near the critical distance the phonon-assisted relaxation remains very slow not only at the resonance but also for any electric field. The phonon-assisted relaxation rate at $D = 7$ nm as a function of electric field at three temperatures ($T = 0$ K, 20 K and 40 K) is shown in Fig. 9.5d. In this case, the results are qualitatively similar to the electron case.

9.3.2 Phonon-Assisted Excitation Transfer

In this section we discuss the evolution of the exciton state in a DQD system coupled to its phonon reservoir. We assume here that the only coupling between

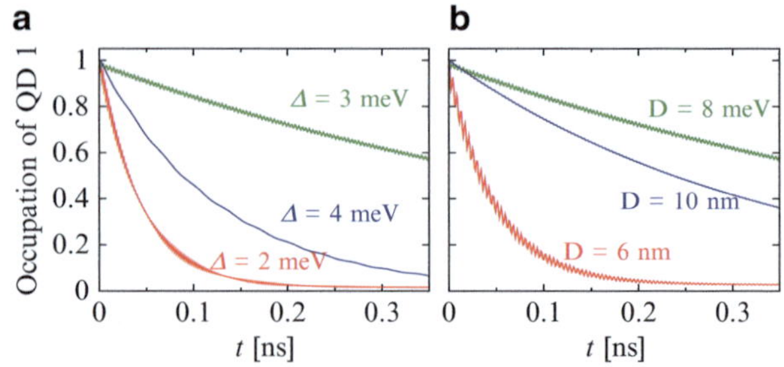

Fig. 9.6 The occupation of the higher energy QD as a function of time at $T = 4$ K: (**a**) $D = 8$ nm and Δ as shown; (**b**) $\Delta = 3$ meV and D as shown [62]

single-exciton states results from the dipole coupling of the Förster type (see Sect. 9.2). Such an interaction opens the possibility of a transfer of occupation between the dots which, in the presence of a dissipation channel, can become irreversible. For confined carriers, the dissipation is dominated by phonon-related mechanisms. Such a process was studied using quantum-kinetic equations [63, 89], perturbative (Fermi golden rule) approaches [90, 91], or Förster phenomenology [92]. Here, we summarize the results of [63].

The system is described by the Hamiltonian $H = H_{\mathrm{DQD}} + H_{\mathrm{ph}} + H_{\mathrm{DQD-ph}}$ [Eqs. (9.3), (9.4), and (9.5)]. We assume that there is one exciton in the system, initially localized in one of the dots and use the simple (Gaussian) model of wave functions. The calculations were originally performed [63] within the correlation expansion approach, outlined in Sect. 9.2.2.3. However, the lowest-order time-convolutionless equations (Sect. 9.2.2.1) yield essentially the same results. As we will see, the interplay of the coupling and phonon-assisted dissipation leads to irreversible excitation transfer between the dots, which is due to a weak Coulomb (Förster) coupling between them.

In general, if the initial state corresponds to the exciton located in one of the dots (state $|01\rangle$ or $|10\rangle$) the evolution is a combination of conservative (unitary) oscillations due to the perturbation induced by the coupling V and a dissipative, irreversible transition towards the lower eigenstate of H_{DQD} [63, 93] induced by the coupling to the phonon continuum. However, in the case of a weak coupling the oscillations are very small and the occupation of the higher-energy dot follows a nearly exponential decay, as shown in Fig. 9.6 (here $l = 4.5$ nm, $l_z = 1$ nm, and the interband dipole moment corresponds to 400 ps radiative lifetime) [62]. An interesting feature visible in Fig. 9.6 is that the rate of the excitation transfer is non-monotonic both in the energy mismatch and in the separation between the dots.

The nearly exponential decay curve suggests that the process can be described in the Markovian approximation. This is possible, since the typical time scale of the decay process is long compared to the transition frequency $V/\hbar$ between the two single-exciton states. It was shown [63] that the formal long-time limit of the TCL equation for this problem in the rotating wave approximation leads to optical Bloch equations describing the system dynamics in the interaction picture and in the rotating frame related to the eigenstates of H_{DQD}.

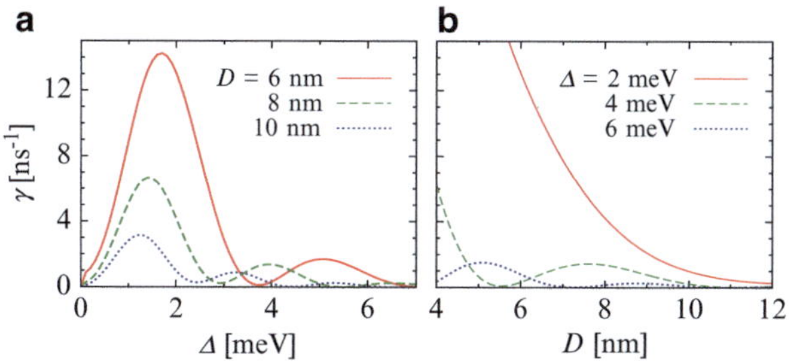

Fig. 9.7 The rate of phonon-assisted excitation transfer as a function of the energy mismatch for a few values of the QD separation D (**a**) and as a function of D for a few values of Δ (**b**) at $T = 4$ K [62]

In general, upon transforming back to the original basis $|01\rangle, |10\rangle$ and to the Schrödinger picture one obtains a complex evolution. However, in the case of $|V| \ll |\Delta|$, which is of particular practical importance, the eigenstates of H_{DQD} are very close to $|10\rangle$ and $|01\rangle$. Moreover, in this limit the energy difference between the eigenstates, $\hbar\Omega = 2\sqrt{V^2+\Delta^2}$ is nearly equal to 2Δ. In this limit, one obtains an exponential excitation transfer with the rate

$$\gamma = 4\pi \left(\frac{V}{\Delta}\right)^2 [R(\Omega) + R(-\Omega)],$$

with the spectral density

$$R(\omega) = \frac{2}{\hbar^2} |n_{\text{B}}(\omega)+1| \sum_{\boldsymbol{k}} \sin^2 \frac{k_z D}{2} |f_{\boldsymbol{k}}|^2 \delta(|\omega| - w_{\boldsymbol{k}}).$$

Thus, the Markovian equations are particularly useful in the limit of weak coupling, where the Markovian dephasing rate may be identified with the rate of irreversible excitation transfer between the dots.

The rate for the phonon-assisted process is governed, on the one hand, by the amplitude of the Förster coupling which decreases roughly as $1/D^3$. On the other hand, it is strongly influenced by the structure of $R(\Omega)$. The latter oscillates as a function of both Ω and D [63] due to the interplay between the wavelength of the emitted phonon and the QD separation in the molecule (phonons are preferentially emitted along the strongest confinement limit, i.e., along the DQD axis). In particular, $R(\Omega)$ has a pronounced minimum whenever Ω is a multiple of $2\pi\hbar c_1/D$ which explains the oscillating dependence on both Δ and D (see Fig. 9.7).

The transfer rate for $\Delta = 4$ meV, $D = 4.5$ nm is $\gamma = 2.3$ ns^{-1}, which is about twice lower than the value of $\Gamma = 5.25$ ns^{-1} deduced from fitting to the photon correlation data [9]. It should be noted, however, that our modeling is based on certain choices of parameters that cannot be uniquely determined. First of all, the magnitude of the Förster coupling for a given inter-dot distance can only be roughly estimated, since the value of the intraband dipole moment in an inhomogeneous, strained structure is not exactly known. Moreover, the carrier–phonon coupling constants in the relevant energy (or wave vector) range are strongly geometry dependent. For instance, by changing the localization widths to $l_e = l_h = 4.0$ nm and $l_z = 0.8$ nm one gets a considerably increased value of $\Gamma = 3.65$ ns^{-1}.

9.3.3 Pure Dephasing Effects

Apart from the real transitions (relaxation, phonon-assisted tunneling, and excitation transfer, as described in the previous sections), carrier–phonon coupling can induce pure dephasing of coherences between various states of a DQD system. These include the decay of optical coherence and of spatial coherences between states localized in different dots, to be discussed in this section, as well as the decay of entanglement, described in Sect. 9.3.4 below.

For single QDs, the phonon response to the ultrafast (sub-picosecond) charge distribution corresponding to the optically generated exciton leads to a partial decay of optical coherence [14–16, 94] due to a kind of "which path" information transfer from the charge subsystem to the phonon reservoir [53]. In DQDs, additional effects appear in the evolution of excitonic coherence due to the interference of phonon packets emitted by the two dots as well as to the mutual impact of the wave packets emitted by one dot on the other dot [58]. As a result, a very small feature appears in the evolution of the modulus of the optical polarization at the time when the wave packet emitted upon excitation of one dot crosses the other one. A more pronounced effect, due to the energy shift resulting from a local compression, is visible in the phase of the optical polarization [58].

In terms of its contribution to the excitonic line shape, the pure dephasing effect dominates over real transitions in individual dots at low temperatures since the excited levels are separated by a large energy distance. In double dots, however, the situation is different: the doublet of states in the ground state sector requires a treatment of both real transitions (off-diagonal carrier–phonon couplings) and dephasing (diagonal couplings) on equal footing. Such a treatment was proposed based on a generalization of the cumulant expansion to a multilevel system [95]. In the second order, virtual transitions to the nearby excited state contribute to the broadening of the excitonic line via effective two-phonon processes [96]. The line width, calculated in this way, is dominated either by real transitions or by virtual transitions, depending on the separation between the levels, with the real transition contribution playing a major role for the energy splitting of a few meV [95].

Another class of dephasing mechanisms results from phonon anharmonicity. In an anharmonic system, a polaron (lattice deformation or polarization field) surrounding the charge distribution confined in a QD serves as a scattering potential for phonons [97,98]. Since the scattering depends on the position of the scatterer, the lattice can extract the information on the presence of an electron and exciton in one or the other dot in the DQD system, hence destroying the spatial coherences present in a delocalized state (a superposition of exciton states located in different dots). The resulting dephasing rate depends extremely strongly on temperature ($\sim T^7$ at low temperatures) and shows dependence on the inter-dot distance ($\sim D^2$ as long as the carrier wave functions overlap and logarithmic for larger distances). As a result, the dephasing time drops from micro-seconds at temperatures of a few Kelvin to picoseconds at 100 K [97].

9.3.4 Entanglement Decay

One of the fundamental qualities of quantum theory is the appearance of non-classical correlations, such as entanglement [99, 100]. Entanglement between two systems manifests itself by the appearance of correlations between the results of appropriately chosen measurements on the entangled subsystems which cannot be accounted for by any classical (realistic and local) theory [101]. Furthermore, entanglement is an important resource in quantum information processing [102], and it is essential for quantum teleportation [103], superdense coding [104], and the distribution of cryptographic keys [105].

Entanglement, as a non-local property of multiple quantum systems, is expected to be more fragile than the phase coherence of individual subsystems. In order to manifest genuinely quantum behavior resulting from entanglement a quantum system must maintain phase relations between the components of its quantum superposition state, involving different states of distinct subsystems. Keeping in mind that the subsystems may be separated by a macroscopic distance, one may expect such a non-local superposition state to be extremely fragile to the dephasing effect of the environment. In fact, it has been shown (for a 2×2 system) that entanglement between two subsystems tends to decay faster than local coherence [19, 21, 106]. As expected, the decay of entanglement is stronger, if the subsystems interact with different environments (which might result from a large spatial separation between them). Furthermore, certain states that show robust entanglement under collective dephasing become disentangled by the interaction with separate environments [21]. It was also shown for two different classes of systems [19,20] that certain states may become separable (completely disentangled) within a finite time under conditions that lead to a usual, exponential decay of local coherence (this phenomenon is known as the "sudden death" of entanglement). Since even partial entanglement of many copies of a bipartite quantum system may be distilled to a smaller number of maximally entangled systems [107], it is essential to understand whether the influence of the environment leads to the appearance of separability in realistic models of dephasing.

The study of entanglement requires an entanglement measure that can be calculated from the system state. For pure states, the von Neumann entropy of one subsystem [108] is a good entanglement measure, but for mixed states a unique entanglement measure has not been found [109, 110]. One choice is to use the entanglement of formation (EOF), defined as the ensemble average of the von

Neumann entropy minimized over all ensemble preparations of the state [110]. Such a measure may be interpreted as the asymptotic number of pure singlets necessary to prepare the state by local operations and classical communication. A practical characterization for mixed state entanglement is available for small systems [111, 112], but an explicit formula for calculating the EOF for a given density matrix is known only for a pair of two-level systems [113, 114]. This is given by

$$\mathrm{EOF}[\rho(t)] = -x_+ \log_2 x_+ - x_- \log_2 x_-,$$

where $x_\pm = (1 \pm \sqrt{1 - C^2[\rho(t)]})/2$, and $C[\rho(t)]$ is the concurrence. The concurrence is an entanglement measure itself and is equal to

$$C[\rho(t)] = \max(0, \lambda_0 - \lambda_1 - \lambda_2 - \lambda_3), \tag{9.18}$$

where λ_i are the square roots of the eigenvalues of the non-Hermitian matrix $\rho(t)(\sigma_y \otimes \sigma_y)\rho^*(t)(\sigma_y \otimes \sigma_y)$ in decreasing order. Here, $\rho(t)$ is the density matrix of the potentially entangled 2×2 system, $\rho^*(t)$ is its complex conjugate, and

$$\sigma_y = \begin{pmatrix} 0 & -i \\ i & 0 \end{pmatrix} \tag{9.19}$$

is one of the Pauli matrices.

Let us study the decay of entanglement under phonon-induced pure dephasing. Since local unitary transformations do not change the amount of entanglement in the system, we can use the density matrix $\tilde{\rho}(t) = e^{-iH_L t}\rho(t)e^{iH_L t}$, with $H_L = E_1(|1\rangle\langle 1| \otimes \mathbb{I}) + E_2(\mathbb{I} \otimes |1\rangle\langle 1|)$, instead of $\rho(t)$ in the formulas for entanglement (in this section, we use the explicit tensor product notation for the states of the two-dot system). Using the method introduced in Sect. 9.2.2.2 and assuming a separable initial system-reservoir state, we find the elements of the density matrix $\tilde{\rho}(t)$. These are equal to

$$[\tilde{\rho}(t)]_{ii} = [\tilde{\rho}_0]_{ii}; \quad [\tilde{\rho}(t)]_{ij} = [\tilde{\rho}_0]_{ij} e^{-iA_{ij}(t) + B_{ij}(t)}, \tag{9.20}$$

with

$$A_{01} = A_{02} = \sum |g_q|^2 \sin w_q t, \tag{9.21}$$

$$A_{03} = 4\sum |g_q|^2 \cos^2(q_z D/2) \sin w_q t - \Delta E t, \tag{9.22}$$

$$A_{12} = 0, \tag{9.23}$$

$$A_{13} = A_{23} = A_{03} - A_{01}, \tag{9.24}$$

$$B_{01} = B_{02} = B_{13} = B_{23} = \sum |g_q|^2 (\cos w_q t - 1)(2n_q + 1), \tag{9.25}$$

$$B_{03} = 4\sum |g_q|^2 \cos^2(q_z D/2)(\cos w_q t - 1)(2n_q + 1), \tag{9.26}$$

$$B_{12} = 4B_{01} - B_{03}, \tag{9.27}$$

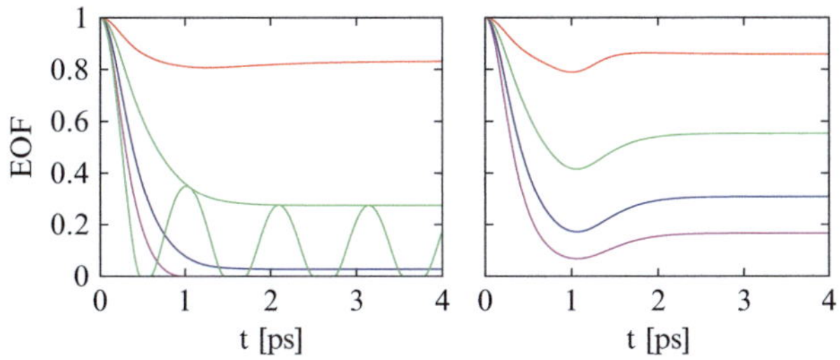

Fig. 9.8 Evolution of entanglement of the two-qubit system at various temperatures (0 K—*red curve*, 40 K—*green curve*, 80 K—*blue curve*, 120 K—*violet curve*) with $D = 6$ nm. *Left panel* corresponds to the initial state (9.28) and the *right panel* corresponds to the initial state (9.29). The green, oscillating curve in the left panel corresponds to $V_B = 6\ \mathrm{ps}^{-1}$ (at $T = 40$ K), the other curves in the left panel correspond to $V_B = 0$

where g_q is defined in Sect. 9.2.2.2. For long times, the factors $\cos w_q t$ and $\sin w_q t$ become quickly oscillating functions of $\boldsymbol{q}$ and their contribution averages to 0. Consequently, the phase damping factors B_{ij} decrease form their initial value of 0 to a certain asymptotic value depending on the material parameters, system geometry and temperature, while the phase shift factors A_{ij} affect the system evolution at small times and then average out to zero. As a result, the off-diagonal elements of the density matrix are reduced and the phase information is partly erased. The following calculations were performed for two identical, self-assembled GaAs/InGaAs QDs, for which the wave function widths were taken equal to 4.4 nm for electrons and 3.6 nm for holes in the in-plane direction, and 1 nm in the growth direction for both.

The calculations of the evolution of the EOF are performed for two initial fully entangled pure states

$$|\psi_0^{(1)}\rangle = \frac{|00\rangle + |01\rangle + |10\rangle - |11\rangle}{2} \tag{9.28}$$

$$|\psi_0^{(2)}\rangle = \frac{|01\rangle - |10\rangle}{\sqrt{2}}. \tag{9.29}$$

The evolution of the EOF of the qubit pair is shown in Figs. 9.8 and 9.9. In the absence of energy shift V_B, entanglement decays on a time scale of a few picoseconds, corresponding to the timescale of the phonon-induced dephasing process. At low temperatures or for overlapping systems, this process resembles the decay of coherences in a single system under the same environmental influence [15]. However, for a sufficiently large separation between the systems and at sufficiently high temperatures the initially maximal entanglement present in the state (9.28) decays completely after a finite time even though the phonon-induced dephasing is always only partial (see left panels of Figs. 9.8 and 9.9). On the other hand, for the singlet initial state [Eq. (9.29)], the destruction of entanglement is always only partial mimicking the pure dephasing process (see right panels of Figs. 9.8 and 9.9).

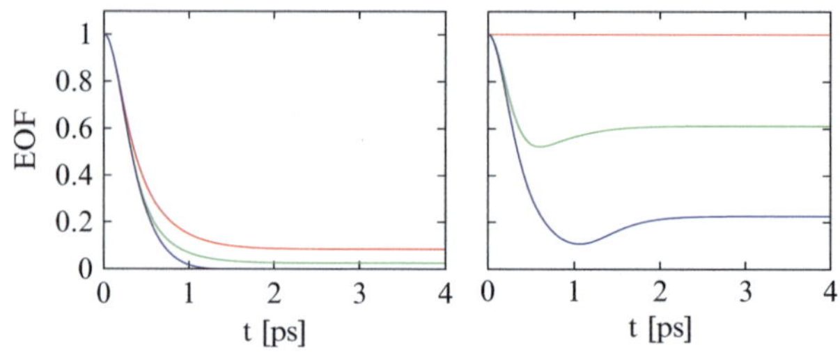

Fig. 9.9 Evolution of entanglement of the two-qubit system for various distances between the dots at $T = 100$ K and $V_B = 0$ (0 nm—*red curve*, 2 nm—*green curve*, 6 nm—*blue curve*). *Left panel* corresponds to the initial state (9.28) and the right panel corresponds to the initial state (9.29)

An important case is that of a nonzero biexcitonic shift, $V_B \neq 0$. Such an energy shift leads to an entanglement-generating evolution. This mechanism is used for performing nontrivial two-qubit gates (controlled-shift) in many proposals for semiconductor-based quantum information processing [115, 116]. As can be seen in Fig. 9.8 (oscillating green line in left panel), in the presence of phonon-induced pure dephasing the cyclic evolution of entanglement is damped and the maximum achievable level of entanglement is reduced. Moreover, extended periods of time appear when the entanglement remains zero.

The appearance of complete disentanglement for some initial states under sufficiently strong partial pure dephasing may be understood with the help of Eq. (9.18). If the completely dephased state (a state where the density matrix is diagonal) has $\lambda_0 - \lambda_1 - \lambda_2 - \lambda_3 < 0$ then, by continuity, it will be surrounded by states with vanishing concurrence, so that entanglement vanishes for sufficiently strongly dephased states, before the complete dephasing is reached. From the Wootters formula for a diagonal density matrix one finds $\lambda_0 - \lambda_1 - \lambda_2 - \lambda_3 = -2\min(\sqrt{\rho_{00}\rho_{33}}, \sqrt{\rho_{11}\rho_{22}})$, meaning that the above condition may only be satisfied, if all four diagonal elements are nonzero, which is the case for the initial state (9.28) but not the case of state (9.29).

The time at which the entanglement of the state (9.28) vanishes completely depends on temperature and on the distance between the systems. This time becomes finite only at a certain temperature (that depends on the coupling strength). Slightly above this critical temperature, complete disentanglement takes place only for strongly separated systems. For higher temperatures the disentanglement time for non-overlapping systems depends very weakly on the distance. It should be stressed that the appearance of complete disentanglement at increased temperatures is only related to stronger dephasing in the system at higher temperatures and, in principle, the state might become separable already at $T = 0$ if the coupling were sufficiently strong.

On the contrary, the dependence on the separation between the two subsystems reflects a more fundamental crossover, from the regime of a common reservoir to that of independent reservoirs [21]. In our model, the off-diagonal element ρ_{12} is unaffected for completely overlapping systems ($D = 0$) which is related to the fact

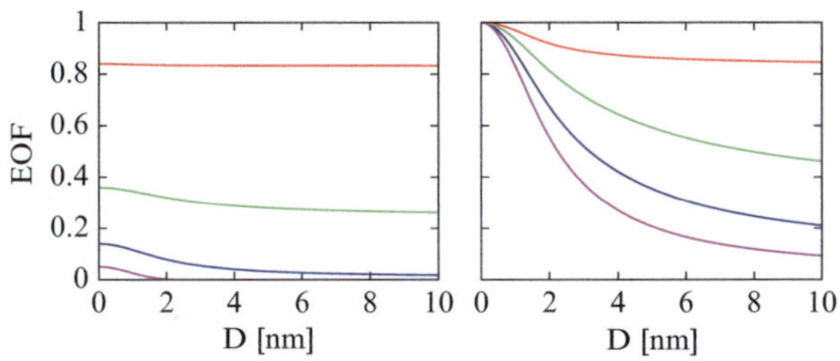

Fig. 9.10 The asymptotic value of the EOF as a function of the distance between the subsystems for different temperatures (0 K—*red curve*, 40 K—*green curve*, 80 K—*blue curve*, 120 K—*violet curve*). *Left panel* corresponds to the initial state (9.28) and the *right panel* corresponds to the initial state (9.29)

that the reservoir can distinguish the subsystems only by their location in space. This single non-vanishing element is enough to prevent complete disentanglement at any temperature and for any coupling strength. Only when the distance between the system grows, dephasing is able to decrease this element to the extent sufficient for the total destruction of entanglement. This distance effect is shown in Fig. 9.10, where we plot the asymptotic (long-time) value of the EOF.

The effect of spatial separation between the systems is very clearly visible for the second initial state [Eq. (9.29)], which involves only this resilient off-diagonal element (see Fig. 9.10, right). Here, the entanglement is robust, if the systems overlap (see also right panel of Fig. 9.9) but becomes fragile as soon as the separation between the systems is comparable to their size. This demonstrates that the distance between the subsystems is the physical parameter that governs the crossover between the two regimes of entanglement decay.

9.4 Collective Effects in the Luminescence

In this section, we analyze purely radiative effects, i.e., we assume that the DQD system interacts only with its photon surrounding. For the wide-gap semiconductors (with energy gap of the order of 1 eV) which we are investigating, the zero-temperature approximation may be used at any reasonable temperature. The Hamiltonian of the system is now

$$H = H_{\mathrm{DQD}} + H_{\mathrm{DQD-rad}},$$

where the exciton Hamiltonian H_{DQD} is defined by the formula (9.3) and $H_{\mathrm{DQD-rad}}$ is the interaction Hamiltonian describing collective coupling of the DQD to radiative reservoir in the Dicke limit (Eq. (9.6)).

Since the ground state $|00\rangle$ and the biexcitonic state $|11\rangle$ are the eigenstates of the carrier Hamiltonian, the nontrivial effects take place only in the single-exciton subspace, where an arbitrary state is a linear combination of localized states $|10\rangle$ and $|01\rangle$, $|\Psi\rangle = c_{10}|10\rangle + c_{01}|01\rangle$, with c_{10} and c_{01} being arbitrary complex numbers such that $|c_{10}|^2 + |c_{01}|^2 = 1$. If a DQD system which is collectively coupled to its radiative surrounding is in state $|\Psi\rangle$, then, according to the Fermi golden rule, the probability that the system releases a photon is proportional to the square of the modulus of the matrix element

$$P \sim \left|\langle 00, k\lambda|H_{\mathrm{rad}}|\Psi\rangle\right|^2 = \left|(c_{10}+c_{01})\, g_{k\lambda}\right|^2 \sim \left|(c_{10}+c_{01})\right|^2 \Gamma$$

where $|00, k\lambda\rangle$ denotes the state with no excitons and with one photon in the mode $k\lambda$, $g_{k\lambda}$ is a coupling constant defined by Eq. (9.7), and Γ is the spontaneous decay rate for a single QD [Eq. (9.16)]. From the point of view of the resulting transition probability two out of the states $|\Psi\rangle$ are particularly interesting. These are the superradiant state $|+\rangle = (|10\rangle + |01\rangle)/\sqrt{2}$, for which the transition probability reaches its maximum value, and the subradiant state $|-\rangle = (|10\rangle - |01\rangle)/\sqrt{2}$, for which the probability of emitting a photon vanishes.

The states $|+\rangle$ and $|-\rangle$ are eigenstates of the carrier Hamiltonian (9.3) only if the two dots form an energetically homogeneous atomic like system ($\Delta = 0$). In this case, an exciton occupation of a DQD system prepared in the superradiant state decays rapidly with a decay rate twice faster (2Γ) than when the electron–hole pair is initially localized in one of the dots. If the DQD is in the subradiant state, then the radiative decoherence does not affect the exciton residing in the system, i.e., the state $|-\rangle$ becomes optically inactive (dark) with ab infinite lifetime. This specific property of the subradiant states makes them useful for quantum information processing, especially for noiseless encoding of quantum information [25]. The superradiant state may be excited from the ground state by weak enough optical pulses which do not allow to excite the higher biexciton state or in QD pairs with sufficiently large biexcitonic shift. It is much more difficult to excite the subradiant state, but recently preparation of this state in a pair of superconducting qubits in a cavity using local qubit control in circuit quantum electrodynamics (QED) was presented [117]. The above-mentioned effects are observed either for coupled ($V \neq 0$) and uncoupled ($V = 0$) dots, since in both cases the states $|\pm\rangle$ are eigenstates of the systems. Although the collective effects have been known for many years and extensively studied in atomic systems [22, 51], the first experimental manifestation of these phenomena in QD ensembles was presented in 2007 [24]. This is mostly because, despite many features in common, there are still properties that differ considerably for both types of structures. The main feature that distinguishes these two systems and affects considerably their coupling to photon surrounding is the inhomogeneity of transition energy $\Delta \neq 0$ of the man-made QD systems. Below we investigate the impact of the energy mismatch Δ and coupling between the dots on effects resulting from collective interaction with electromagnetic field.

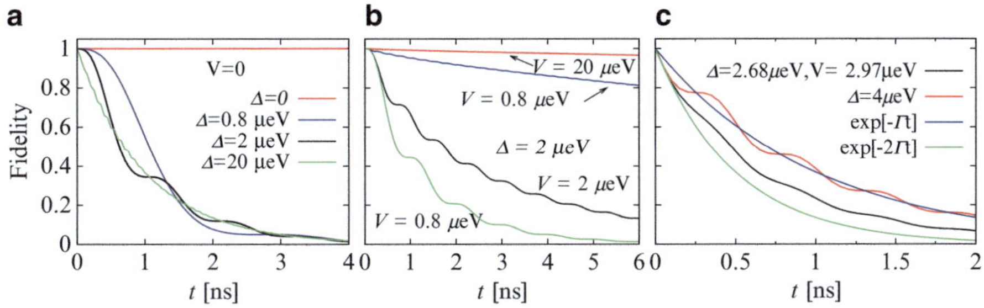

Fig. 9.11 (**a**) and (**b**) The fidelity for a subradiant initial state for a system of uncoupled QDs for a few values of the energy mismatch (**a**) and for a coupled system with the energy mismatch $\Delta = 2\,\mu\text{eV}$. (**c**) The fidelity for a superradiant initial state for a constant energy splitting $\mathcal{E} = 8\,\mu\text{eV}$ compared to the exponential decay with decay rates Γ and 2Γ [61]

9.4.1 Stability of Sub- and Superradiant States

The dynamics of a quantum system is governed by the Schrödinger equation which, for a superposition state, leads to a system of equations for amplitudes of the contributing states. The Hamiltonian in the rotating wave approximation conserves the number of excitations (excitons plus photons), thus the evolution of a system initially prepared in a sub- or superradiant state is governed by a system of three equations for single exciton amplitudes of the localized states $|10\rangle$ and $|01\rangle$ and an amplitude of the state $|00,k\lambda\rangle$. This system of equations may be solved using the Weisskopf–Wigner method [118] which is equivalent to description with the Master equation in the Lindblad form discussed in Sect. 9.2.2.4. For more details, see [119] and [61].

Using the Weisskopf–Wigner method we will solve the evolution equations for a DQD system and test the stability of sub- and superradiant states spanned in an inhomogeneous DQD. In order to do this we will analyze the fidelity $F = \sqrt{\langle\Phi|\rho|\Phi\rangle}$ between the actual state $\rho(t)$ of the exciton subsystem and the pure state $|\Phi(t)\rangle$ evolving from the initial state in the absence of the electromagnetic field ($\Gamma = 0$). In Fig. 9.11a we show the evolution of the fidelity for DQD system initially prepared in the subradiant state $|-\rangle$ in the limit of vanishing coupling between the dots (sufficiently distant dots) [61]. For atomic-like systems ($\Delta = 0$) the fidelity is stable, since the state $|-\rangle$ is the eigenstate of the exciton Hamiltonian of identical dots. This property is extremely sensitive to the homogeneity of the transition energy and is destroyed for energy splittings of the order of transition line width ($\hbar\Gamma = 0.658\,\mu\text{eV}$). As can be seen, the energy mismatch induces transitions between sub- and superradiant regimes which lead to oscillations around the exponential (uncorrelated) decay. The state $|-\rangle$ maintains its stable character until $t \sim \pi\hbar/(2\Delta)$, and then the system enters the superradiant regimes, i.e., the fidelity rapidly drops below the exponential decay to return to the subradiant phase after time $t \sim \pi\hbar/\Delta$. While the energy mismatch increases the number of oscillations also increases, but its amplitude decreases, so in the limit of large energy mismatches the transitions become very frequent, their amplitude becomes small and, as expected for systems interacting with disjoint frequency ranges of the photon reservoir, the decay of fidelity closely follows the exponential function which is characteristic for uncorrelated systems.

The coupling between the single-exciton states of the QDs becomes effective when the distance between the systems is close enough. Contrary to the previous case, where the localized states $|10\rangle$ and $|01\rangle$ were the eigenstates of the system, for coupled dots the eigenstates are linear combinations of the states $|10\rangle$ and $|01\rangle$, i.e.,

$$|\Psi_+\rangle = \cos\frac{\theta}{2}|10\rangle + \sin\frac{\theta}{2}|01\rangle, \tag{9.30}$$

$$|\Psi_-\rangle = -\sin\frac{\theta}{2}|10\rangle + \cos\frac{\theta}{2}|01\rangle, \tag{9.31}$$

where the mixing angle $\theta = \text{arctg}(V/\Delta)$ takes values in the range $[-\pi, \pi)$. If the coupling exceeds the energy splitting, then these states become relatively close to the sub- and superradiant states (they become the sub- and superradiant states only in the ideal case of identical dots, where they correspond to eigenvalues $\pm V$). In this case, the transition to the superradiant regime is suppressed and, as can be seen in Fig. 9.11b, the superradiant character of the evolution is recovered to a great extent. Even if the coupling is not sufficiently strong to rebuild the stability, it is clear from Fig. 9.11b that its existence reduces the decay rate.

If a pair of uncoupled QDs with different transition energies is prepared in a superradiant state (red line in Fig. 9.11c), then the system maintains its superradiant character and decays with a decay rate 2Γ until $t \sim \pi\hbar/(2\Delta)$ and then enters the subradiant regime. As previously, the fidelity oscillates around the exponential decay and the collective character of the evolution, in this case this means a decay with the rate 2Γ, may be recovered for a sufficiently strong coupling between the dot. Apart from the limiting case of vanishing and strong coupling, the decay is non-exponential and its modulation yields information about the origin of the energy splitting in the system. It is clearly seen in Fig. 9.11c, where we compare a decay of the fidelity for two systems with the same energy splitting $\mathscr{E} = 2\sqrt{\Delta^2 + V^2}$, originating only from the difference between the two dots (red curve) or from the system properties and coupling between the single-exciton states (black curve).

As discussed above, the collective character of evolution is destroyed, if the sub- and superradiant states are spanned in inhomogeneous ensembles with energy mismatches of the order of μeV. In realistic DQDs the transition energy splitting is rather on the order of meV than μeV; therefore, once again we will consider the dynamics of the rapidly decaying state $|+\rangle$ and optically inactive state $|-\rangle$, but this time in terms of the exciton occupation of the system and for the technologically achievable $\Delta = 1$ meV. As can be seen in Fig. 9.12, due to the large value of Δ, the decay of the exciton occupation shows no oscillations. In the limit of weak coupling, $V \ll \Delta$, regardless of whether the system was prepared initially in the sub- or superradiant state, the exciton occupation decays exponentially with a decay

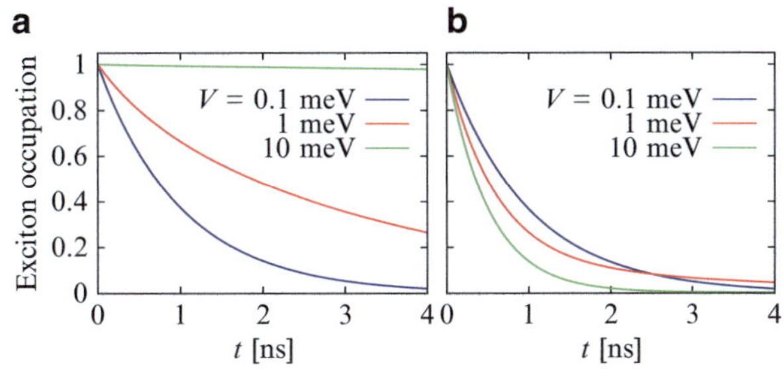

Fig. 9.12 The exciton occupation for sub- (**a**) and superradiant initial state (**b**) for a DQD with technologically achievable energy mismatch $\Delta = 1$ meV, for few values of the coupling between the dots [61]

rate Γ, as expected for dots interacting with photon modes of different energy. For increasing coupling between the dots, the evolution of the excitation becomes non-exponential until the opposite limiting case $V \gg \Delta$ is reached, where the state $|-\rangle$ becomes stable and state $|+\rangle$ decays with a twice larger rate.

The evolution of a superradiant state, which is particularly relevant for optical experiments, is discussed in more detail in [61].

9.4.2 Vacuum-Induced Coherence

A consequence of the existence of sub- and superradiant states is a vacuum-induced coherence effect which is related to spontaneous excitation transfer between emitters and leads to population trapping in a decoherence resistant state [121]. The initial state of the analyzed system is a localized single-exciton state, $|10\rangle$ or $|01\rangle$, which cannot be excited optically, but can appear naturally, e.g., as an effect of incoherent trapping or controlled tunnel injection of carriers in an injection structure similar to that studied in [122], but with two QDs instead of one. The localized initial states may be spit into linear superpositions of dark and bright states,

$$|10\rangle = \frac{1}{\sqrt{2}}(|+\rangle + |-\rangle),$$

$$|01\rangle = \frac{1}{\sqrt{2}}(|+\rangle - |-\rangle),$$

with equal distribution of excitation between these two states. If the investigated system consists of two identical QDs ($\Delta = 0$), then coupling to the photon reservoir induces emission only from the superradiant state, since the subradiant one is unaffected by the electromagnetic surrounding. This leads to the coherent excitation transfer from the initially occupied QD to the initially empty one until occupations of both dots stabilize at the same level, which reflects equal contribution of the localized states $|10\rangle$ and $|01\rangle$ to the superradiant one (Fig. 9.13a). While the photon emission process is taking place, coherence builds up spontaneously in the system (Fig. 9.13b). Finally the DQD system is blocked in a decoherence resistant state

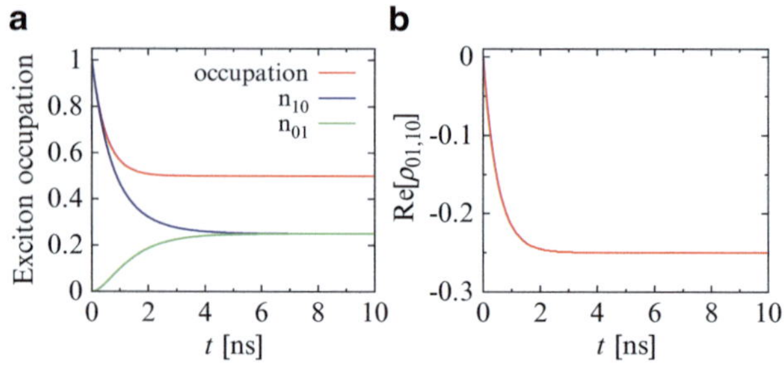

Fig. 9.13 Vacuum-induced coherence effect in a system of two identical QDs. (**a**) Exciton occupation of the DQD system and occupations of the localized states $|10\rangle$ (n_{10}) and $|01\rangle$ (n_{01}). (**b**) The off-diagonal density matrix element $\rho_{01,10}$ [120]

with stable density matrix elements. Irrespective of the choice of the initial state, the coherences are always equal $-0{,}25$, while the occupations of both dots are 0.25, i.e., the total occupation of the system is reduced by half, compared to the initial state. Thus, in a pair of closely spaced QDs, the existence of a nearby empty dot strongly affects the evolution of the excitation of the other QD: The occupation of a single dot decays exponentially while the presence of another dot leads to the vacuum-induced coherence effect and population trapping.

In spite of rapid technological progress, "on demand" manufacturing of DQDs consisting of two identical emitters is still not feasible. In Sect. 9.4.1 we have shown that in the realistic case of nonzero energy mismatch, the collective character of the sub- and superradiant states is lost, and the emission of light from both the sub- and superradiant states takes place. Due to the lack of a stable state in which the system might be trapped, the vacuum-induced coherence effect is destroyed and already for the energy splitting of the order of μeV quenching of the exciton occupation is observed (Fig. 9.14). The coupling to photon reservoir maintains its collective character until $t \sim \pi\hbar/2\Delta$; therefore, the initial character of the exciton occupation evolution does not differ considerably between identical and nonidentical dots and in both cases the occupation of the initial state decreases while the number of excitons confined in the initially empty dot increases, but later the population of both dots starts to decay due to the emission from the subradiant state. For relatively small energy mismatches it is clearly seen that after the emission from the short-living state excitation is transferred to the subradiant state, the occupation of both dots reaches the same number (due to equal, up to a sign, contributions of single-exciton states to the subradiant state) and the following decay of both occupation numbers is indistinguishable (Fig. 9.14b and c).

The effect of vacuum-induced coherence is strongly affected by the coupling (Förster or tunnel) between the single-exciton states. If the spatial separation of the dots is small enough and the coupling exceeds the energy mismatch by far, then the eigenstates of the carrier Hamiltonian (9.30) and (9.31) are relatively close to the sub- and superradiant states, and thus a transition from a localized single-exciton state to a nearly stable delocalized one is possible even for systems

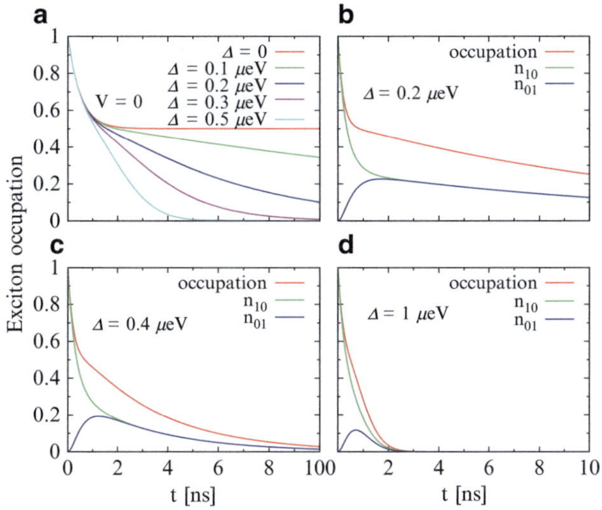

Fig. 9.14 Vacuum-induced coherence effect in a system of two QDs. (**a**) Exciton occupation of the DQD system. (**b**)–(**d**) Exciton occupation of the DQD system and occupations of the localized states $|10\rangle$ (n_{10}) and $|01\rangle$ (n_{01}) for different energy mismatches [123]

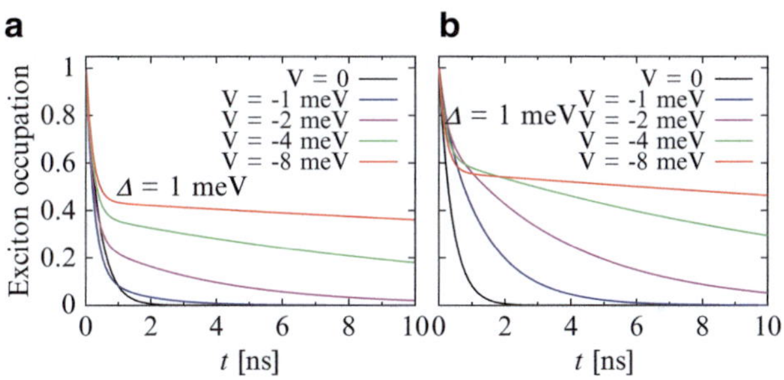

Fig. 9.15 The exciton occupation for a localized initial state $|01\rangle$ (**a**) and $|10\rangle$ (**b**) for a technologically achievable energy mismatch $\Delta = 1$ meV and for few values of coupling between the dots [123]

with a technologically achievable transition energy splitting on the order of meV. Full stabilization of the exciton trapping is impossible for nonidentical systems; therefore, even in strongly coupled DQDs the quenching of the population always takes place, but the decay of the "trapped" state is considerably reduced. For a weaker coupling the decay of the exciton occupation becomes faster, but still the exciton lifetime is much longer than for uncoupled dots (Fig. 9.15).

In contrast to the case of identical QDs, the evolution of coupled dots depends on the choice of the initially occupied dot. For non-vanishing coupling between the dots, the initial localized states may be expressed as non-equal combinations of the eigenstates of the exciton Hamiltonian (9.30) and (9.31),

$$|10\rangle = \cos\frac{\theta}{2}|\Psi_+\rangle - \sin\frac{\theta}{2}|\Psi_-\rangle,$$

$$|01\rangle = \sin\frac{\theta}{2}|\Psi_+\rangle + \cos\frac{\theta}{2}|\Psi_-\rangle,$$

respectively, where the contribution from each eigenstate depends on the rate between the coupling and energy mismatch ($\tan\theta = V/\Delta$). For the negative amplitude of coupling V the initial state $|01\rangle$ (exciton in the lower energy dot) has a stronger overlap with the state $|\Psi_-\rangle$ which has a partly superradiant character. Therefore the contribution from the short-living state to the initial state $|10\rangle$ is larger than from the long-living state and the initial decay of occupation is stronger. On the other hand, if the exciton is initially confined in the higher energy dot, then the initial state $|01\rangle$ has a larger overlap with a partly subradiant state $|\Psi_+\rangle$ and quenching of the occupation is slowed down compared to the previous case.

9.4.3 Optical Response to an Ultrafast Laser Excitation

In this section, we identify the signatures of collective coupling to electromagnetic environment in the linear optical response to the laser excitation. We assume that the DQD, prepared initially in the ground state $|00\rangle\langle 00|$, is excited with an instantaneous laser pulse which is spectrally broad enough in order not to discriminate between the two dots in the structure. Due to a small (sub-wavelength) distance between the dots, they cannot be resolved spatially, either. Therefore, the pulse induces optical polarizations symmetrically and independently in both dots. In most optical experiments, the measured quantity is proportional to the square modulus of the polarization $P(t)$, thus in order to determine the signatures of the superradiant emission we will compare the evolution of polarization induced by the DQD system interacting with common electromagnetic field in the Dicke limit with the response of a hypothetical system consisting of two dots interacting with independent reservoirs. In the first case, after the excitation the DQD evolves according to the Lindblad equation discussed in the Sect. 9.2.2.4. In the second case, the system dynamics are also governed by the Master equation in the Markov approximation (Lindblad) where the Lindblad dissipator is a sum of two elements referring to different dots. For a more detailed discussion, see [124].

9.4.3.1 Uncoupled Quantum Dots

First, we will analyze the optical response induced by a systems of uncoupled quantum dots ($V = V_{\mathrm{B}} = 0$). In this case, the system evolution is determined by the interplay of the energy mismatch Δ and the recombination rate Γ. In Fig. 9.16,

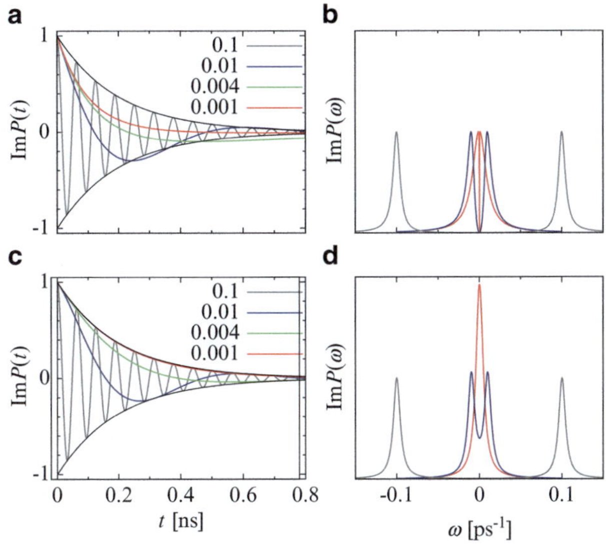

Fig. 9.16 (**a**),(**c**) The optical polarization as a function of time after an ultrafast excitation for uncoupled dots $V = V_B = 0$ in the linear-response limit. (**b**),(**d**) The corresponding spectrum. (**a**), (**b**): QDs interacting with a common reservoir in the Dicke limit; (**c**), (**d**): QDs radiating into independent reservoirs. Here, $\Gamma = 0.01$ ps^{-1} and the values of $\Delta/\hbar$ are as shown in the figure (in ps^{-1}). Line definitions in (**b**) and (**d**) are the same as in (**a**) and (**c**). *Black lines* in (**a**) and (**c**) show the envelope $\pm\exp(-\Gamma t/2)$. The vertical scale in (**b**) and (**d**) is the same [124]

we show the linear optical response in the time and frequency domains for a fixed value of the recombination rate. The imaginary part of the Fourier transform of the polarization is proportional to the absorption spectrum.

For $\Delta \gg \hbar\Gamma$, there is no noticeable difference between systems interacting with common and separate reservoirs as well in the time domain (gray lines in Fig. 9.16a and c) as in the frequency one (gray lines in Fig. 9.16b and d). In both cases the emitted signal is dominated by optical beats due to the interference of fields emitted from the two dots which is manifested in the absorption spectrum as a sum of two Lorentzians centered around distant frequencies. This is not surprising since systems with considerably different transition energies emit into disjoint frequency ranges of the electromagnetic field and thus essentially interact with different photon reservoirs.

The effects of different types of coupling to the electromagnetic environment become essential with decreasing energy mismatch and are manifested by the decrease of the frequency of the oscillations in the collective case (more detailed discussion in [124]). This does not lead to a considerable difference in the time domain (Fig. 9.16a and c), but the effect is clearly seen in the frequency domain (Fig. 9.16b). It is clearly seen in Fig. 9.16b that for a DQD interacting with photon modes in the Dicke limit the absorption spectrum differs considerably from a sum of two identical Lorentzians when the latter overlap.

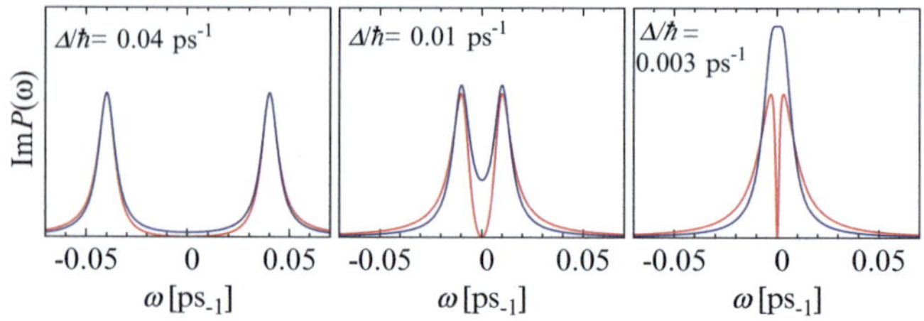

Fig. 9.17 The transition from the "different dots" regime to the "identical dots" regime as manifested by the form of the absorption spectrum of a DQD system. Here, $\Gamma = 0.01\ \mathrm{ps}^{-1}$ and the values of $\Delta/\hbar$ are shown in the figures. *Red lines*: the actual system; *blue lines*: the hypothetical system made of two dots interacting with separate reservoirs [124]

For $\Delta < \hbar\Gamma/2$, $\Omega_+ = \sqrt{(\Delta/\hbar)^2 - \Gamma^2/4}$ is imaginary and the absorption spectrum can be written as

$$\mathrm{Im}P(\omega|\alpha = 0, V = 0, \Delta < \hbar\Gamma/2) = \frac{\Gamma_+}{|\Omega_+|}\frac{\Gamma_+}{4\omega^2 + \Gamma_+^2} - \frac{\Gamma_-}{|\Omega_+|}\frac{\Gamma_-}{4\omega^2 + \Gamma_-^2},$$

where $\Gamma_\pm = \Gamma \pm 2|\Omega_+|$. In the limit of $\Delta \to 0$ the decay rates $\Gamma_+ \to 2\Gamma$ and $\Gamma_- \to 0$, and thus the width of the first Lorentzian becomes twice larger than in the case of emission to separated reservoirs while the width of the second one vanishes. Since the transition from a single exponential decay rate to the twice larger one is a signature of the existence of a short-living state, the first Lorentzian corresponds to the superradiant component of the evolution. On the other hand, the transition which leads to quenching of the decay rate is a fingerprint of subradiant character of the second (negative) component. As can be seen in Fig. 9.16b and d, the amplitudes of the two Lorentzians are always opposite and thus the condition $P(\omega = 0) = 0$ is fulfilled. These spectral features are reflected in the time domain. In the intermediate range of Δ (Δ slightly below $\hbar\Gamma/2$) the decay of polarization becomes a sum of two exponential factors: one positive, large and short living and the other one, negative, small and long living. As seen in Fig. 9.16a and c, also in this range it is hard to distinguish systems interacting with common reservoirs from the ones emitting to disjoint energy ranges where the oscillations are strongly damped. The analysis of the emitted signal in the time domain allows to specify the coupling only in the case of nearly identical dots ($\Delta \ll \hbar\Gamma/2$) where for the collectively coupled systems the initial response is dominated by the superradiant component and thus the polarization decays with a twice as large rate compared to the dots coupled to separate reservoirs (red lines in Fig. 9.16a and c).

In Fig. 9.17, we compare the transition between the limiting cases of non-identical and identical quantum dots for systems coupled to common and separate reservoirs. As long as the fundamental transition energies of the two dots differ considerably ($\Delta \gg \hbar\Gamma/2$) the absorption spectrum of both types of systems is indistinguishable, since the dots emit into disjoint frequency ranges. The situation changes with decreasing energy mismatch. In the intermediate range of $\Delta \sim \hbar\Gamma/2$, the spectrum is non-Lorentzian and difference between quantum dots coupled to common and separated reservoirs is clearly seen. Then, with decreasing Δ, the spectrum of collectively coupled systems switches to the sum of two Lorentzians with different widths and opposite signs centered around the zero frequency (only in this case the polarization has two exponential components). In the hypothetical case of quantum dots emitting to separate reservoirs, the spectrum is a sum of two identical Lorentzians.

9.4.3.2 Coupled Quantum Dots

Below we analyze the role of the coupling between single-exciton states ($V \neq 0$, $V_B = 0$) in the optical response to a laser pulse excitation. In this case, the transition to the collective regime takes place for the energy mismatches of the order of the coupling between the systems which we assume to be larger than the relaxation rate, $V \gg \hbar\Gamma$. Thus the decay rate Γ becomes the smallest frequency parameter of the system and one can simplify the discussion by retaining only terms linear in Γ or assume $\Gamma = 0$ (for more detailed discussion, see [124]). With these assumptions, a decay of the linear polarization is described by a sum of two exponential components with decay rates $\Gamma_{\pm} = [1 \mp 2V/\mathscr{E}]\Gamma$, where $\mathscr{E}$ is the energy splitting. One of these rates, subradiant, decays slowly with a rate $\Gamma_{-}/2 < \Gamma/2$ and its amplitude vanishes in the limit of $\Delta \ll V$ while the second, superradiant, component decays rapidly with a rate $\Gamma_{+}/2 > \Gamma/2$ and dominates the evolution in the limit of strongly coupled dots.

In Fig. 9.18a and c, we show the evolution of the envelopes of the optical beats for three sets of parameters corresponding to the same energy splitting $\mathscr{E} = 6\,\mathrm{ps}^{-1}$. One effect that is seen in the dynamics of the systems coupled to common as well as to separate reservoirs is the decrease of the beating amplitude. Although the excitation pulse acts symmetrically on both dots, the occupation of one dot has a larger overlap with one of the eigenstates and thus the two systems emit radiation with different amplitudes which leads to the reduction of the beating amplitude of the total signal.

The second effect, i.e., the change in the decay rate, is observed only in systems collectively coupled to electromagnetic environment (Fig. 9.18a). For large energy mismatches ($\Delta \gg V$) the polarization decays with a characteristic for single dots decay rate $\Gamma/2$. With the increasing coupling (and decreasing Δ at the same time) the polarization decay is transferred from a single exponential decay to a sum of two exponential factors with different rates (long- and short-living terms). In the limit of strong coupling ($V \gg \Delta$) only the superradiant state is excited (it coincides with the optically active symmetric superposition) and thus the signal decays with twice larger rate than in the case of $\Delta \gg V$. On the contrary, the signal emitted by a pair of dots coupled to separated reservoirs always decays with a single decay rate.

Also in this case, the collective effects are more clearly seen in the absorption spectrum shown in Fig. 9.18b, as well as the difference between common and

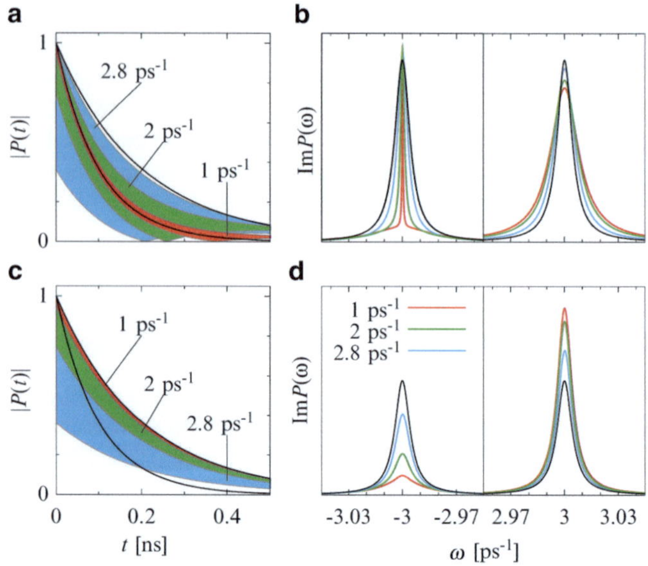

Fig. 9.18 *Left panels*: envelope of the optical beats for two coupled QDs emitting to a common reservoir and to separate reservoirs. For each value of $\Delta/\hbar$ shown in the plot, the coupling V is adjusted so that $\mathscr{E} = 6\ \mathrm{ps}^{-1}$ in each case. *Right panels*: the corresponding absorption spectra. Plots (**a**) and (**b**) show the properties of a DQD emitting to a common reservoir, while plots (**c**) and (**d**) correspond to two dots emitting to separate reservoirs. *Black lines* in (**a**) and (**c**) show exponential decay curves with the rates $\Gamma/2$ and Γ, while in (**b**) and (**d**) they show Lorentzians of equal weight and width $\Gamma/2$ [124]

separated reservoirs. In both cases the increase of the coupling (and decrease of Δ) induces a similar transfer of line weight from one line to the other, but the way the line shape changes is substantially different for the two types of interaction. In the case of independent reservoirs, only the line amplitude changes while the width of the two Lorentzians remains constant. If the dots interact with common electromagnetic field, then the line amplitudes change only slightly, but the shape of the two components allows to distinguish the subradiant factor for which the width of the Lorentzian decreases with increasing coupling and the superradiant factor for which the width increases while the energy mismatch tends to zero. It is interesting to note that since the weight of each line behaves almost in the same way in both cases no difference can be observed if the absorption spectrum is averaged over an inhomogeneous ensemble of DQDs for which the transition energies are distributed on the wider interval than the width of a single line. The static dipole interaction which induces the biexcitonic shift is relevant only for third and higher order contributions to the optical response of collectively coupled dots. It leads to the biexcitonic beats which show two-rate decay in the intermediate range of parameters ($\Delta \sim \Gamma\hbar$) which becomes dominated by the superradiant component in the limit of $\Delta \to 0$ [124].

9.4.4 Four-Wave Mixing

Effects resulting from collective coupling to electromagnetic field are also visible in the nonlinear response of DQD samples. Below we will show how this kind of coupling affects the four-wave mixing response of an ensemble of DQDs. The modeled experiment consists of exciting the system of many DQDs with two ultrafast laser pulses delayed in time by τ. To obtain the optical response of the whole ensemble, we first calculate the polarization induced by a single DQD. As in the previous case, we assume that the laser pulses act symmetrically on both dots in the pair. After the first excitation, arriving at time $t = -\tau$, the system evolves according to the Lindblad equation discussed in Sect. 9.2.2.4. The second laser pulse arrives at time $t = 0$. Later, the system again evolves according to the Lindblad equation. In order to extract the signal measured in the four-wave mixing experiment we pick out only the terms containing the phase factor $\exp[i(2\phi_2 - \phi_1)]$, where the $\phi_{1,2}$ are the phases of the exciting pulses. The signals emitted by ensembles of DQDs are usually very weak and thus their detection is based on a heterodyne technique [125], where the signal emitted by the system of QDs is superimposed on the reference signal. Therefore, in the next step, we calculate the overlap of the signal induced by quantum dots and the reference signal which arrives at time $t = t_0$ and has a Gaussian envelope. Since we want to calculate the signal emitted by an ensemble of DQDs we average the polarization induced by a single pair with a Gaussian distribution for the fundamental transition energies E_1 and E_2 which, expressed in terms of the average transition energy and the energy mismatch, becomes a product of two Gaussian functions, $g(E_1, E_2) = g(E, \Delta) = g(E)g(\Delta)$, where $g(E)$ and $g(\Delta)$ are the Gaussian distributions of the mean transition energy and the energy mismatch,

$$g(E) = \frac{1}{\sqrt{2\pi}\sigma_E} \exp\left[-\frac{(E-\bar{E})^2}{2\sigma_E^2}\right] \quad \text{and} \quad g(\Delta) = \frac{1}{\sqrt{2\pi}\sigma_\Delta} \exp\left[-\frac{(\Delta-\bar{\Delta})^2}{2\sigma_\Delta^2}\right],$$

respectively, with the corresponding variances σ_E and σ_Δ. Following this procedure and performing some approximations, which are discussed in detail in [126], we find the signal in the form

$$F(t_0, \tau) = \sum_{n=1}^{3} F_n(t_0, \tau),$$

where

$$F_n(t_0, \tau) = \frac{i}{8} \sin(\alpha_1) \cos^2\left(\frac{\alpha_1}{2}\right) \sin^2(\alpha_2) \int d\Delta g_\Delta(\Delta)[\mathscr{F}_{n+}(\Delta) + \mathscr{F}_{n-}(\Delta)]. \quad (9.32)$$

Here, $\alpha_{1,2}$ are the areas of the exciting pulses and

$$\mathscr{F}_{n\pm}(\Delta) = e^{i\overline{E}(t_0-\tau)/\hbar} \int dt P_{n\pm}^{(+)}(t,\Delta) E_{\mathrm{ref}}^{(-)}(t) = \frac{1}{4\sqrt{1+\tau_{\mathrm{ref}}^2 \sigma_E^2/\hbar}} \tag{9.33}$$

$$\times \exp\left[-\frac{\sigma_E^2 (t_0-\tau)^2}{2(\hbar^2+\tau_{\mathrm{ref}}^2\sigma_E^2)}\right] \exp\left[\mp i \frac{\mathscr{E}(t_0-\tau)}{2(\hbar+\tau_{\mathrm{ref}}^2\sigma_E^2/\hbar)}\right] \Phi_{n\pm}(\Delta),$$

where τ_{ref} is the variance of the reference signal and

$$\Phi_{1\pm}(\Delta) = \beta_\pm^2 \exp\left[-\frac{\mathscr{E}^2\tau_{\mathrm{ref}}^2}{8(\hbar^2+\tau_{\mathrm{ref}}^2\sigma_E^2)}\right] \exp[-\beta_\pm \Gamma \tau], \tag{9.34}$$

$$\Phi_{2\pm}(\Delta) = \beta_+\beta_- \exp\left[-\frac{\mathscr{E}^2\tau_{\mathrm{ref}}^2}{8(\hbar^2+\tau_{\mathrm{ref}}^2\sigma_E^2)}\right] \exp\left[\mp\frac{i\mathscr{E}\tau}{\hbar}\right] \exp[-\Gamma\tau], \tag{9.35}$$

$$\Phi_{3\pm}(\Delta) = -\beta_\mp \exp\left[-\frac{(\mathscr{E}/2 \pm V_{\mathrm{B}})^2\tau_{\mathrm{ref}}^2}{2(\hbar^2+\tau_{\mathrm{ref}}^2\sigma_E^2)}\right] \exp\left[-i\frac{(V_{\mathrm{B}}\pm\mathscr{E})\tau}{\hbar} - \frac{iV_{\mathrm{B}}(t_0-\tau)}{(\hbar+\tau_{\mathrm{ref}}^2\sigma_E^2/\hbar)}\right]$$
$$\times \exp[-(1+\beta_\mp)\Gamma\tau], \tag{9.36}$$

with $\beta_\pm = 1 \pm 2V/\mathscr{E}$.

To calculate the nonlinear response from the ensemble of DQDs in the time domain (FWM signal), we integrate Eq. (9.33) numerically with the weight factor $g(\Delta)$. Another integration, over the arrival time t_0 of the reference pulse, gives the time integrated four-wave mixing signal (TIFWM) as a function of the delay time τ which is used to characterize phase decoherence.

Due to the Gaussian term in the formula (9.32), the signal is restricted to short ranges of delay times τ, of the width $\sim \hbar/\sigma_{\mathrm{E}}$, centered around t_0 and thus the DQD response shows "photon echo" type behavior. The contribution $F_1(t_0,\tau)$ [Eqs. (9.32) and (9.34)] has a different structure of phase factors $\exp(i\mathscr{E}\tau/\hbar)$ which results in different [compared to $F_2(t_0,\tau)$ and $F_3(t_0,\tau)$] character of evolution. The energy mismatch Δ and the energy splitting $\mathscr{E}$ vary across the ensemble and thus the terms $\exp(i\mathscr{E}\tau/\hbar)$ interfere destructively when the signal emitted by different DQDs is summed. The relevant phase factor appearing in the first component F_1 depends only on $t_0-\tau$ which is limited to the width $\sim\hbar/\sigma_{\mathrm{E}}$ by the Gaussian term in (9.32); thus, the spread of the phase factors is also limited and independent of τ. Therefore, the phase factor induces only oscillations in $\mathrm{Im}F_1(t_0,\tau)$ (Fig. 9.19a and e) which are always in phase with the center of the echo peak and thus the area of the echo pulse (time-integrated four-wave mixing signal) is constant.

The contributions $F_2(t_0,\tau)$ and $F_3(t_0,\tau)$ contain phase terms proportional to $\mathscr{E}\tau/\hbar$ and $(V_{\mathrm{B}}\pm\mathscr{E})\tau/\hbar$ which induce variations of the phase of the signal at $t_0=\tau$, with the delay time τ, and subsequently variations in the shape and magnitude of the echo signal. The strong dependence on the delay times τ is clearly seen, if one compares Fig. 9.19b and c with Fig. 9.19f and g. Consequently, the total measured

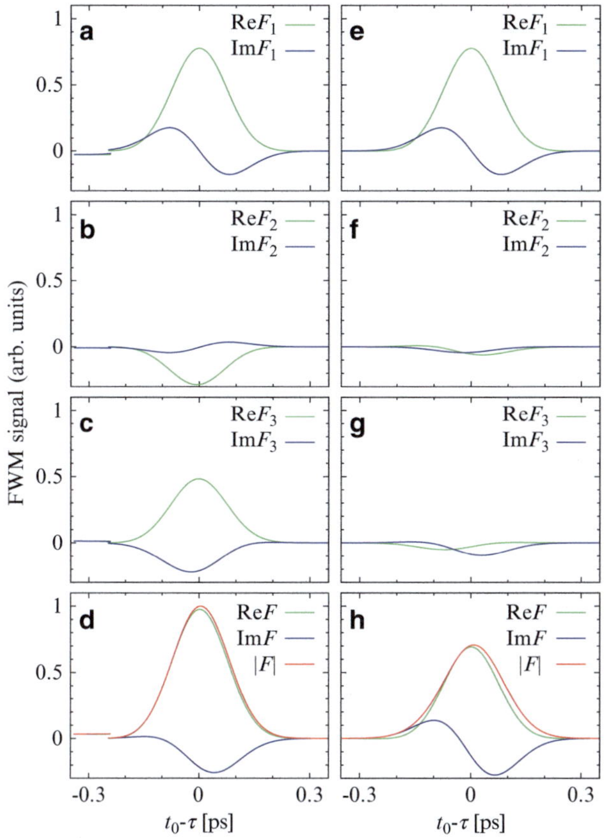

Fig. 9.19 (**a**)–(**c**) Real and imaginary parts (*green and blue lines*, respectively) of the three contributions $F_n(t_0, \tau)$ to the FWM echo $F(t_0, \tau)$ for short delays, with $\tau = 0.23$ ps, $V = 2$ meV, $V_B = 1$ meV, and $\sigma_\Delta = 1$ meV. (**e**)–(**g**) As previously, but at $\tau = 0.61$ ps. (**d**) and (**h**) Real and imaginary parts of the total signal $F(t_0, \tau)$, as well as its amplitude, which corresponds to the measured signal, for the two values of τ [126]

signal also varies with time τ on the picosecond scale (Fig. 9.19d and h). This leads to oscillations in the time-integrated four-wave mixing signal (equal to the area of the echo pulse) at short (picosecond) time scales which are a manifestation of the optical beats between the two dots forming all the quantum dot pairs (Fig. 9.20).

The character of the initial evolution of the time-integrated signal depends on the interplay between σ_Δ and $\overline{\Delta}$. If $\sigma_\Delta \ll \overline{\Delta}$, then the measured signal is a superposition of the fields emitted by all DQDs whose energy mismatches Δ are spread roughly within a range $(\overline{\Delta} - \sigma_\Delta; \overline{\Delta} + \sigma_\Delta)$. Since the effective energy mismatch depends on Δ, the inhomogeneity of fundamental transition energies leads to the spread of the phase factors, $\exp(i\mathscr{E}\tau/\hbar)$, which increases with increasing delay time. Due to this phase distribution the signals emitted by different DQDs interfere destructively

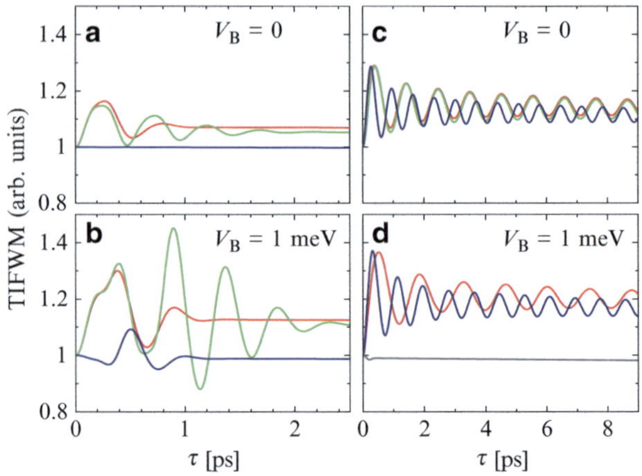

Fig. 9.20 (**a**) and (**b**) The evolution of the time-integrated signal for short delays without (**a**) and with (**b**) a biexcitonic shift for $\sigma_\Delta \ll \bar{\Delta}$ (no resonant dots in the ensemble). *Red lines*: $V = 2$ meV and $\sigma_\Delta = 1$ meV; *green lines*: $V = 2$ meV and $\sigma_\Delta = 0.5$ meV; *blue lines*: $V = 0$ and $\sigma_\Delta = 1$ meV. (**c**) and (**d**) As in (**a**) and (**b**), respectively, but for σ_Δ comparable with $\bar{\Delta}$. *Red lines*: $V = 2$ meV and $\sigma_\Delta = 3$ meV; *green line*: $V = 2$ meV and $\sigma_\Delta = 5$ meV; *blue lines*: $V = 3$ meV and $\sigma_\Delta = 3$ meV; and *gray line*: $V = 0$ and $\sigma_\Delta = 3$ meV [126]

which leads to quenching of the contributions $F_2(t_0, \tau)$ and $F_3(t_0, \tau)$ at delay times $\tau \sim \hbar\overline{\mathscr{E}}/(\Delta\sigma_{\overline{\Delta}})$ and thus a decay of the oscillations (Fig. 9.20a and b). This effect is a result of the symmetric action of the second exciting pulse which can invert only the dephasing process induced by the inhomogeneous distribution of the average transition energies, but not that resulting from the inhomogeneity of the energy mismatches in the ensemble.

If $\sigma_\Delta \sim \overline{\Delta}$, then the analyzed sample contains a subensemble of resonant QDs which have nearly identical fundamental transition energies. The fraction of identical DQDs emits radiation in phase because the frequency $\mathscr{E}/\hbar$ has a minimum at $\Delta/\hbar = 0$, which corresponds to a stationary point of the phase distribution over the ensemble. Therefore, the signal emitted by these resonant dots may dominate the contributions $F_2(t_0, \tau)$ and $F_3(t_0, \tau)$ even, if the subensemble of identical dots constitutes a minority in the ensemble. As can be seen in Fig. 9.20c and d, the beats induced by the resonant DQDs show much slower damping than those originating from the majority of pairs, but still they vanish on time scales of several picoseconds.

For the resonant subensemble $\Delta \sim 0$ and thus the frequency of the beats is very close to $2V/\hbar$ or $(2V - V_B)/\hbar$ and does not depend on the energy mismatch. In the limit of $\Delta \to 0$ the weight factor β_- associated with the frequency $(2V + V_B)/\hbar$ in the formula (9.36) vanishes and thus the term with the frequency $(2V - V_B)/\hbar$ dominates the response (beats). It is essential because in this way the time-integrated four-wave mixing signal allows one to characterize the coupling between the dots and gives the direct access to the properties of the resonant dots even, if they are only a minority of the investigated ensemble.

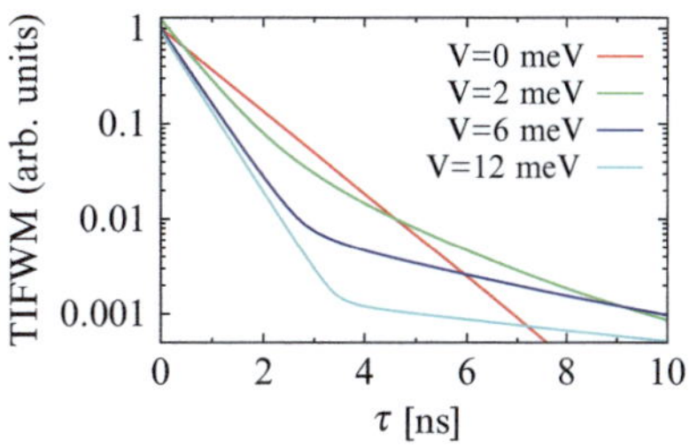

Fig. 9.21 Dependence on the coupling strength V of the time-integrated FWM signal for $\sigma_\Delta = 3$ meV [126]

For longer delay times of the order of nanoseconds, the two components $F_2(t_0, \tau)$ and $F_3(t_0, \tau)$ vanish and the total measured signal is described only by the first component $F_1(t_0, \tau)$ which is split into two terms. One of the terms decays with the rate $\Gamma_+ = \beta_+ \Gamma > \Gamma$ which reaches 2Γ for $V \gg \Delta$ and thus shows a superradiant character. The other, subradiant, term has the decay rate $\Gamma_- = \beta_- \Gamma < \Gamma$ and vanishes in the limit of strongly coupled dots.

In Fig. 9.21, we show a transition of the time-integrated four-wave mixing signal from the regime of uncoupled QD pairs to strongly coupled systems. In the limit of vanishing coupling, both decay rates Γ_+ and Γ_- are equal to Γ and one observes single exponential decay. As the coupling between the dots increases the decay rates, Γ_+ and Γ_- of the two components become different and the decay of the measured optical response becomes non-exponential. In the intermediate range of parameters, the decay rate sets between Γ and 2Γ, but the decay quickly deviates from being exponential. If the coupling between the dots is sufficiently strong, the superradiant term dominates in the optical response (in this limit the relative amplitude of the two components is $\beta_+/\beta_- > 1$), but this is a short-living term and decays very fast, therefore for the longer delay times the measured signal is dominated by the slowly decaying (subradiant) contribution (Eq. (9.21)).

Interestingly, in the leading (third) order of the response, the biexcitonic shift appears only in the short-living term $F_3(t_0, \tau)$ which vanishes on delay times τ of the order of several picoseconds. At longer times, this type of coupling does not affect the four-wave mixing signal and thus the beats from the molecular biexcitons do not appear in the measured signal.

9.5 Interplay of Phonon-Related and Radiative Effects

A real QD system interacts with its phonon environment as well as with the radiation field. In this section, we present the results of theoretical modeling of a DQD system interacting with both these reservoirs. First, we describe the effect of phonon-induced transitions and thermalization of occupations on the kinetics of exciton recombination (spontaneous emission). Next, we revisit the problem of entanglement decay and extend the description by including spontaneous emission effects. Finally, we study the phonon impact on the process of generating vacuum-induced coherence.

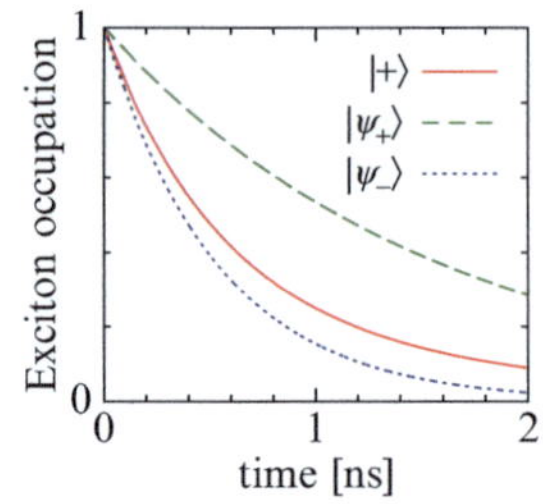

Fig. 9.22 The decay of the exciton occupation without carrier–phonon interaction for three different initial states at constant temperature $T = 4$ K, $D = 6$ nm, and $\theta = -\pi/6$ [52]

9.5.1 *The Role of Phonons in the Collective Luminescence from Double Quantum Dots*

In this section, we present and discuss the results of numerical simulations of the spontaneous emission from DQD systems under the influence of phonons [52]. We study theoretically the kinetics of radiative recombination of excitons confined in a system consisting of two coupled self-assembled QDs.

As discussed in Sect. 9.4.1, the coupling between the dots leads to the formation of two delocalized eigenstates $|\Psi_{\pm}\rangle$ with different decay rates. We assume negative coupling ($V \leq 0$); hence, the mixing angle of the single-exciton states (defined in Sect. 9.4.1) is also negative, $\theta \leq 0$. In this case, the state $|\Psi_{+}\rangle$ is darker and decays slower while the state $|\Psi_{-}\rangle$ is brighter and decays faster. Figure 9.22 presents the evolution of the exciton occupation in the DQD system without carrier–phonon coupling (in this case, the resulting decay depends only on the mixing angle θ). Two limiting cases, the dark and bright state, are shown with the green dashed and blue dotted lines, respectively. For a general initial state, including the state $|+\rangle = (|01\rangle + |10\rangle)/\sqrt{2}$ (red solid line), the decay is not simply exponential. Note that this state is particularly interesting because it corresponds to optical excitation by a laser pulse.

When the carrier–phonon interaction is taken into account, one observes changes of the recombination kinetics. Typically, carrier–phonon interaction leads to the redistribution of the occupations on a time scale much shorter than the exciton life time, hence the dependence on the initial state observed in Fig. 9.22 is suppressed. As shown in the left panel of Fig. 9.24, for strong phonon coupling, corresponding to the first maximum of the spectral density (Fig. 9.23), the radiative decay is nearly insensitive to the initial state. When the energy splitting corresponds to the first minimum of the spectral density, the situation is still very similar in spite of a much weaker coupling, as shown in the right panel of Fig. 9.24.

An important effect of phonon interaction in radiative recombination is a strong temperature dependence of the emission process. This is shown in Fig. 9.25 (left panel) for the initial optically generated state $|+\rangle$ at various temperatures and system parameters. In this case, phonon-induced relaxation is efficient as the energy splitting $2\mathcal{E}$ corresponds to the first maximum of the spectral density. With increasing temperature the occupation of the dark state increases at the

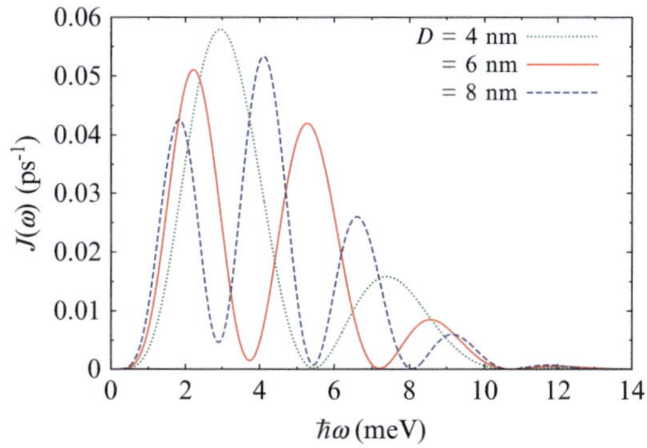

Fig. 9.23 The phonon spectral density for a few values of the distance between the dots [52]

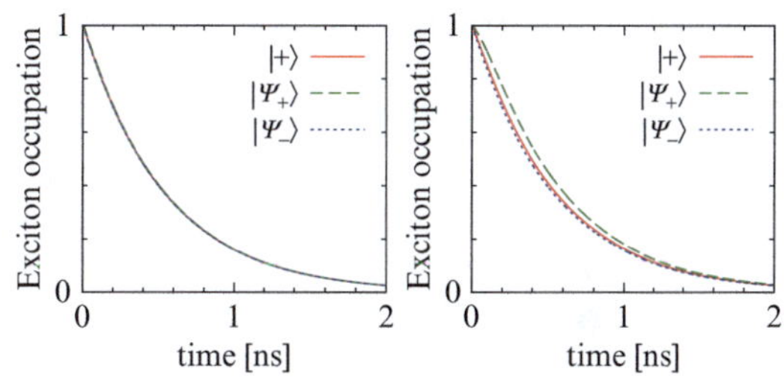

Fig. 9.24 The decay of exciton occupation with phonons (*left*: $2\mathcal{E} = 1.77$ meV; *right*: $2\mathcal{E} = 3.55$ meV) for the same initial conditions as in Fig. 9.22 [52]

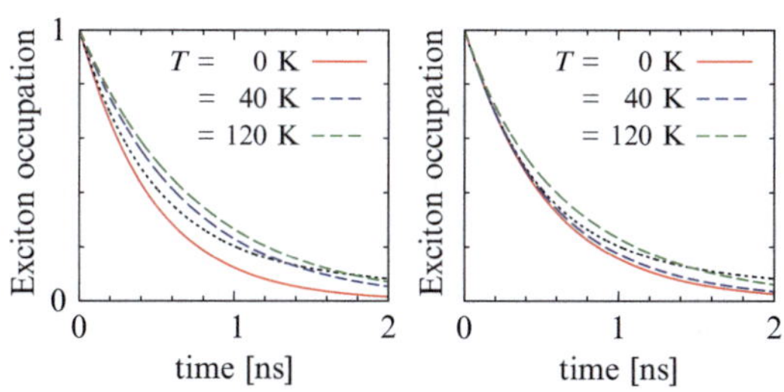

Fig. 9.25 The decay of exciton occupation for the initial state ($|+\rangle$) for the distance $D = 6$ nm at a few temperatures, dominated by relaxation ($\theta = -\pi/3$). *Left*: with strong carrier–phonon coupling ($2\mathcal{E} = 1.77$ meV); *right*: with a weaker coupling ($2\mathcal{E} = 7.1$ meV) [52]

thermal quasi-equilibrium and the resulting decay becomes slower. This effect of the resulting redistribution of occupations can be observed also for higher values of $2\mathcal{E}$ close to the second minimum of the spectral density (see Fig. 9.25, right panel), but the effect is weaker and appears at higher temperatures.

The effect of thermal extension of the exciton lifetime can be directly reflected in the dynamics of photoluminescence (PL), which may be observed in a time-resolved experiment. In such an experiment, the sample is excited with a light pulse and the decay of PL can be measured as a function of time. We calculate the PL signal by multiplying the occupations of the eigenstates $|\Psi_+\rangle$ and $|\Psi_-\rangle$, by the corresponding decay rates $\Gamma_\pm = \Gamma_0(1 \pm \sin\theta)$, where Γ_0 is the spontaneous decay rate for a single dot. The PL depends on the occupations of the states (dark and

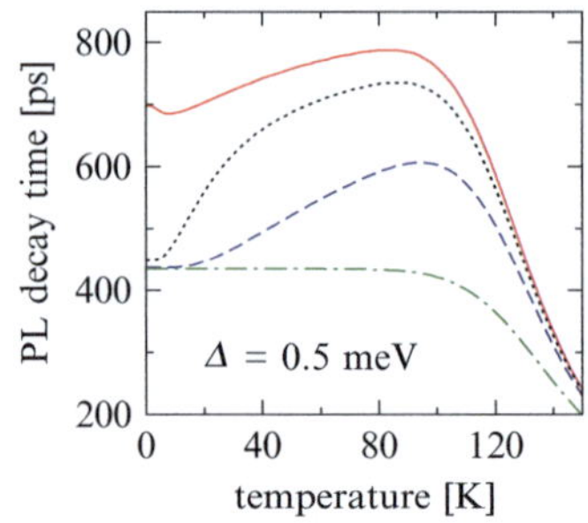

Fig. 9.26 The decay time of the photoluminescence from a single DQD with $\Delta = 0.5$ meV and three different values of the coupling: $V = -0.2$ meV (*red solid line*), $V = -1$ meV (*black dotted line*), $V = -3$ meV (*blue dashed line*), and $V = -8$ meV (*green dash-dotted line*) [52]

bright), so we study the evolution of the intensity of the PL, $I_{\rm PL} = n_+\Gamma_+ + n_-\Gamma_-$, where $n_\pm$ are the occupations of the state $|\Psi_+\rangle$ and $|\Psi_-\rangle$, respectively. Additionally, the values obtained from the numerical simulations are suppressed by the factor $\exp(-\Gamma_{\rm nr}t)$, corresponding to thermally activated non-radiative processes [127]. We set $\Gamma_{\rm nr} = \Gamma_{\rm nr}^{(0)} \exp[-E_{\rm a}/(k_{\rm B}T)]$, corresponding to the exciton escape [127]. The PL decay time is determined as the time when the PL intensity is reduced by the factor $1/e$ (the PL decay is not exactly exponential).

In the case of a single DQD structure, we assume that the system under study is inhomogeneous and optically prepared in the initial state $|+\rangle$ although, as we mentioned previously, the initial conditions are not very important, if the phonon-induced thermalization is fast enough. In Fig. 9.26, we choose a fixed value of the energy mismatch, $\Delta = 0.5$ meV, and plot the results for several values of the inter-dot coupling. At low temperatures, the observed rate is dominated by radiative recombination, while at higher temperatures non-radiative suppression plays a major role. The red solid line corresponds to a very weak coupling between the dots, so that the energy mismatch dominates and the two dots are almost equally bright. Hence, their radiative rates are similar. Therefore, thermal redistribution of their occupations leads only to marginal changes in the PL decay time with temperature. With increasing coupling values (black dotted and blue dashed lines), the ground state has a considerably increased decay rate; hence, the PL decay is much faster at low temperatures. At moderate temperatures, the increasing thermal occupation of the higher, dark state leads to a considerable growth of the PL decay time until the non-radiative effects become important, which leads to the formation of a pronounced maximum. This situation changes when the coupling strengths are higher: in this case, there is no maximum since the energy splitting between the two states (roughly equal to $2V$) is large enough to prevent a non-negligible occupation of the dark state in the relevant temperature range.

In a more feasible case, the PL signal is collected from an inhomogeneous ensemble composed of DQDs with different fundamental transition energies. The results for a model in which the values of Δ vary across the ensemble according to the Gaussian distribution with the mean value $\overline{\Delta} = 0$ and the standard deviation σ are shown in Fig. 9.27 (for the initial state $|+\rangle$). In the left panel, we set $\sigma = 0.5$ meV and show the results for a few coupling strengths V, in analogy with Fig. 9.26.

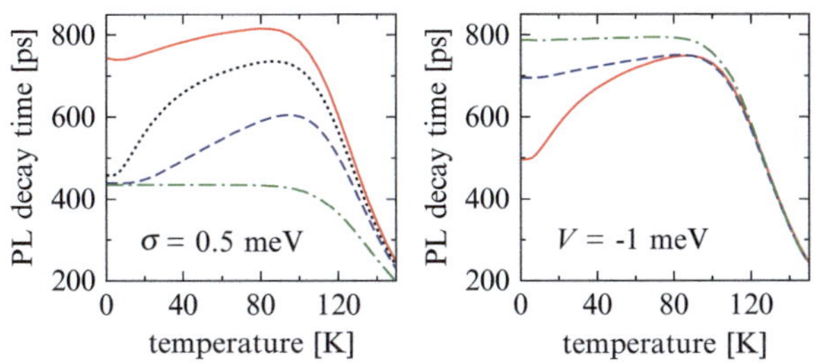

Fig. 9.27 The decay time of the photoluminescence from an ensemble of DQDs with the average energy mismatch $\overline{\Delta} = 0$. *Left*: the standard deviation of Δ is equal to $\sigma = 0.5$ meV and the values of the coupling V are the same as in Fig. 9.26. *Right*: $V = 1$ meV, $\sigma = 1$ meV (*red solid line*), 3 meV (*blue dashed line*), and 10 meV (*green dash-dotted line*) [52]

The resulting temperature dependence is very similar to that observed for a single dot (Fig. 9.26). In the right panel of Fig. 9.27, the results for larger values of the standard deviation σ are shown. Now, the shape of temperature dependence changes significantly. When $\sigma \gg |V|$, the ensemble is dominated by DQDs that are composed of very different dots. This is the situation when $\Delta \gg |V|$ for which the collective effects are negligible. Hence, one does not observe a decrease of the PL decay time at low temperatures for such ensembles.

9.5.2 Entanglement Decay

In Sect. 9.3.4, we discussed the evolution of entanglement in the DQD system in the case of a restricted Hamiltonian which can be diagonalized exactly. In the following, the full model will be taken into account, including both the phonon reservoir and coupling to the radiation vacuum, and we will discuss the role of radiative recombination for long-time entanglement decay, as well as the effects on the evolution of entanglement induced by the phonon-assisted excitation transfer, and the results of the interplay between the interactions of the DQD system with the phonon and photon environments [62].

The extended model is much more complicated and presents many new features, on which we wish to focus. To this end, we will restrict the discussion to one of the previously discussed, fully entangled initial states, namely the "singlet" state of Eq. (9.29). As discussed previously, the zero-temperature approximation may be used for the radiation reservoir at any reasonable temperature, so there is no interaction in the system which can lead to the creation of the biexcitonic state and the state space is restricted to the three lower-energy DQD states, $|00\rangle$, $|01\rangle$, and $|10\rangle$. In this case, the problem of calculating the concurrence and subsequently the EOF is greatly reduced since the concurrence is always proportional to the

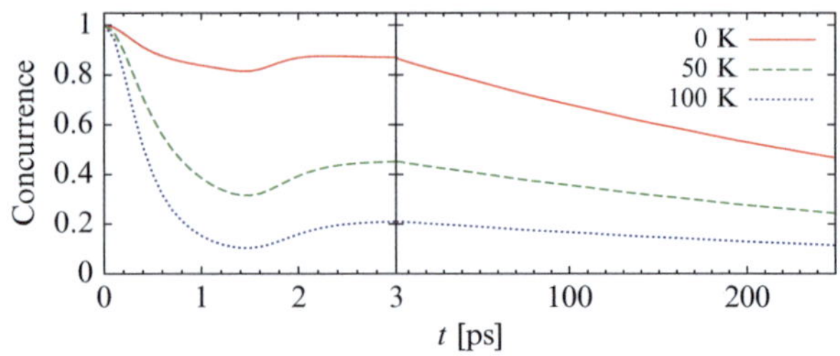

Fig. 9.28 Evolution of entanglement of the initial singlet state under the influence of radiative recombination without the tunnel coupling ($V = 0$) at various temperatures [62]

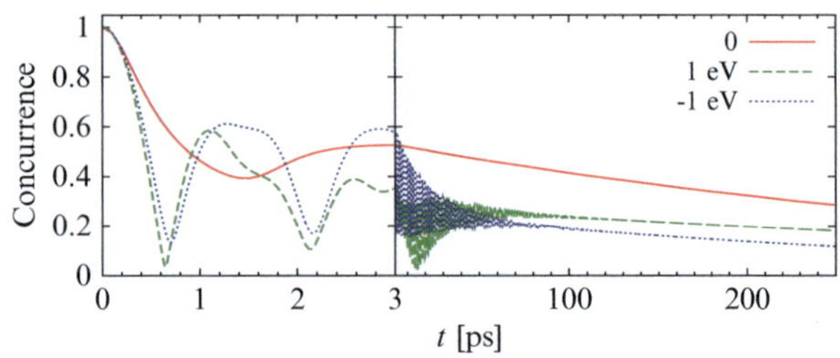

Fig. 9.29 Evolution of entanglement for tunnel-coupled dots ($V = \pm 1$ meV, as shown) compared to the uncoupled case ($V = 0$) at $T = 40$ K [62]

amplitude of the off-diagonal element of the density matrix between the states $|01\rangle$ and $|10\rangle$, and is now equal to

$$C[\rho(t)] = 2|\langle 01|\rho(t)|10\rangle|. \tag{9.37}$$

The evolution of the entanglement between the dots is shown in Fig. 9.28. At short (picosecond) time scales, phonon-induced dephasing leads to a drop of concurrence, ending with a temperature-dependent plateau level, analogously to what is seen in the right panel of Fig. 9.8. In the absence of radiative recombination, the entanglement would remain constant after this initial dephasing stage. However, in the presence of carrier–photon coupling the exciton lifetime becomes finite, which leads to an exponential population decay and, in consequence, to decay of entanglement on the time scales $\sim 1/\Gamma$. The situation becomes much more complicated, if the dots are coupled by a transfer-type interaction (that is, $V \neq 0$). The evolution in this case is shown in Fig. 9.29. The interaction leads to the appearance of a few new effects. The most striking feature are the oscillations of concurrence on a picosecond timescale. Since the coupling is comparable to the energy mismatch, the system performs rotations in the single-exciton subspace, coming close to the separable states, $|01\rangle$ or $|10\rangle$, every half-period. These oscillations are damped on a time scale of tens of picoseconds, as the excitation is dissipatively transferred to the lower-energy eigenstate of H_{DQD} by a process discussed in the previous section. Depending on the sign of the interaction, this eigenstate (which can still be entangled) can have either subradiant or superradiant character (for $V > 0$ and $V < 0$, respectively). This property manifests itself on long timescales, leading to the visible difference in the decay rates of entanglement for $V = 1$ meV and $V = -1$ meV (see right panel of Fig. 9.29).

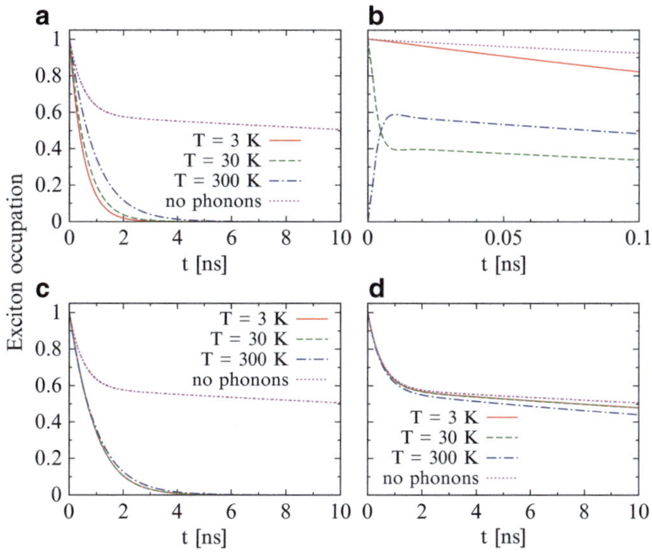

Fig. 9.30 Phonon impact on VIC process: time dependence of the exciton occupation at a few temperatures compared to the evolution without carrier–phonon coupling for $D = 8$ nm for $\mathscr{E} =$ 4 meV (**a**,**b**), 8.07 meV (**c**), and 13.36 meV (**d**). In (**b**), a short-time section of (**a**) is shown, with the *red solid line* representing the total exciton occupation, the *magenta dotted line* corresponding to the total occupation without phonons, and the *green dashed and blue dash-dotted lines* showing the occupations of the two dots [120]

9.5.3 The Role of Phonons in the Vacuum-Induced Coherence Effect

For a system of vertically stacked quantum dots, the amplitudes of the coupling between single-exciton states due to both the Förster and the tunneling mechanism are always negative, thus the darker eigenstate corresponds to the higher energy eigenvalue. The lifetime of this state is strongly reduced due to coupling to lattice vibrations which induce phonon-assisted excitation transfer from the higher to the lower energy eigenstate. As a result, there is no dark or slowly decaying eigenstate in which the exciton occupation could be trapped and thus the vacuum-induced coherence effect is destroyed, unless special parameter choices are made. In Fig. 9.30, we show the impact of phonon dynamics on the process of spontaneous buildup of coherence [120]. We keep the mixing angle θ (i.e., the ratio V/Δ) constant and use the energy splitting $\mathscr{E} = 2\sqrt{\Delta^2 + V^2}$ as a parameter. As can be seen in Fig. 9.30a and c, phonon-related relaxation indeed strongly suppresses the effect of exciton occupation trapping and induces a rapid decay of occupation. The role of phonons in this process is clearly seen in Fig. 9.30b, where we show the first 100 ps of the evolution presented in Fig. 9.30a. A very fast, compared to the purely radiative case (Sect. 9.4.2), phonon-assisted redistribution of occupations takes place, after which the thermal quasi-equilibrium of occupations of the dark and bright states is maintained. This transfer of occupation to the short-living lower energy state makes the dark state unstable and precludes occupation trapping.

The only way to avoid the phonon-induced quenching of exciton occupation is to adjust the system parameters in such a way that the thermal processes become slow compared to the radiative emission. This is possible due to the oscillating form of the spectral density (Fig. 9.23). If the energy parameters Δ and V are adjusted in such a way that the energy splitting $\mathcal{E}$ corresponds to a sufficiently low minimum of the spectral density, the phonon effects become weak. As can be seen by comparing Fig. 9.30c and d, the phonon-induced processes at the third minimum of the spectral density (Fig. 9.30c) are still to strong for the vacuum-induced coherence to be restored but for the energy $\mathcal{E}$ corresponding to the fifth minimum phonon effects become very weak and thus the slow decay of the final occupation is observed again.

Interestingly, the character of temperature dependence of the evolution is different for strong and weak phonon coupling. In the former case, the decay rate decreases with increasing temperatures, while in the latter case the dependence is opposite. This can be explained as follows. If the phonon dynamic is fast, the occupation of the two single-exciton states remains in a quasi-equilibrium, which means that the occupation of the higher energy (darker) eigenstate increases with temperature. On the contrary, if the phonon dynamic is slow compared to the radiative processes, the long-living tail is present, due to the excitation trapping in the darker eigenstate. Since the probability of phonon-assisted escape to the brighter state grows with temperature so does the emission intensity.

9.6 Conclusions

The presented overview of phonon-related and radiative effects in DQD structures shows that the physics of these systems is much richer than that of a single QD.

Already the coupling to phonons alone can lead to new phenomena, including phonon-assisted tunneling of carriers or Coulomb-mediated dissipative excitation transfer. Also the processes that are known from QD systems attain new aspects. In particular, new effects appear in the pure dephasing process, where, e.g., the mutual impact of phonon packets emitted by the other dot plays a role and the decay of spatial coherence occurs. Moreover, due to the presence of an energetically close doublet of levels, the broadening of the fundamental line is determined by an interplay of real and virtual phonon scattering. Phonon-induced dephasing of exciton states in DQDs leads also to strong suppression of entanglement between the two dots.

The linear and nonlinear optical response of DQDs also show features characteristic of these systems. The essential effect here is the formation of bright and dark states and the interplay of the coupling and transition energy mismatch in the two dots. It turns out that sufficiently strong coupling can stabilize the superradiance-like and subradiance-like effects even for DQD systems composed of very different dots. This is reflected in the luminescence of individual DQDs and four-wave mixing signals from DQD ensembles. One of such manifestations is a two-rate decay of the PL and FWM signal, due to the presence of two eigenstates with different decay rates. Another manifestation of collective spontaneous emission dynamics is occupation trapping and spontaneous appearance of inter-dot coherence.

The interplay of radiative and phonon-related effects results in a nontrivial dynamics due to the phonon-induced redistribution of occupations between the bright and dark excitonic eigenstates. As a result, the exciton lifetime becomes temperature dependent. Under certain conditions this dependence may be non-monotonic and counter-intuitive, with longer life times at higher temperatures.

Many of the special properties of DQD systems predicted by the theory are manifested in the results of appropriate optical measurements, which may allow one to characterize DQD systems by analyzing the peculiarities of experimental data. Phonon-induced and radiative decoherence is also an essential limiting factor for quantum-coherent applications, e.g. in the field of quantum information processing. Finally, understanding the fundamental physical mechanisms that govern the optical properties of double-dot systems is an important step towards understanding of the collective emission from larger QD ensembles [24].

Acknowledgements This work was supported in parts by the Polish National Centre for Science (DEC-2011/01/B/ST3/02415) and by the TEAM programme of the Foundation for Polish Science, co-financed from the European Regional Development Fund.

References

1. Rolon, J.E., Ulloa, S.E.: Phys. Rev. B **82**(11), 115307 (2010). DOI 10.1103/PhysRevB.82.115307
2. Unold, T., Mueller, K., Lienau, C., Elsaesser, T., Wieck, A.D.: Phys. Rev. Lett. **94**(13), 137404 (2005)
3. Gywat, O., Burkard, G., Loss, D.: Phys. Rev. B **65**, 205329 (2002)
4. Goryca, M., Kazimierczuk, T., Nawrocki, M., Golnik, A., Gaj, J.A., Kossacki, P., Wojnar, P., Karczewski, G.: Phys. Rev. Lett. **103**(8), 087401 (2009). DOI 10.1103/PhysRevLett.103.087401
5. Xie, Q., Madhukar, A., Chen, P., Kobayashi, N.P.: Phys. Rev. Lett. **75**(13), 2542 (1995)
6. Solomon, G.S., Trezza, J.A., Marshall, A.F., Harris, J.S. Jr.: Phys. Rev. Lett. **76**(6), 952 (1996)
7. Gerardot, B., Shtrichman, I., Hebert, D., Petroff, P.: J. Cryst. Growth **252**(1–3), 44 (2003)
8. Fafard, S., Wasilewski, Z.R., Allen, C.N., Picard, D., Spanner, M., McCaffrey, J.P., Piva, P.G.: Phys. Rev. B **59**(23), 15368 (1999)
9. Gerardot, B.D., Strauf, S., de Dood, M.J.A., Bychkov, A.M., Badolato, A., Hennessy, K., Hu, E.L., Bouwmeester, D., Petroff, P.M.: Phys. Rev. Lett. **95**(13), 137403 (2005)
10. Ortner, G., Yugova, I., Baldassarri Höger von Högersthal, G., Larionov, A., Kurtze, H., Yakovlev, D.R., Bayer, M., Fafard, S., Wasilewski, Z., Hawrylak, P., Lyanda-Geller, Y.B., Reinecke, T.L., Babinski, A., Potemski, M., Timofeev, V.B., Forchel, A.: Phys. Rev. B **71**(12), 125335 (2005)
11. Szafran, B., Chwiej, T., Peeters, F.M., Bednarek, S., Adamowski, J., Partoens, B.: Phys. Rev. B **71**, 205316 (2005)
12. Szafran, B.: Acta Phys. Polon. A **114**(5), 1013 (2008)

13. Borri, P., Langbein, W., Schneider, S., Woggon, U., Sellin, R.L., Ouyang, D., Bimberg, D.: Phys. Rev. Lett. **87**(15), 157401 (2001)
14. Vagov, A., Axt, V.M., Kuhn, T., Langbein, W., Borri, P., Woggon, U.: Phys. Rev. B **70**(20), 201305(R) (2004)
15. Krummheuer, B., Axt, V.M., Kuhn, T.: Phys. Rev. B **65**(19), 195313 (2002)
16. Jacak, L., Machnikowski, P., Krasnyj, J., Zoller, P.: Eur. Phys. J. D **22**, 319 (2003)
17. Grodecka, A., Machnikowski, P.: Phys. Rev. B **73**, 125306 (2006)
18. Diósi, L.: In: Benatti, F., Floreanini, R. (eds.) Irreversible Quantum Dynamics (Lecture Notes in Physics, vol. 622), pp. 157–163. Springer, Berlin (2003). Quant-ph/0301096
19. Yu, T., Eberly, J.H.: Phys. Rev. Lett. **93**(14), 140404 (2004)
20. Dodd, P.J., Halliwell, J.J.: Phys. Rev. A **69**(5), 052105 (2004)
21. Yu, T., Eberly, J.H.: Phys. Rev. B **68**(16), 165322 (2003)
22. Gross, M., Haroche, S.: Phys. Rep. **93**(5), 301 (1982)
23. Skribanowitz, N., Herman, I.P., MacGilvray, J.C., Feld, M.S.: Phys. Rev. Lett. **30**(8), 309 (1973)
24. Scheibner, M., Schmidt, T., Worschech, L., Forchel, A., Bacher, G., Passow, T., Hommel, D.: Nat. Phys. **3**, 106 (2007)
25. Zanardi, P., Rasetti, M.: Phys. Rev. Lett. **79**, 3306 (1997)
26. Zanardi, P., Rossi, F.: Phys. Rev. Lett. **81**, 4752 (1998)
27. Bryant, G.W.: Phys. Rev. B **47**(3), 1683 (1993)
28. Schliwa, A., Stier, O., Heitz, R., Grundmann, M., Bimberg, D.: Phys. Stat. Sol. (b) **224**(2), 405 (2001)
29. Szafran, B., Bednarek, S., Adamowski, J.: Phys. Rev. B **64**, 125301 (2001)
30. Bayer, M., Hawrylak, P., Hinzer, K., Fafard, S., Korkusinski, M., Wasilewski, Z.R., Stern, O., Forchel, A.: Science **291**, 451 (2001)
31. Ortner, G., Bayer, M., Larionov, A., Timofeev, V.B., Forchel, A., Lyanda-Geller, Y.B., Reinecke, T.L., Hawrylak, P., Fafard, S., Wasilewski, Z.: Phys. Rev. Lett. **90**(8), 086404 (2003)
32. Ortner, G., Bayer, M., Lyanda-Geller, Y., Reinecke, T.L., Kress, A., Reithmaier, J.P., Forchel, A.: Phys. Rev. Lett. **94**(15), 157401 (2005)
33. Krenner, H.J., Sabathil, M., Clark, E.C., Kress, A., Schuh, D., Bichler, M., Abstreiter, G., Finley, J.J.: Phys. Rev. Lett. **94**, 057402 (2005)
34. Lovett, B.W., Reina, J.H., Nazir, A., Briggs, G.A.D.: Phys. Rev. B **68**, 205319 (2003)
35. Ahn, K.J., Förstner, J., Knorr, A.: Phys. Rev. B **71**(15), 153309 (2005)
36. Förster, T.: Ann. Phys. (Leipzig) **2**, 55 (1948)
37. Dexter, D.L.: J. Chem. Phys. **21**(5), 836 (1953)
38. Heitz, R., Mukhametzhanov, I., Chen, P., Madhukar, A.: Phys. Rev. B **58**(16), R10151 (1998)
39. Tackeuchi, A., Kuroda, T., Mase, K., Nakata, Y., Yokoyama, N.: Phys. Rev. B **62**(3), 1568 (2000)
40. Rodt, S., Turck, V., Heitz, R., Guffarth, F., Engelhardt, R., Pohl, U.W., Straßburg, M., Dworzak, M., Hoffmann, A., Bimberg, D.: Phys. Rev. B **67**(23), 235327 (2003)
41. Ortner, G., Oulton, R., Kurtze, H., Schwab, M., Yakovlev, D.R., Bayer, M., Fafard, S., Wasilewski, Z., Hawrylak, P.: Phys. Rev. B **72**(16), 165353 (2005)
42. Nakaoka, T., Clark, E.C., Krenner, H.J., Sabathil, M., Bichler, M., Arakawa, Y., Abstreiter, G., Finley, J.J.: Phys. Rev. B **74**(12), 121305(R) (2006)
43. Pryor, C., Kim, J., Wang, L.W., Williamson, A.J., Zunger, A.: J. Appl. Phys. **83**(5), 2548 (1998)
44. Winkler, R.: Spin-Orbit Coupling Effects in Two-Dimensional Electron and Hole Systems, Springer Tracts in Modern Physics, vol. 191. Springer, Berlin (2003)
45. Wójs, A., Hawrylak, P., Fafard, S., Jacak, L.: Phys. Rev. B **54**, 5604 (1996)
46. Messiah, A.: Quantum Mechanics. North-Holland, Amsterdam (1966)
47. Machnikowski, P., Grodecka, A., Weber, C., Knorr, A.: Mater. Sci. (Poland) **26**, 851 (2008). ArXiv:0706.0276v1 [cond-mat.mes-hall]

48. Grodecka, A., Jacak, L., Machnikowski, P., Roszak, K.: In: Ling, P.A. (ed.) Quantum Dots: Research Developments, p. 47. Nova Science, NY (2005). Cond-mat/0404364
49. Mahan, G.D.: Many-Particle Physics. Kluwer, New York (2000)
50. Mahan, G.D.: In: Devreese, J.T. (ed.) Polarons in Ionic Crystals and Polar Semiconductors, p. 553. North-Holland, Amsterdam (1972)
51. Dicke, R.H.: Phys. Rev. **93**(1), 99 (1954)
52. Karwat, P., Sitek, A., Machnikowski, P.: Phys. Rev. B **84**, 195315 (2011). DOI 10.1103/PhysRevB.84.195315
53. Roszak, K., Machnikowski, P.: Phys. Lett. A **351**(4–5), 251 (2006). Cond-mat/0504135
54. Rossi, F., Kuhn, T.: Rev. Mod. Phys. **74**(3), 895 (2002)
55. Förstner, J., Weber, C., Danckwerts, J., Knorr, A.: Phys. Rev. Lett. **91**, 127401 (2003)
56. Krügel, A., Axt, V.M., Kuhn, T., Machnikowski, P., Vagov, A.: Appl. Phys. B **81**, 897 (2005)
57. Krügel, A., Axt, V.M., Kuhn, T.: Phys. Rev. B **73**(3), 035302 (2006)
58. Huneke, J., Krügel, A., Kuhn, T., Vagov, A., Axt, V.M.: Phys. Rev. B **78**(8), 085316 (2008)
59. Roszak, K., Grodecka, A., Machnikowski, P., Kuhn, T.: Phys. Rev. B **71**(19), 195333 (2005)
60. Roszak, K., Machnikowski, P.: Phys. Rev. A **73**, 022313 (2006). Quant-ph/0507027
61. Sitek, A., Machnikowski, P.: Phys. Rev. B **75**(3), 035328 (2007)
62. Machnikowski, P., Roszak, K., Sitek, A.: Acta Phys. Polon. A **116**(5), 818 (2009)
63. Rozbicki, E., Machnikowski, P.: Phys. Rev. Lett. **100**(2), 027401 (2008)
64. Kazimierczuk, T., Suffczyński, J., Golnik, A., Gaj, J.A., Kossacki, P., Wojnar, P.: Phys. Rev. B **79**(15), 153301 (2009). DOI 10.1103/PhysRevB.79.153301
65. Reischle, M., Beirne, G.J., Roßbach, R., Jetter, M., Schweizer, H., Michler, P.: Phys. Rev. B **76**(8), 085338 (2007). DOI 10.1103/PhysRevB.76.085338
66. Nishibayashi, K., Kawazoe, T., Ohtsu, M., Akahane, K., Yamamoto, N.: Appl. Phys. Lett. **93**(4), 042101 (2008). DOI 10.1063/1.2945289
67. Park, J.H., Choi, D.G., Lee, T.K., Oh, E., Lee, S., Furdyna, J.K.: Appl. Phys. Lett. **90**(20), 201916 (2007). DOI 10.1063/1.2740484
68. Bajracharya, P., Nguyen, T.A., Mackowski, S., Smith, L.M., Wagner, H.P., Pohl, U.W., Bimberg, D., Strassburg, M.: Phys. Rev. B **75**(3), 035321 (2007). DOI 10.1103/PhysRevB.75.035321
69. de Sales, F.V., da Silva, S.W., Cruz, J.M.R., Monte, A.F.G., Soler, M.A.G., Morais, P.C., da Silva, M.J., Quivy, A.A.: Phys. Rev. B **70**(23), 235318 (2004). DOI 10.1103/PhysRevB.70.235318
70. Mazur, Y.I., Wang, Z.M., Tarasov, G.G., Xiao, M., Salamo, G.J., Tomm, J.W., Talalaev, V., Kissel, H.: Appl. Phys. Lett. **86**(6), 063102 (2005). DOI 10.1063/1.1861980
71. Seufert, J., Obert, M., Bacher, G., Forchel, A., Passow, T., Leonardi, K., Hommel, D.: Phys. Rev. B **64**(12), R121303 (2001). DOI 10.1103/PhysRevB.64.121303
72. Chang, W.H., Lin, H., Wang, S.Y., Lin, C.H., Cheng, S.J., Lee, M.C., Chen, W.Y., Hsu, T.M., Hsieh, T.P., Chyi, J.I.: Phys. Rev. B **77**(24), 245314 (2008). DOI 10.1103/PhysRevB.77.245314
73. Nakaoka, T., Tatebayashi, J., Arakawa, Y., Saito, T.: J. Appl. Phys. **96**(1), 150 (2004). DOI 10.1063/1.1755857
74. Wang, B.R., Sun, B.Q., Ji, Y., Dou, X.M., Xu, Z.Y., Wang, Z.M., Salamo, G.J.: J. Appl. Phys. **93**(1), 011107 (2008). DOI 10.1063/1.2957466
75. Müller, K., Bechtold, A., Ruppert, C., Zecherle, M., Reithmaier, G., Bichler, M., Krenner, H.J., Abstreiter, G., Holleitner, A.W., Villas-Boas, J.M., Betz, M., Finley, J.J.: Phys. Rev. Lett. **108**, 197402 (2012). DOI 10.1103/PhysRevLett.108.197402
76. Gawarecki, K., Pochwała, M., Grodecka-Grad, A., Machnikowski, P.: Phys. Rev. B **81**, 245312 (2010)
77. Gawarecki, K., Machnikowski, P.: Phys. Rev. B **85**, 041305(R) (2012). DOI 10.1103/PhysRevB.85.041305
78. Wu, Z.J., Zhu, K.D., Yuan, X.Z., Jiang, Y.W., Zheng, H.: Phys. Rev. B **71**(20), 205323 (2005)
79. Vorojtsov, S., Mucciolo, E.R., Baranger, H.U.: Phys. Rev. B **71**(20), 205322 (2005)
80. Stavrou, V.N., Hu, X.: Phys. Rev. B **72**(7), 075362 (2005)

81. López-Richard, V., Oliveira, S.S., Hai, G.Q.: Phys. Rev. B **71**, 075329 (2005)
82. Grodecka, A., Machnikowski, P., Förstner, J.: Phys. Rev. B **78**(8), 085302 (2008)
83. Grodecka-Grad, A., Förstner, J.: Phys. Rev. B **81**, 115305 (2010)
84. Wijesundara, K.C., Rolon, J.E., Ulloa, S.E., Bracker, A.S., Gammon, D., Stinaff, E.A.: Phys. Rev. B **84**, 081404 (2011). DOI 10.1103/PhysRevB.84.081404
85. Bockelmann, U., Bastard, G.: Phys. Rev. B **42**, 8947 (1990)
86. Jaskólski, W., Zieliński, M.: Acta Phys. Polon. A **106**(2), 193 (2004)
87. Jaskolski, W., Zielinski, M., Bryant, G.W., Aizpurua, J.: Phys. Rev. B **74**(19), 195339 (2006). DOI 10.1103/PhysRevB.74.195339
88. Climente, J.I., Korkusinski, M., Goldoni, G., Hawrylak, P.: Phys. Rev. B **78**, 115323 (2008). DOI 10.1103/PhysRevB.78.115323. URL http://link.aps.org/doi/10.1103/PhysRevB.78.115323
89. Richter, M., Ahn, K.J., Knorr, A., Schliwa, A., Bimberg, D., Madjet, M.E.A., Renger, T.: Phys. Stat. Sol. (b) **243**(10), 2302 (2006)
90. Govorov, A.O.: Phys. Rev. B **71**(15), 155323 (2005)
91. Thilagam, A., Lohe, M.A.: Phonon-assisted decoherence in coupled quantum dots (2007). ArXiv:cond-mat/0702428
92. Govorov, A.O.: Phys. Rev. B **68**(7), 075315 (2003)
93. Rozbicki, E., Machnikowski, P.: Acta Phys. Pol. A **112**, 197 (2007). ArXiv:0706.0862 [cond-mat.mes-hall]
94. Krummheuer, B., Axt, V.M., Kuhn, T.: Phys. Rev. B **72**(24), 245336 (2005)
95. Muljarov, E.A., Takagahara, T., Zimmermann, R.: Phys. Rev. Lett. **95**(17), 177405 (2005)
96. Muljarov, E.A., Zimmermann, R.: Phys. Rev. Lett. **93**, 237401 (2004)
97. Machnikowski, P.: Phys. Rev. Lett. **96**(14), 140405 (2006)
98. Machnikowski, P.: Phys. Stat. Sol. (b) **243**(10), 2247 (2006)
99. Einstein, A., Podolsky, B., Rosen, N.: Phys. Rev. **47**, 777 (1935)
100. Schrödinger, E.: Proc. Cambridge Philos. Soc. **31**, 555 (1935)
101. Bell, J.S.: Physics **1**, 195 (1964)
102. Nielsen, M.A., Chuang, I.L.: Quantum Computation and Quantum Information. Cambridge University Press, Cambridge (2000)
103. Bennett, C.H., Brassard, G., Crépeau, C., Jozsa, R., Peres, A., Wootters, W.K.: Phys. Rev. Lett. **70**, 1895 (1993)
104. Bennett, C.H., Wiesner, S.J.: Phys. Rev. Lett. **69**(20), 2881 (1992)
105. Ekert, A.K.: Phys. Rev. Lett. **67**(6), 661 (1991)
106. Yu, T., Eberly, J.H.: Phys. Rev. B **66**(19), 193306 (2002)
107. Horodecki, M., Horodecki, P., Horodecki, R.: Phys. Rev. Lett. **78**(4), 574 (1997)
108. Bennett, C.H., Bernstein, H.J., Popescu, S., Schumacher, B.: Phys. Rev. A **53**(7), 2046 (1996)
109. Bennett, C.H., Brassard, G., Popescu, S., Schumacher, B., Smolin, J.A., Wootters, W.K.L Phys. Rev. Lett. **76**(5), 722 (1996)
110. Bennett, C.H., DiVincenzo, D.P., Smolin, J.A., Wootters, W.K.: Phys. Rev. A **54**(5), 3824 (1996)
111. Peres, A.: Phys. Rev. Lett. **77**(8), 1413 (1996)
112. Horodecki, M., Horodecki, P., Horodecki, R.: Phys. Lett. A **223**, 1 (1996)
113. Hill, S., Wootters, W.K.: Phys. Rev. Lett. **78**(26), 5022 (1997)
114. Wootters, W.K.: Phys. Rev. Lett. **80**(10), 2245 (1998)
115. Biolatti, E., Iotti, R.C., Zanardi, P., Rossi, F.: Phys. Rev. Lett. **85**, 5647 (2000)
116. Pazy, E., Biolatti, E., Calarco, T., D'Amico, I., Zanardi, P., Rossi, F., Zoller, P.: Europhys. Lett. **62**, 175 (2003)
117. Filipp, S., van Loo, A.F., Baur, M., Steffen, L., Wallraff, A.: Phys. Rev. A **84**, 061805(R) (2011). DOI 10.1103/PhysRevA.84.061805
118. Weisskopf, V., Wigner, E.: Z. Phys. **63**, 54 (1930)
119. Scully, M.O., Englert, B.G., Walther, H.: Nature **351**, 111 (1991)
120. Sitek, A., Machnikowski, P.: Phys. Rev. B **86**, 205315 (2012). DOI 10.1103/PhysRevB.86.205315

121. Agarwal, G.S.: In: Höhler, G. (ed.) Quantum Statistical Theories of Spontaneous Emission and Their Relation to Other Approaches, Springer Tracts in Modern Physics, vol. 70. Springer, Berlin (1974)
122. Rudno-Rudzinski, W., Kudrawiec, R., Podemski, P., Sek, G., Misiewicz, J., Somers, A., Schwertberger, R., Reithmaier, J.P., Forchel, A.: Appl. Phys. Lett. **89**(3), 031908 (2006). DOI 10.1063/1.2226503
123. Sitek, A., Machnikowski, P.: Phys. Stat. Sol. (b) **248**(4), 847 (2011). DOI 10.1002/pssb.201000848
124. Sitek, A., Machnikowski, P.: Phys. Rev. B **80**(11), 115319 (2009)
125. Borri, P., Langbein, W., Mørk, J., Hvam, J.M., Heinrichsdorff, F., Mao, M.H., Bimberg, D.: Phys. Rev. B **60**(11), 7784 (1999)
126. Sitek, A., Machnikowski, P.: Phys. Rev. B **80**(11), 115301 (2009)
127. Bardot, C., Schwab, M., Bayer, M., Fafard, S., Wasilewski, Z., Hawrylak, P.: Phys. Rev. B **72**(3), 035314 (2005)

Chapter 10
Förster Resonant Energy Transfer Signatures in Optically Driven Quantum Dot Molecules

Juan E. Rolon and Sergio E. Ulloa

Abstract The present chapter discusses the optical signatures of Förster resonant energy transfer (FRET) in optically pumped and electrically gated quantum dot molecules (QDMs). To this end, an excitonic dressed Hamiltonian is constructed and the level occupation of each exciton is calculated as function of the pump laser energy and applied electric field. Level occupation maps can offer a systematic way of identifying FRET signatures through the analysis of the spectral weight and level anticrossing behavior of each exciton that is pumped in the QDM. The resulting level occupation maps show a clear splitting of the spatially-direct excitons and nontrivial satellites following the spectral lines of the spatially-indirect excitons. These lines are clearly visible starting at the molecular resonance regime up to a regime where charge tunneling is suppressed. In this sense, FRET induces a nontrivial behavior on the spatially-indirect excitons, which is reflected by a robust signature that can be coherently controlled to avoid the detrimental effects of charge tunneling and direct exciton recombination. In addition, our work suggests that FRET optical signatures in QDMs can be addressed via pump-probe differential transmission or level anticrossing PL spectroscopy.

J.E. Rolon (✉)
Department of Physics and Astronomy, University of North Carolina at Chapel Hill, Chapel Hill, NC 27593-3255, USA
e-mail: rolon@live.unc.edu

S.E. Ulloa
Department of Physics and Astronomy and Nanoscale and Quantum Phenomena Institute, Ohio University, Athens, OH 45701-2979, USA
e-mail: ulloa@ohio.edu

J. Wu and Z.M. Wang (eds.), *Quantum Dot Molecules*, Lecture Notes in Nanoscale Science and Technology 14, DOI 10.1007/978-1-4614-8130-0_10,

10.1 Introduction

A prototypical quantum dot molecule (QDM) [1] consists of a stacked pair of single quantum dots (QDs) coupled by charge tunneling (i.e., electron or hole tunneling) [2]. Excitons in QDMs can be coherently excited and controlled by pulsed electromagnetic fields. At the same time, for self-assembled systems embedded in Schottky structures, an external electric field applied along the growth direction can be used to localize charges in either dot. The molecular excitonic states of a QDM can be labeled with either exciton spin and charge isospin degrees of freedom. The electrically controlled interdot coupling effectively rotates the system effective isospin, giving rise to a quantum superposition (molecular states) of the individual excitons [3, 4]. This exciting behavior allows for the implementation and coherent manipulation of excitons in QDMs. If such delicate control of the interdot coupling is achieved, QDs and coupled QD arrays offer a robust and scalable architecture for the next generation of optical sensing and quantum information devices.

A fundamental process that occurs in optically excited molecular systems is Förster (fluorescence) resonant energy transfer (FRET). This process occurs between a donor (D) excited state and an acceptor (A) state of a molecule in the ground state. In the case of a QDM the fluorescence occurs upon the recombination of an exciton on each single QD. A basic requirement for this process is the spectral overlap between the donor emission and the acceptor absorption spectra, respectively. Within a QDM, FRET effectively couples two single QDs via long-range dipole–dipole interactions between the donor and acceptor excitonic levels. Since FRET plays a fundamental role in the charge dynamics in molecular interactions, key questions are whether FRET could be a relevant and controllable coupling mechanism for QDs, and which types of optical signatures can be expected in realistic self-assembled QDMs.

10.2 Förster Resonant Energy Transfer

Excitation energy transfer within a molecular complex is a directed transfer of electrostatic energy in the absence of charge transfer. This process arises from non-radiative multi-pole Coulomb interactions among the conformational states of the molecule involving optically excited states. Furthermore, it produces excitation delocalization within the whole complex, which might be interpreted as exciton hopping between distant sites the molecule [5]. Excitation transfer kinetics can be modeled as a coherent process, or as an open quantum systems problem within the context of the equation of motion for the excitonic density matrix. Let us introduce the notion of excitation energy transfer among two molecules Q_1 and Q_2. Consider each molecule as a two-level quantum system, with the levels being the ground state and the first excited state of each molecule, g_1, e_1 and g_2, e_2, respectively, see Fig. 10.1.

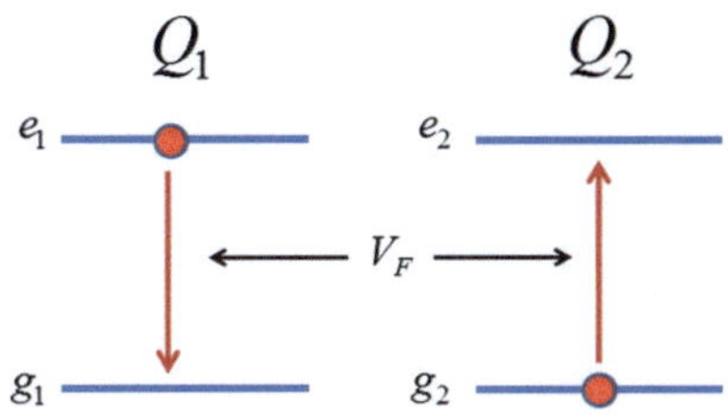

Fig. 10.1 Excitation energy transfer between molecular complexes Q_1 and Q_2. A laser pulse pumps the excited state e_1, upon de-excitation electrostatic energy is transferred to Q_2; absorption on Q_2 leads to the excited state e_2. V_F is the dipole–dipole Coulomb interaction responsible for this process

In addition, let us suppose that molecule Q_1 (donor) is initially on its electronic excited state e_1 after being pumped with an external laser pulse, whereas Q_2 (acceptor) is on its ground state g_2. The Coulomb interaction among the two molecules causes de-excitation of Q_1 with electrostatic energy being transferred to Q_2 causing its respective excitation. This process is effectively equivalent to the recombination of an exciton on site Q_1 and its subsequent creation at site Q_2, or an exciton hopping from Q_1 to Q_2. This processes is known on the literature as FRET mechanism, based on the early works of Förster and Dexter [6, 7]. FRET is involved in a great variety of physical chemistry phenomena, such as superradiance, long-range delocalization, and exciton transport. The most fascinating manifestation of these phenomena occurs in photosynthesis, which relies on the directed excitation transfer of the energy collected by chromophores and photosynthetic light harvesting antennas [8].

10.3 FRET Coupling in QDMs

In a QDM the roles of the donor and acceptor complexes are played by individual QDs [9–11]. Let's consider the two vertically and dissimilar disk-shaped QDs, QD1 and QD2, separated by a distance $|\mathbf{R}|$, each having radius $|\mathbf{r}_1|$ and $|\mathbf{r}_2|$, respectively, see Fig. 10.2.

The first step on deriving the interdot coupling leading to FRET is the calculation of the interband transition dipole moments of exciton levels in each dot. This is followed by the calculation of the Coulomb matrix element connecting exciton wave functions at different dots. The Coulomb matrix element can be approximated by a dipole–dipole interaction, V_F, following the standard procedure of Förster and Dexter [6, 7, 11]. In the limit where the length-scale of exciton wave functions are small in comparison with $|\mathbf{R}|$, we can write,

$$V_F = -\frac{e^2}{4\pi\varepsilon_0\varepsilon_r R^3}\left(\langle\mathbf{r}_1\rangle\cdot\langle\mathbf{r}_2\rangle - \frac{3(\langle\mathbf{r}_1\rangle\cdot\mathbf{R})(\langle\mathbf{r}_2\rangle\cdot\mathbf{R})}{R^2}\right). \quad (10.1)$$

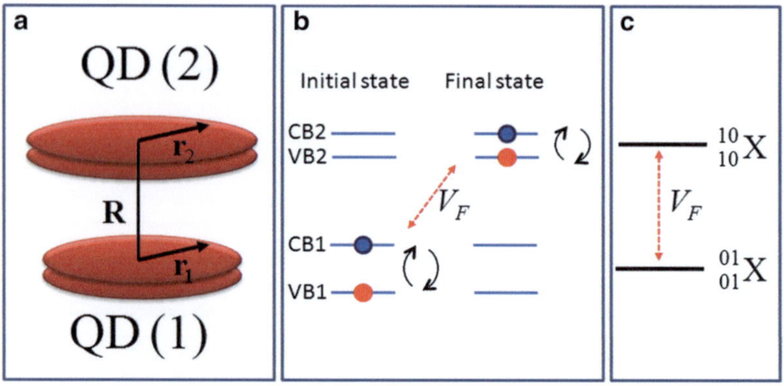

Fig. 10.2 FRET process in a QDM (**a**) QDM schematics; two dissimilar *disk-shaped* vertically stacked QDs. (**b**) Ground state excitons in each QD are represented by an electronic state (*blue*) and a hole state (*red*) in the conduction band (CB) and valence band (VB), respectively. A FRET interaction, V_F, de-excites an exciton on QD1 and transfers its energy to create an exciton on QD2. (**c**) The FRET process represented as a single transition between two excitons ${}^{10}_{10}X$ and ${}^{01}_{01}X$. Here an exciton is denoted by ${}^{e_1e_2}_{h_1h_2}X$, with h_i, e_i being the occupation numbers on the ith QD

Within the envelope wave function approximation, the single carrier wave functions are given by the products $\psi_i(\mathbf{r}) = \phi_i(\mathbf{r})u_i(\mathbf{r})$. The envelope part, $\phi_i(\mathbf{r})$, contains the slow variation of the wave function amplitude over the QD, and most of the important properties of single carrier states. $u_i(\mathbf{r})$ is the Bloch wave function which possesses the lattice periodicity, and it is important for deriving the optical transition dipole moments, and many-body interactions. Therefore,

$$\langle \mathbf{r}_1 \rangle = \int d^3\mathbf{r}_1 \psi^*_{e,n}(\mathbf{r}_1)\mathbf{r}_1 \psi_{h,n}(\mathbf{r}_1) \tag{10.2}$$

$$\langle \mathbf{r}_2 \rangle = \int d^3\mathbf{r}_2 \psi^*_{h,n}(\mathbf{r}_2)\mathbf{r}_2 \psi_{e,n}(\mathbf{r}_2), \tag{10.3}$$

where n is the label for single particle states.

Using the results above, it is possible to separate the contributions arising from the envelope and Bloch parts of the wave function. This results in a simplified expression for the Förster coupling,

$$V_F = -\frac{e^2}{4\pi\varepsilon_0\varepsilon_r R^3} O_1 O_2 [\mu_1 \cdot \mu_2 - \frac{3}{R^2}(\mu_1 \cdot \mathbf{R})(\mu_2 \cdot \mathbf{R})] \tag{10.4}$$

where $O_{1(2)}$ are the overlap integrals of the envelope wave functions for QD1 and QD2, respectively,

$$O_i = \int d^3\mathbf{r}\phi_e(\mathbf{r}_i)\phi_h(\mathbf{r}_i). \tag{10.5}$$

The inter-band dipole matrix elements, connecting the electron and hole Bloch wave functions, $u_{e(h)}(\mathbf{r})$, are given by

$$\mu_i = e \int u_e(\mathbf{r}_i)\mathbf{r}_i u_h(\mathbf{r}_i) d^3\mathbf{r}_i . \tag{10.6}$$

Using the axial symmetry of the problem, and assuming QDs with transition dipole moments along the in-plane direction, the expression for V_F reduces to

$$V_F = -\frac{\mu_1 \mu_2}{4\pi\varepsilon_0\varepsilon_r R^3}\kappa . \tag{10.7}$$

In the model, μ_1 and μ_2 are not calculated explicitly, since more accurate values are readily available from experiments [12–14]. The interaction strength of V_F depends on the degree of exciton confinement within each QD (via the electron–hole wave function overlap). It also depends on the relative orientation of their dipole moments, $\kappa = \frac{\mu_1 \cdot \mu_2}{\mu_1 \mu_2}$; assuming in-plane and parallel permanent dipoles, $\kappa \sim 1$. Furthermore, in this approximation the FRET exciton hopping rate, $K_{\mathrm{DA}} = \frac{2\pi}{\hbar} V_F^2 \Theta$, depends on the spectral overlap Θ between the donor emission and acceptor absorption [15, 16]. However, the excitonic levels in self-assembled QDs are spectrally narrow, which typically makes FRET signatures difficult to observe in experiments. On the other hand, the energy mismatch, $\Delta_{X_{12}}$, between the bottom $({}^{10}_{10}X)$ and top $({}^{01}_{01}X)$ exciton levels, depends on growth conditions, so QDs might be grown to have near resonant exciton transitions [17, 18]. For QDMs such that $\Delta_{X_{12}} \leq V_F$, the resonant energy transfer is nearly coherent "exciton hopping"; on the other hand, when $\Delta_{X_{12}} \gg V_F$, the transfer is incoherent, requiring phonon-assisted transitions via auxiliary excited levels [19].

10.4 Effective Hamiltonian

In what follows, we consider a QDM formed by vertically stacking two self-assembled InGaAs/GaAs QDs, a top QD (T) and a bottom QD (B) separated by a distance d. We assume QDs grown by the Stranski–Krastanov method on a GaAs host matrix, so the barrier material separating the dots is mostly GaAs with InGaAs/GaAs alloying on the QD boundaries [20]. Exciton transition energies in single QDs are critically affected by strain-induced modifications to the single particle conduction and valence band offsets. In addition, the QDM has an intrinsic broken inversion symmetry, which leads to a difference between the hole confinement potential experienced at the base of the top dot from that experienced at the base of the bottom dot [21]. This difference suppresses interdot hole tunneling, which typically is measured to be an order of magnitude less than electron tunneling. Atomistic pseudopotential calculations of exciton wave functions on strained QDMs have shown that holes are mostly heavy-hole like, experiencing a higher interdot potential barrier. Thus, the interdot barrier experienced by the heavy-holes changes with the width of the interdot GaAs barrier, d, leading to strong modifications of the exciton transition energies. Our model uses input numerical parameters from atomistic simulations carried out by Bester et al. [21]. In doing this, we explicitly incorporate the effects of changing d, since the exciton energies were obtained by configuration interaction fitted to tight binding parameters. The d-dependent input parameters include single particle energies, electron–hole Coulomb interactions, and electron and hole tunneling energies.

We construct the basis for a two-site Hamiltonian representing the QDM, as products of electron and hole states, $|e_T\rangle|h_T\rangle$, $|e_T\rangle|h_B\rangle$, $|e_B\rangle|h_T\rangle$, and $|e_B\rangle|h_B\rangle$, where the subindex indicates the QD in which the single particle is localized; all single particle states are assumed to be the ground state in each QD/band. On this basis, the non-interacting Hamiltonian is given by

$$H_0 = \begin{pmatrix} E_{e_T h_T} & t_h & t_e & 0 \\ t_h & E_{e_T h_B} & 0 & t_e \\ t_e & 0 & E_{e_B h_T} & t_h \\ 0 & t_e & t_h & E_{e_B h_B} \end{pmatrix}. \tag{10.8}$$

where the diagonal matrix elements of H_0 are the neutral exciton transition energies, given in terms of the single particle energies, ε_i, and the electron hole Coulomb interaction, U^{eh}. Explicitly,

$$\begin{aligned} E_{e_T h_T} &= \varepsilon_{e_T} - \varepsilon_{h_T} + U^{eh}_{TT} \\ E_{e_T h_B} &= \varepsilon_{e_T} - \varepsilon_{h_B} + U^{eh}_{TB} \\ E_{e_B h_T} &= \varepsilon_{e_B} - \varepsilon_{h_T} + U^{eh}_{BT} \\ E_{e_B h_B} &= \varepsilon_{e_B} - \varepsilon_{h_B} + U^{eh}_{BB}, \end{aligned} \tag{10.9}$$

For simplicity we have only shown the exciton manifold corresponding to single neutral excitons. A similar construction can be made when including the biexciton manifold or charged exciton states.

10.5 Effective Exciton Hamiltonian

In our model exciton pumping is generated by a coherent external laser field, which results in coherent Rabi oscillations among all different exciton states [22,23]. Now, lets us introduce a notation for exciton states derived from the optically perturbed Hamiltonian, H. This corresponds to constructing an optically pumped excitonic basis ${}^{e_B e_T}_{h_B h_T}X$, where $h_{B(T)}$ is the occupation number of holes in the B(T) QDs, and $e_{B(T)}$ the number of electrons, respectively [4]. The QDM is pumped by broadband laser pulse of frequency ω, which pumps nearby exciton states. We assume a rectangular pulse width long enough to capture several Rabi oscillations of the excitonic level populations. When, $e_{B(T)}, h_{B(T)} = \{0,1,2\}$, the resulting Hamiltonian basis contains a total of 14 neutral exciton states. This corresponds to: the vacuum ${}^{00}_{00}X$; a pair of monoexcitons ${}^{10}_{10}X, {}^{01}_{01}X$; a pair of spatially indirect monoexcitons, ${}^{10}_{01}X, {}^{01}_{10}X$; direct biexciton states ${}^{20}_{20}X, {}^{02}_{02}X$; indirect biexciton states ${}^{20}_{02}X, {}^{02}_{20}X$; a delocalized biexciton ${}^{11}_{11}X$ and a remaining set of trion-like states, ${}^{02}_{11}X, {}^{20}_{11}X, {}^{11}_{02}X, {}^{11}_{20}X$.

Up to single occupancy of holes and electrons, $e_{B(T)}, h_{B(T)} = \{0,1\}$, the perturbed Hamiltonian corresponding to the monoexciton manifold in (10.8), is given by

$$H_{\text{tot}} = H_D + H_{\text{tun}} + H_{V_F} + H_{\text{opt}}\,. \tag{10.10}$$

The Hamiltonian in (10.10) includes the diagonal contribution of (10.8) plus the vacuum state $|0\rangle$, the tunneling matrix elements of electrons and holes, H_{tun}, the resonant energy transfer interaction, V_F, and the laser excitation or optical perturbation, H_{opt}. Explicitly,

$$\begin{aligned}
H_D &= |0\rangle\langle 0| + \sum_i E_i |i\rangle\langle i| \\
H_{\text{tun}} &= t_e\left(|{}^{01}_{01}X\rangle\langle{}^{01}_{10}X| + \text{H.c}\right) + t_h\left(|{}^{01}_{01}X\rangle\langle{}^{01}_{10}X| + \text{H.c}\right) \\
H_{V_F} &= V_F\left(|{}^{01}_{01}X\rangle\langle{}^{10}_{10}X| + \text{H.c}\right) \\
H_{opt} &= \Omega_T\left(e^{-i\omega t}|0\rangle\langle{}^{01}_{01}X| + e^{i\omega t}|{}^{01}_{01}X\rangle\langle 0|\right) + \Omega_B\left(e^{-i\omega t}|0\rangle\langle{}^{10}_{10}X| + e^{i\omega t}|{}^{10}_{10}X\rangle\langle 0|\right).
\end{aligned} \tag{10.11}$$

In the model, only spatially direct monoexcitons are coupled to the radiation field by $\Omega_T(t) = \langle 0|\mu_T \cdot \mathbf{E}(t)|{}^{01}_{01}X\rangle$ and $\Omega_B(t) = \langle 0|\mu_B \cdot \mathbf{E}(t)|{}^{10}_{10}X\rangle$, for QD(T) and QD(B), respectively. On the other hand, the spatially indirect excitons have typically a much weaker oscillator strength; its inclusion in the model is straightforward but will be ignored in what follows. Here, $\mu_\mathbf{i}$ are the interband transition dipole moments and $\mathbf{E}(t)$ is the electric field component of the radiation pulse amplitude. In order to simplify the equation of motion, we have performed the rotating wave approximation (RWA) and a unitary transformation that removes the time-dependent fast oscillatory terms of the matter–radiation interaction [24]. The latter approximation is carried out by the following transformation,

$$\begin{aligned}
\tilde{H} &= \Lambda^\dagger H \Lambda + i\hbar \frac{\partial \Lambda^\dagger}{\partial t}\Lambda \\
\Lambda &= \sum_k e^{-i\xi_k t}|k\rangle\langle k|\,,
\end{aligned} \tag{10.12}$$

with the constraint $\xi_i - \xi_j = 0$. The transformed Hamiltonian has diagonal matrix elements containing the detunings of the exciton transition energies with respect to the laser excitation energy, $\tilde{H}_D = \sum_i \delta_i |i\rangle\langle i|$, with $\delta_i = E_i - \hbar\omega$, whereas the couplings to the radiation field are striped of the $e^{\pm i\omega t}$ terms.

10.6 Electrically Tunable Molecular Coupling

Level anticrossing spectroscopy (LACS) is an invaluable tool for probing the optical signatures of quantum coupling in QD aggregates. It consists of performing cryogenic resonant or non-resonant photoluminescence spectroscopy while applying external perturbations which shift the exciton levels of the system [3, 4, 25]. In this manner, all the different possible resonances of connecting different transitions give a direct measure of the electronic and molecular structure of the system. In single QDs and QDMs, this is typically achieved by embedding them in an $n-i$ Schottky junction. In this way the QD system is subject to an external axial electric field controlled by a bias voltage between the n^- contact and the Schottky gate, see Fig. 10.3.

The strongest signature of molecular coupling arises from interdot tunneling coupling [3, 4, 26]. This manifests as an anticrossing in the eigenvalue spectrum and indicates the onset of carrier delocalization over the two QDs. Figure 10.4 shows the hallmark of this delocalization, a QDM under optical excitation generates an exciton, or electron (blue) hole (red) pair; initially, emission would arise from recombination of a single neutral direct exciton as in Fig. 10.4a; as we vary the electric field, the conduction band levels of both QDs become resonant via the Stark shift, allowing electron tunneling, as shown in Fig. 10.4b. In this regime, the QDM

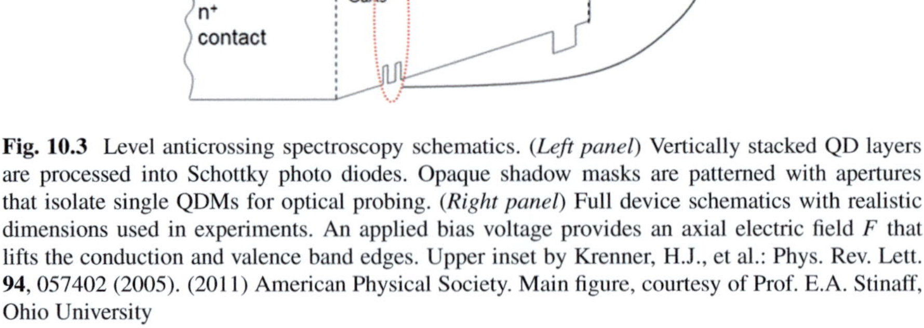

Fig. 10.3 Level anticrossing spectroscopy schematics. (*Left panel*) Vertically stacked QD layers are processed into Schottky photo diodes. Opaque shadow masks are patterned with apertures that isolate single QDMs for optical probing. (*Right panel*) Full device schematics with realistic dimensions used in experiments. An applied bias voltage provides an axial electric field F that lifts the conduction and valence band edges. Upper inset by Krenner, H.J., et al.: Phys. Rev. Lett. **94**, 057402 (2005). (2011) American Physical Society. Main figure, courtesy of Prof. E.A. Stinaff, Ohio University

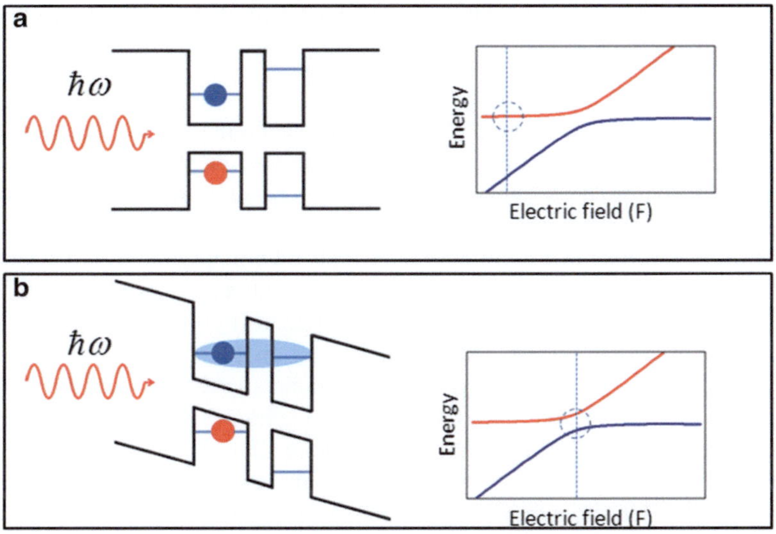

Fig. 10.4 A QDM coupled by electron tunneling. (**a**) At zero bias the laser field pumps a spatially direct exciton, where both hole and electron are localized in the same QD; this corresponds to a plateau on the F-dependent eigenvalue spectrum. (**b**) The field tilts the band edges, inducing a resonance among the conduction bands of the two QDs; electron tunneling occurs, mixing the direct with the spatially indirect exciton with an anticrossing opening in the spectrum

exciton states are superpositions of a spatially direct and indirect states and are therefore truly molecular in nature [3, 4].

The effect of the external axial electric field is to induce, on the one hand, a quantum confined Stark effect (QCSE) shift of the exciton levels within each individual QD. On the other hand, it exerts a DC Stark shift of excitons with spatially indirect character, effectively separating electron and holes in opposite directions. In what follows we ignore QCSE since it will be rather small for the set of parameters used in our simulations. The Stark shift on spatially indirect exciton shifts the energy by $\Delta_S = eFd$. With this in mind, we can finally write the interacting effective QDM excitonic Hamiltonian as

$$\tilde{H} = \begin{pmatrix} \delta_0 & \Omega_T & 0 & 0 & \Omega_B \\ \Omega_T & \delta_{01}^{01} & t_e & t_h & V_F \\ 0 & t_e & \delta_{01}^{10} + \Delta_S & 0 & t_h \\ 0 & t_h & 0 & \delta_{10}^{01} - \Delta_S & t_e \\ \Omega_B & V_F & t_h & t_e & \delta_{10}^{10} \end{pmatrix}, \tag{10.13}$$

where the columns are associated with the states $|0\rangle$, $|_{01}^{01}X\rangle$, $|_{01}^{10}X\rangle$, $|_{10}^{01}X\rangle$, and $|_{10}^{10}X\rangle$, where the third and fourth columns represent spatially indirect excitons.

10.7 Coherent Exciton Dynamics and Level Anticrossing Population Maps

A level anticrossing map can be generated by calculating the average population of all exciton levels, as function of the laser excitation energy and applied electric field. A given coordinate on these maps, $(F,\hbar\omega)$, represents the long time dynamics of the QDM. The contribution of a pumped exciton is given by an amplitude p_i. Two or more excitons can contribute with their own amplitude to the coordinate $(F,\hbar\omega)$, only when having non-vanishing components in a dressed state [27–29]. Using the maps corresponding to each exciton state, it is possible to reassemble the entire spectrum of the system. On the other hand, mapping the population of the vacuum state ${}^{00}_{00}X = |0\rangle$, one can obtain the complete LACS spectra, such that all coordinates $(F,\hbar\omega)$ on this map, represents an exciton that is being *depopulated* into the vacuum, or redistributing its weight among all remaining states [30].

The first step of this procedure consists in the calculation of the eigenvalue spectrum of the Hamiltonian in (10.13) and the calculation of the unitary dynamics of the system. To justify unitarity we calculate the typical Rabi period, T_R, associated with a molecular resonance, which depends on all couplings of the Hamiltonian. To work in the coherent dynamics regime, this period should be of the order of a few ps [22], so that $T_R \ll \tau_X \sim 1ns$, the radiative exciton recombination time found in experiments [31, 32]. An additional requirement is the strong coupling limit for the resonant transfer processes, $V_F \gg \Delta_{X_{TB}}, \hbar/\tau_X$, so that the dynamics of the system is coherent for times t, such that, $T_R < t \ll \tau_X$. Therefore if the time evolution is unitary, the propagator is given by $U(t) = \exp(\frac{-iHt}{\hbar})$, so that the population of an exciton state $|i\rangle$ is given at time t by

$$P_i(t) = |\langle i|U(t)|0\rangle|^2 . \tag{10.14}$$

If the initial condition is full occupation of the vacuum $|0\rangle$ (i.e., an "empty" QDM state), then the averaged population for long integration times is given by

$$p_i = (1/t_\infty) \int_0^{t_\infty} P_i(t)dt , \tag{10.15}$$

where t_∞ is the pulse duration for a broadband square pulse, which is long enough to accommodate several Rabi oscillations of the excitonic populations, $T_R \ll t_\infty < \tau_X$. Only a few Rabi oscillations are enough to compute well-converged long-time population averages. A lower bound for the initial condition is the time for which transient effects have elapsed, and an upper bound for the time to end the simulation would correspond to a value such that damping starts becoming important.

10.8 Competing Effects of Tunneling and Förster Energy Transfer in Monoexcitons

Let us consider the general signatures of charge tunneling and investigate what are the overall features of FRET according to the level diagram in Fig. 10.5a. To this end we have chosen the QD structural parameters (see Appendix Table 10.2), such that excitons ${}^{01}_{01}X$ and ${}^{10}_{10}X$ are initially in near resonance (strongly coupled).

Figure 10.5a shows the level diagram corresponding to the effective Hamiltonian in (10.13). With this Hamiltonian we can construct a level anticrossing population map for the vacuum state ${}^{00}_{00}X$. Figure 10.5b shows the LACS results for a QDM with an interdot distance of $d = 8.4$ nm, assuming negligible Förster coupling among ${}^{10}_{10}X$ and ${}^{01}_{01}X$ exciton states ($V_F = 0$). With these assumptions, the electron and hole tunnelings, t_e and t_h, are obviously the dominant interdot couplings. On the other hand, coupling to the radiation field, $\Omega_{T(B)}$, has the effect of dressing all exciton states, effectively coupling the direct states very weakly via higher order tunneling

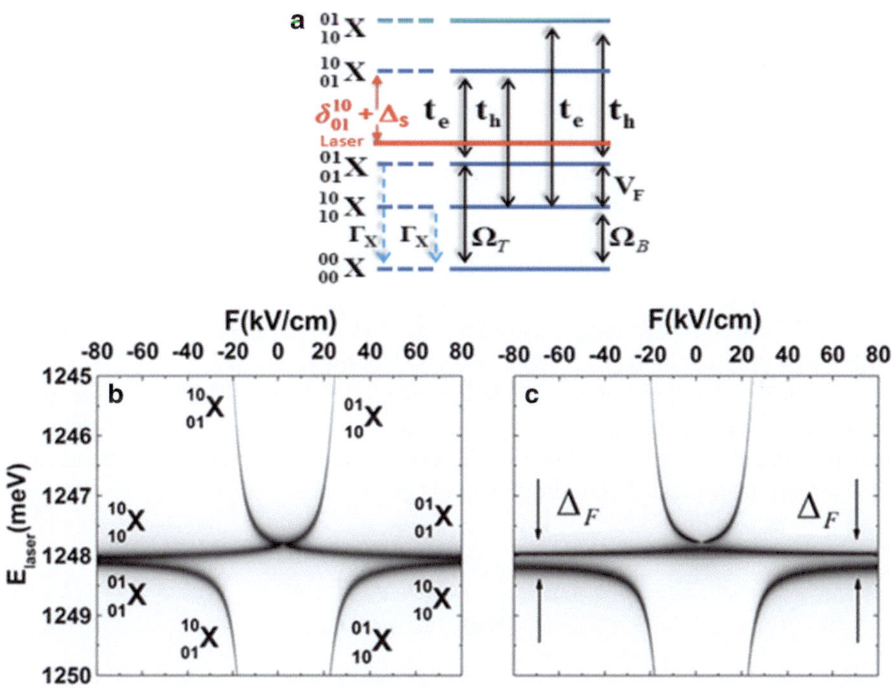

Fig. 10.5 (**a**) Single-exciton energy level diagram showing the relevant couplings with the empty QDM state, ${}^{00}_{00}X$. (**b**) and (**c**) Show occupation maps of vacuum state ${}^{00}_{00}X$ in (10.13), exhibiting features as indicated, where direct, ${}^{10}_{10}X$, ${}^{01}_{01}X$, and indirect excitons, ${}^{01}_{10}X$, ${}^{10}_{01}X$, have nonzero occupation; for vanishing (in **b**) and nonzero V_F (**c**). Notice sizeable Δ_F splitting at high fields, away from tunneling anticrossings. Rolon, J.E., Ulloa, S.E.: Phys. Rev. B **79**, 245309 (2009). (2011) American Physical Society

and excitation processes. A molecular resonance occurs at ($F \simeq \pm 20\,\text{kV/cm}$), resulting in a clear tunneling-induced anticrossing. Here, the applied field F Stark shifts the spatially indirect exciton states by $\pm\Delta_S$, tuning them into resonance with the direct states. For simplicity we have neglected QCSE effects which also shift weakly the direct exciton states, being a nonessential element in our discussion. Lets us consider now the case when the direct excitons are near resonant, thus strongly coupled by V_F. In this case FRET splits considerably the direct exciton spectral lines by $\Delta_F = 0.16\,\text{meV}$, Fig. 10.5c for $\hbar\omega \simeq 1248\,\text{meV}$. On the other hand, the tunneling remains unchanged, since its anticrossing is dominated by $t_e \gg V_F$.

A more detailed view of the FRET optical signature is seen in Fig. 10.6, which shows the level anticrossing map of the ${}^{10}_{10}X$ exciton state. The splitting of the direct state for $\omega \simeq 1248\,\text{meV}$ appears as a plateau satellite far from the tunneling anticrossing, see Fig. 10.6. However, for $F \simeq (-35,-20)\,\text{kV/cm}$, the satellite curvature increases, following the (diabatic) spectral line of ${}^{10}_{01}X$. Interestingly, the FRET mechanism competes strongly with electron tunneling, making the indirect exciton to "light up," by receiving some of the population of the direct state. Eventually the FRET signature quenches at the anticrossing where electron tunneling dominates, which demonstrates that tunneling is detrimental to FRET at low values of F. We emphasize that a proper description of the dynamics of such system needs to take into account the entire set of exciton states, since the direct coupling terms in the Hamiltonian and the various higher order virtual processes make the decoupling of direct and indirect exciton subspaces not possible. For completeness, we show in the right column panel of Fig. 10.6 the results of our simulation when expanding the Hamiltonian to include the biexciton manifold, expanding up to 14 the elements of the excitonic basis; clearly there is no qualitative difference between the LACS maps arising from the long time averaged dynamics. Figure 10.6e, f compares the populations for fixed values of the axial electric field, $F = -70$ and $-35\,\text{kV/cm}$. The FRET satellite peak amplitude increases at stronger fields, suggesting that the FRET signature strength can be controlled by electrical means.

In order to understand the changing behavior of the FRET satellite peak amplitudes, we can truncate the Hamiltonian (10.13) such that its off-diagonal matrix elements connect only the two direct excitons $|{}^{10}_{10}X\rangle, |{}^{01}_{01}X\rangle$, and the vacuum, $|{}^{00}_{00}X\rangle$ state; the truncated Hamiltonian H_D is given by,

$$H_D = \begin{pmatrix} 0 & \Omega_T & \Omega_B \\ \Omega_T & \delta^{01}_{01} & V_F \\ \Omega_B & V_F & \delta^{10}_{10} \end{pmatrix}. \tag{10.16}$$

The level diagram corresponding to the Hamiltonian (10.16) is shown in Fig. 10.7a; Fig. 10.7b shows the corresponding eigenvalue spectrum as function of the direct state laser detuning, δ^{10}_{10}. For $\Omega_B \simeq \Omega_T = \Omega$ and $\delta^{10}_{10} \simeq \delta^{01}_{01}$, the spectrum shows an anticrossing at $\delta^{10}_{10} = -V_F$ and a level crossing at $\delta^{10}_{10} = \frac{(V_F^2 - \Omega^2)}{V_F} = V_X$, with a relative separation $\Delta_F = 2V_F - \left(\frac{\Omega^2}{V_F}\right)$. As a consequence, if we tune the

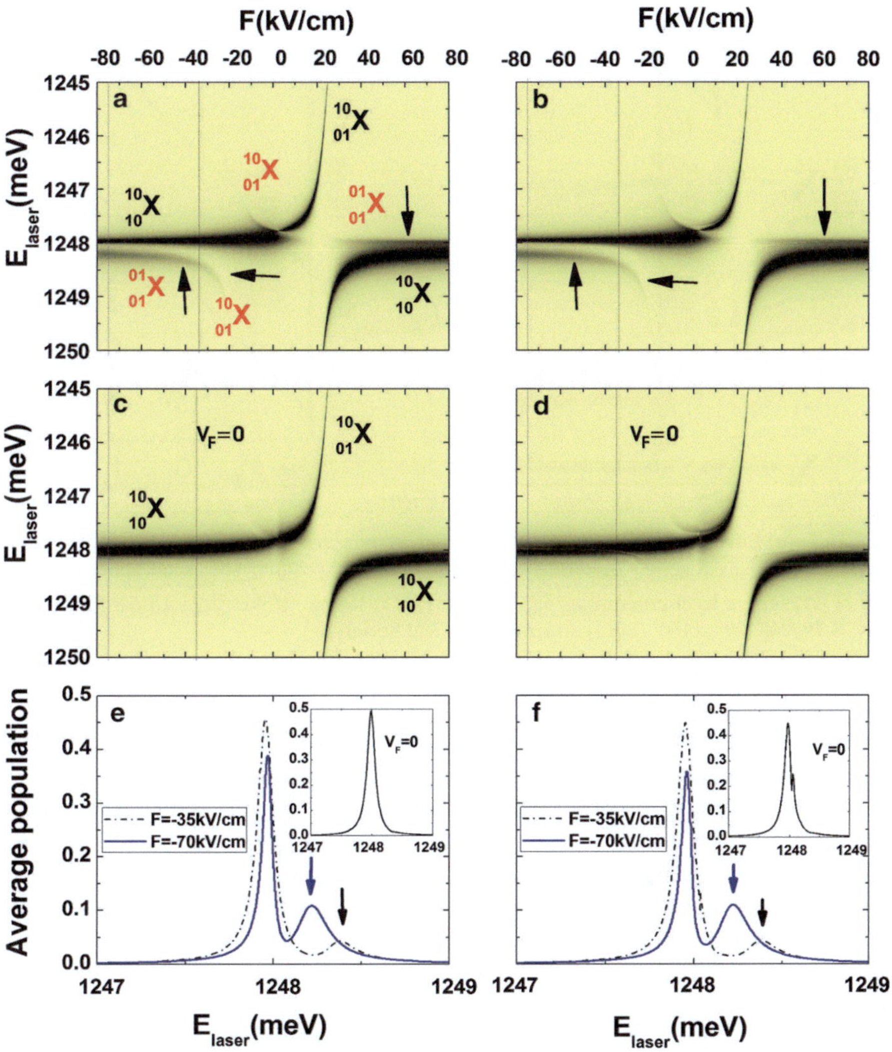

Fig. 10.6 LACS population maps of the monoexciton ${}^{10}_{10}X$ for a system described by (10.13) and parameters given in Appendix Table 10.2 (**a**) Shows FRET-induced weak satellite line (*black arrows*), parallel up to the tunneling anticrossing region. (**b**) Signature persists, even when taking all biexciton transitions into account (all 14 states). The respective population maps (**c**) and (**d**) show no satellites for $V_F = 0$ (no FRET). (**e**) Acceptor population for two fixed values of $F = -35$ (*black dashed line*) and $F = -70$ kV/cm (*blue solid*). (**f**) Acceptor population as in (**e**) but for full system in (**b**). Notice the absence of features when $V_F = 0$ in insets. Rolon, J.E., Ulloa, S.E.: Phys. Rev. B **79**, 245309 (2009). (2011) American Physical Society

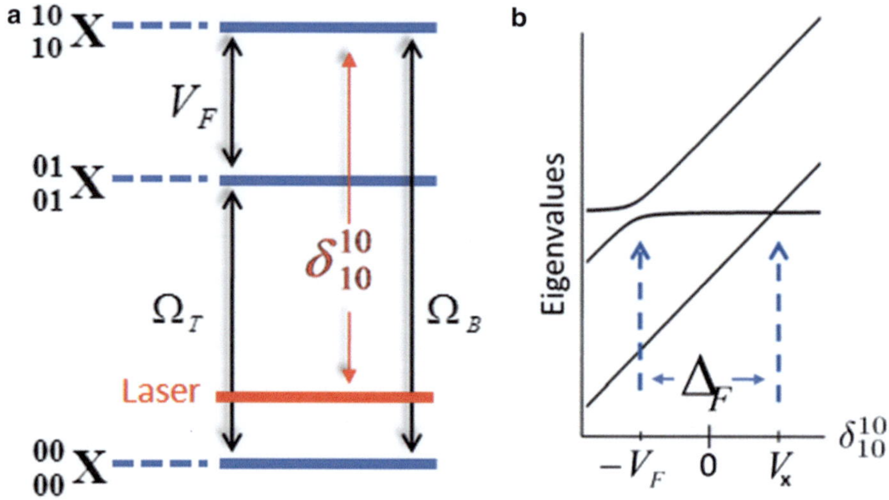

Fig. 10.7 (**a**) Effective level diagram corresponding to truncated Hamiltonian (10.16). (**b**) Eigenvalue spectrum as function of energy detuning $\delta_{X^{10}_{10}}$, the spectrum shows an anticrossing at $\delta^{10}_{10} = -V_F$ and a level crossing at $\delta^{10}_{10} = V_F - (\Omega^2/V_F) = V_X$. Rolon, J.E., Ulloa, S.E.: Phys. Rev. B **79**, 245309 (2009). (2011) American Physical Society

laser excitation energy to $\hbar\omega = E_{{}^{10}_{10}X} + V_F$, the system would be strongly coupled by the energy transfer mechanism V_F, which is manifest on the strong population of the exciton state ${}^{10}_{10}X$. However, when $\hbar\omega \approx E_{{}^{10}_{10}X} - V_X$, the coupling is weaker and the population lower; with V_X having positive or negative values depending on the relative magnitude of V_F and Ω. Hence, the exciton spectral line splits into a doublet; one high peaked line separated and a weak narrower satellite spectral line, both separated from each other by Δ_F (for $|F| \gg 1$). This result implies that a better spectral resolution for the FRET signature is achieved for low excitation power, $\Omega \ll V_F$ and when $\Delta_F \simeq 2V_F$. For the parameters used in our simulations, the splitting in Fig. 10.5b is about 0.25 meV, with $2V_F = 0.14$ meV, since the actual gap is slightly magnified by perturbative corrections arising from charge tunneling and power broadening (not included in this simple algebraic analysis but included in the numerical simulations), even for values of $F \simeq -80$ kV/cm.

A proof of the arguments given above arises when one calculates the time evolution of the population (probability density) of the direct exciton states. Upon diagonalization of the truncated Hamiltonian (10.16), the time-dependent populations are given by

$$|\psi^{10}_{10}(t)|^2 = |\psi^{01}_{01}(t)|^2 = \left(\frac{2\Omega^2}{\Omega_R^2}\right)(1 - \cos(\Omega_R t)) , \tag{10.17}$$

where the Rabi frequency in the regime where, $\delta_{10}^{10} \simeq -V_F$, is given by

$$\Omega_R = \sqrt{(\delta_{10}^{10} + V_F)^2 + 8\Omega^2}, \tag{10.18}$$

with the width of the gap at the anticrossing in Fig. 10.7b given by $2\sqrt{2}\Omega$. Integration of the time-dependent probability density gives the average exciton population, $p_{10}^{10} = 2\Omega^2/\Omega_R^2$, which has a maximum at $\delta_{10}^{10} = -V_F$. We readily notice that p_{10}^{10} increases monotonically with laser power for different values of the detuning. The dependence of p_{10}^{10} on the laser detuning and excitation power suggests a high degree of tunability of the FRET optical signatures via optical parameters only. In this sense, our results suggest that the identification of FRET signatures in QDMs, using monoexciton PL experiments, is attainable for realistic parameters and experimental conditions.

10.9 FRET in Biexciton Optical Signatures

Increasing the QDM laser excitation power results in the possibility of pumping additional exciton states outside the direct exciton manifold [33]. On the one hand, excited states of single electron–hole pairs exist a few meV above the lowest transition energy and correspond to excited states of holes or electrons [25, 34]. On the other hand, longitudinal optical (LO) phonons might be resonant at ~35 meV above the lowest transition; for the parameters considered in the model, and for a sufficiently narrow laser line-width and narrow exciton transitions due to cryogenic temperatures, $T \leq 4$ K, render those processes negligible under suitable experimental conditions [35, 36]. More probable situations arise when considering relaxation and exciton dephasing due to longitudinal acoustic (LA) phonons and pumping of charged excitons and biexcitons [32, 37, 38]. As we would prove in Chap. 4 of this dissertation, indirect exciton states would result to be more prone to LA phonon exciton dephasing in QDMs, rather than their direct counterparts, which are mostly involved in FRET processes. Charged excitons, on the other hand, can be sufficiently suppressed by controlling carrier capture from the back contacts of the Schottky structure. However, biexciton transitions in QDMs, ${}_{20}^{20}X$, ${}_{02}^{02}X$ and ${}_{11}^{11}X$, are more favorable due to the existence of more intermediate transition cascades leading to their formation. Direct transitions such as ${}_{00}^{00}X \rightarrow {}_{20}^{20}X\,({}_{02}^{02}X)$ are usually forbidden by selection rules. The most probable pathways for pumping biexciton in single QDs involves the transitions

$$\begin{gathered} {}_{00}^{00}X \rightarrow {}_{10}^{10}X \rightarrow {}_{20}^{20}X \\ {}_{00}^{00}X \rightarrow {}_{01}^{01}X \rightarrow {}_{02}^{02}X \\ {}_{00}^{00}X \rightarrow {}_{01}^{01}X\,({}_{01}^{01}X) \rightarrow {}_{11}^{11}X, \end{gathered} \tag{10.19}$$

where the parenthesis indicates an alternative route for the preceding transition. However, in QDMs electron and hole tunneling expands the number of pathways leading to biexcitons, which should involve indirect excitons as intermediate states. For example,

$$ {}^{00}_{00}X \rightarrow^{01}_{01} X \rightarrow^{10}_{01} X \rightarrow^{11}_{02} X \rightarrow^{02}_{02} X({}^{11}_{11}X) \rightarrow^{01}_{01} X({}^{10}_{10}X), \tag{10.20} $$

which, among many other pathways, interferes with the FRET transition ${}^{01}_{01}X \rightarrow^{10}_{10} X$. On the other hand, biexciton recombination times are typically faster than monoexciton recombination times[32], $\tau_{XX^0} \simeq \frac{1}{4}\tau_{X^0}$, which might cause biexciton transition spectral lines to broaden significatively in comparison. Moreover, binding Coulomb interactions typically produce a red shift with respect to the monoexciton transition, reducing their relative detuning from as low as 1.7 meV up to 3.5 meV [39]. For these reasons biexciton resonances cannot be ruled out from the dynamics involving monoexcitons as just given by the Hamiltonian in (10.13).

In optically driven QDMs the pumping of biexcitons, ${}^{20}_{20}X$, ${}^{02}_{02}X$ and ${}^{11}_{11}X$, strongly affects the time evolution of the monoexcitons, as the additional biexcitonic manifold redistributes the monoexciton probability densities [40]. However, as we have shown in the previous section, the monoexcitonic FRET satellite in the molecular doublet spectral line is quite robust. The right panel of Fig. 10.6 shows the results of simulations using the full Hamiltonian with 14 basis states, including monoexciton and biexciton manifolds. The main biexcitonic effects appear as a small attenuation of the central peak in Fig. 10.6f, with the satellite peak keeping its relative amplitude. Interestingly, our simulation shows that FRET signatures manifest on biexcitonic complexes by themselves, such as the spatially indirect biexcitons, ${}^{11}_{02}X$, ${}^{20}_{11}X$, ${}^{02}_{11}X$ and ${}^{11}_{20}X$, as seen in Fig. 10.8. These spatially indirect states possess an effective dipole moment (due to their spatially direct component), such that they would be connected by the FRET mechanism. Additionally, these neutral biexcitons can be thought of being composed by a trion localized in one QD together with a single charge (electron or hole) localized in the second QD.

Let's focus our attention on the biexciton manifold spawning the transition cascade ${}^{10}_{01}X \leftrightarrow^{20}_{11} X \leftrightarrow^{02}_{11} X$, as shown in Fig. 10.8c. Truncation of the full Hamiltonian restricted to such states yields

$$ H_{XXI} = \begin{pmatrix} \Delta_S & \Omega_T & \Omega_B \\ \Omega_T & \delta^{11}_{02} + \Delta_S & V_F \\ \Omega_B & V_F & \delta^{20}_{11} + \Delta_S \end{pmatrix}. \tag{10.21} $$

A crucial symmetry of this Hamiltonian makes its eigenvalues invariant against variations in the applied electric field F, since all three states possess the same Stark shift Δ_S. This symmetry makes the eigenvalue problem of (10.21) formally equivalent to that of (10.16); in this case FRET optical signatures appear as a bright exciton spectral doublet with the satellite peak following the Stark shifted spectral

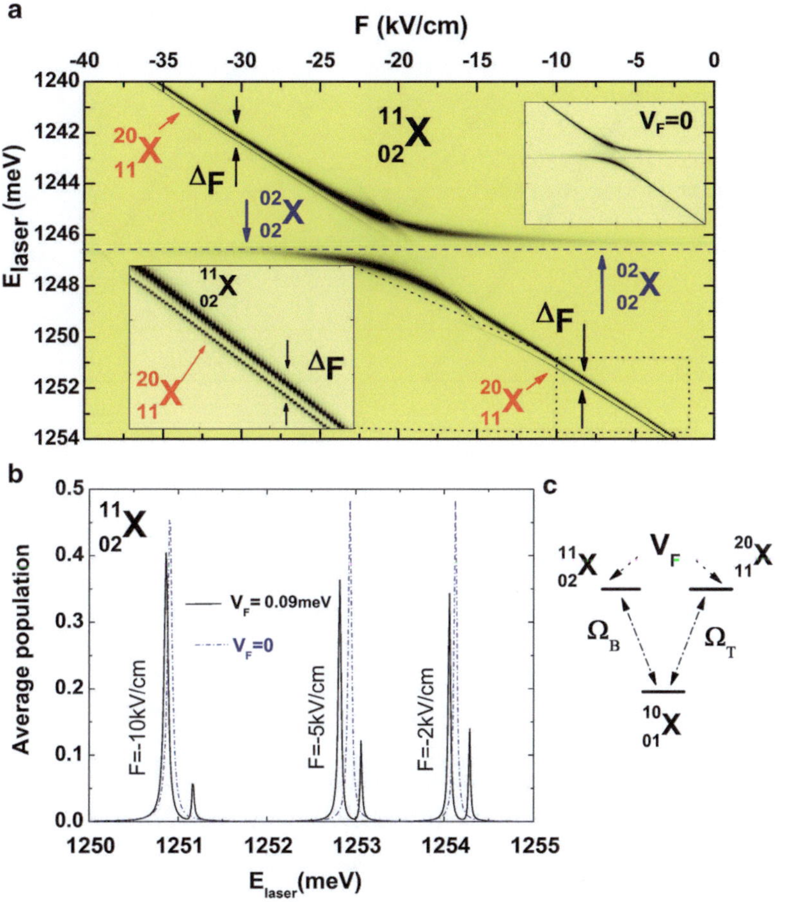

Fig. 10.8 FRET signatures on the biexciton state ${}^{11}_{02}X$ according to parameters given in Table 10.3. (**a**) LACS population map showing splitting of indirect exciton line over large F range. *Blue dashed line* indicates the asymptotic position of the direct biexciton ${}^{02}_{02}X$ mixing with the acceptor level. *Top right inset*: map with no FRET. *Bottom left inset*: magnification of the indirect line splitting. (**b**) Average population for three values of F, showing invariance of indirect split-off energy: satellite line in (**a**) tracks the brighter exciton line nearly parallel. (**c**) Reduced level diagram describing the system for high values of F. Rolon, J.E., Ulloa, S.E.: Phys. Rev. B **79**, 245309 (2009). (2011) American Physical Society

line associated with the spatially indirect component of the trion-like biexciton. This is precisely seen to be demonstrated by our simulated LACS population map for the ${}^{11}_{02}X$ state in Fig. 10.8a. For $F \in (-15, 0)$ kV/cm, the split-feature signature is evident (see lower inset); as expected, the splitting disappears at the tunneling anticrossings where ${}^{20}_{11}X$ hybridizes with the spatially direct biexciton ${}^{02}_{02}X$. Interestingly, the width of the splitting does not change appreciably as we increase the value of the electric field F, as seen in Fig. 10.8b, which shows different cuts along the laser energy axis

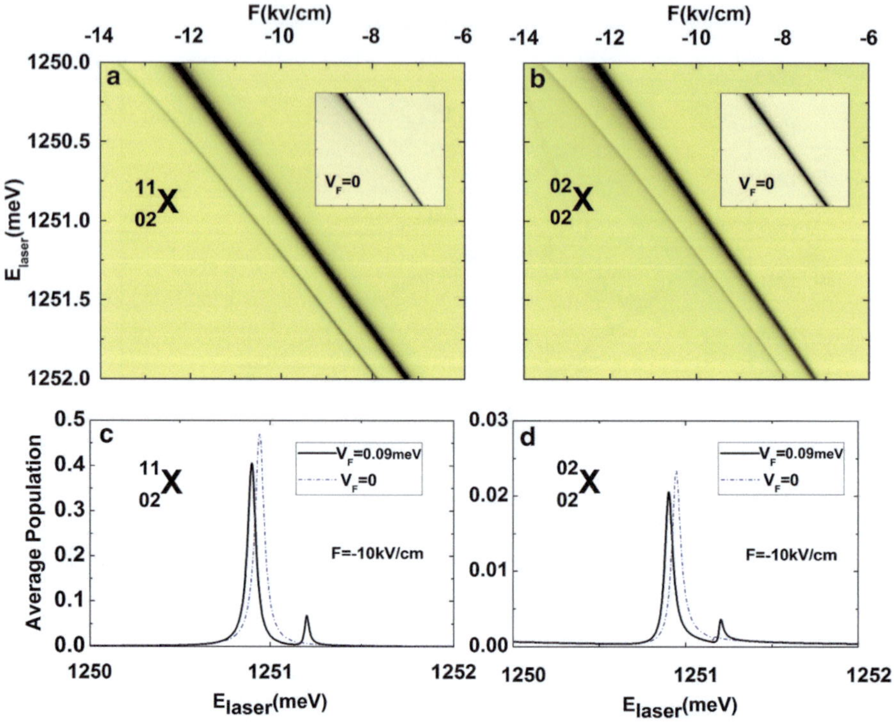

Fig. 10.9 (**a**) The trion-like state $^{11}_{02}X$ has satellite after mixing with the biexciton resonance $^{02}_{02}X$, as shown in (**b**). Notice inset shows no splitting when $V_F = 0$. Comparison of (**c**) and (**d**) confirms mixing of FRET satellites between direct and trion-like biexciton states; FRET signature on direct biexciton state is weaker by an order of magnitude. Simulation parameters according to Appendix Table 10.3. Rolon, J.E., Ulloa, S.E.: Phys. Rev. B **79**, 245309 (2009). (2011) American Physical Society

within the region enclosed by the dashed box. Most remarkably, the amplitude of the FRET peak steadily increases as the field is swept away from the molecular anticrossing, which corresponds to a further projection of the exciton dynamics into the relevant subspace spanned by the FRET coupled states $^{11}_{02}X$ and $^{20}_{11}X$. These results indicate that in experiments one can enhance the FRET signal (obtaining a better contrast) by just increasing the field.

In order to support more our arguments we constructed LACS population maps of both of the hybridized states, at fixed $F \simeq -10\,\text{kV/cm}$, within the region shown in dashed box at the bottom of Fig. 10.8a. These maps appear in Fig. 10.9, which shows almost identical level anticrossing signatures around $\hbar\omega \sim 1251$, for both $^{11}_{02}X$ and $^{20}_{11}X$ states, with the sum of the population at the peak maxima, in Fig. 10.9c, d, plus the amplitude of the vacuum, adding up to near unity.

10.10 Conclusions

In this chapter we have presented theoretical results regarding detection of resonant energy transfer processes in self-assembled InGaAs QDMs. We have shown that LACS experiments would be a suitable methodology for FRET detection. In this regard, FRET optical signatures can be detectable in spite of the strong effects of charge tunneling. As interdot tunnel mixing is controlled by the application of an external axial electric field, enhancement of FRET optical signatures can be achieved in the same way.

Although the effects we discuss have yet to be identified in experiments, the main message of this work is to point out that under suitable conditions Förster interaction may give rise to clearly identifiable signatures in PL or differential transmission spectroscopy. In fact, to date most workers in this field see the Förster interaction as difficult to detect in self-assembled QDMs, partly because clear signatures in experiments have not been identified from a theoretical point of view. We emphasize, especially on Fig. 10.6, for example, that this interaction can give rise to clearly defined split-off satellites which become stronger as an electric field is applied across the molecule ("away" from the region where the electronic tunneling is dominant). This should be easily testable in experiments. For example, the QDM can be explored via pump-probe differential transmission spectroscopy measurements of exciton populations. These could in principle exhibit signatures as those discussed above. For example, the observation of dressed exciton states in QDs using this technique has appeared recently [27–29]. An exciton state can be pumped into the first QD via pulsed broadband laser excitation, creating a collection of dressed states in the system. A second weak probe pulse can be sent into resonance with the excitonic transition in the second QD, which has a slightly different transition energy with respect to the QD, measuring in this way the transient differential transmission, which reflects the exciton population in the second QD.

10.11 Appendix: Simulation Parameters

Table 10.1 Constant parameters used in simulations

Name	Symbol	Value
Interband transition moments	$\mu_{T(B)}$	6.2 eÅ
Exciton recombination time	τ_X	1 ns
Biexciton recombination time	τ_{XX}	0.5 ns

Table 10.2 Parameters used in simulations corresponding to Figs. 10.5–10.7

Name	Symbol	Value
Interdot distance	d	8.4 nm
Förster coupling strength	V_F	0.08 meV
Electron tunneling strength	t_e	2.0 meV
Hole tunneling strength	t_h	0.1 meV
Radiation field coupling	Ω	0.4 meV
T QD direct exciton bare energy	$E_{{}^{01}_{01}X}$	1248.1 meV
B QD direct exciton bare energy	$E_{{}^{10}_{10}X}$	$E_{{}^{01}_{01}X}$+10 μeV
Indirect exciton $e_B h_T$	$E_{{}^{10}_{01}X}$	1263.11 meV
Indirect exciton $e_B h_T$	$E_{{}^{01}_{10}X}$	1267.12 meV

Table 10.3 Parameters used in simulations corresponding to Figs. 10.8 and 10.9

Name	Symbol	Value
Interdot distance	d	8.25 nm
Förster coupling strength	V_F	0.09 meV
Electron tunneling strength	t_e	2.1 meV
Hole tunneling strength	t_h	0.12 meV
Radiation field coupling	Ω	1.4 meV
T QD direct exciton bare energy	$E_{{}^{01}_{01}X}$	1248.62 meV
B QD direct exciton bare energy	$E_{{}^{10}_{10}X}$	1248.79 meV
Indirect exciton $e_B h_T$	$E_{{}^{10}_{01}X}$	1263.49 meV
Indirect exciton $e_B h_T$	$E_{{}^{01}_{10}X}$	1267.60 meV
T QD direct biexciton bare energy	$E_{{}^{02}_{02}X}$	2492.85 meV
Trion like biexciton energy	$E_{{}^{20}_{11}X}$	$E_{{}^{02}_{02}X} + 17$ meV
Trion like biexciton energy	$E_{{}^{11}_{02}X}$	$E_{{}^{20}_{11}X} - 20$ μeV

References

1. Bayer, M., et al.: Science **291**, 451 (2001)
2. Wang, L., et al.: Adv. Mater. **21**, 2601 (2009)
3. Krenner, H.J., et al.: Phys. Rev. Lett. **94**, 057402 (2005)
4. Stinaff, E.A., et al.: Science **311**, 636 (2006)
5. May, V., Kühn, O.: Charge and Energy Transfer Dynamics in Molecular Systems, 2nd edn. Wiley, Berlin (2004)
6. Förster, T.: Discuss. Faraday Soc. **27**, 7 (1959)
7. Dexter, J.: Chem. Phys. **21**, 836 (1953)
8. Burghardt, I., et al.: Energy Transfer Dynamics in Biomaterial Systems. Springer, Berlin (2009)
9. Govorov, A.O.: Phys. Rev. B **68**, 075315 (2003)
10. Govorov, A.O.: Phys. Rev. B **71**, 155323 (2005)
11. Nazir, A., et al.: Phys. Rev. B **71**, 045334 (2005)
12. Warburton, R.J., et al.: Phys. Rev. B **65**, 113303 (2002)
13. Silverman, K.L., et al.: Appl. Phys. Lett. **82**, 4552 (2003)
14. Muller, A., et al.: Appl. Phys. Lett. **84**, 981 (2004)
15. Al-Ahmadi, A.N., Ulloa, S.E.: Appl. Phys. Lett. **88**, 043110 (2006)
16. Lacowicz, J.R.: Principles of Fluorescence Spectroscopy. Springer, New York (2006)
17. Schumann, O., et al.: Phys. Rev. B **71**, 245316 (2005)
18. Gerardot, B.D., et al.: J. Cryst. Growth **252**, 44 (2003)

19. Leegwater, J.A.: J. Phys. Chem. **100**, 14403 (1996)
20. Ledentsov, N.N., et al.: Phys. Rev. B **54**, 8743 (1996)
21. Bester, G., et al.: Phys. Rev. B **71**, 075325 (2005)
22. Stievater, T.H., et al.: Phys. Rev. Lett. **87**, 133603 (2001)
23. Zrenner, A., et al.: Nature **418**, 612 (2002)
24. Cohen-Tannoudji, C., et al.: Atom-Photon Interactions. Wiley, New York (1992)
25. Scheibner, M., et al.: Nat. Phys. **4**, 291 (2008)
26. Villas-Boas, J.M., et al.: Phys. Rev. B **69**, 125342 (2004)
27. Muller, A., et al.: Phys. Rev. Lett. **101**, 027401 (2008)
28. Jundt, G., et al.: Phys. Rev. Lett. **100**, 177401 (2008)
29. Boyle, S.J., et al.: Phys. Rev. Lett. **102**, 207401 (2009)
30. Rolon, J.E., Ulloa, S.E.: Phys. Rev. B **79**, 245309 (2009)
31. Bardot, C., et al.: Phys. Rev. B **72**, 035314 (2005)
32. Narvaez, G.A., et al.: Phys. Rev. B **72**, 245318 (2005)
33. Xu, X., et al.: Science **317**, 929 (2007)
34. Williamson, A.J., et al.: Phys. Rev. B **62**, 12963 (2000)
35. Stock, E., et al.: Phys. Rev. B **83**, 041304 (2011)
36. Ortner, G., et al.: Phys. Rev. B **72**, 165353 (2005)
37. Cheche, T.O.: Europhys. Lett. **86**, 67011 (2009)
38. Bayer, M., et al.: Phys. Rev. B **58**, 4740 (1998)
39. Dalgarno, P.A., et al.: Phys. Rev. B **77**, 245311 (2008)
40. Klimov, V., et al.: Phys. Rev. B **50**, 8110 (1994)

Chapter 11
Stark Effect and the Measurement of Electric Fields with Quantum Dot Molecules

Eric Stinaff, Kushal Wijesundara, Mauricio Garrido, and Swati Ramanathan

Abstract Using the physically separated electron and hole of an interdot exciton in a quantum dot molecule we have studied local electric fields with extremely high resolution. By monitoring the interdot exciton energy we have measured an electric field generated through non-resonant excitation in a Schottky device. A maximum optically generated field of ~3.25 kV/cm was observed which corresponds to 5.04% of the total applied field. The time decay of the field was found to be in the range of 110–140 μs while the onset of the field was shorter than our experimental resolution (7–8 μs).

11.1 Introduction

Since the first observations of molecular wavefunction formation between coupled quantum dots (QDs), these quantum dot molecule (QDM) structures have been the subject of increasing interest [1–11]. The potential use of QDMs for spintronics and quantum information applications has been supported by observations such as electric field tunable *g*-factors and exciton lifetimes [12–14], coherent manipulation, and multiple spin entanglement [15]. In this chapter we will focus on the origin and uses of the Stark effect in QDMs. The situation in QDMs is quite intriguing as there are not only intradot exciton states, where the electron and hole predominantly reside within a single QD, but there are also quite prominent interdot exciton states [2, 4]. Here, since the electron and hole are localized in separate dots, the exciton displays a dramatic linear dependence on the local electric field. This Stark shift of the interdot exciton energy can be controlled through sample growth and may provide a sensitive probe for local electric fields [16].

E. Stinaff (✉) • K. Wijesundara • M. Garrido • S. Ramanathan
Clippinger Labs 364, Department of Physics and Astronomy, Ohio University, Athens, OH 45701, USA
e-mail: stinaff@ohio.edu; kw195305@ohio.edu; mg180004@ohio.edu; sr218305@ohio.edu

J. Wu and Z.M. Wang (eds.), *Quantum Dot Molecules*, Lecture Notes in Nanoscale Science and Technology 14, DOI 10.1007/978-1-4614-8130-0_11,

The chapter is divided into three sections. First we will discuss the sample structures used in this investigation along with the experimental details and an explanation of the important spectroscopic features. This will be followed by a brief explanation of the Stark effect in QDMs. We will then conclude with a specific example where an optically generated field, arising from photovoltaic effects, is characterized in detail using the Stark effect in QDMs.

11.2 Sample Structure

The samples described in this chapter consist of two sequential layers of InAs QDs grown using molecular beam epitaxy [17, 18]. All of the samples discussed in this chapter were grown on n+-doped GaAs substrates to facilitate the fabrication of Schottky diode structures. In a typical sample an 80 nm thick layer of undoped GaAs is deposited on the doped substrate. The first layer of QDs is then formed due to the strain induced by the lattice mismatch between the GaAs and the deposited InAs material. This involves the standard growth process of the formation of a two-dimensional wetting layer (WL) on which, after a critical thickness, the QDs begin to nucleate. The QD layer is then partially capped with GaAs and an Indium flush technique [19], where the temperature is increased to redistribute and partially remove the exposed InAs, is used to control the height, and therefore the bound state energies, of the QDs. After the first layer has been truncated another layer of undoped GaAs is deposited. This layer forms the potential barrier between the QDs and the thickness, and therefore the coupling, can be precisely controlled. The strain-driven Stranski–Krastanow growth mode is extremely well suited for the formation of QDMs as after the barrier growth, the QDs formed in the second layer will preferentially nucleate above those in the first layer. The second layer of QDs can then be truncated in a manner similar to the first. The remainder of the device consists of 230 nm of undoped GaAs followed by a 40 nm thick layer of $Al_{0.3}Ga_{0.7}As$ and a 10 nm GaAs capping layer. To complete the device structure a semi-transparent layer of titanium, 5 nm thick, and an aluminum shadow mask, 100 nm thick, are deposited to form the Shottky contact. Using negative tone photoresist and e-beam lithography an aluminum shadow mask is created with micron sized apertures and alignment marks to facilitate the spectroscopic study of individual QDMs (Fig. 11.1).

11.3 Experimental Setup

The majority of the measurements were done with a standard photoluminescence (PL) setup. The excitation was provided by a tunable Spectra-Physics Ti:Sapphire laser. Typical excitation was non-resonant (800–900 nm), into, or higher in energy than, the quasi-continuum of the InAs WL. After relaxation of the electron–hole

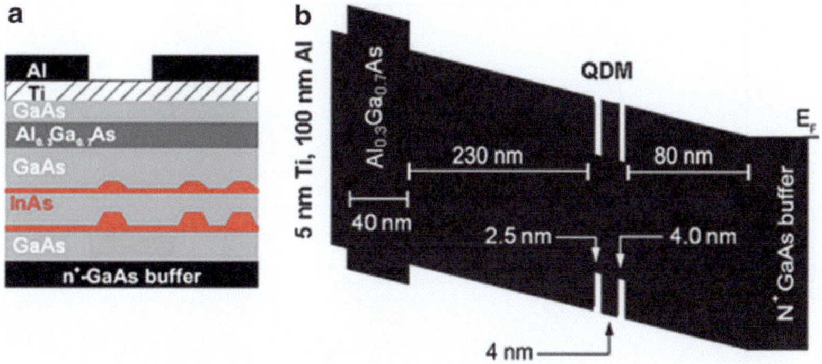

Fig. 11.1 (**a**) QDM device layer sequence. (**b**) Band edge diagram with applied electric field

(e–h) pairs to the lowest available energy states, recombination of the resulting excitons generates photons that can be used to map the charge and spin states of the QDMs. The detection of PL is made using a Princeton Instrument's Trivista triple grating spectrometer along with a Spec-10 CCD detector with a quantum efficiency of 50–60% over the wavelength range where the QDMs emit. The samples were kept at cryogenic temperatures using Advanced Research System's closed cycle cryostat. This system is designed using helium gas to provide the heat exchange between the expander and the sample, reducing vibrations of less than a micron and allowing for the spectroscopic study of individual QDMs.

11.4 Field-Dependent Photoluminescence

To observe the various states within the QDM and their behavior with applied field we use PL as our primary tool. We display the data in a "biasmap" where we plot the intensity as a function of applied field and PL energy. This is done by taking spectra at different electric fields and stacking them next to each other to form a contour plot of electric field vs. PL energy (Fig. 11.2), where the colors represent the intensity at a given field and energy. In a typical experiment, laser excitation of the QDM device creates e–h pairs that may then relax, recombine, or be ionized by the applied field. If the e–h pairs relax into the QD states within a QDM, they may then recombine. In these samples, with n+-doped substrates, the typical electric field range over which PL can be observed is 20–80 kV/cm. On the high field side the loss of luminescence is due to the electric field-induced ionization of the optically created e–h pairs and on the low field side the QDs become saturated with electrons from the substrate and, for excitation above the WL, the PL becomes dominated by the WL emission.

Within the region where PL is visible, signatures of neutral, charged, and multi-exciton states are observed. Identification of the various excitonic states is aided by comparison with the field-dependent spectra of a typical, well-understood, single

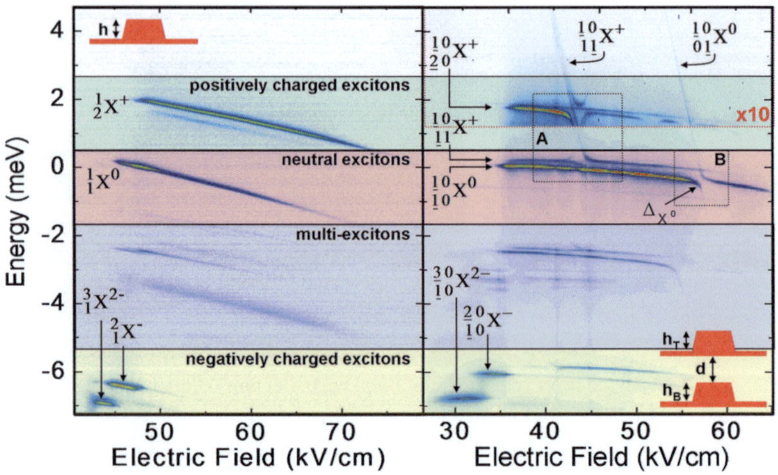

Fig. 11.2 PL as a function of applied electric field for a single QD (*left*) and QDM (*right*). Identification of the QDM states is aided by the similarities between the positive, negative, neutral, and multi-exciton states [2]. *Panel B* shows the region of neutral exciton anticrossing indicating the formation of molecular bonding and antibonding orbitals for the hole states. Detail from a positively charged exciton in the QDM is displayed in *panel A*, where the additional structure is due to the fact that anticrossings occur in both the initial and final states. The nomenclature is as described in the text

QD [20–23]. To facilitate the discussion of the various exciton states we will introduce the following nomenclature, ${}^{e_Be_T}_{h_Bh_T}X^q$, where $e_{B(T)}$ represents the number of electrons in the bottom (top) dot and similarly $h_{B(T)}$ represents the number of holes in the bottom (top) and where q is the charge state of the exciton. The position of the electron and hole participating in the recombination are underlined, for example a positively charged exciton, where the two holes reside in separate QDs, may recombine via the hole in either the bottom (${}^{10}_{\underline{1}1}X^+$) or top (${}^{10}_{1\underline{1}}X^+$) dot.

Perhaps the most obvious similarity is the onset of the emission from negatively charged exciton states. This is indicated by the decrease of the intensity of the neutral exciton emission (${}^{1}_{1}X^0$ and ${}^{10}_{\underline{1}0}X^0$) at ∼36 kV/cm and the subsequent increase of the negative trion emission (${}^{2}_{1}X^-$ and ${}^{20}_{\underline{1}0}X^-$) approximately 6 meV lower in energy. This well-understood shift in energy due to Coulomb interactions is observed in both spectra and provides a good reference point from which to identify the additional lines. For example, the lines which extend to high electric fields along with, but a few meV higher in energy than, the neutral exciton are identified as arising from positively charged exciton configurations (${}^{1}_{2}X^+$ and ${}^{10}_{\underline{2}0}X^+$). It is in these states that the most obvious differences between the single QD and the QDM can be observed. In the QDM a signature x-pattern is seen which arises from the symmetric and antisymmetric molecular hole wavefunctions and the anticrossing of these levels with applied electric field. These anticrossings occur in both the ground and excited states providing the rich features observed.

Details of the optical signatures of QDMs are discussed in detail in references [2, 4, 24]. As previously mentioned, one of the important features for the current chapter is that the excitons in a QDM may recombine as either intradot transitions (electron and hole recombine within the same dot, e.g. ${}^{10}_{1\underline{1}}X^{+}$ and ${}^{10}_{1\underline{0}}X^{0}$) or interdot transitions (electron and hole recombine between the two dots, e.g., ${}^{\overline{1}0}_{1\underline{1}}X^{+}$ and ${}^{1\overline{0}}_{0\underline{1}}X^{0}$). For the sample structure used in the current study the intradot transitions show up as highly field-dependent lines and interdot as approximately field-independent. It is the large field dependence, discussed in the next section, which provides a sensitive probe to measure changes in the local fields.

11.5 Stark Effect

The ability to tune optical transitions with an applied electric field in single photon emitting structures such as QDs has potential applications in secure communication, quantum information processing, optical switches, and tunable emitters [25–27]. The electric field-dependent energy shifts, due to the Stark effect, are well known from atoms and molecules. In semiconductor heterostructures where the dimensions are on the order of the Bohr radius of the charge carriers, such as in quantum wells (QWs) and QDs, this takes the form of the quantum confined Stark effect (QCSE). Here, perturbations of the confining potential and electrostatic interactions of the applied field and the electron and hole result in changes of the energies with applied field. The QCSE has been well studied in single QDs, however, due to the size of the QD, and therefore the confining potential, the magnitude is often small, resulting in a narrow tuning range of the optical transition [25]. Due to the molecular nature of the carriers in a QDM a significant enhancement in the magnitude of the QCSE can be realized [28–30].

In a QDM, as with coupled QWs, the formation of molecular wavefunctions (Fig. 11.3) may result in both an intradot and an interdot exciton configuration. For the intradot exciton the measured field-dependent PL leads to a Stark shift of $\sim$0.13 meV/kV/cm, and a maximum tunable PL energy of $\sim$2 meV over the field range in which emission is visible. Therefore, the PL emission energy can be varied over a range of three orders of magnitude larger than the homogenous linewidth of the intradot exciton. This range is even larger for the interdot exciton where the electron and hole are in separate dots and therefore the dipole moment is on the order of the separation between, plus half the heights, of the constituent dots in the QDM. The interdot exciton PL shows a large, linear bias dependence up to an order of magnitude larger than the largest intradot exciton. For the largest studied barrier (6 nm) the Stark shift is found to be $\sim$0.97 meV/kV/cm, corresponding to a tunable PL emission energy of $\sim$20 meV over a field range in which the emission is visible. In this chapter we are primarily concerned with the interdot exciton state where the large dipole moment can be used to investigate changes in local electric fields.

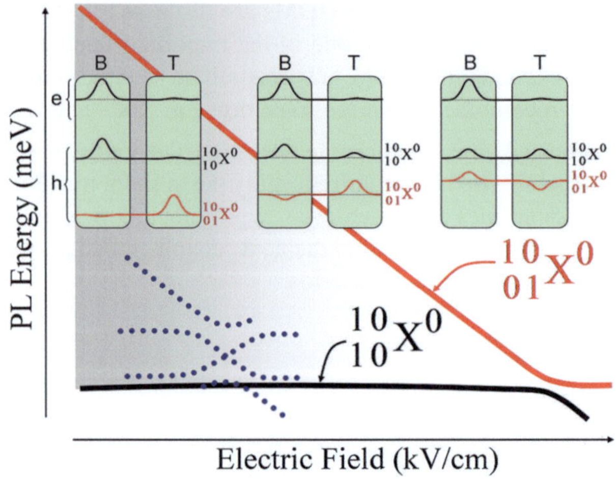

Fig. 11.3 Representation of the interdot ($^{10}_{01}X^0$, *red*) and intradot ($^{10}_{10}X^0$, *black*) exciton states. The *blue dotted lines* represent the positively charged exciton states. The qualitative behavior of the electron and hole molecular wavefunctions is shown in the sequence of *green boxes*. As the QDM is tuned, with decreasing electric field, the holes become increasingly localized. We also note the linear field dependence of the interdot configuration

There is a continuous change in the QCSE of excitons as a function of the separation between the constituent QDs (Fig. 11.4). This can be understood in terms of carrier separation, the formation of molecular wavefunctions, Coulomb interactions, and the differing effective masses for the electron and hole [30]. In a QDM the electron and hole become increasingly delocalized with smaller barrier resulting in a change in both the linear (dipole moment) and quadratic (polarizability) terms of the QCSE for the exciton. For the current discussion we will neglect the quadratic term and show results for a simple fit of the PL biasmap data to only a linear component [31, 32]. Even with this simple approximation the effects of molecular wavefunction formation can be seen in the deviation from the approximation where the electron and hole remain fixed as the barrier is reduced (dashed lines in Fig. 11.4). Though a detailed discussion of these effects including the quadratic contribution is beyond the scope of this chapter, a simple picture is that as the QDs are brought closer together the electron and hole symmetric molecular wavefunctions will shift toward the center of the molecule [30]. Due to its smaller effective mass, this will occur to a larger degree for the electron leading to a reversal of the typical dipole moment associated with a single QD for the intradot exciton and a deviation from the linear barrier-dependent behavior for the interdot exciton. As the barrier is increased, the interdot exciton oscillator strength will decrease leading to a reduction in the observed PL intensity. Therefore, in the remainder of the chapter we will focus on the barrier separation of 4 nm. In these samples we find an excellent balance of a strongly linear interdot Stark shift and reasonable intensity.

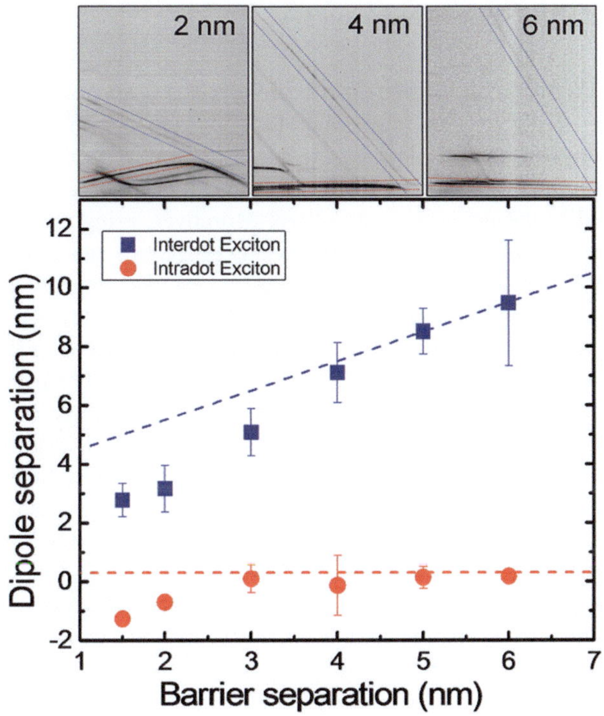

Fig. 11.4 *Top panels* show representative spectra from samples with 2, 4, and 6 nm barriers where the direct and indirect neutral exciton lines are highlighted. The *plot* shows the measured dipole separation for the direct (*red circles*) and indirect (*blue squares*) exciton with a simple linear dipole approximation. The *thin dashed lines* show the single QD limit with large barrier

In general, the Stark shift is dependent on, among other details, indium concentration, strain, and barrier profile within the QDM, and could potentially be used as a spectroscopic tool to gain insight into these properties [31–34]. In this section, a continuous shift was observed in the dipole of the intradot exciton with barrier size that shows that at a low barrier, there is a reversal in the dipole separation, which arises from highly molecular behavior of the two dots in the QDM. Starting with nearly single dot behavior for the large barriers, the bottom dot shows a reversal of its dipole as the barrier between the two dots is decreased, consistent with a simple picture of the shift of the symmetric molecular wavefunction toward the center of the molecule which depends on the carrier effective mass. This ability to engineer the electric field dependence of the excitonic PL in QDMs could prove to be useful for multiple applications requiring tunable single photon emitters or absorbers. Here we will focus on utilizing this as a sensitive probe of local electric fields.

11.6 Optically Generated Electric Field

The ability to modulate an applied electric field is necessary for many experiments but can often be difficult to implement [35–37]. One possibility would be to modulate the applied electric field by means of an optically created electric field. Using photoionization of optically created excitons it is possible to rapidly generate an electric field in the devices studied here. Moreover, using the interdot exciton recombination it is possible to precisely characterize this field. In the following sections we will use this tool to investigate the detailed dependence on energy, power, applied field, and time of the photogenerated field in these QDM devices.

As discussed earlier, by embedding the QDMs in a Schottky diode structure, the bound energy levels of an individual QD are easily tuned with respect to each other by means of an applied electric field [2]. By monitoring the interdot exciton states, we can directly observe the effect of the optically created electric field as shown in Fig. 11.5b. In contrast, for an intradot exciton, we observe only a small shift, if any, due to the QCSE. For the samples used in these measurements the GaAs barrier between the QDs within the molecule was 4 nm and the average interdot Stark shift was 0.74 meV/kV/cm, up to 100 times larger than that of the intradot exciton. As discussed in the previous section, this large enhancement comes from the displacement of the electron and hole and is proportional to the distance between the two dots.

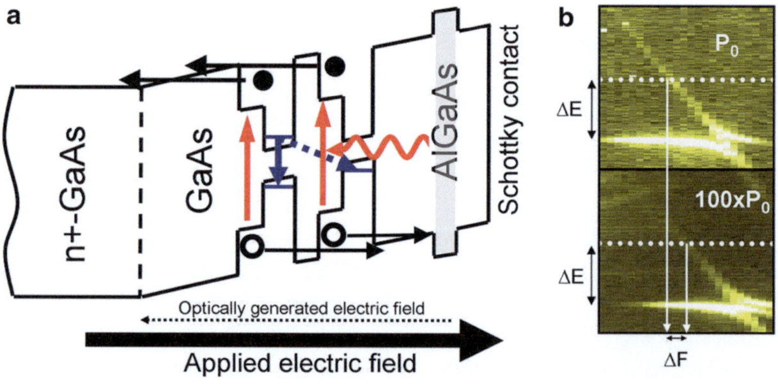

Fig. 11.5 (**a**) Schematic representation of the device structure and the interdot (*blue dashed arrow*) and intradot (*blue solid arrow*) exciton recombination. Once the laser energy is higher than that of the WL (*red arrows*) the absorbed photons may create large numbers of e–h pairs which may ionize. These ionized charges may then be trapped within the device resulting in an electric field which opposes the applied field. (**b**) Example data showing the observed shift in the interdot PL energy. If we measure the field at which the interdot PL is a fixed energy (ΔE) away from the intradot PL we find that this occurs at a different field (ΔF) depending on the laser power. We could conversely measure the change in PL energy at a fixed applied field

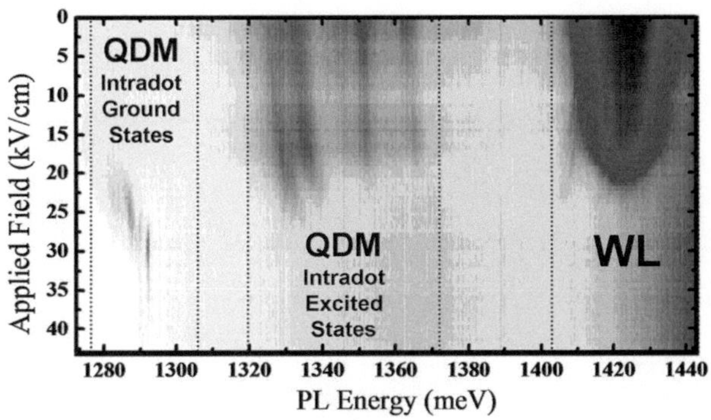

Fig. 11.6 Low resolution biasmap showing the QDM ground and excited state PL emission as well as the broad WL emission. Note that at this resolution and intensity scale the interdot lines are not visible

Measurements similar to this could be done using coupled QWs; however, comparing the linewidth of the ground state PL in the QDMs to that of the WL (Fig. 11.6), we can see the advantage of using the interdot PL to measure small changes. Even the relatively narrow QW-like emission of the WL is one to two orders of magnitude broader than the interdot linewidth. Therefore, as seen in Fig. 11.5b, even very small changes in the electric field will result in a measurable shift in the emission energy of the interdot lines. We could either measure the shift of the interdot line PL energy at a given applied field or measure the shift of the field at a given interdot PL energy. Either way, changes in the electric field down to a few tenths of a kilovolt per cm can be measured. The interdot PL peaks in the spectra for different applied fields were fitted to a Gaussian, which was found to provide a moderately better fit than the expected Lorentzian possibly due to the sensitivity to random charge fluctuations. These peaks were then fitted to a line which was used to calculate the shift in electric field for an arbitrary PL energy. This was done as a function of excitation energy, excitation power, and applied electric field. All of the results, which will be discussed in the following sections, are consistent with the effect of photovoltaic band flattening [38, 39]. Here, the ionization of the photogenerated e–h pairs within the Schottky diode can induce an opposing field if the ionized electrons and holes separate and remain within the device region. This trapping most likely occurs at impurity sites and material boundaries (e.g., the GaAs/AlGaAs interface, the WLs, or the doped/intrinsic GaAs interface), while carriers not trapped will contribute to the measured photocurrent. The most likely source of trapping is at the GaAs/AlGaAs interface. As can be seen in Fig. 11.5a, for this device structure this interface forms a two-dimensional potential well for the holes.

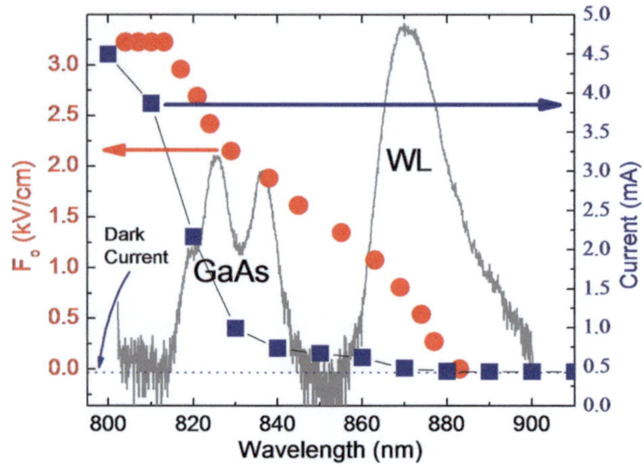

Fig. 11.7 Dependence of the optically generated field (F_o) on laser excitation energy. A typical PL spectrum is displayed in *gray* with intensity on a log scale. The *red circles* indicate the measured strength of F_o and the *blue squares* are the measured photocurrent. Only when the laser energy becomes resonant with, and then higher in, energy than the WL does F_o become measurable. F_o is found to saturate at 3.25 kV/cm above the GaAs band edge

11.6.1 Energy Dependence

The dependence of the optically generated field (F_o) on laser power, laser energy, and the applied field is a complicated function of the various tunneling, recombination, and generation rates, as well as the type and density of traps within the device. We can begin to deduce some qualitative trends as we measure the effects of F_o as a function of laser energy. Here we perform photoluminescence excitation (PLE) experiments and monitor the shift of the interdot line as a function of laser energy. To provide additional insight into the processes involved we also monitor the photocurrent. The PLE intensity and the photocurrent are proportional to the absorption of photons and tunneling of carriers out of the device and as the excitation energy becomes resonant with the WL, a shift of the electric field-dependent spectra is observed (Fig. 11.5b). The shift increases with increasing laser energy, and once the energy is above the GaAs band edge, no further shift is observed. Since these measurements were done at a fixed power (2 W/cm^2) this could be interpreted as arising from an increased absorption probability with increasing laser energy. In the next section we will discuss the dependence on power and what additional insight we can gain. In Fig. 11.7 we see that at $\lambda_e = 880$ nm there is a PLE signal but nearly no photocurrent, suggesting that a fraction of the absorbed photons, those which do not relax and recombine, are ionized and trapped, thereby contributing to F_o. As the laser energy is increased we observe an increase in PLE, photocurrent, and the optically generated field. Above the GaAs both the PLE and photocurrent increase dramatically indicating efficient absorption, ionization, and tunneling.

11.6.2 Power Dependence

Power dependence curves show that, at a given laser energy, as the photon flux (laser power) is increased the electric field shift also increases and ultimately saturates. The saturation demonstrates a limit to the optically generated electric field most likely due to the saturation of the trapping of the ionized carriers. At a fixed excitation wavelength, just above the WL ($\lambda_e = 860.1$ nm), this saturation was measured to be around 1.4 kV/cm and occurred for a power of 0.5 W/cm^2. Reproducing the same measurement on eight different QDMs within the same sample we found an average saturation of 1.12 kV/cm ($\lambda_e = 860.1$ nm).

An interesting observation is that at a laser energy of $\lambda_e = 860.1$ nm (above the WL), the field shift appears to saturate at a higher limit than at $\lambda_e = 892.0$ nm (below the WL). This could be interpreted as a large decrease in absorption and possibly a lower trapping rate. F_o does still appear to saturate and it is possible that at even higher power the maximum F_o would approach that seen above the WL. It is also interesting to note that the saturation of F_o at 1.12 kV/cm for excitation at 860.1 nm is below the maximum F_o measured for excitation above the GaAs band edge. It is possible that either again at higher powers F_o would approach the maximum measured above the GaAs band edge. It could also be that the rate at which carriers leave the device region, and therefore no longer contribute to F_o, is such that the number of traps filled in steady state is below the maximum. Unfortunately at higher powers the spectra become dominated by background PL making unambiguous measurements of the optically induced shifts impossible with the current samples. It is possible that with larger apertures we may be able to observe uniform saturation for all wavelengths and future experiments are in the works.

11.6.3 Applied Field Dependence

Not surprisingly, the optically generated field also displays a dependence on the externally applied field itself. We can measure this effect by using interdot lines associated with different charge states. As mentioned earlier in the chapter, due to the Coulomb interaction, different charge states are shifted in energy. This results in the interdot versions of these lines appearing more prominently at different applied fields.

In Fig. 11.8 we plot the power dependence of the shift of two such indirect lines (${}^{1\underline{0}}_{\underline{0}1}X^0$ and ${}^{10}_{\underline{1}1}X^+$) which are observed at different applied electric field values (48.0 and 31.7 kV/cm, respectively). With a given laser power and energy, for example 0.5 W/cm^2 and 860.1 nm, we find that for an applied electric field of $F_A = 31.7$ kV/cm we have an optically created electric field value of $F_o = 1.3$ kV/cm; while for a higher applied field of $F_A = 48.0$ kV/cm we find a lower value of $F_o = 0.74$ kV/cm. This is a relatively small effect, only 1% of the applied field, and therefore we find no measureable deviation from linearity for the

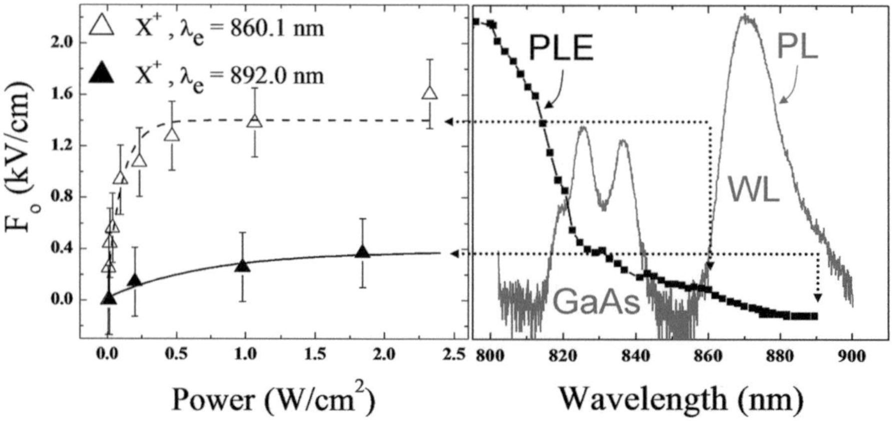

Fig. 11.8 Power dependence of F_o measured using the ${}^{10}_{11}X^+$ state as the probe. When exciting above the WL peak PL energy there is a rapid saturation of F_o at around 1.4 kV/cm. This saturation is found to occur much slower with power and at an apparently lower field for excitation below the WL energy

field-dependent PL of the interdot exciton. If the effect were larger, we would expect a noticeable quadratic dependence on the interdot lines.

We attribute this effect to the applied field dependence of the tunneling and trapping rates of the electron and hole. At higher applied electric field, charges are less easily trapped and tunnel more easily out of the device, therefore reducing the optically created field. Though the saturation point of F_o appears to be nearly the same, it is possible that there is a dependence of the maximum F_o which could be attributed to the field dependence of the potential well created at the GaAs/AlGaAs interface. As the field is increased the number of states available would decrease leading to a reduction of the maximum F_o (Fig. 11.9).

11.6.4 Time Dependence

In this section we investigate the dynamics of F_o with time-resolved measurements. As previously discussed, with laser excitation above the WL energy, we optically generate an electric field within a device containing the QDMs and observe the field strength through a shift in the interdot PL spectra. The excitation above the WL energy was seen to generate a shift of up to 3.5 kV/cm in the electric field dependent PL spectra of QDMs, whereas for excitation below the WL energy only a negligible shift was observed [16]. By modulating the above WL excitation, oscillations of the optically generated electric field can be created. Monitoring the PL energy of the interdot excitonic emission at different time delays from the creation to the

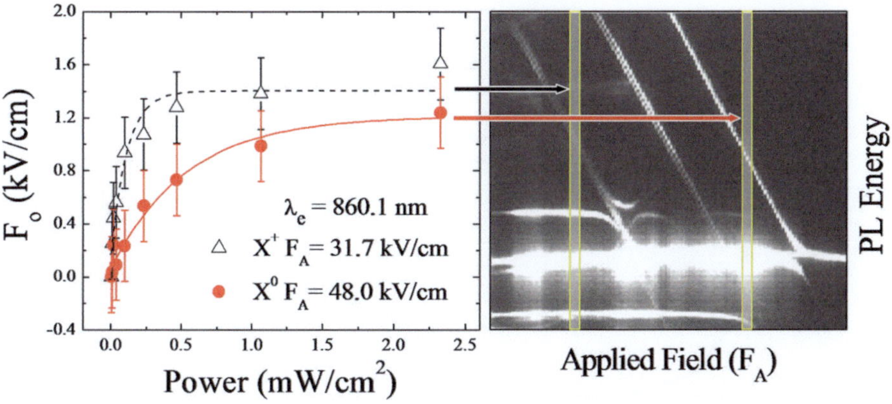

Fig. 11.9 Comparison of the power dependence of F_o at two different applied field. This is accomplished by monitoring the $^{10}_{0\underline{1}}X^0$state and the $^{10}_{1\underline{1}}X^+$ state simultaneously

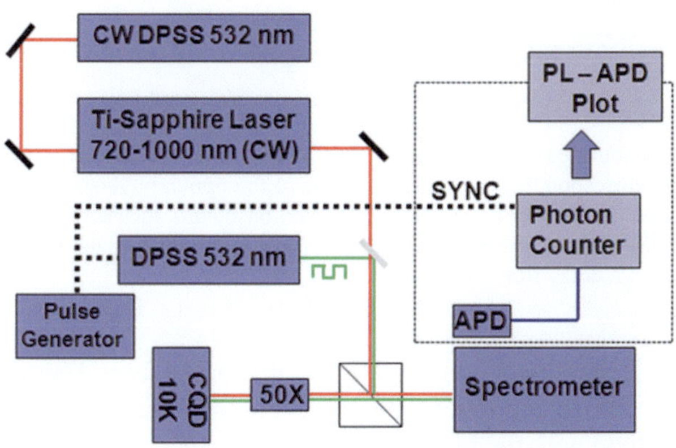

Fig. 11.10 Schematic representation of the quasi-pump probe setup. A modulated (250 Hz) diode pumped solid states (DPSS) laser is used to generate the optical field while a CW laser is used to probe and monitor the indirect exciton

extinction of the optically generated electric field we were able to determine the decay rate ofthe optically generated electric field within the device.

To determine the dynamics of the optically created field, time-resolved measurements were performed in which the position of the interdot line was monitored relative to a modulated excitation laser. A diode pumped solid-state (DPSS) laser with energy above the WL (532 nm), modulated at 250 Hz, was used to create the optically generated field (Fig. 11.10). However, due to the relatively short PL lifetime of the exciton states in QDMs, typically on the order of nanoseconds, we cannot use the interdot PL generated by the 532 nm excitation to monitor the temporal behavior of F_o. To be able to continue to observe the interdot exciton PL we used the CW Ti:Sapphire laser with energy below the WL, around 920 nm, such

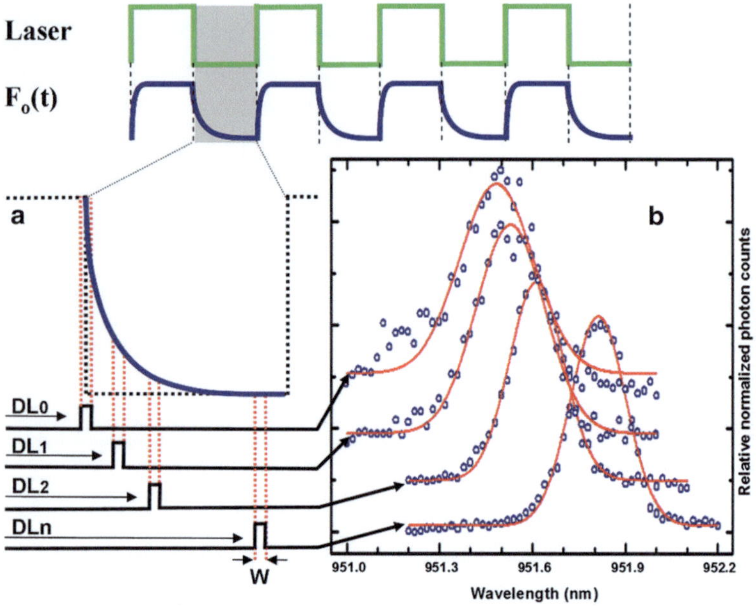

Fig. 11.11 Gated photon counting technique. Modulated laser excitation (250 Hz) energy above WL (532 nm) and corresponding charging behavior of the optically generated electric field $F_0(t)$ schematically shown in the *top figure*. (**a**) Gated photon counts measured by the single photon counting module (SPCM) for different delay times (DL) with a fixed gate width (W). (**b**) Sample spectra for an example delay times and gate width shown

that no optically generated field is created by this probe laser. Using this two-color approach we are able to continuously monitor the local electric field of QDM while modulating this field with the high energy laser.

For the detection we used the same PL setup as discussed earlier in this chapter while replacing the CCD with the avalanche photodiode single photon counting module (SPCM) attached to an SR-400 photon counter. To isolate only the photons within a given time interval, a gated photon counting technique was used to discriminate against photons that arrive outside of the time interval of interest. The process by which gated photon counting was carried out is schematically illustrated in Fig. 11.11. For example, at delay DL0 the SPCM registers only the photons which occur within the gate width of W. To monitor the energy of the indirect PL emission for a given gate delay the spectrometer is scanned while equal numbers of gate periods are counted. By then increasing the gate delay and repeating the spectrometer scan the entire time decay of F_0 could be measured. An example spectrum for a specific gate width, $W = 100$ μs, is shown in Fig. 11.11b and fits to additional spectra showing the decay of the field are plotted as a function of delay time (DL). The experiment was performed with different gate widths when determining the decay time constant in order to eliminate any possible influence of the gate width. There was no measurable difference when varying the gate width

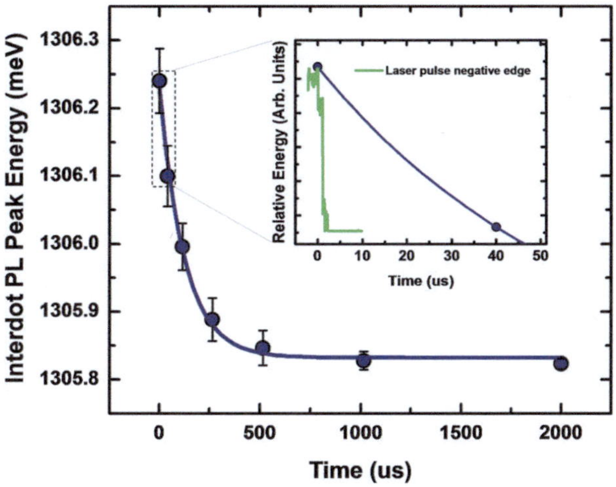

Fig. 11.12 Temporal response of the optically generated electric field, F_o. The decay of F_o is measured from the plot of energy of the photon peak with time delay using the gated photon counting technique. The *inset* shows the turn-off of the triggering laser pulse relative to the decay of F_o

from 1 to 100 μs. Therefore the data reported here all had the width set to 100 μs to maximize the signal to noise.

To understand the time dependence associated with F_o in our QDM device, the peak photon count corresponding to each delay time was evaluated by fitting the data to a Gaussian. We then plotted these peaks as a function of the time delay as shown in Fig. 11.12 and fitted to a first order exponential decay function. From eight sets of decay plots the average decay time constant was calculated to be 125 ± 18 μs. The negative edge of the laser pulse that triggered the onset of the decay of F_o is also shown in the inset of Fig. 11.12. The negative edge of the laser pulse had a turn-off time of 2–3 μs, which was much shorter than that of the decay rate of F_o. Hence convolution effects were minimal and can be neglected on the measured temporal response.

As mentioned earlier, F_o is dependent on electron and hole generation, tunneling, and recombination rates, as well as the type and density of traps. Typically for bulk GaAs the generation rate, tunneling rate, and recombination rates are on the order of nanoseconds [40, 41]. The longer decay rate of 110–140 μs for F_o is due to the excitons ionizing and the charges remaining within the device region of the Schottky diode. This happens through trapping of these charge carriers at impurity sites and material boundaries such as the GaAs/AlGaAs interface, doped or intrinsic GaAs interface, and WLs.

The formation of F_o, however, should be significantly faster. As we excite above the WL the tunneling and relaxation times should result in a build-up of charge on the order of nano to picoseconds. We conducted the same experiment as shown in Fig. 11.10, and instead observed the rising edge of the optically generated electric

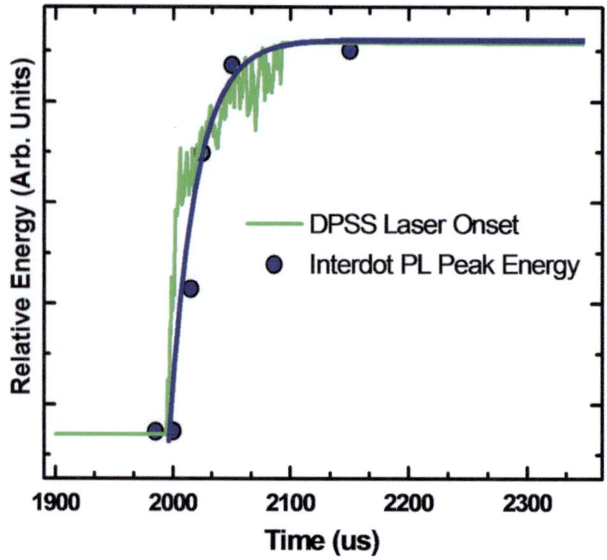

Fig. 11.13 Generation rate of F_o along with the rising edge of the modulated laser pulse

field as opposed to the falling edge as shown in Fig. 11.11a. With our data analysis we observed that the rise time of the optically generated electric field followed the positive edge of the laser pulse which has a rise time of 7–8 μs as shown in Fig. 11.13. Therefore, unfortunately, with our current experimental setup we were unable to determine the rise time of the optically generated electric filed, but we can put an upper bound of 7–8 μs. This can be interpreted as the generation of the optically generated electric field having a frequency approaching MHz, whereas the decay frequency falls within the KHz range.

We also measured the decay time of the optically generated electric field for above the WL laser excitation powers of 0.3 and 0.6 mW (Fig. 11.14). As we increase the laser excitation power more e–h pairs contribute to the creation of F_o, which can be seen as a larger variation of the peak energy values for the first two data points, corresponding to a delay time of 0 μs. But, as the charge carriers tunnel out of the device the separation of the data points becomes smaller and subsequently ends up at the same end point. However, the decay rate was found to remain in the range of 125–137 μs. This indicates that for the powers measured, the number of electrons and holes created does not effect the response time of F_o.

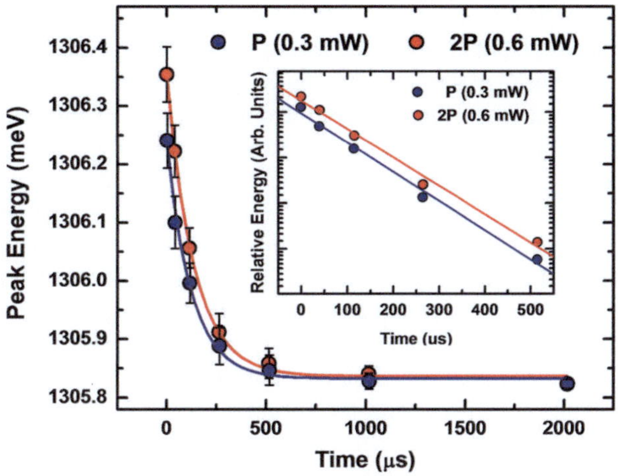

Fig. 11.14 Decay of the optically generated electric field with laser excitation above the WL for two different powers of P (0.3 mW) and 2P (0.6 mW). *Inset* shows the semi-logarithmic plot of the power dependence

11.7 Summary

In this chapter we have briefly discussed the Stark effect observed in exciton emission in QDMs. The observed energy shifts with electric field are a result of carrier separation, the formation of molecular wavefunctions, Coulomb interactions, and the differing effective masses for the electron and hole which results in a measureable effect. The intradot exciton can be engineered to display Stark shifts of up to ~0.13 meV/kV/cm, an order of magnitude larger than their intrinsic linewidths. The interdot exciton Stark shift is predominantly due to the barrier separation between the QDs and can therefore result in extremely large field dependencies of up to ~0.97 meV/kV/cm, an order of magnitude larger than the intradot shifts. It is worth noting that if one could resolve the lifetime limited linewidth of the interdot exciton that, with the Stark shifts measured in these devices, it would be theoretically possible to detect the field due to a single electron on the order of microns away.

We then used the large interdot Stark shift, from a QDM with a 4 nm barrier, to study the optically generated electric field, which was found to have a maximum value of ~3.25 kV/cm, corresponding to 5.04% of the total field. The observed effects are consistent with photovoltaic band flattening, which is produced by the ionization of the photogenerated e–h pairs within the Schottky diode. Basically, the shift of the PL in electric field indicates a local electric field which is the result of creating electron–hole pairs by means of optical absorption by the sample and consequent tunneling of charges to opposite sides of the device. This will create charge accumulation which will generate a local electric field opposing the applied electric field. To induce an opposing field, however, the ionized electrons and holes must not only separate but also remain within the device region. In our device, this is most likely done through trapping at impurity sites and materials boundaries (e.g., the GaAs/AlGaAs interface, the WLs, or the doped/intrinsic GaAs interface), while those not trapped will contribute to the measured photocurrent.

Using a gated photon counting technique, the associated temporal response of F_0 was measured using a modulated laser excitation with energy above the WL while continuously monitoring the PL through a CW laser that was below the WL energy. We observed a relatively long decay time on the order of 110–140 μs for the temporal response of F_0. This can be related to the trapping of carriers at material interfaces and impurity sites and was consistent with carrier lifetimes in III–V materials as well as estimates of tunneling rates of the holes through the AlGaAs barrier. Experiments associated with the variations in the laser power signified that the decay rate of F_0 does not have an influence on the number of generated charge carrier densities. The onset of the optically generated electric field was limited by the temporal resolution of the experiment (7–8 μs). However, the observed results of the generation rate of F_0 yielded a frequency response on the order of MHz, which can possibly be useful for future applications of fast, non-contact, electric field modulation techniques.

Acknowledgments The authors would like to thank Dan Gammon and Allan Bracker for helpful discussions. This work was supported by the Ohio University CMSS program and NSF grant number DMR-1005525.

References

1. Bayer, M., Hawrylak, P., Hinzer, K., Fafard, S., Korkusinski, M., Wasilewski, Z.R., Stern, O., Forchel, A.: Science **291**, 451 (2001)
2. Stinaff, E.A., Scheibner, M., Bracker, A.S., Ponomarev, I.V., Korenev, V.L., Ware, M.E., Doty, M.F., Reinecke, T.L., Gammon, D.: Science **311**, 636 (2006)
3. Koppens, F.H.L., Folk, J.A., Elzerman, J.M., Hanson, R., Willems van Beveren, L.H., Vink, I.T., Tranitz, H.P., Wegscheider, W., Kouwenhoven, L.P., Vandersypen, L.M.K.: Science **309**, 1346 (2005)
4. Krenner, H.J., Sabathil, M., Clark, E.C., Kress, A., Schuh, D., Bichler, M., Abstreiter, G., Finley, J.J.: Phys. Rev. Lett. **94**, 057402 (2005)
5. Ortner, G., Bayer, M., Lyanda-Geller, Y., Reinecke, T.L., Kress, A., Reithmaier, J.P., Forchel, A.: Phys. Rev. Lett. **94**, 157401 (2005)
6. Gerardot, B.D., Strauf, S., de Dood, M.J.A., Bychkov, A.M., Badolato, A., Hennessy, K., Hu, E.L., Bouwmeester, D., Petroff, P.M.: Phys. Rev. Lett. **95**, 137403 (2005)
7. Nakaoka, T., Clark, E.C., Krenner, H.J., Sabathil, M., Bichler, M., Arakawa, Y., Abstreiter, G., Finley, J.J.: Phys. Rev. B **74**, 121305 (2006)
8. Türeci, A.E., Taylor, J.M., Imamoglu, A.: Phys. Rev. B **75**, 235313 (2007)
9. Troiani, F., Wilson-Rae, I., Tejedor, C.: Appl. Phys. Lett. **90**, 144103 (2007)
10. Degani, M.H., Maialle, M.Z.: Phys. Rev. B **75**, 115322 (2007)
11. He, L., Zunger, A.: Phys. Rev. B **75**, 075330 (2007)

12. Doty, M.F., Scheibner, M., Ponomarev, I.V., Stinaff, E.A., Bracker, A.S., Korenev, V.L., Reinecke, T.L., Gammon, D.: Phys. Rev. Lett. **97**, 197202 (2006)
13. Scheibner, M., Doty, M.F., Ponomarev, I.V., Bracker, A.S., Stinaff, E.A., Korenev, V.L., Reinecke, T.L., Gammon, D.: Phys. Rev. B **75**, 245318 (2007)
14. Wijesundara, K.C., Rolon, J.E., Ulloa, S.E., Bracker, A.S., Gammon, D., Stinaff, E.A.: Phys. Rev. B **84**, 081404(R) (2011)
15. Greilich, A., Carter, S.G., Kim, D., Bracker, A.S., Gammon, D.: Nat. Photonics **5**, 703 (2011)
16. Garrido, M., Wijesundara, K.C., Ramanathan, S., Bracker, A.S., Gammon, D., Stinaff, E.A.: Appl. Phys. Lett. **96**, 211115 (2010)
17. Xie, Q., Madhukar, A., Chen, P., Kobayashi, N.P.: Phys. Rev. Lett. **75**, 2542 (1995)
18. Bracker, A.S., Scheibner, M., Doty, M.F., Stinaff, E.A., Ponomarev, I.V., Kim, J.C., Whitman, L.J., Reinecke, T.L., Gammon, D.: Appl. Phys. Lett. **89**, 233110 (2006)
19. Wasilewski, R., Fafard, S., McCaffrey, J.P.: J. Cryst. Growth **201/202**, 1131 (1999)
20. Warburton, R.J., Schäflein, C., Haft, D., Bickel, F., Lorke, A., Karrai, K., Garcia, J.M., Schoenfeld, W., Petroff, P.M.: Nature **405**, 926 (2000)
21. Dekel, E., Gershoni, D., Ehrenfreund, E., Spektor, D., Garcia, J.M., Petroff, P.M.: Phys. Rev. Lett. **80**, 4991 (1998)
22. Regelman, D.V., Dekel, E., Gershoni, D., Ehrenfreund, E., Williamson, A.J., Shumway, J., Zunger, A., Schoenfeld, W.V., Petroff, P.M.: Phys. Rev. B **64**, 165301 (2001)
23. Besombes, L., Kheng, K., Marsal, L., Mariette, H.: Phys. Rev. B **65**, 121314 (2002)
24. Scheibner, M., Bracker, A.S., Kim, D., Gammon, D.: Solid State Commun. **149**, 1427 (2009)
25. Miller, D.A.B., Chemla, D.S., Damen, T.C., Gossard, A.C., Wiegmann, W., Wood, T.H., Burrus, C.A.: Phys. Rev. Lett. **53**, 2173 (1984)
26. Capasso, F., Sirtori, C., Cho, A.Y.: IEEE J. Quantum Electron. **30**, 1313 (1994)
27. Wood, T.H., Burrus, C.A., Gnauck, A.H., Wiesenfeld, J.M., Miller, D.A.B., Chemla, D.S., Damen, T.C.: Appl. Phys. Lett. **47**, 190 (1985)
28. Le, H.Q., Zayhowski, J.J., Goodhue, W.D.: Appl. Phys. Lett. **50**, 1518 (1987)
29. Chen, Y.J., Koteles, E.S., Elman, B.S., Armiento, C.A.: Phys. Rev. B **36**, 4562 (1987)
30. Ramanathan, S., Petersen, G., Wijesundara, K., Thota, R., Stinaff, E.A., Kerfoot, M.L., Scheibner, M., Bracker, A.S., and Gammon D.: Quantum-confined Stark effects in coupled InAs/GaAs quantum dots. Appl. Phys. Lett. **102**, 213101 (2013)
31. Sheng, W., Leburton, J.-P.: Phys. Rev. Lett. **88**, 167401 (2002)
32. Szafran, B., Peeters, F.M., Bednarek, S.: Phys. Rev. B **75**, 115303 (2007)
33. Fry, P.W., Itskevich, I.E., Mowbray, D.J., Skolnick, M.S., Finley, J.J., Barker, J.A., O'Reilly, E.P., Wilson, L.R., Larkin, I.A., Maksym, P.A., Hopkinson, M., Al-Khafaji, M., David, J.P.R., Cullis, A.G., Hill, G., Clark, J.C.: Phys. Rev. Lett. **84**, 733 (2000)
34. Itskevich, I.E., Rybchenko, S.I., Tartakovskii, I.I., Stoddart, S.T., Levin, A., Main, P.C., Eaves, L., Henini, M., Parnell, S.: Appl. Phys. Lett. **76**, 3932 (2000)
35. Huggard, P.G., Shaw, C.J., Andrews, S.R., Cluff, J.A., Grey, R.: Phys. Rev. Lett. **84**, 1023 (2000)
36. Santori, C., Fattal, D., Vučković, J., Solomon, G.S., Yamamoto, Y.: Nature **419**, 594 (2002)
37. Smith, G.H., Novak, D., Lim, C., Wu, K.: Electron. Lett. **33**, 1159–1160 (1997)
38. Hecht, M.H.: J. Vac. Sci. Technol. B **8**, 1018 (1990)
39. Hecht, M.H.: Phys. Rev. B **41**, 7918–7921 (1990)
40. Alexandrou, A., Kash, J.A., Mendez, E.E., Zachau, M., Hong, J.M., Fukuzawa, T., Hase, Y.: Phys. Rev. B **42**, 9225 (1990)
41. Dumke, W.P.: Phys. Rev. **132**, 1998 (1963)

Index

J. Wu and Z.M. Wang (eds.), *Quantum Dot Molecules*, Lecture Notes in Nanoscale Science and Technology 14, DOI 10.1007/978-1-4614-8130-0,

MIX
Papier aus verantwortungsvollen Quellen
Paper from responsible sources
FSC® C105338

If you have any concerns about our products,
you can contact us on
ProductSafety@springernature.com

In case Publisher is established outside the EU,
the EU authorized representative is:
Springer Nature Customer Service Center GmbH
Europaplatz 3, 69115 Heidelberg, Germany

Printed by Libri Plureos GmbH
in Hamburg, Germany